INTRODUCTION TO VASSILIEV KNOT INVARIANTS

With hundreds of worked examples, exercises and illustrations, this detailed exposition of the theory of Vassiliev knot invariants opens the field to students with little or no knowledge in this area. It also serves as a guide to more advanced material.

The book begins with a basic and informal introduction to knot theory, giving many examples of knot invariants before the class of Vassiliev invariants is introduced. This is followed by a detailed study of the algebras of Jacobi diagrams and 3-graphs, and the construction of functions on these algebras via Lie algebras. The authors then describe two constructions of a universal invariant with values in the algebra of Jacobi diagrams: via iterated integrals and via the Drinfeld associator, and extend the theory to framed knots. Various other topics are then discussed, such as Gauss diagram formulae, the diagrammatic version of the Duflo isomorphism, connection to the theory of nilpotent groups and more. The book ends with Vassiliev's original construction.

S. CHMUTOV is Associate Professor in the Department of Mathematics at Ohio State University, Mansfield.

S. DUZHIN is a Senior Researcher in the St. Petersburg Department of the Steklov Institute of Mathematics.

J. MOSTOVOY is Professor in the Department of Mathematics at the Centre for Research and Advanced Studies of the National Polytechnic Institute (CINVESTAV-IPN), Mexico City.

INTRODUCTION TO VASSILIEV KNOT INVARIANTS

S. CHMUTOV
Ohio State University, Mansfield

S. DUZHIN
Steklov Institute of Mathematics, St Petersburg

J. MOSTOVOY
*Centre for Research and Advanced Studies of the National
Polytechnic Institute (CINVESTAV-IPN), Mexico City*

CAMBRIDGE
UNIVERSITY PRESS

CAMBRIDGE
UNIVERSITY PRESS

University Printing House, Cambridge CB2 8BS, United Kingdom

One Liberty Plaza, 20th Floor, New York, NY 10006, USA

477 Williamstown Road, Port Melbourne, VIC 3207, Australia

314-321, 3rd Floor, Plot 3, Splendor Forum, Jasola District Centre, New Delhi - 110025, India

103 Penang Road, #05-06/07, Visioncrest Commercial, Singapore 238467

Cambridge University Press is part of the University of Cambridge.

It furthers the University's mission by disseminating knowledge in the pursuit of education, learning and research at the highest international levels of excellence.

www.cambridge.org
Information on this title: www.cambridge.org/9781107020832

© S. Chmutov, S. Duzhin and J. Mostovoy 2012

First published 2012

A catalogue record for this publication is available from the British Library

ISBN 978-1-107-02083-2 Hardback

To the memory of V. I. Arnold

Contents

Preface

This book is a detailed introduction to the theory of finite type (Vassiliev) knot invariants, with a stress on its combinatorial aspects. It is intended to serve both as a textbook for readers with no or little background in this area, and as a guide to some of the more advanced material. Our aim is to lead the reader to understanding by means of pictures and calculations, and for this reason we often prefer to convey the idea of the proof on an instructive example rather than give a complete argument. While we have made an effort to make the text reasonably self-contained, an advanced reader is sometimes referred to the original papers for the technical details of the proofs.

Historical remarks

The notion of a finite type knot invariant was introduced by Victor Vassiliev (Moscow) in the end of the 1980s and first appeared in print in his paper (1990a). Vassiliev, at the time, was not specifically interested in low-dimensional topology. His main concern was the general theory of discriminants in the spaces of smooth maps, and his description of the space of knots was just one, though the most spectacular, application of a machinery that worked in many seemingly unrelated contexts. It was V. I. Arnold (1992) who understood the importance of finite type invariants, coined the name "Vassiliev invariants" and popularized the concept; since that time, the term "Vassiliev invariants" has become standard.

A different perspective on the finite type invariants was developed by Mikhail Goussarov (St. Petersburg). His notion of n-equivalence, which first appeared in print in Goussarov (1993), turned out to be useful in different situations, for example, in the study of the finite type invariants of 3-manifolds.[1] Nowadays some people use the expression "Vassiliev–Goussarov invariants" for the finite type invariants.

---

[1] Goussarov cites Vassiliev's works in his earliest paper (1991). Nevertheless, according to O. Viro, Goussarov first mentioned finite type invariants in a talk at the Leningrad topological seminar as early as in 1987.

Vassiliev's definition of finite type invariants is based on the observation that knots form a topological space and the knot invariants can be thought of as the locally constant functions on this space. Indeed, the space of knots is an open subspace of the space M of all smooth maps from S^1 to $\mathbb{R}^3$; its complement is the so-called discriminant Σ which consists of all maps that fail to be embeddings. Two knots are isotopic if and only if they can be connected in M by a path that does not cross Σ.

Using simplicial resolutions, Vassiliev constructs a spectral sequence for the homology of Σ. After applying the Alexander duality, this spectral sequence produces cohomology classes for the space of knots $M - \Sigma$; in dimension zero these are precisely the Vassiliev knot invariants.

Vassiliev's approach, which is technically rather demanding, was simplified by J. Birman and X.-S. Lin (1993). They explained the relation between the Jones polynomial and finite type invariants[2] and emphasized the role of the algebra of chord diagrams. M. Kontsevich (1993) showed that the study of real-valued Vassiliev invariants can, in fact, be reduced entirely to the combinatorics of chord diagrams. His proof used an analytic tool (the Kontsevich integral) which is, essentially, a power series encoding all the finite type invariants of a knot. Kontsevich also defined a coproduct on the algebra of chord diagrams which turns it into a Hopf algebra.

D. Bar-Natan was the first to give a comprehensive treatment of Vassiliev knot and link invariants. In his preprint (1991a) and PhD thesis (1991b) he found the relationship between finite type invariants and the topological quantum field theory developed by his thesis advisor E. Witten (1989, 1995). Bar-Natan's paper (1995a) (whose preprint edition appeared in 1992) is still the most authoritative source on the fundamentals of the theory of Vassiliev invariants. About the same time, T. Le and J. Murakami (1996a), relying on V. Drinfeld's work (1989, 1990), proved the rationality of the Kontsevich integral.

Among further developments in the area of finite type knot invariants, let us mention:

- The existence of non-Lie-algebraic weight systems (Vogel 1997, Lieberum 1999) and an interpretation of all weight systems as Lie algebraic weight systems in a suitable category (Hinich and Vaintrob 2002);
- J. Kneissler's analysis (2000, 2001a, 2001b) of the structure of the algebra Λ introduced by P. Vogel (1997);
- The proof by Goussarov (1998b) that Vassiliev invariants are polynomials in the gleams for a fixed Turaev shadow;

---

[2]independently from Goussarov, who was the first to discover this relation in Goussarov (1991).

- Gauss diagram formulae of M. Polyak and O. Viro (1994) and the proof by M. Goussarov (1998a) that all finite type invariants can be expressed by such formulae;

- Habiro's theory of claspers (Habiro 2000) (see also Goussarov 2001);

- V. Vassiliev's papers (2001, 2005) where a general technique for deriving combinatorial formulae for cohomology classes in the complements to discriminants, and in particular, for finite type invariants, is proposed;

- Explicit formulae for the Kontsevich integral of some knots and links (Bar-Natan, Le and Thurston 2003; Bar-Natan and Lawrence 2004; Rozansky 2003; Kricker 2000; Marché 2004; Garoufalidis and Kricker 2004);

- The interpretation of the Vassiliev spectral sequence in terms of the Hochschild homology of the Poisson operad by V. Turchin (Tourtchine 2004);

- The alternative approaches to the topology of the space of knots via configuration spaces and the Goodwillie calculus (Sinha 2009).

One serious omission in this book is the connection between the Vassiliev invariants and the Chern–Simons theory. This connection motivates much of the interest in finite type invariants and gives a better understanding of the nature of the Kontsevich integral. Moreover, it suggests another form of the universal Vassiliev invariant, namely, the configuration space integral. There are many texts that explain this connection with great clarity; the reader may start, for instance, with Labastida (1999), Sawon (2006) or Polyak (2005). The original paper of Witten (1989) has not lost its relevance and, while it does not deal directly with the Vassiliev invariants (which were not defined at the time), it still is one of the indispensable references.

An important source of information on finite type invariants is the online *Bibliography of Vassiliev invariants* started by D. Bar-Natan and currently located at

> http://www.pdmi.ras.ru/~duzhin/VasBib/

On January 19, 2011 it contained 641 items and this number is increasing. The study of finite type invariants is ongoing. However, notwithstanding all efforts, the most important question put forward in 1990:

Is it true that Vassiliev invariants distinguish knots?

is still open. At the moment it is not even known whether the Vassiliev invariants can detect knot orientation. A number of open problems related to finite type invariants are listed in Ohtsuki (2004b).

Prerequisites

We assume that the reader has a basic knowledge of calculus on manifolds (vector fields, differential forms, Stokes' theorem), general algebra (groups, rings, modules, Lie algebras, fundamentals of homological algebra), linear algebra (vector spaces, linear operators, tensor algebra, elementary facts about representations) and topology (topological spaces, homotopy, homology, Euler characteristic). Some of this and more advanced algebraic material (bialgebras, free algebras, universal enveloping algebras etc.), which is of primary importance in this book, can be found in the appendix at the end of the book. No knowledge of knot theory is presupposed, although it may be useful.

Contents

The book consists of fifteen chapters, which can logically be divided into four parts.

The first part, Chapters 1–4, opens with a short introduction into the theory of knots and their classical polynomial invariants and closes with the definition of Vassiliev invariants.

In Chapters 5–7, we systematically study the graded Hopf algebra naturally associated with the filtered space of Vassiliev invariants, which appears in three different guises: as the algebra of framed chord diagrams $\mathscr{A}$, as the algebra of closed Jacobi diagrams $\mathscr{C}$ and as the algebra of open Jacobi diagrams $\mathscr{B}$. After that, we study the auxiliary algebra Γ generated by regular trivalent graphs and closely related to the algebras $\mathscr{A}$, $\mathscr{B}$, $\mathscr{C}$ as well as to Vogel's algebra Λ. In Chapter 7 we discuss the weight systems defined by Lie algebras, both universal and depending on a chosen representation.

Chapters 8–10 are dedicated to a detailed exposition of the Kontsevich integral; it contains the proof of the main theorem of the theory of Vassiliev knot invariants that reduces their study to combinatorics of chord diagrams and related algebras. Chapters 8 and 9 treat the Kontsevich integral from the analytic point of view. Chapter 10 is dedicated to the Drinfeld associator and the combinatorial construction of the Kontsevich integral.

The last part of the book, Chapters 11–15, is devoted to various topics left out in the previous exposition. Chapter 11 contains some additional material on the Kontsevich integral: the wheels formula, the Rozansky rationality conjecture etc. This is followed in Chapters 12–14 by discussions of the Vassiliev invariants for braids, Gauss diagram formulae, the Melvin–Morton conjecture, the Goussarov–Habiro theory, the size of the space of Vassiliev invariants, etc. The book closes with a description of Vassiliev's original construction for the finite type invariants.

Chapter dependence

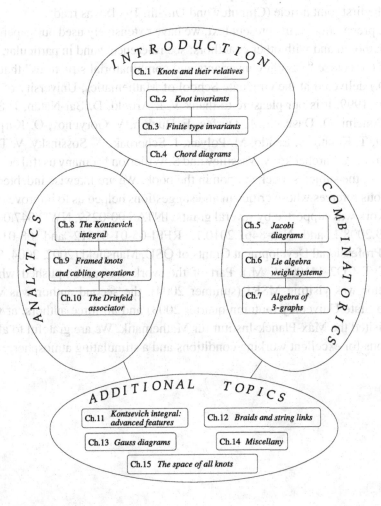

The book is intended to be a textbook, so we have included many exercises. Some exercises are embedded in the text; the others appear in a separate section at the end of each chapter. Open problems are marked with an asterisk.

Acknowledgements

The work of the first two authors on this book actually began in August 1992, when our colleague Inna Scherbak returned to Pereslavl-Zalessky from the First European Mathematical Congress in Paris and brought a photocopy of Arnold's lecture notes about the newborn theory of Vassiliev knot invariants. We spent several months filling our waste-paper baskets with pictures of chord diagrams, before the first joint article (Chmutov and Duzhin 1994) was ready.

In the preparation of the present text, we have extensively used our papers (joint, single-authored and with other coauthors; see references) and in particular, lecture notes of the course "Vassiliev invariants and combinatorial structures" that one of us (S. D.) delivered at the Graduate School of Mathematics, University of Tokyo, in Spring 1999. It is our pleasure to thank V. I. Arnold, D. Bar-Natan, J. Birman, C. De Concini, O. Dasbach, A. Durfee, F. Duzhin, V. Goryunov, O. Karpenkov, T. Kerler, T. Kohno, S. Lando, M. Polyak, I. Scherbak, A. Sossinsky, V. Turchin, A. Vaintrob, A. Varchenko, V. Vassiliev and S. Willerton for many useful comments concerning the subjects touched upon in the book. We are likewise indebted to the anonymous referees whose criticism and suggestions helped us to improve the text.

Our work was supported by several grants: INTAS 00-0259, NWO 047.008.005, NSh-709.2008.1 and Nsh-8462.2010.1, RFFI-05-01-01012 and 08-01-00379 (S. D.), Professional Development Grants of OSU, Mansfield (2002, 2004, S. Ch.), CONACyT CO2-44100 (J. M.). Part of the work was accomplished when the first author was visiting MSRI (summer 2004), the second author was visiting the Ohio State University (autumn quarter 2003) and all three authors, at various times, visited the Max-Planck-Institut für Mathematik. We are grateful to all these institutions for excellent working conditions and a stimulating atmosphere.

1

Knots and their relatives

This book is about knots. It is, however, hardly possible to speak about knots without mentioning other one-dimensional topological objects embedded into the three-dimensional space. Therefore, in this introductory chapter we give basic definitions and constructions pertaining to knots and their relatives: links, braids and tangles.

The table of knots provided in Table 1.1 will be used throughout the book as a source of examples and exercises.

1.1 Definitions and examples

1.1.1 Knots

A *knot* is a closed non-self-intersecting curve in 3-space. In this book, we shall mainly study smooth oriented knots. A precise definition can be given as follows.

Definition 1.1. A *parametrized knot* is an embedding of the circle S^1 into the Euclidean space $\mathbb{R}^3$.

Recall that an *embedding* is a smooth map which is injective and whose differential is nowhere zero. In our case, the non-vanishing of the differential means that the tangent vector to the curve is non-zero. In the above definition and everywhere in the sequel, the word *smooth* means *infinitely differentiable*.

A choice of an orientation for the parametrizing circle

$$S^1 = \{(\cos t, \sin t) \mid t \in \mathbb{R}\} \subset \mathbb{R}^2$$

gives an orientation to all the knots simultaneously. We shall always assume that S^1 is oriented counterclockwise. We shall also fix an orientation of the 3-space; each time we pick a basis for $\mathbb{R}^3$ we shall assume that it is consistent with the orientation.

If coordinates x, y, z are chosen in $\mathbb{R}^3$, a knot can be given by three smooth periodic functions of one variable $x(t)$, $y(t)$, $z(t)$.

Example 1.2. The simplest knot is represented by a plane circle:

$$
\begin{aligned}
x &= \cos t, \\
y &= \sin t, \\
z &= 0.
\end{aligned}
$$

Example 1.3. The curve that goes 3 times around and 2 times across a standard torus in $\mathbb{R}^3$ is called the *(left) trefoil knot*, or the $(2, 3)$-*torus knot*:

$$
\begin{aligned}
x &= (2 + \cos 3t) \cos 2t, \\
y &= (2 + \cos 3t) \sin 2t, \\
z &= \sin 3t.
\end{aligned}
$$

Exercise. Give the definition of a (p, q)-torus knot. What are the appropriate values of p and q for this definition?

It will be convenient to identify knots that only differ by a change of a parametrization. An *oriented knot* is an equivalence class of parametrized knots under orientation-preserving diffeomorphisms of the parametrizing circle. Allowing *all* diffeomorphisms of S^1 in this definition, we obtain *unoriented knots*. Alternatively, an unoriented knot can be defined as the *image* of an embedding of S^1 into $\mathbb{R}^3$; an oriented knot is then an image of such an embedding together with the choice of one of the two possible directions on it.

We shall distinguish oriented/unoriented knots from parametrized knots in the notation: oriented and unoriented knots usually will be denoted by capital letters, while for the individual embeddings lowercase letters will be used. As a rule, the word "knot" will mean "oriented knot," unless it is clear from the context that we deal with unoriented knots, or consider a specific choice of parametrization.

1.1.2 Isotopy

The study of parametrized knots falls within the scope of differential geometry. The *topological* study of knots requires an equivalence relation which would not only discard the specific choice of parametrization, but also model the physical transformations of a closed piece of rope in space.

By a *smooth family of maps*, or a *map smoothly depending on a parameter*, we understand a smooth map $F : S^1 \times I \to \mathbb{R}^3$, where $I \subset \mathbb{R}$ is an interval. Assigning a fixed value a to the second argument of F, we get a map $f_a : S^1 \to \mathbb{R}^3$.

Definition 1.4. A smooth isotopy of a knot $f : S^1 \to \mathbb{R}^3$, is a smooth family of knots f_u, with u a real parameter, such that for some value $u = a$ we have $f_a = f$.

For example, the formulae

$$
\begin{aligned}
x &= (u + \cos 3t) \cos 2t, \\
y &= (u + \cos 3t) \sin 2t, \\
z &= \sin 3t,
\end{aligned}
$$

where $u \in (1, +\infty)$, represent a smooth isotopy of the trefoil knot which corresponds to $u = 2$. In the pictures below, the space curves are shown by their projection to the (x, y) plane:

$u = 2$ $u = 1.5$ $u = 1.2$ $u = 1$

For any $u > 1$ the resulting curve is smooth and has no self-intersections, but as soon as the value $u = 1$ is reached we get a singular curve with three coinciding cusps[1] corresponding to the values $t = \pi/3$, $t = \pi$ and $t = 5\pi/3$. This curve is not a knot.

Definition 1.5. Two parametrized knots are said to be *isotopic* if one can be transformed into another by means of a smooth isotopy. Two oriented knots are isotopic if they represent the classes of isotopic parametrized knots; the same definition is valid for unoriented knots.

Example 1.6. This picture shows an isotopy of the figure eight knot into its mirror image:

There are other notions of knot equivalence, namely, *ambient equivalence* and *ambient isotopy*, which, for smooth knots, are the same thing as isotopy. Here are the definitions. A proof that they are equivalent to our definition of isotopy can be found in Burde and Zieschang (2003).

---

[1] A *cusp* of a spatial curve is a point where the curve can be represented as $x = s^2$, $y = s^3$, $z = 0$ in some local coordinates.

Definition 1.7. Two parametrized knots, f and g, are *ambient equivalent* if there is a commutative diagram

$$
\begin{array}{ccc}
S^1 & \xrightarrow{\ f\ } & \mathbb{R}^3 \\
{\scriptstyle \varphi}\downarrow & & \downarrow{\scriptstyle \psi} \\
S^1 & \xrightarrow{\ g\ } & \mathbb{R}^3
\end{array}
$$

where φ and ψ are orientation-preserving diffeomorphisms of the circle and the 3-space, respectively.

Definition 1.8. Two parametrized knots, f and g, are *ambient isotopic* if there is a smooth family of diffeomorphisms of the 3-space $\psi_t : \mathbb{R}^3 \to \mathbb{R}^3$ with $\psi_0 = \mathrm{id}$ and $\psi_1 \circ f = g$.

Definition 1.9. A knot, equivalent to the plane circle on page 2 is referred to as a *trivial knot*, or an *unknot*.

Sometimes, it is not immediately clear from a diagram of a trivial knot that it is indeed trivial:

Trivial knots

There are algorithmic procedures to detect whether a given knot diagram represents an unknot. One of them, based on W. Thurston's ideas, is implemented in J. Weeks' computer program `SnapPea`; see Weeks (2010); another algorithm, due to I. Dynnikov, is described in Dynnikov (2006).

Here are several other examples of knots.

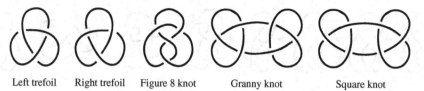

| Left trefoil | Right trefoil | Figure 8 knot | Granny knot | Square knot |

1.1.3 Links

Knots are a special case of links.

Definition 1.10. A *link* is a smooth embedding $S^1 \sqcup \cdots \sqcup S^1 \to \mathbb{R}^3$, where $S^1 \sqcup \cdots \sqcup S^1$ is the disjoint union of several circles.

| Trivial 2-component link | Hopf link | Whitehead link | Borromean rings |

Equivalence of links is defined in the same way as for knots – with the exception that now one may choose whether or not to distinguish between the components of a link and thus speak about the equivalence of links with *numbered* or *unnumbered* components.

In the future, we shall often say "knot (link)" instead of "equivalence class," or "topological type of knots (links)."

1.2 Plane knot diagrams

1.2.1 Knot diagrams

Knots are best represented graphically by means of *knot diagrams*. A knot diagram is a plane curve whose only singularities are transversal double points (crossings), together with the choice of one branch of the curve at each crossing. The chosen branch is called an *overcrossing*; the other branch is referred to as an *undercrossing*. A knot diagram is thought of as a projection of a knot along some "vertical" direction; overcrossings and undercrossings indicate which branch is "higher" and which is "lower." To indicate the orientation, an arrow is added to the knot diagram.

Theorem 1.11 (Reidemeister 1948, proofs can be found in Prasolov and Sossinsky 1997; Burde and Zieschang 2003; and Murasugi 1996). *Two unoriented knots, K_1 and K_2, are equivalent if and only if a diagram of K_1 can be transformed into a diagram of K_2 by a sequence of isotopies of the plane and local moves of the following three types:*

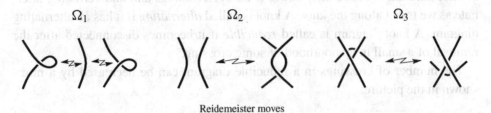

Reidemeister moves

To adjust the assertion of this theorem to the oriented case, each of the three Reidemeister moves has to be equipped with orientations in all possible ways. Smaller sufficient sets of oriented moves exist; one such set will be given later in terms of Gauss diagrams (see Section 1.7.3).

Exercise. Determine the sequence of Reidemeister moves that relates the two diagrams of the trefoil knot below:

1.2.2 Local writhe

Crossing points on a diagram come in two species, positive and negative:

Positive crossing   Negative crossing

Although this sign is defined in terms of the knot orientation, it is easy to check that it does not change if the orientation is reversed. For *links* with more than one component, the choice of orientation is essential.

The *local writhe* of a crossing is defined as $+1$ or -1 for positive or negative points, respectively. The *writhe* (or *total writhe*) of a diagram is the sum of the writhes of all crossing points, or, equivalently, the difference between the number of positive and negative crossings. Of course, the same knot may be represented by diagrams with different total writhes. In Chapter 2 we shall see how the writhe can be used to produce knot invariants.

1.2.3 Alternating knots

A knot diagram is called *alternating* if its overcrossings and undercrossing alternate as we travel along the knot. A knot is called *alternating* if it has an alternating diagram. A knot diagram is called *reducible* if it becomes disconnected after the removal of a small neighbourhood of some crossing.

The number of crossings in a reducible diagram can be decreased by a move shown in the picture:

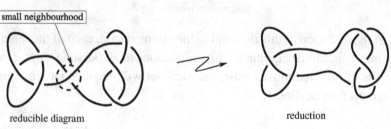

reducible diagram   reduction

A diagram which is not reducible is called *reduced*. As there is no immediate way to simplify a reduced diagram, the following conjecture naturally arises (Tait, 1898).

The Tait conjecture. *A reduced alternating diagram has the minimal number of crossings among all diagrams of the given knot.*

This conjecture stood open for almost 100 years. It was proved only in 1986 (using the newly invented Jones polynomial) simultaneously and independently by L. Kauffman (1987b), K. Murasugi (1987) and M. Thistlethwaite (1987) (see Exercise 2.27).

1.3 Inverses and mirror images

Change of orientation (taking the inverse) and taking the mirror image are two basic operations on knots which are induced by orientation reversing smooth involutions on S^1 and $\mathbb{R}^3$ respectively. Every such involution on S^1 is conjugate to the reversal of the parametrization; on $\mathbb{R}^3$ it is conjugate to a reflection in a plane mirror.

Let K be a knot. Composing the parametrization reversal of S^1 with the map $f : S^1 \to \mathbb{R}^3$ representing K, we obtain the *inverse K^** of K. The *mirror image* of K, denoted by $\overline{K}$, is a composition of the map $f : S^1 \to \mathbb{R}^3$ with a reflection in $\mathbb{R}^3$. Both change of orientation and taking the mirror image are involutions on the set of (equivalence classes of) knots. They generate a group isomorphic to $\mathbb{Z}_2 \oplus \mathbb{Z}_2$; the symmetry properties of a knot K depend on the subgroup that leaves the knot invariant. The group $\mathbb{Z}_2 \oplus \mathbb{Z}_2$ has five (not necessarily proper) subgroups, which give rise to five symmetry classes of knots.

Definition 1.12. A knot is called:

- *invertible*, if $K^* = K$,
- *plus-amphicheiral*, if $\overline{K} = K$,
- *minus-amphicheiral*, if $\overline{K} = K^*$,
- *fully symmetric*, if $K = K^* = \overline{K} = \overline{K}^*$,
- *totally asymmetric*, if all knots $K, K^*, \overline{K}, \overline{K}^*$ are different.

The word *amphicheiral* means either plus- or minus-amphicheiral. For invertible knots, this is the same. Amphicheiral and non-amphicheiral knots are also referred to as *achiral* and *chiral* knots, respectively.

The five symmetry classes of knots are summarized in the following table. The word "minimal" means "with the minimal number of crossings"; σ and τ denote the involutions of taking the mirror image and the inverse respectively. The notation for concrete knots in the last column will be explained in the next section.

Stabilizer	Orbit	Symmetry type	Min example
$\{1\}$	$\{K, \overline{K}, K^*, \overline{K}^*\}$	totally asymmetric	9_{32}, 9_{33}
$\{1, \sigma\}$	$\{K, K^*\}$	+amphicheiral, non-inv	12^a_{427}
$\{1, \tau\}$	$\{K, \overline{K}\}$	invertible, chiral	3_1
$\{1, \sigma\tau\}$	$\{K, K^*\}$	−amphicheiral, non-inv	8_{17}
$\{1, \sigma, \tau, \sigma\tau\}$	$\{K\}$	fully symmetric	4_1

Example 1.13. The trefoil knots are invertible, because the rotation through $180°$ around an axis in $\mathbb{R}^3$ changes the direction of the arrow on the knot.

The existence of non-invertible knots was first proved by H. Trotter (1964). The simplest instance of Trotter's theorem is a *pretzel knot with parameters* $(3, 5, 7)$:

Among the knots with up to eight crossings (see Table 1.1) there is only one non-invertible knot: 8_{17}, which is, moreover, minus-amphicheiral. These facts were proved by A. Kawauchi (1979).

Example 1.14. The trefoil knots are not amphicheiral, hence the distinction between the left and the right trefoil. A proof of this fact, based on the calculation of the Jones polynomial, will be given in Section 2.4.

Remark 1.15. Knot tables only list knots up to taking inverses and mirror images. In particular, there is only one entry for the trefoil knots. Either of them is often referred to as *the trefoil*.

Example 1.16. The figure eight knot is amphicheiral. The isotopy between this knot and its mirror image is shown in Example 1.6.

Among the 35 knots with up to eight crossings shown in Table 1.1, there are exactly seven amphicheiral knots: 4_1, 6_3, 8_3, 8_9, 8_{12}, 8_{17}, 8_{18}, out of which 8_{17} is minus-amphicheiral, the rest, as they are invertible, are both plus- and minus-amphicheiral.

The simplest totally asymmetric knots appear in nine crossings, they are 9_{32} and 9_{33}. The following are all non-equivalent:

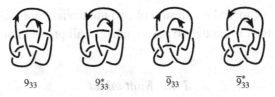

$$9_{33} \qquad 9_{33}^* \qquad \overline{9}_{33} \qquad \overline{9}_{33}^*$$

Here is the simplest plus-amphicheiral non-invertible knot, together with its inverse:

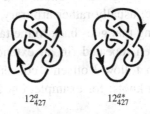

$$12_{427}^a \qquad 12_{427}^{a*}$$

In practice, the easiest way to find the symmetry type of a given knot or link is by using the computer program Knotscape (Hoste and Thistlethwaite 1999), which can handle link diagrams with up to 49 crossings.

1.4 Knot tables

1.4.1 Connected sum

There is a natural way to fuse two knots into one: cut each of the two knots at some point, then connect the two pairs of loose ends. This must be done with some caution: first, by a smooth isotopy, both knots should be deformed so that for a certain plane projection they look as shown in the picture below on the left, then they should be changed inside the dashed disk as shown on the right:

The connected sum makes sense only for oriented knots. It is well-defined and commutative on the equivalence classes of knots. The connected sum of knots K_1 and K_2 is denoted by $K_1 \# K_2$.

Definition 1.17. A knot is called *prime* if it cannot be represented as the connected sum of two nontrivial knots.

Each knot is a connected sum of prime knots, and this decomposition is unique (see Crowell and Fox (1963) for a proof). In particular, this means that a trivial knot cannot be decomposed into a sum of two nontrivial knots. Therefore, in order to classify all knots, it is enough to have a table of all prime knots.

1.4.2 Knot tables

Prime knots are tabulated according to the minimal number of crossings that their diagrams can have. Within each group of knots with the same crossing number, knots are numbered in some, usually rather arbitrary, way. In Table 1.1, we use the widely adopted numbering that goes back to the table compiled by Alexander and Briggs (1926/1927), then repeated (in an extended and modified way) by D. Rolfsen (1976). We also follow Rolfsen's conventions in the choice of the version of non-amphicheiral knots: for example, our 3_1 is the left, not the right, trefoil.

Rolfsen's table of knots, authoritative as it is, contained an error. It is the famous *Perko pair* (knots 10_{161} and 10_{162} in Rolfsen) – two equivalent knots that were thought to be different for 75 years since 1899:

The equivalence of these two knots was established in 1973 by K. A. Perko (Perko 1973), a lawyer from New York who studied mathematics at Princeton in 1960–1964 (Perko 2002) but later chose jurisprudence to be his profession.[2]

Complete tables of knots are currently known up to crossing number 16 (Hoste, Thistlethwaite and Weeks 1998). For knots with 11 through 16 crossings it is nowadays customary to use the numbering of Knotscape (Hoste and Thistlethwaite 1999) where the tables are built into the software. For each crossing number, Knotscape has a separate list of alternating and non-alternating knots. For example, the notation 12_{427}^a used in Section 1.3, refers to item number 427 in the list of alternating knots with 12 crossings.

1.5 Algebra of knots

Denote by $\mathcal{K}$ the set of the equivalence classes of knots. It forms a commutative monoid (semigroup with a unit) under the connected sum of knots, and therefore we can construct the monoid algebra $\mathbb{Z}\mathcal{K}$ of $\mathcal{K}$. By definition, elements of $\mathbb{Z}\mathcal{K}$ are formal finite linear combinations $\sum \lambda_i K_i$, $\lambda_i \in \mathbb{Z}$, $K_i \in \mathcal{K}$, the product

---

[2] The combination of a professional lawyer and an amateur mathematician in one person is not new in the history of mathematics (think of Pierre Fermat!).

Table 1.1 *Prime knots, up to orientation and mirror images, with at most 8 crossings. Amphicheiral knots are marked by 'a', the (only) non-invertible minus-amphicheiral knot by 'na–'.*

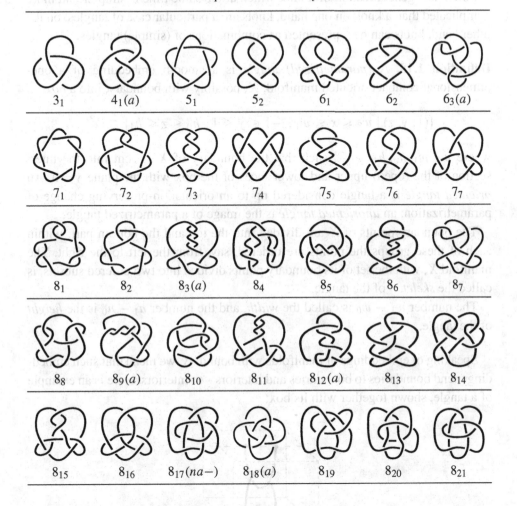

is defined by $(K_1, K_2) \mapsto K_1 \# K_2$ on knots and then extended by linearity to the entire space $\mathbb{Z}\mathcal{K}$. This algebra $\mathbb{Z}\mathcal{K}$ will be referred to as the *algebra of knots*.

The algebra of knots provides a convenient language for the study of *knot invariants* (see the next chapter): in these terms, a knot invariant is nothing but a linear functional on $\mathbb{Z}\mathcal{K}$. Ring homomorphisms from $\mathbb{Z}\mathcal{K}$ to some ring are referred to as *multiplicative invariants*; later, in Section 4.3, we shall see the importance of this notion.

In the sequel, we shall introduce more operations in this algebra, as well as in the dual algebra of knot invariants. We shall also study a filtration on $\mathbb{Z}\mathcal{K}$ that will give us the notion of a finite type knot invariant.

1.6 Tangles, string links and braids

1.6.1 The definition of a tangle

A *tangle* is a generalization of a knot which at the same time is simpler and more complicated than a knot: on one hand, knots are a particular case of tangles; on the other hand, knots can be represented as combinations of (simple) tangles.

Definition 1.18. A *(parametrized) tangle* is a smooth embedding of a one-dimensional compact oriented manifold, X, possibly with boundary, into a *box*

$$\{(x, y, z) \mid w_0 \leqslant x \leqslant w_1, \, -1 \leqslant y \leqslant 1, \, h_0 \leqslant z \leqslant h_1\} \subset \mathbb{R}^3,$$

where $w_0, w_1, h_0, h_1 \in \mathbb{R}$, such that the boundary of X is sent into the intersection of the (open) upper and lower faces of the box with the plane $y = 0$. An *oriented tangle* is a tangle considered up to an orientation-preserving change of parametrization; an *unoriented tangle* is the image of a parametrized tangle.

The boundary points of X are divided into the top and the bottom part; within each of these groups the points are ordered, say, from the left to the right. The manifold X, with the set of its boundary points divided into two ordered subsets, is called the *skeleton* of the tangle.

The number $w_1 - w_0$ is called the *width*, and the number $h_1 - h_0$ is the *height* of the tangle.

Speaking of embeddings of manifolds with boundary, we mean that such embeddings send boundaries to boundaries and interiors – to interiors. Here is an example of a tangle, shown together with its box:

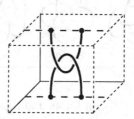

Usually the boxes will be omitted in the pictures.

We shall always identify tangles obtained by translations of boxes. Further, it will be convenient to have two notions of equivalences for tangles. Two tangles will be called *fixed-end isotopic* if one can be transformed into the other by a boundary-fixing isotopy of its box. We shall say that two tangles are simply *isotopic*, or
· *equivalent* if they become fixed-end isotopic after a suitable re-scaling of their boxes of the form

$$(x, y, z) \rightarrow (f(x), y, g(z)),$$

where f and g are strictly increasing functions.

1.6.2 Operations

In the case when the bottom of a tangle T_1 coincides with the top of another tangle T_2 of the same width (for oriented tangles we require the consistency of orientations, too), one can define the *product* $T_1 \cdot T_2$ by putting T_1 on top of T_2 (and, if necessary, smoothing out the corners at the joining points):

$$T_1 = \left(\begin{array}{c} \rule{0pt}{1.5em} \end{array}\right); \qquad T_2 = \begin{array}{c} \rule{0pt}{1.5em} \end{array}; \qquad T_1 \cdot T_2 = \left(\begin{array}{c} \rule{0pt}{1.5em} \end{array}\right).$$

Another operation, *tensor product*, is defined by placing one tangle next to the other tangle of the same height:

$$T_1 \otimes T_2 = \left(\begin{array}{c} \rule{0pt}{1.5em} \end{array}\right).$$

Both operations give rise to products on equivalence classes of tangles. The product of two equivalence classes is defined whenever the bottom of one tangle and the top of the other consist of the same number of points (with matching orientations in the case of oriented tangles), the tensor product is defined for any pair of equivalence classes.

1.6.3 Special types of tangles

Knots, links and braids are particular cases of tangles. For example, an n-component link is just a tangle whose skeleton is a union of n circles (and whose box is disregarded).

Let us fix n distinct points p_i on the top boundary of a box of unit width and let q_i be the projections of the p_i to the bottom boundary of the box. We choose the points p_i (and, hence, the q_i) to lie in the plane $y = 0$.

Definition 1.19. A *string link on n strings* (or *strands*) is an (unoriented) tangle whose skeleton consists of n intervals, the ith interval connecting p_i with q_i. A string link on one string is called a *long knot*.

Definition 1.20. A string link on n strings whose tangent vector is never horizontal is called a *pure braid on n strands*.

One difference between pure braids and string links is that the components of a string link can be knotted. However, there are string links with unknotted strands that are not equivalent to braids.

Let σ be a permutation of the set of n elements.

Definition 1.21. A *braid on n strands* is an (unoriented) tangle whose skeleton consists of n intervals, the ith interval connecting p_i with $q_{\sigma(i)}$, with the property that the tangent vector to it is never horizontal.

Pure braids are a specific case of braids with σ the identity permutation. Note that with our definition of equivalence, an isotopy between two braids can pass through tangles with points where the tangent vector is horizontal. Often, in the definition of the equivalence for braids it is required that an isotopy consist entirely of braids; the two approaches are equivalent.

The above definitions are illustrated by the following pictures:

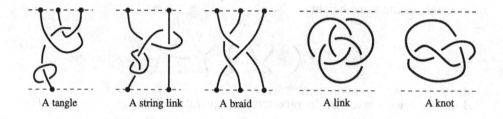

| A tangle | A string link | A braid | A link | A knot |

1.6.4 Braids

Braids are useful in the study of links, because any link can be represented as a *closure* of a braid (Alexander 1923):

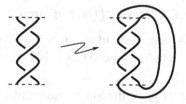

Braids are in many respects easier to work with, as they form groups under tangle multiplication: the set of equivalence classes of braids on n strands is the *braid group* denoted by B_n. A convenient set of generators for the group B_n consists of the elements $\sigma_i, i = 1, \ldots, n-1$:

$$[\cdots \times \cdots]_{i \ \ i+1}$$

which satisfy the following complete set of relations.

Far commutativity, $\sigma_i \sigma_j = \sigma_j \sigma_i$,   for $|i-j| > 1$.

Braiding relation, $\sigma_i \sigma_{i+1} \sigma_i = \sigma_{i+1} \sigma_i \sigma_{i+1}$,   for $i = 1, 2, \dots, n-2$.

Assigning to each braid in B_n the corresponding permutation σ, we get an epi-morphism $\pi : B_n \to S_n$ of the braid group on n strands onto the symmetric group on n letters. The kernel of π consists of pure braids and is denoted by P_n.

Theorem 1.22 (Markov 1935; Birman 1974). *Two closed braids are equivalent (as links) if and only if the braids are related by a finite sequence of the following Markov moves:*

(M1) $b \longleftrightarrow aba^{-1}$ *for any* $a, b \in B_n$;

(M2) $B_n \ni$ $\longleftrightarrow$ $\in B_{n+1}$, $\longleftrightarrow$.

1.6.5 Elementary tangles

A link can be cut into several *simple* tangles by a finite set of horizontal planes, and the link is equal to the product of all such tangles. Every simple tangle is a tensor product of the following *elementary* tangles.

Unoriented case:

$$\mathrm{id} := \; \big| \big| , \quad X_+ := \; \rlap{X}\; , \quad X_- := \; \rlap{X}\; , \quad \max := \; \cap , \quad \min := \; \cup .$$

Oriented case:

$$\mathrm{id} := \; \uparrow , \quad \mathrm{id}^* := \; \downarrow , \quad X_+ := \; \rlap{X}\; , \quad X_- := \; \rlap{X}\; ,$$

$$\overrightarrow{\max} := \; \cap , \quad \overleftarrow{\max} := \; \cap , \quad \overrightarrow{\min} := \; \cup , \quad \overleftarrow{\min} := \; \cup .$$

For example, the generator $\sigma_i \in B_n$ of the braid group is a simple tangle represented as the tensor product, $\sigma_i = \mathrm{id}^{\otimes(i-1)} \otimes X_+ \otimes \mathrm{id}^{\otimes(n-i-1)}$.

Exercise. Decompose the tangle 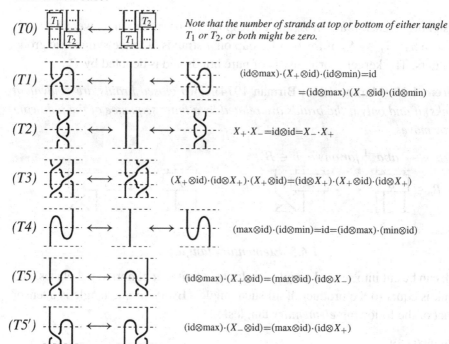 into elementary tangles.

1.6.6 The Turaev moves

Having presented a tangle as a product of simple tangles, it is natural to ask for an analogue of Reidemeister's Theorem 1.11 and Markov's Theorem 1.22, that is, a criterion for two such presentations to give isotopic tangles. Here is the answer.

Theorem 1.23 (Turaev 1990). *Two products of simple tangles are isotopic if and only if they are related by a finite sequence of the following Turaev moves.*

Unoriented case:

(T0)   Note that the number of strands at top or bottom of either tangle T_1 or T_2, or both might be zero.

(T1) $(\mathrm{id}\otimes\max)\cdot(X_+\otimes\mathrm{id})\cdot(\mathrm{id}\otimes\min)=\mathrm{id}$
    $=(\mathrm{id}\otimes\max)\cdot(X_-\otimes\mathrm{id})\cdot(\mathrm{id}\otimes\min)$

(T2) $X_+\cdot X_-=\mathrm{id}\otimes\mathrm{id}=X_-\cdot X_+$

(T3) $(X_+\otimes\mathrm{id})\cdot(\mathrm{id}\otimes X_+)\cdot(X_+\otimes\mathrm{id})=(\mathrm{id}\otimes X_+)\cdot(X_+\otimes\mathrm{id})\cdot(\mathrm{id}\otimes X_+)$

(T4) $(\max\otimes\mathrm{id})\cdot(\mathrm{id}\otimes\min)=\mathrm{id}=(\mathrm{id}\otimes\max)\cdot(\min\otimes\mathrm{id})$

(T5) $(\mathrm{id}\otimes\max)\cdot(X_+\otimes\mathrm{id})=(\max\otimes\mathrm{id})\cdot(\mathrm{id}\otimes X_-)$

(T5') $(\mathrm{id}\otimes\max)\cdot(X_-\otimes\mathrm{id})=(\max\otimes\mathrm{id})\cdot(\mathrm{id}\otimes X_+)$

Oriented case:

(T0) *Same as in the unoriented case with arbitrary orientations of participating strings.*

(T1 – T3) *Same as in the unoriented case with orientations of all strings from bottom to top.*

(T4) $(\overrightarrow{\max}\otimes\mathrm{id})\cdot(\mathrm{id}\otimes\min)=\mathrm{id}=(\mathrm{id}\otimes\overrightarrow{\max})\cdot(\min\otimes\mathrm{id})$

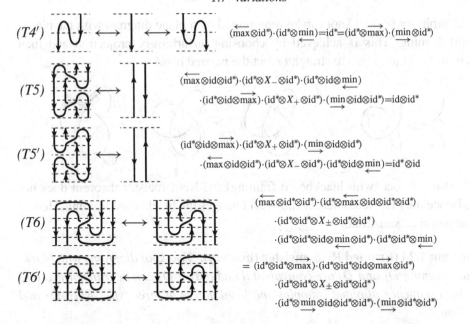

(T4') $\quad$ $(\overleftarrow{\max}\otimes\mathrm{id}^*)\cdot(\mathrm{id}^*\otimes\overleftarrow{\min})=\mathrm{id}^*=(\mathrm{id}^*\otimes\overrightarrow{\max})\cdot(\overrightarrow{\min}\otimes\mathrm{id}^*)$

(T5) $\quad$ $(\overleftarrow{\max}\otimes\mathrm{id}\otimes\mathrm{id}^*)\cdot(\mathrm{id}^*\otimes X_-\otimes\mathrm{id}^*)\cdot(\mathrm{id}^*\otimes\mathrm{id}\otimes\overleftarrow{\min})$
$\cdot(\mathrm{id}^*\otimes\mathrm{id}\otimes\overrightarrow{\max})\cdot(\mathrm{id}^*\otimes X_+\otimes\mathrm{id}^*)\cdot(\overrightarrow{\min}\otimes\mathrm{id}\otimes\mathrm{id}^*)=\mathrm{id}\otimes\mathrm{id}^*$

(T5') $\quad$ $(\mathrm{id}^*\otimes\mathrm{id}\otimes\overrightarrow{\max})\cdot(\mathrm{id}^*\otimes X_+\otimes\mathrm{id}^*)\cdot(\overrightarrow{\min}\otimes\mathrm{id}\otimes\mathrm{id}^*)$
$\cdot(\overleftarrow{\max}\otimes\mathrm{id}\otimes\mathrm{id}^*)\cdot(\mathrm{id}^*\otimes X_-\otimes\mathrm{id}^*)\cdot(\mathrm{id}^*\otimes\mathrm{id}\otimes\overleftarrow{\min})=\mathrm{id}^*\otimes\mathrm{id}$

(T6) $\quad$ $(\overleftarrow{\max}\otimes\mathrm{id}^*\otimes\mathrm{id}^*)\cdot(\mathrm{id}^*\otimes\overleftarrow{\max}\otimes\mathrm{id}\otimes\mathrm{id}^*\otimes\mathrm{id}^*)$
$\cdot(\mathrm{id}^*\otimes\mathrm{id}^*\otimes X_\pm\otimes\mathrm{id}^*\otimes\mathrm{id}^*)$
$\cdot(\mathrm{id}^*\otimes\mathrm{id}^*\otimes\mathrm{id}\otimes\overleftarrow{\min}\otimes\mathrm{id}^*)\cdot(\mathrm{id}^*\otimes\mathrm{id}^*\otimes\overleftarrow{\min})$

(T6') $\quad$ $=(\mathrm{id}^*\otimes\mathrm{id}^*\otimes\overrightarrow{\max})\cdot(\mathrm{id}^*\otimes\mathrm{id}^*\otimes\mathrm{id}\otimes\overrightarrow{\max}\otimes\mathrm{id}^*)$
$\cdot(\mathrm{id}^*\otimes\mathrm{id}^*\otimes X_\pm\otimes\mathrm{id}^*\otimes\mathrm{id}^*)$
$\cdot(\mathrm{id}^*\otimes\overrightarrow{\min}\otimes\mathrm{id}\otimes\mathrm{id}^*\otimes\mathrm{id}^*)\cdot(\overrightarrow{\min}\otimes\mathrm{id}^*\otimes\mathrm{id}^*)$

1.7 Variations

1.7.1 Framed knots

A *framed knot* is a knot equipped with a *framing*, that is, a smooth family of non-zero vectors perpendicular to the knot. Two framings are considered as equivalent, if one can be transformed to another by a smooth deformation. Up to this equivalence relation, a framing is uniquely determined by one integer: the linking number between the knot itself and the curve formed by a small shift of the knot in the direction of the framing. This integer, called the *self-linking number*, can be arbitrary. The framing with self-linking number n will be called the *n-framing* and a knot with the *n*-framing will be referred to as *n-framed*.

One way to choose a framing is to use the *blackboard framing*, defined by a plane knot projection, with the vector field everywhere parallel to the projection plane, for example:

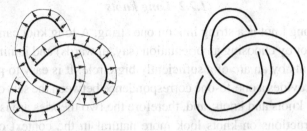

A framed knot can also be visualized as a *ribbon knot*, that is, a narrow knotted strip (see the right picture above).

An arbitrary framed knot can be represented by a plane diagram with the blackboard framing. This is achieved by choosing an arbitrary projection and then performing local moves to straighten out the twisted band:

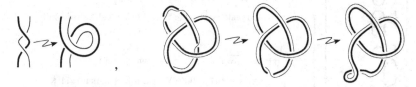

For framed knots (with blackboard framing) the Reidemeister theorem does not hold since the first Reidemeister move Ω_1 changes the blackboard framing. Here is an appropriate substitute.

Theorem 1.24 (Framed Reidemeister theorem). *Two knot diagrams with blackboard framing D_1 and D_2 are equivalent if and only if D_1 can be transformed into D_2 by a sequence of plane isotopies and local moves of three types $F\Omega_1$, Ω_2 and Ω_3, where*

$$F\Omega_1 : \qquad$$

and Ω_2 and Ω_3 are the ordinary Reidemeister moves.

One may also consider framed tangles. These are defined in the same manner as framed knots, with the additional requirement that at each boundary point of the tangle the normal vector is equal to $(\varepsilon, 0, 0)$ for some $\varepsilon > 0$. Framed tangles can be represented by tangle diagrams with blackboard framing. For such tangles there is an analogue of Turaev's theorem (Theorem 1.23): the Turaev move (T1) should be replaced by its framed version that mimics the move $F\Omega_1$.

1.7.2 Long knots

Recall that a long knot is a string link on one string. A long knot can be converted into a usual knot by choosing an orientation (say, upwards) and joining the top and the bottom points by an arc of a sufficiently big circle. It is easy to prove that this construction provides a one-to-one correspondence between the sets of equivalence classes of long knots and knots, and, therefore the two theories are isomorphic.

Some constructions on knots look more natural in the context of long knots. For example, the cut and paste procedure for the connected sum becomes a simple concatenation.

1.7.3 *Gauss diagrams and virtual knots*

Plane knot diagrams are convenient for presenting knots graphically, but for other purposes, such as coding knots in a computer-recognizable form, *Gauss diagrams* are better suited.

Definition 1.25. A Gauss diagram is an oriented circle with a distinguished set of distinct points divided into ordered pairs, each pair carrying a sign ± 1.

Graphically, an ordered pair of points on a circle can be represented by a chord with an arrow connecting them and pointing, say, to the second point. Gauss diagrams are considered up to orientation-preserving homeomorphisms of the circle. Sometimes, an additional basepoint is marked on the circle and the diagrams are considered up to homeomorphisms that keep the basepoint fixed. In this case, we speak of *based Gauss diagrams*.

To a plane knot diagram one can associate a Gauss diagram as follows. Pairs of points on the circle correspond to the values of the parameter where the diagram has a self-intersection, each arrow points from the overcrossing to the undercrossing and its sign is equal to the local writhe at the crossing.

Here is an example of a plane knot diagram and the corresponding Gauss diagram:

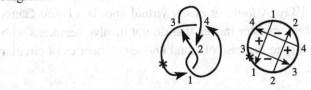

Exercise. What happens to a Gauss diagram, if (a) the knot is mirrored, (b) the knot is reversed?

A knot diagram can be uniquely reconstructed from the corresponding Gauss diagram. We call a Gauss diagram *realizable*, if it comes from a knot. Not every Gauss diagram is realizable, the simplest example being

with arbitrary signs of the arrows.

As we know, two oriented knot diagrams give the same knot type if and only if they are related by a sequence of oriented Reidemeister moves. The corresponding moves translated into the language of Gauss diagrams look as follows:

$V\Omega_1:$

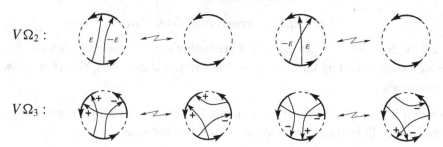

In fact, the two moves $V\Omega_3$ do not exhaust all the possibilities for representing the third Reidemeister move on Gauss diagrams. It can be shown, however, that all the other versions of the third move are combinations of the moves $V\Omega_2$ and $V\Omega_3$; see Exercises 1.24–1.26 for examples and Östlund (2001) for a proof.

These moves, of course, have a geometric meaning only for realizable diagrams. However, they make sense for all Gauss diagrams, whether realizable or not. In particular, a realizable diagram may be equivalent to non-realizable one:

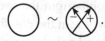

Definition 1.26. A *virtual knot* is a Gauss diagram considered up to the Reidemeister moves $V\Omega_1$, $V\Omega_2$, $V\Omega_3$. A *long*, or *based* virtual knot is a based Gauss diagram, considered up to Reidemeister moves that do not involve segments with the basepoint on them. Contrary to the case of usual knots, the theories of circular and long virtual knots differ.

It can be shown that the isotopy classes of knots form a subset of the set of virtual knots. In other words, if there is a chain of Reidemeister moves connecting two realizable Gauss diagrams, we can always modify it so that it goes only though realizable diagrams.

Virtual knots were introduced by L. Kauffman (1999). Almost at the same time, they turned up in the work of M. Goussarov, M. Polyak and O. Viro (2000). There are various geometric interpretations of virtual knots. Many knot invariants are known to extend to invariants of virtual knots.

1.7.4 Knots in arbitrary manifolds

We have defined knots as embeddings of the circle into the Euclidean space $\mathbb{R}^3$. In this definition $\mathbb{R}^3$ can be replaced by the 3-sphere S^3, since the one-point compactification $\mathbb{R}^3 \to S^3$ establishes a one-to-one correspondence between the equivalence classes of knots in both manifolds. Going further and replacing $\mathbb{R}^3$ by an arbitrary 3-manifold M, we can arrive to a theory of knots in M which may well be different from the usual case of knots in $\mathbb{R}^3$; see, for instance, Kalfagianni (1998) and Vassiliev (1998).

If the dimension of the manifold M is bigger than 3, then all knots in M that represent the same element of the fundamental group $\pi_1(M)$, are isotopic. It does not mean, however, that the theory of knots in M is trivial: the space of all embeddings $S^1 \to M$ may have non-trivial higher homology groups. These homology groups are certainly of interest in dimension 3 too; see Vassiliev (1998). Another way of doing knot theory in higher-dimensional manifolds is studying multidimensional knots, like embeddings $S^2 \to \mathbb{R}^4$; see, for example, Rolfsen (1976). An analogue of knot theory for 2-manifolds is Arnold's theory of immersed curves (Arnold 1990).

Exercises

1.1 Find the following knots in the knot table (Table 1.1):

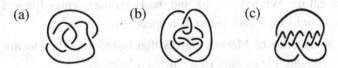

(a)　　　　　　(b)　　　　　　(c)

1.2 Can you find the following links in the picture on page 5?

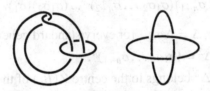

1.3 Borromean rings (see page 5) have the property that after deleting any component the remaining two-component link becomes trivial. Links with such property are called *Brunnian*. Find a Brunnian link with 4 components.

1.4 Table 1.1 shows 35 topological types of knots up to change of orientation and taking the mirror images. How many distinct knots do these represent?

1.5 Find an isotopy that transforms the knot 6_3 into its mirror image $\overline{6_3}$.

1.6 Repeat Perko's achievement: find an isotopy that transforms one of the knots of the Perko pair into another one.

1.7 Let G_n be the *Goeritz diagram* (Goeritz 1934) with $2n + 5$ crossings, as in the figure below.

1. Show that G_n represents a trivial knot.

$$G_n = $$ *n crossings*　　*n+1 crossings*

2. Prove that for $n \geqslant 3$ in any sequence of the Reidemeister moves transforming G_n into the plane circle there is an intermediate knot diagram with more than $2n + 5$ crossings.

3. Find a sequence of 23 Reidemeister moves (containing the Ω_1 move 5 times, the Ω_2 move 7 times and the Ω_3 move 11 times), transforming G_3 into the plane circle. See the picture of G_3 on page 4.

1.8 Decompose the knot on the right into a connected sum of prime knots.

1.9 Show that by changing some crossings from overcrossing to undercrossing or vice versa, any knot diagram can be transformed into a diagram of the unknot.

1.10 (Adams 1994) Show that by changing some crossings from overcrossing to undercrossing or vice versa, any knot diagram can be made alternating.

1.11 Represent the knots $4_1, 5_1, 5_2$ as closed braids.

1.12 Analogously to the braid closure, one can define the closure of a string link. Represent the Whitehead link and the Borromean rings (page 5) as closures of string links on 2 and 3 strands respectively.

1.13 Find a sequence of Markov moves that transforms the closure of the braid $\sigma_1^2\sigma_2^3\sigma_1^4\sigma_2$ into the closure of the braid $\sigma_1^2\sigma_2\sigma_1^4\sigma_2^3$.

1.14 Garside's *fundamental braid* $\Delta \in B_n$ is defined as

$$\Delta := (\sigma_1\sigma_2\ldots\sigma_{n-1})(\sigma_1\sigma_2\ldots\sigma_{n-2})\ldots(\sigma_1\sigma_2)(\sigma_1).$$

$\Delta = $

1. Prove that $\sigma_i\Delta = \Delta\sigma_{n-i}$ for every standard generator $\sigma_i \in B_n$.

2. Prove that $\Delta^2 = (\sigma_1\sigma_2\ldots\sigma_{n-1})^n$.

3. Check that Δ^2 belongs to the centre $Z(B_n)$ of the braid group.

4. Show that any braid can be represented as a product of a certain power (possibly negative) of Δ and a *positive braid*, that is, a braid that contains only positive powers of standard generators σ_i.

In fact, for $n \geqslant 3$, the centre $Z(B_n)$ is the infinite cyclic group generated by Δ^2. The word and conjugacy problems in the braid group were solved by F. Garside (1969). The structure of positive braids that occur in the last statement was studied in Adyan (1984) and El-Rifai and Morton (1994).

1.15 1. Prove that the sign of the permutation corresponding to a braid b is equal to the parity of the number of crossings of b, that is $(-1)^{\ell(b)}$, where $\ell(b)$ is the length of b as a word in generators $\sigma_1, \ldots, \sigma_{n-1}$.

2. Prove that the subgroup P_n of pure braids is generated by the braids A_{ij} linking the ith and jth strands with each other behind all other strands.

$A_{ij} = $

1.16 Let V be a vector space of dimension n with a distinguished basis $e_1, \ldots, e_n$, and let Ξ_i be the counterclockwise $90°$ rotation in the plane $\langle e_i, e_{i+1} \rangle$: $\Xi_i(e_i) = e_{i+1}$, $\Xi_i(e_{i+1}) = -e_i$, $\Xi_i(e_j) = e_j$ for $j \neq i, i+1$. Prove that sending each elementary generator $\sigma_i \in B_n$ to Ξ_i we get a representation $B_n \to GL_n(\mathbb{R})$ of the braid group.

1.17 **Burau representation.** Consider the free module over the ring of Laurent polynomials $\mathbb{Z}[x^{\pm 1}]$ with a basis $e_1, \ldots, e_n$. The *Burau representation* $B_n \to GL_n(\mathbb{Z}[x^{\pm 1}])$ sends $\sigma_i \in B_n$ to the linear operator that transforms e_i into $(1 - x)e_i + e_{i+1}$, and e_{i+1} into $x e_i$.

1. Prove that it is indeed a representation of the braid group.
2. The Burau representation is reducible. It splits into the trivial one-dimensional representation and an $(n-1)$-dimensional irreducible representation which is called the *reduced Burau representation*.

   Find a basis of the reduced Burau representation where the matrices have the form

$$
\sigma_1 \mapsto \begin{pmatrix} -x & x & \ldots & 0 \\ 0 & 1 & \ldots & 0 \\ \vdots & \vdots & \ddots & \vdots \\ 0 & 0 & \ldots & 1 \end{pmatrix}, \quad \sigma_i \mapsto \begin{pmatrix} 1 & & & \\ & \ddots & & \\ & 1 & 0 & 0 \\ & 1 & -x & x \\ & 0 & 0 & 1 \\ & & & & \ddots \\ & & & & & 1 \end{pmatrix}, \quad \sigma_{n-1} \mapsto \begin{pmatrix} 1 & \ldots & 0 & 0 \\ \vdots & \ddots & \vdots & \vdots \\ 0 & \ldots & 1 & 0 \\ 0 & \ldots & 1 & -x \end{pmatrix}.
$$

Answer to 2. $\{xe_1 - e_2, xe_2 - e_3, \ldots, xe_{n-1} - e_n\}$.

The Burau representation is faithful for $n \leqslant 3$ (Birman 1974), and not faithful for $n \geqslant 5$ (Bigelow 1999). The case $n = 4$ remains open.

1.18 **Lawrence–Krammer–Bigelow representation.** Let V be a free $\mathbb{Z}[q^{\pm 1}, t^{\pm 1}]$ module of dimension $n(n-1)/2$ with a basis $e_{i,j}$ for $1 \leqslant i < j \leqslant n$. The *Lawrence–Krammer–Bigelow representation* can be defined via the action of $\sigma_k \in B_n$ on V:

$$
\sigma_k(e_{i,j}) = \begin{cases} e_{i,j} & \text{if } k < i-1 \text{ or } k > j, \\ e_{i-1,j} + (1-q)e_{i,j} & \text{if } k = i-1, \\ tq(q-1)e_{i,i+1} + qe_{i+1,j} & \text{if } k = i < j-1, \\ tq^2 e_{i,j} & \text{if } k = i = j-1, \\ e_{i,j} + tq^{k-i}(q-1)^2 e_{k,k+1} & \text{if } i < k < j-1, \\ e_{i,j-1} + tq^{j-i}(q-1)e_{j-1,j} & \text{if } i < k = j-1, \\ (1-q)e_{i,j} + qe_{i,j+1} & \text{if } k = j. \end{cases}
$$

Prove that this assignment determines a representation of the braid group. It was shown by Bigelow (2001) and Krammer (2002) that this representation is faithful for any $n \geqslant 1$. Therefore the braid group is a linear group.

1.19 Represent the knots $4_1, 5_1, 5_2$ as products of simple tangles.

1.20 Consider the following two knots given as products of simple tangles:

$$
(\overleftarrow{\max} \otimes \overrightarrow{\max}) \cdot (\mathrm{id}^* \otimes X_+ \otimes \mathrm{id}^*) \cdot (\mathrm{id}^* \otimes X_+ \otimes \mathrm{id}^*) \cdot (\mathrm{id}^* \otimes X_+ \otimes \mathrm{id}^*) \cdot (\overrightarrow{\min} \otimes \overleftarrow{\min})
$$

and

$$\overrightarrow{\max} \cdot (\mathrm{id} \otimes \overrightarrow{\max} \otimes \mathrm{id}^*) \cdot (X_+ \otimes \mathrm{id}^* \otimes \mathrm{id}^*) \cdot (X_+ \otimes \mathrm{id}^* \otimes \mathrm{id}^*) \cdot (X_+ \otimes \mathrm{id}^* \otimes \mathrm{id}^*) \cdot (\mathrm{id} \otimes \overleftarrow{\min} \otimes \mathrm{id}^*) \cdot \overleftarrow{\min}$$

1. Show that these two knots are equivalent.
2. Indicate a sequence of the Turaev moves that transforms one product into another.
3. Forget about the orientations and consider the corresponding unoriented tangles. Find a sequence of unoriented Turaev moves that transforms one product into another.

1.21 Represent the oriented tangle move on the right as a sequence of oriented Turaev moves from Theorem 1.23.

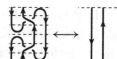

1.22 **Whitney trick.** Show that the move $F\Omega_1$ in the framed Reidemeister theorem (Theorem 1.24) can be replaced by the move shown on the right.

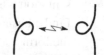

1.23 The group $\mathbb{Z}_2^{k+1}$ acts on oriented k-component links, changing the orientation of each component and taking the mirror image of the link. How many different links are there in the orbit of an oriented Whitehead link under this action?

1.24 Show that each of the moves $V\Omega_3$ can be obtained as a combination of the moves $V\Omega_2$ with the moves $V\Omega_3'$ below:

$V\Omega_3'$:

Conversely, show that the moves $V\Omega_3'$ can be obtained as combinations of the moves $V\Omega_2$ and $V\Omega_3$.

1.25 Show that the following moves are equivalent modulo $V\Omega_2$.

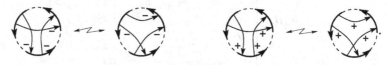

This means that either one can be obtained as a combination of another one with the $V\Omega_2$ moves.

1.26 (Östlund 2001) Show that the second version of $V\Omega_2$:

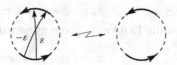

is redundant. It can be obtained as a combination of the first version,

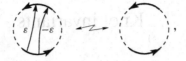

with the moves $V\Omega_1$ and $V\Omega_3$.

1.27 (Polyak 2010) Show that the following moves

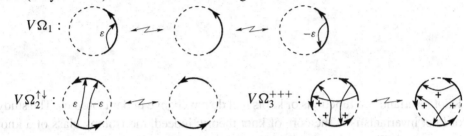

are sufficient to generate all Reidemeister moves $V\Omega_1$, $V\Omega_2$, $V\Omega_3$.

2

Knot invariants

Knot invariants are functions of knots that do not change under isotopies. The study of knot invariants is at the core of knot theory; indeed, the isotopy class of a knot is, tautologically, a knot invariant.

2.1 Definition and first examples

Let $\mathcal{K}$ be the set of all equivalence classes of knots. A *knot invariant with values in a set S* is a map from $\mathcal{K}$ to S. In the same way one can speak of invariants of links, framed knots, etc.

2.1.1 Crossing number

Any knot can be represented by a plane diagram in infinitely many ways.

Definition 2.1. The *crossing number* $c(K)$ of a knot K is the minimal number of crossing points in a plane diagram of K.

Exercise. Prove that if $c(K) \leqslant 2$, then the knot K is trivial.

It follows that the minimal number of crossings required to draw a diagram of a nontrivial knot is at least 3. A little later we shall see that the trefoil knot is indeed nontrivial.

Obviously, $c(K)$ is a knot invariant taking values in the set of non-negative integers.

2.1.2 Unknotting number

Another integer-valued invariant of knots which admits a simple definition is the unknotting number.

Represent a knot by a plane diagram. The diagram can be transformed by plane isotopies, Reidemeister moves and crossing changes:

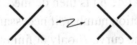

As we know, modifications of the first two kinds preserve the topological type of the knot, and only crossing switches can change it.

Definition 2.2. The *unknotting number* $u(K)$ of a knot K is the minimal number of crossing changes in a plane diagram of K that convert it to a trivial knot, provided that any number of plane isotopies and Reidemeister moves is also allowed.

Exercise. What is the unknotting number of the knots 3_1 and 8_3?

Finding the unknotting number, if it is greater than 1, is a difficult task; for example, the second question of the previous exercise was answered only in 1986 (Kanenobu and Murakami 1986).

2.1.3 Knot group

The *knot group* is the fundamental group of the complement to the knot in the ambient space: $\pi(K) = \pi_1(\mathbb{R}^3 \setminus K)$. The knot group is a very strong invariant. For example, a knot is trivial if and only if its group is infinite cyclic. More generally, two prime knots with isomorphic fundamental groups are isotopic. For a detailed discussion of knot groups see Lickorish (1997).

Exercise. Prove that

1. the group of the trefoil is generated by two elements x, y with one relation $x^2 = y^3$;
2. this group is isomorphic to the braid group B_3 (in terms of x, y find another pair of generators a, b that satisfy $aba = bab$).

2.2 Linking number

2.2.1 Definition

The *linking number* is an example of a Vassiliev invariant of two-component links; it has an analog for framed knots, called *self-linking number*.

Intuitively, the linking number $lk(A, B)$ of two oriented spatial curves A and B is the number of times A winds around B. To give a precise definition, choose an oriented disk D_A immersed in space so that its oriented boundary is the curve A

(this means that the ordered pair consisting of an outward-looking normal vector to A and the orienting tangent vector to A gives a positive basis in the tangent space to D_A). The linking number $lk(A, B)$ is then defined as the intersection number of D_A and B. To find the intersection number, if necessary, make a small perturbation of D_A so as to make it meet the curve B only at finitely many points of transversal intersection. At each intersection point, define the sign to be equal to ± 1 depending on the orientations of D_A and B at this point. More specifically, let (e_1, e_2) be a positive pair of tangent vectors to D_A, while e_3 a positively directed tangent vector to B at the intersection point; the sign is set to $+1$ if and only if the frame (e_1, e_2, e_3) defines a positive orientation of $\mathbb{R}^3$. Then the linking number $lk(A, B)$ is the sum of these signs over all intersection points $p \in D_A \cap B$. One can prove that the result does not depend on the choice of the surface D_A and that $lk(A, B) = lk(B, A)$.

Example 2.3. The two curves shown in the picture

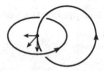

have their linking number equal to -1.

Given a plane diagram of a two-component link, there is a simple combinatorial formula for the linking number. Let I be the set of crossing points involving branches of both components A and B (crossing points involving branches of only one component are irrelevant here). Then I is the disjoint union of two subsets I_B^A (points where A passes over B) and I_A^B (where B passes over A).

Proposition 2.4.

$$lk(A, B) = \sum_{p \in I_B^A} w(p) = \sum_{p \in I_A^B} w(p) = \frac{1}{2} \sum_{p \in I} w(p)$$

where $w(p) = \pm 1$ is the local writhe of the crossing point.

Proof Crossing changes at all points $p \in I_A^B$ make the two components unlinked. Call the new curves A' and B', then $lk(A', B') = 0$. It is clear from the pictures below that each crossing switch changes the linking number by $-w$ where w is the local writhe:

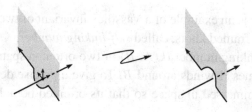

Therefore, $lk(A, B) - \sum_{p \in I_A^B} w(p) = 0$, and the assertion follows. □

Example 2.5. For the two curves below, both ways to compute the linking number give $+1$:

There are various integral formulae for the linking number. The most famous

2.2.2 Integral formulae

There are various integral formulae for the linking number. The most famous formula was found by Gauss (see Spivak 1979 for a proof).

Theorem 2.6. *Let A and B be two non-intersecting curves in $\mathbb{R}^3$, parametrized, respectively, by the smooth functions $\alpha, \beta : S^1 \to \mathbb{R}^3$. Then*

$$lk(A, B) = \frac{1}{4\pi} \int_{S^1 \times S^1} \frac{(\beta(v) - \alpha(u), du, dv)}{|\beta(v) - \alpha(u)|^3},$$

where the parentheses in the numerator stand for the mixed product of 3 vectors.

Geometrically, this formula computes the degree of the Gauss map from $A \times B = S^1 \times S^1$ to the 2-sphere S^2, that is, the number of times the normalized vector connecting a point on A to a point on B goes around the sphere.

A different integral formula for the linking number will be stated and proved in Chapter 8. It represents the simplest term of the *Kontsevich integral*, which encodes all Vassiliev invariants.

2.2.3 Self-linking

Let K be a framed knot and let K' be the knot obtained from K by a small shift in the direction of the framing.

Definition 2.7. The *self-linking number* of K is the linking number of K and K'.

Note, by the way, that the linking number is the same if K is shifted in the direction *opposite* to the framing.

Proposition 2.8. *The self-linking number of a framed knot given by a diagram D with blackboard framing is equal to the total writhe of the diagram D.*

Proof Indeed, in the case of blackboard framing, the only crossings of K with K' occur near the crossing points of K. The neighbourhood of each crossing point looks like

The local writhe of the crossing where K passes over K' is the same as the local writhe of the crossing point of the knot K with itself. Therefore, the claim follows from the combinatorial formula for the linking number on page 28.   □

2.3 The Conway polynomial

2.3.1 Definition

In what follows we shall usually consider invariants with values in a commutative ring. Of special importance in knot theory are *polynomial knot invariants* taking values in the rings of polynomials (or Laurent polynomials[1]) in one or several variables, usually with integer coefficients.

Historically, the first polynomial invariant for knots was the *Alexander polynomial A(K)* (Alexander 1928). See Crowell and Fox (1963); Lickorish (1997); and Rolfsen (1976) for a discussion of the beautiful topological theory related to the Alexander polynomial. In 1970 J. Conway (1970) found a simple recursive construction of a polynomial invariant $C(K)$ which differs from the Alexander polynomial only by a change of variable, namely, $A(K) = C(K) \mid_{t \mapsto x^{1/2} - x^{-1/2}}$. In this book, we only use Conway's normalization. Conway's definition, given in terms of plane diagrams, relies on crossing point resolutions that may take a knot diagram into a link diagram; therefore, we shall speak of links rather than knots.

Definition 2.9. The Conway polynomial C is an invariant of oriented links (and, in particular, an invariant of oriented knots) taking values in the ring $\mathbb{Z}[t]$ and defined by the two properties:

---

[1] A *Laurent polynomial* in x is a polynomial in x and x^{-1}.

$$C\left(\bigcirc\right) = 1,$$

$$C\left(\overset{\nearrow}{\times}\right) - C\left(\overset{\nwarrow}{\times}\right) = tC\left(\smile\frown\right).$$

Here $\bigcirc$ stands for the unknot (trivial knot) while the three pictures in the second line stand for three diagrams that are identical everywhere except for the fragments shown. The second relation is referred to as *Conway's skein relation*. Skein relations are equations on the values of some functions on knots (links, etc.) represented by diagrams that differ from each other by local changes near a crossing point. These relations often give a convenient way to work with knot invariants.

2.3.2 Calculations

It is not quite trivial to prove the existence of an invariant satisfying this definition, but as soon as this fact is established, the computation of the Conway polynomial becomes fairly easy.

Example 2.10.

(i) $\quad C\left(\bigcirc\bigcirc\right) = \dfrac{1}{t}C\left(\infty\right) - \dfrac{1}{t}C\left(\infty\right) = 0,$

because the two knots on the right are equivalent (both are trivial).

(ii) $\quad C\left(\bigcirc\!\!\bigcirc\right) = C\left(\bigcirc\!\!\bigcirc\right) - tC\left(\bigcirc\!\!\bigcirc\right)$

$$= C\left(\bigcirc\bigcirc\right) - tC\left(\bigcirc\right) = -t.$$

(iii) $\quad C\left(\text{trefoil}\right) = C\left(\text{knot}\right) - tC\left(\text{link}\right)$

$$= C\left(\bigcirc\right) - tC\left(\bigcirc\!\!\bigcirc\right) = 1 + t^2.$$

Table 2.1 *Conway polynomials of knots with up to 8 crossings*

K	$C(K)$	K	$C(K)$	K	$C(K)$
3_1	$1+t^2$	7_6	$1+t^2-t^4$	8_{11}	$1-t^2-2t^4$
4_1	$1-t^2$	7_7	$1-t^2+t^4$	8_{12}	$1-3t^2+t^4$
5_1	$1+3t^2+t^4$	8_1	$1-3t^2$	8_{13}	$1+t^2+2t^4$
5_2	$1+2t^2$	8_2	$1-3t^4-t^6$	8_{14}	$1-2t^4$
6_1	$1-2t^2$	8_3	$1-4t^2$	8_{15}	$1+4t^2+3t^4$
6_2	$1-t^2-t^4$	8_4	$1-3t^2-2t^4$	8_{16}	$1+t^2+2t^4+t^6$
6_3	$1+t^2+t^4$	8_5	$1-t^2-3t^4-t^6$	8_{17}	$1-t^2-2t^4-t^6$
7_1	$1+6t^2+5t^4+t^6$	8_6	$1-2t^2-2t^4$	8_{18}	$1+t^2-t^4-t^6$
7_2	$1+3t^2$	8_7	$1+2t^2+3t^4+t^6$	8_{19}	$1+5t^2+5t^4+t^6$
7_3	$1+5t^2+2t^4$	8_8	$1+2t^2+2t^4$	8_{20}	$1+2t^2+t^4$
7_4	$1+4t^2$	8_9	$1-2t^2-3t^4-t^6$	8_{21}	$1-t^4$
7_5	$1+4t^2+2t^4$	8_{10}	$1+3t^2+3t^4+t^6$		

The values of the Conway polynomial on knots with up to 8 crossings are given in Table 2.1. Note that the Conway polynomials of the inverse knot K^* and the mirror knot $\overline{K}$ coincide with that of knot K.

For every n, the coefficient c_n of t^n in C is a numerical invariant of the knot.

The behaviour of the Conway polynomial under the change of orientation of one component of a link does not follow any known rules. Here is an example.

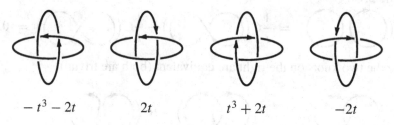

$$-t^3-2t \qquad\qquad 2t \qquad\qquad t^3+2t \qquad\qquad -2t$$

2.4 The Jones polynomial

2.4.1 The Kauffman bracket and the Jones polynomial

The invention of the Jones polynomial (Jones 1985) produced a genuine revolution in knot theory. The original construction of V. Jones was given in terms of state sums and von Neumann algebras. It was soon noted, however, that the Jones polynomial can be defined by skein relations, in the spirit of Conway's definition in Section 2.3.1.

Instead of simply giving the corresponding formal equations, we explain, following L. Kauffman (1987b), *how* this definition could be invented. As with the Conway polynomial, the construction given below requires that we consider invariants on the totality of all links, not only knots, because the transformations used may turn a knot diagram into a link diagram with several components.

Suppose that we are looking for an invariant of unoriented links, denoted by angular brackets, that has a prescribed behaviour with respect to the resolution of diagram crossings and the addition of a disjoint copy of the unknot:

$$\langle \times \rangle = a\langle \rangle\langle \rangle + b\langle \asymp \rangle,$$

$$\langle L \sqcup \bigcirc \rangle = c\langle L \rangle,$$

where a, b and c are certain fixed coefficients.

For the bracket $\langle\,,\,\rangle$ to be a link invariant, it must be stable under the three Reidemeister moves Ω_1, Ω_2, Ω_3 (see Section 1.2).

Exercise. Show that the bracket $\langle\,,\,\rangle$ is Ω_2-invariant if and only if $b = a^{-1}$ and $c = -a^2 - a^{-2}$. Prove that Ω_2-invariance in this case implies Ω_3-invariance.

Exercise. Suppose that $b = a^{-1}$ and $c = -a^2 - a^{-2}$. Check that the behaviour of the bracket with respect to the first Reidemeister move is described by the equations

$$\langle \text{ } \rangle = -a^{-3}\langle \text{ } \rangle,$$

$$\langle \text{ } \rangle = -a^{3}\langle \text{ } \rangle.$$

In the assumptions $b = a^{-1}$ and $c = -a^2 - a^{-2}$, the bracket polynomial $\langle L \rangle$ normalized by the initial condition

$$\langle \bigcirc \rangle = 1$$

is referred to as the *Kauffman bracket* of L. We see that the Kauffman bracket changes only under the addition (or deletion) of a small loop, and this change depends on the local writhe of the corresponding crossing. It is easy, therefore, to write a formula for a quantity that would be invariant under all three Reidemeister moves:

$$J(L) = (-a)^{-3w}\langle L \rangle,$$

where w is the total writhe of the diagram (the difference between the number of positive and negative crossings).

The invariant $J(L)$ is a Laurent polynomial called the *Jones polynomial* (in *a-normalization*). The more standard *t-normalization* is obtained by the substitution $a = t^{-1/4}$. Note that the Jones polynomial is an invariant of an oriented link, although in its definition we use the Kauffman bracket which is determined by a diagram without orientation.

Exercise. Check that the Jones polynomial is uniquely determined by the skein relation

$$t^{-1} J\left(\overset{\nearrow\;\nwarrow}{\times}\right) - t J\left(\overset{\nwarrow\;\nearrow}{\times}\right) = (t^{1/2} - t^{-1/2}) J\left(\right) \left(\right) \qquad (2.1)$$

and the initial condition

$$J\left(\bigcirc\right) = 1. \qquad (2.2)$$

Example 2.11. Let us compute the value of the Jones polynomial on the left trefoil 3_1. The calculation requires several steps, each consisting of one application of the rule (2.1) and some applications of rule (2.2) and/or using the results of the previous steps. We leave the details to the reader.

(i) $\quad J\left(\bigcirc\bigcirc\right) = -t^{1/2} - t^{-1/2}.$

(ii) $\quad J\left(\right) = -t^{-5/2} - t^{-1/2}.$

(iii) $\quad J\left(\right) = -t^{-4} + t^{-3} + t^{-1}.$

Exercise. Repeat the previous calculation for the right trefoil and prove that $J(\overline{3_1}) = t + t^3 - t^4.$

We see that the Jones polynomial J can tell apart two knots which the Conway polynomial C cannot. This does not mean, however, that J is stronger than C. There are pairs of knots, for example, $K_1 = 10_{71}$, $K_2 = 10_{104}$ such that $J(K_1) = J(K_2)$, but $C(K_1) \neq C(K_2)$ (see, for instance, the knot atlas or tables at A. Stoimenow's homepage).

2.4.2 The values of the Jones polynomial

The values of the Jones polynomial on standard knots with up to 8 crossings are given in Table 2.2. The Jones polynomial does not change when the knot is inverted (this is no longer true for links), see Exercise 2.25. The behaviour of the Jones polynomial under mirror reflection is described in Exercise 2.24.

Table 2.2 *Jones polynomials of knots with up to 8 crossings*

3_1	$-t^{-4} + t^{-3} + t^{-1}$
4_1	$t^{-2} - t^{-1} + 1 - t + t^2$
5_1	$-t^{-7} + t^{-6} - t^{-5} + t^{-4} + t^{-2}$
5_2	$-t^{-6} + t^{-5} - t^{-4} + 2t^{-3} - t^{-2} + t^{-1}$
6_1	$t^{-4} - t^{-3} + t^{-2} - 2t^{-1} + 2 - t + t^2$
6_2	$t^{-5} - 2t^{-4} + 2t^{-3} - 2t^{-2} + 2t^{-1} - 1 + t$
6_3	$-t^{-3} + 2t^{-2} - 2t^{-1} + 3 - 2t + 2t^2 - t^3$
7_1	$-t^{-10} + t^{-9} - t^{-8} + t^{-7} - t^{-6} + t^{-5} + t^{-3}$
7_2	$-t^{-8} + t^{-7} - t^{-6} + 2t^{-5} - 2t^{-4} + 2t^{-3} - t^{-2} + t^{-1}$
7_3	$t^2 - t^3 + 2t^4 - 2t^5 + 3t^6 - 2t^7 + t^8 - t^9$
7_4	$t - 2t^2 + 3t^3 - 2t^4 + 3t^5 - 2t^6 + t^7 - t^8$
7_5	$-t^{-9} + 2t^{-8} - 3t^{-7} + 3t^{-6} - 3t^{-5} + 3t^{-4} - t^{-3} + t^{-2}$
7_6	$-t^{-6} + 2t^{-5} - 3t^{-4} + 4t^{-3} - 3t^{-2} + 3t^{-1} - 2 + t$
7_7	$-t^{-3} + 3t^{-2} - 3t^{-1} + 4 - 4t + 3t^2 - 2t^3 + t^4$
8_1	$t^{-6} - t^{-5} + t^{-4} - 2t^{-3} + 2t^{-2} - 2t^{-1} + 2 - t + t^2$
8_2	$t^{-8} - 2t^{-7} + 2t^{-6} - 3t^{-5} + 3t^{-4} - 2t^{-3} + 2t^{-2} - t^{-1} + 1$
8_3	$t^{-4} - t^{-3} + 2t^{-2} - 3t^{-1} + 3 - 3t + 2t^2 - t^3 + t^4$
8_4	$t^{-5} - 2t^{-4} + 3t^{-3} - 3t^{-2} + 3t^{-1} - 3 + 2t - t^2 + t^3$
8_5	$1 - t + 3t^2 - 3t^3 + 3t^4 - 4t^5 + 3t^6 - 2t^7 + t^8$
8_6	$t^{-7} - 2t^{-6} + 3t^{-5} - 4t^{-4} + 4t^{-3} - 4t^{-2} + 3t^{-1} - 1 + t$
8_7	$-t^{-2} + 2t^{-1} - 2 + 4t - 4t^2 + 4t^3 - 3t^4 + 2t^5 - t^6$
8_8	$-t^{-3} + 2t^{-2} - 3t^{-1} + 5 - 4t + 4t^2 - 3t^3 + 2t^4 - t^5$
8_9	$t^{-4} - 2t^{-3} + 3t^{-2} - 4t^{-1} + 5 - 4t + 3t^2 - 2t^3 + t^4$
8_{10}	$-t^{-2} + 2t^{-1} - 3 + 5t - 4t^2 + 5t^3 - 4t^4 + 2t^5 - t^6$
8_{11}	$t^{-7} - 2t^{-6} + 3t^{-5} - 5t^{-4} + 5t^{-3} - 4t^{-2} + 4t^{-1} - 2 + t$
8_{12}	$t^{-4} - 2t^{-3} + 4t^{-2} - 5t^{-1} + 5 - 5t + 4t^2 - 2t^3 + t^4$
8_{13}	$-t^{-3} + 3t^{-2} - 4t^{-1} + 5 - 5t + 5t^2 - 3t^3 + 2t^4 - t^5$
8_{14}	$t^{-7} - 3t^{-6} + 4t^{-5} - 5t^{-4} + 6t^{-3} - 5t^{-2} + 4t^{-1} - 2 + t$
8_{15}	$t^{-10} - 3t^{-9} + 4t^{-8} - 6t^{-7} + 6t^{-6} - 5t^{-5} + 5t^{-4} - 2t^{-3} + t^{-2}$
8_{16}	$-t^{-6} + 3t^{-5} - 5t^{-4} + 6t^{-3} - 6t^{-2} + 6t^{-1} - 4 + 3t - t^2$
8_{17}	$t^{-4} - 3t^{-3} + 5t^{-2} - 6t^{-1} + 7 - 6t + 5t^2 - 3t^3 + t^4$
8_{18}	$t^{-4} - 4t^{-3} + 6t^{-2} - 7t^{-1} + 9 - 7t + 6t^2 - 4t^3 + t^4$
8_{19}	$t^3 + t^5 - t^8$
8_{20}	$-t^{-5} + t^{-4} - t^{-3} + 2t^{-2} - t^{-1} + 2 - t$
8_{21}	$t^{-7} - 2t^{-6} + 2t^{-5} - 3t^{-4} + 3t^{-3} - 2t^{-2} + 2t^{-1}$

2.5 Algebra of knot invariants

Knot invariants with values in a given commutative ring $\mathcal{R}$ form an algebra $\mathcal{I}$ over that ring with respect to usual pointwise operations on functions

$$(f + g)(K) = f(K) + g(K),$$
$$(fg)(K) = f(K)g(K).$$

Extending knot invariants by linearity to the whole algebra of knots we see that

$$\mathscr{I} = \mathrm{Hom}_{\mathbb{Z}}(\mathbb{Z}\mathscr{K}, \mathscr{R}).$$

In particular, as an $\mathscr{R}$-module (or a vector space, if $\mathscr{R}$ is a field) $\mathscr{I}$ is dual to the algebra $\mathscr{R}\mathscr{K} := \mathbb{Z}\mathscr{K} \otimes \mathscr{R}$, where $\mathbb{Z}\mathscr{K}$ is the algebra of knots introduced in Section 1.5. It turns out (see Section 4.3.1) that the product on $\mathscr{I}$ corresponds under this duality to the *coproduct* on the algebra $\mathscr{R}\mathscr{K}$ of knots.

2.6 Quantum invariants

The subject of this section is not entirely elementary. However, we are not going to develop here a full theory of quantum groups and corresponding invariants, confining ourselves to some basic ideas which can be understood without going deep into complicated details. The reader will see that it is possible to use quantum invariants without even knowing what a quantum group is!

2.6.1 Background

The discovery of the Jones polynomial inspired many people to search for other skein relations compatible with Reidemeister moves and thus defining knot polynomials. This led to the introduction of the HOMFLY (Freyd *et al.* 1985; Przytycki and Traczyk 1988) and Kauffman's (1987a; 1990) polynomials. It soon became clear that all these polynomials are the first members of a vast family of knot invariants called *quantum invariants*.

The original idea of quantum invariants (in the case of 3-manifolds) was proposed by E. Witten (1989). Witten's approach coming from physics was not completely justified from the mathematical viewpoint. The first mathematically impeccable definition of quantum invariants of links and 3-manifolds was given by Turaev (1988) and Reshetikhin and Turaev (1990), who used in their construction the notion of *quantum groups* introduced shortly before that by V. Drinfeld (1985) (see also Drinfeld 1987) and M. Jimbo (1985). In fact, a quantum group is not a group at all. Instead, it is a family of algebras, more precisely, of *Hopf algebras* (see Section A.2.6), depending on a complex parameter q and satisfying certain axioms. The quantum group $U_q\mathfrak{g}$ of a semisimple Lie algebra $\mathfrak{g}$ is a deformation of the universal enveloping algebra (see Section A.1.6) of $\mathfrak{g}$ (which corresponds to the value $q = 1$) in the class of Hopf algebras.

In this section, we show how the Jones polynomial J can be obtained by the techniques of quantum groups, following the approach of Reshetikhin and Turaev. It turns out that J coincides, up to normalization, with the quantum invariant corresponding to the Lie algebra $\mathfrak{g} = \mathfrak{sl}_2$ in its standard two-dimensional representation (see Section A.1.4). Later in the book, we shall sometimes refer to the ideas illustrated in this section. For detailed expositions of quantum groups, we refer the interested reader to Jantzen (1996), Kassel (1995) and Kassel *et al.* (1997).

2.6.2 The R-matrix for the Lie algebra $\mathfrak{sl}_2$

Let $\mathfrak{g}$ be a semisimple Lie algebra and let V be its finite-dimensional representation. One can view V as a representation of the universal enveloping algebra $U(\mathfrak{g})$ (see Section A.1.6). It is remarkable that this representation can also be deformed with parameter q to a representation of the quantum group $U_q\mathfrak{g}$. The vector space V remains the same, but the action now depends on q. For a generic value of q all irreducible representations of $U_q\mathfrak{g}$ can be obtained in this way. However, when q is a root of unity, the representation theory is different and resembles the representation theory of $\mathfrak{g}$ in finite characteristic. It can be used to derive quantum invariants of 3-manifolds. For the purposes of knot theory it is enough to use generic values of q, that is, those which are not roots of unity.

An important property of quantum groups is that every representation gives rise to a solution R of the *quantum Yang–Baxter equation*

$$(R \otimes \mathrm{id}_V)(\mathrm{id}_V \otimes R)(R \otimes \mathrm{id}_V) = (\mathrm{id}_V \otimes R)(R \otimes \mathrm{id}_V)(\mathrm{id}_V \otimes R)$$

where R (the *R-matrix*) is an invertible linear operator $R : V \otimes V \to V \otimes V$, and both sides of the equation are understood as linear transformations $V \otimes V \otimes V \to V \otimes V \otimes V$.

Exercise. Given an R-matrix, construct a representation of the braid group B_n in the space $V^{\otimes n}$.

There is a procedure to construct an R-matrix associated with a representation of a Lie algebra. We are not going to describe it in general, confining ourselves just to one example: the Lie algebra $\mathfrak{g} = \mathfrak{sl}_2$ and its standard two-dimensional representation V (for $\mathfrak{sl}_N$ case see Exercise 2.38). In this case the associated R-matrix has the form

$$R : \begin{cases} e_1 \otimes e_1 & \mapsto & q^{1/4} e_1 \otimes e_1 \\ e_1 \otimes e_2 & \mapsto & q^{-1/4} e_2 \otimes e_1 \\ e_2 \otimes e_1 & \mapsto & q^{-1/4} e_1 \otimes e_2 + (q^{1/4} - q^{-3/4}) e_2 \otimes e_1 \\ e_2 \otimes e_2 & \mapsto & q^{1/4} e_2 \otimes e_2 \end{cases}$$

for an appropriate basis $\{e_1, e_2\}$ of the space V. The inverse of R (we shall need it later) is given by the formulae

$$R^{-1}: \begin{cases} e_1 \otimes e_1 & \mapsto & q^{-1/4} e_1 \otimes e_1 \\ e_1 \otimes e_2 & \mapsto & q^{1/4} e_2 \otimes e_1 + (-q^{3/4} + q^{-1/4}) e_1 \otimes e_2. \\ e_2 \otimes e_1 & \mapsto & q^{1/4} e_1 \otimes e_2 \\ e_2 \otimes e_2 & \mapsto & q^{-1/4} e_2 \otimes e_2 \end{cases}$$

Exercise. Check that this operator R satisfies the quantum Yang–Baxter equation.

2.6.3 The construction of quantum invariants

The general procedure of constructing quantum invariants is organized as follows (see details in Ohtsuki 2002). Consider a knot diagram in the plane and take a generic horizontal line. To each intersection point of the line with the diagram assign either the representation space V or its dual V^* depending on whether the orientation of the knot at this intersection is directed upwards or downwards. Then take the tensor product of all such spaces over the whole horizontal line. If the knot diagram does not intersect the line, then the corresponding vector space is the ground field $\mathbb{C}$.

A portion of a knot diagram between two such horizontal lines represents a tangle T (see the general definition in Section 1.6). We assume that this tangle is framed by the blackboard framing. With T we associate a linear transformation $\theta^{fr}(T)$ from the vector space corresponding to the bottom of T to the vector space corresponding to the top of T. The following three properties hold for the linear transformation $\theta^{fr}(T)$:

- $\theta^{fr}(T)$ is an invariant of the isotopy class of the framed tangle T;
- $\theta^{fr}(T_1 \cdot T_2) = \theta^{fr}(T_1) \circ \theta^{fr}(T_2)$;
- $\theta^{fr}(T_1 \otimes T_2) = \theta^{fr}(T_1) \otimes \theta^{fr}(T_2)$.

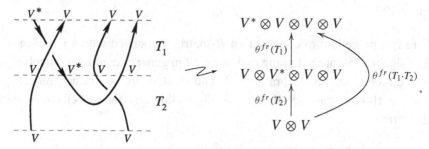

Now we can define a knot invariant $\theta^{fr}(K)$ regarding the knot K as a tangle between the two lines below and above K. In this case $\theta^{fr}(K)$ would be a linear transformation from $\mathbb{C}$ to $\mathbb{C}$, that is, multiplication by a number. Since our linear transformations depend on the parameter q, this number is actually a function of q.

Because of the multiplicativity property $\theta^{fr}(T_1 \cdot T_2) = \theta^{fr}(T_1) \circ \theta^{fr}(T_2)$ it is enough to define $\theta^{fr}(T)$ only for elementary tangles T such as a crossing, a minimum or a maximum point. This is precisely where quantum groups come in. Given a quantum group $U_q\mathfrak{g}$ and its finite-dimensional representation V, one can associate certain linear transformations with elementary tangles in a way consistent with the Turaev oriented moves from Section 1.6.6. The R-matrix appears here as the linear transformation corresponding to a positive crossing, while R^{-1} corresponds to a negative crossing. Of course, for a trivial tangle consisting of a single string connecting the top and bottom, the corresponding linear operator should be the identity transformation. So we have the following correspondence valid for all quantum groups:

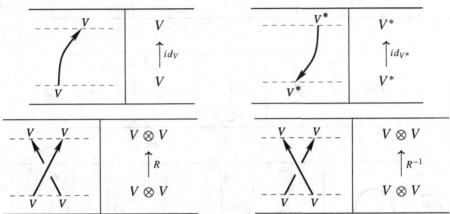

Using this we can easily check that the invariance of a quantum invariant under the third Reidemeister move is nothing else but the quantum Yang–Baxter equation:

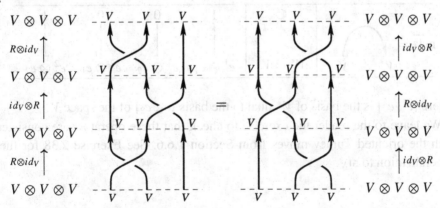

$$(R \otimes id_V) \circ (id_V \otimes R) \circ (R \otimes id_V) = (id_V \otimes R) \circ (R \otimes id_V) \circ (id_V \otimes R)$$

Similarly, the fact that we assigned mutually inverse operators (R and R^{-1}) to positive and negative crossings implies the invariance under the second Reidemeister move. (The first Reidemeister move is treated in Exercise 2.37.)

To complete the construction of our quantum invariant we should assign appropriate operators to the minimum and maximum points. These depend on all the data involved: the quantum group, the representation and the R-matrix. For the quantum group $U_q\mathfrak{sl}_2$, its standard two-dimensional representation V and the R-matrix chosen in Section 2.6.2 these operators are:

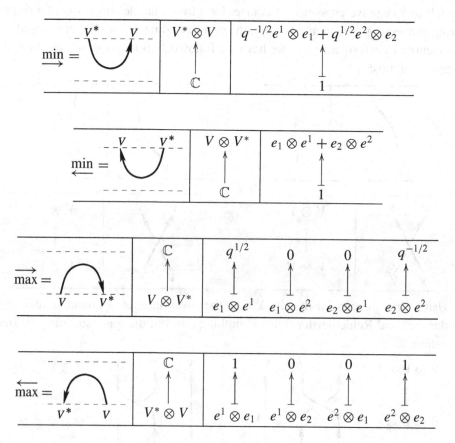

where $\{e^1, e^2\}$ is the basis of V^* dual to the basis $\{e_1, e_2\}$ of the space V.

We leave to the reader the exercise to check that these operators are consistent with the oriented Turaev moves from Section 1.6.6. See Exercise 2.38 for their generalization to $\mathfrak{sl}_N$.

2.6.4 Some examples

Example 2.12. Let us compute the $\mathfrak{sl}_2$-quantum invariant of the unknot. Represent the unknot as a product of two tangles and compute the composition of the corresponding transformations:

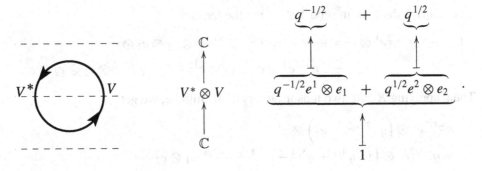

So $\theta^{fr}(\text{unknot}) = q^{1/2} + q^{-1/2}$. Therefore, in order to normalize our invariant so that its value on the unknot is equal to 1, we must divide $\theta^{fr}(\cdot)$ by $q^{1/2} + q^{-1/2}$. We denote the normalized invariant by $\widetilde{\theta}^{fr}(\cdot) = \frac{\theta^{fr}(\cdot)}{q^{1/2} + q^{-1/2}}$.

Example 2.13. Let us compute the quantum invariant for the left trefoil. Represent the diagram of the trefoil as follows:

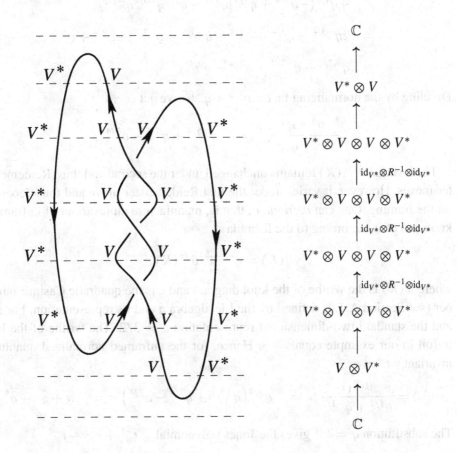

Two maps at the bottom send $1 \in \mathbb{C}$ into the tensor

$$1 \;\mapsto\; q^{-1/2}e^1 \otimes e_1 \otimes e_1 \otimes e^1 \quad + \quad q^{-1/2}e^1 \otimes e_1 \otimes e_2 \otimes e^2$$
$$+ \;\; q^{1/2}e^2 \otimes e_2 \otimes e_1 \otimes e^1 \quad + \quad q^{1/2}e^2 \otimes e_2 \otimes e_2 \otimes e^2.$$

Then, applying R^{-3} to two tensor factors in the middle, we get

$$q^{-1/2}e^1 \otimes \left(q^{-3/4}e_1 \otimes e_1\right) \otimes e^1$$
$$+ q^{-1/2}e^1 \otimes \left((-q^{9/4}+q^{5/4}-q^{1/4}+q^{-3/4})e_1 \otimes e_2\right.$$
$$+ (-q^{7/4}-q^{3/4}-q^{-1/4})e_2 \otimes e_1\Big) \otimes e^2$$
$$+ q^{1/2}e^2 \otimes \left((q^{7/4}-q^{3/4}+q^{-1/4})e_1 \otimes e_2 + (-q^{5/4}+q^{1/4})e_2 \otimes e_1\right) \otimes e^1$$
$$+ q^{1/2}e^2 \otimes \left(q^{-3/4}e_2 \otimes e_2\right) \otimes e^2.$$

Finally, the two maps at the top contract the whole tensor into a number

$$\theta^{fr}(3_1) \;=\; q^{-1/2}q^{-3/4}q^{1/2}+q^{-1/2}(-q^{9/4}+q^{5/4}-q^{1/4}+q^{-3/4})q^{-1/2}$$
$$+q^{1/2}(-q^{5/4}+q^{1/4})q^{1/2}+q^{1/2}q^{-3/4}q^{-1/2}$$
$$=\; 2q^{-3/4}-q^{5/4}+q^{1/4}-q^{-3/4}+q^{-7/4}-q^{9/4}+q^{5/4}$$
$$=\; q^{-7/4}+q^{-3/4}+q^{1/4}-q^{9/4}.$$

Dividing by the normalizing factor $q^{1/2}+q^{-1/2}$ we get

$$\frac{\theta^{fr}(3_1)}{q^{1/2}+q^{-1/2}} \;=\; q^{-5/4}+q^{3/4}-q^{7/4}.$$

The invariant $\theta^{fr}(K)$ remains unchanged under the second and third Reidemeister moves. However, it varies under the first Reidemeister move and thus depends on the framing. One can *deframe* it, that is, manufacture an invariant of unframed knots out of it, according to the formula

$$\theta(K) = q^{-\frac{c \cdot w(K)}{2}}\theta^{fr}(K),$$

where $w(K)$ is the writhe of the knot diagram and c is the quadratic Casimir number (see Section A.1.4) defined by the Lie algebra $\mathfrak{g}$ and its representation. For $\mathfrak{sl}_2$ and the standard two-dimensional representation $c = 3/2$. The writhe of the left trefoil in our example equals -3. Hence, for the unframed normalized quantum invariant we have

$$\widetilde{\theta}(3_1) = \frac{\theta(3_1)}{q^{1/2}+q^{-1/2}} \;=\; q^{9/4}\left(q^{-5/4}+q^{3/4}-q^{7/4}\right) \;=\; q+q^3-q^4.$$

The substitution $q = t^{-1}$ gives the Jones polynomial $\quad t^{-1}+t^{-3}-t^{-4}$.

2.7 Two-variable link polynomials

2.7.1 HOMFLY polynomial

The *HOMFLY polynomial* $P(L)$ is an unframed link invariant. It is defined as the Laurent polynomial in two variables a and z with integer coefficients satisfying the following skein relation and the initial condition:

$$a P\left(\times\right) - a^{-1} P\left(\times\right) = z P\left(\right) \left(\right) ; \qquad P\left(\bigcirc\right) = 1.$$

The existence of such an invariant is a difficult theorem. It was established simultaneously and independently by five groups of authors (Freyd *et al.* 1985; Przytycki and Traczyk 1988) (see also Lickorish 1997). The HOMFLY polynomial is equivalent to the collection of quantum invariants associated with the Lie algebra $\mathfrak{sl}_N$ and its standard N-dimensional representation for all values of N (see Exercise 2.38 for details).

Important properties of the HOMFLY polynomial are contained in the following exercises.

Exercise.

1. Prove the uniqueness of such an invariant. In other words, prove that the relations above are sufficient to compute the HOMFLY polynomial.
2. Compute the HOMFLY polynomial for the knots 3_1, 4_1 and compare your results with those given in Table 2.3.

$$C = \left(\right)$$

3. Compare the HOMFLY polynomials of the Conway and Kinoshita–Terasaka knots on the right (see, for instance, Sossinsky 2002).

$$KT = \left(\right)$$

Exercise. Prove that the HOMFLY polynomial of a link is preserved when the orientation of all components is reversed.

Exercise (Lickorish 1997). Prove that

1. $P(\overline{L}) = \overline{P(L)}$, where $\overline{L}$ is the mirror reflection of L and $\overline{P(L)}$ is the polynomial obtained from $P(L)$ by substituting $-a^{-1}$ for a;
2. $P(K_1 \# K_2) = P(K_1) \cdot P(K_2)$;
3. $P(L_1 \sqcup L_2) = \dfrac{a - a^{-1}}{z} \cdot P(L_1) \cdot P(L_2)$, where $L_1 \sqcup L_2$ means the split union of links (that is, the union of L_1 and L_2 such that each of these two links is contained inside its own ball, and the two balls do not have common points);

Table 2.3 *HOMFLY polynomials of knots with up to 8 crossings*

3_1 $(2a^2 - a^4) + a^2 z^2$

4_1 $(a^{-2} - 1 + a^2) - z^2$

5_1 $(3a^4 - 2a^6) + (4a^4 - a^6)z^2 + a^4 z^4$

5_2 $(a^2 + a^4 - a^6) + (a^2 + a^4)z^2$

6_1 $(a^{-2} - a^2 + a^4) + (-1 - a^2)z^2$

6_2 $(2 - 2a^2 + a^4) + (1 - 3a^2 + a^4)z^2 - a^2 z^4$

6_3 $(-a^{-2} + 3 - a^2) + (-a^{-2} + 3 - a^2)z^2 + z^4$

7_1 $(4a^6 - 3a^8) + (10a^6 - 4a^8)z^2 + (6a^6 - a^8)z^4 + a^6 z^6$

7_2 $(a^2 + a^6 - a^8) + (a^2 + a^4 + a^6)z^2$

7_3 $(a^{-4} + 2a^{-6} - 2a^{-8}) + (3a^{-4} + 3a^{-6} - a^{-8})z^2 + (a^{-4} + a^{-6})z^4$

7_4 $(2a^{-4} - a^{-8}) + (a^{-2} + 2a^{-4} + a^{-6})z^2$

7_5 $(2a^4 - a^8) + (3a^4 + 2a^6 - a^8)z^2 + (a^4 + a^6)z^4$

7_6 $(1 - a^2 + 2a^4 - a^6) + (1 - 2a^2 + 2a^4)z^2 - a^2 z^4$

7_7 $(a^{-4} - 2a^{-2} + 2) + (-2a^{-2} + 2 - a^2)z^2 + z^4$

8_1 $(a^{-2} - a^4 + a^6) + (-1 - a^2 - a^4)z^2$

8_2 $(3a^2 - 3a^4 + a^6) + (4a^2 - 7a^4 + 3a^6)z^2 + (a^2 - 5a^4 + a^6)z^4 - a^4 z^6$

8_3 $(a^{-4} - 1 + a^4) + (-a^{-2} - 2 - a^2)z^2$

8_4 $(a^4 - 2 + 2a^{-2}) + (a^4 - 2a^2 - 3 + a^{-2})z^2 + (-a^2 - 1)z^4$

8_5 $(4a^{-2} - 5a^{-4} + 2a^{-6}) + (4a^{-2} - 8a^{-4} + 3a^{-6})z^2$
       $+ (a^{-2} - 5a^{-4} + a^{-6})z^4 - a^{-4}z^6$

8_6 $(2 - a^2 - a^4 + a^6) + (1 - 2a^2 - 2a^4 + a^6)z^2 + (-a^2 - a^4)z^4$

8_7 $(-2a^{-4} + 4a^{-2} - 1) + (-3a^{-4} + 8a^{-2} - 3)z^2 + (-a^{-4} + 5a^{-2} - 1)z^4$
       $+ a^{-2}z^6$

8_8 $(-a^{-4} + a^{-2} + 2 - a^2) + (-a^{-4} + 2a^{-2} + 2 - a^2)z^2 + (a^{-2} + 1)z^4$

8_9 $(2a^{-2} - 3 + 2a^2) + (3a^{-2} - 8 + 3a^2)z^2 + (a^{-2} - 5 + a^2)z^4 - z^6$

8_{10} $(-3a^{-4} + 6a^{-2} - 2) + (-3a^{-4} + 9a^{-2} - 3)z^2 + (-a^{-4} + 5a^{-2} - 1)z^4$
       $+ a^{-2}z^6$

8_{11} $(1 + a^2 - 2a^4 + a^6) + (1 - a^2 - 2a^4 + a^6)z^2 + (-a^2 - a^4)z^4$

8_{12} $(a^{-4} - a^{-2} + 1 - a^2 + a^4) + (-2a^{-2} + 1 - 2a^2)z^2 + z^4$

8_{13} $(-a^{-4} + 2a^{-2}) + (-a^{-4} + 2a^{-2} + 1 - a^2)z^2 + (a^{-2} + 1)z^4$

8_{14} $1 + (1 - a^2 - a^4 + a^6)z^2 + (-a^2 - a^4)z^4$

8_{15} $(a^4 + 3a^6 - 4a^8 + a^{10}) + (2a^4 + 5a^6 - 3a^8)z^2 + (a^4 + 2a^6)z^4$

8_{16} $(-a^4 + 2a^2) + (-2a^4 + 5a^2 - 2)z^2 + (-a^4 + 4a^2 - 1)z^4 + a^2 z^6$

8_{17} $(a^{-2} - 1 + a^2) + (2a^{-2} - 5 + 2a^2)z^2 + (a^{-2} - 4 + a^2)z^4 - z^6$

8_{18} $(-a^{-2} + 3 - a^2) + (a^{-2} - 1 + a^2)z^2 + (a^{-2} - 3 + a^2)z^4 - z^6$

8_{19} $(5a^{-6} - 5a^{-8} + a^{-10}) + (10a^{-6} - 5a^{-8})z^2 + (6a^{-6} - a^{-8})z^4 + a^{-6}z^6$

8_{20} $(-2a^4 + 4a^2 - 1) + (-a^4 + 4a^2 - 1)z^2 + a^2 z^4$

8_{21} $(3a^2 - 3a^4 + a^6) + (2a^2 - 3a^4 + a^6)z^2 - a^4 z^4$

4. $P(8_8) = P(\overline{10_{129}})$.

These knots can be distinguished by the two-variable Kauffman polynomial defined below.

2.7.2 Two-variable Kauffman polynomial

L. Kauffman (1994) found another invariant Laurent polynomial $F(L)$ in two variables a and z. First, for an unoriented link diagram D we define a polynomial $\Lambda(D)$ which is invariant under Reidemeister moves Ω_2 and Ω_3 and satisfies the relations

$$\Lambda\left(\!\left(\!\includegraphics[scale=0.1]{}\!\right)\!\right) + \Lambda\left(\!\left(\!\includegraphics[scale=0.1]{}\!\right)\!\right) = z\left(\Lambda\left(\!\left(\!\includegraphics[scale=0.1]{}\!\right)\!\right) + \Lambda\left(\!\left(\!\includegraphics[scale=0.1]{}\!\right)\!\right)\right),$$

$$\Lambda\left(\!\left(\!\includegraphics[scale=0.1]{}\!\right)\!\right) = a\Lambda\left(\!\left(\!\includegraphics[scale=0.1]{}\!\right)\!\right), \qquad \Lambda\left(\!\left(\!\includegraphics[scale=0.1]{}\!\right)\!\right) = a^{-1}\Lambda\left(\!\left(\!\includegraphics[scale=0.1]{}\!\right)\!\right),$$

and the initial condition $\Lambda\left(\bigcirc\right) = 1$.

Now, for any diagram D of an oriented link L we put

$$F(L) := a^{-w(D)}\Lambda(D).$$

It turns out that this polynomial is equivalent to the collection of the quantum invariants associated with the Lie algebra $\mathfrak{so}_N$ and its standard N-dimensional representation for all values of N (see Turaev 1990).

As in the previous section, we conclude with a series of exercises with additional information on the Kauffman polynomial.

Exercise. Prove that the defining relations are sufficient to compute the Kauffman polynomial.

Exercise. Compute the Kauffman polynomial for the knots $3_1, 4_1$ and compare the results with those given in Table 2.4.

Exercise. Prove that the Kauffman polynomial of a knot is preserved when the knot orientation is reversed.

Exercise (Lickorish 1997). Prove that

1. $F(\overline{L}) = \overline{F(L)}$, where $\overline{L}$ is the mirror reflection of L, and $\overline{F(L)}$ is the polynomial obtained from $F(L)$ by substituting a^{-1} for a;
2. $F(K_1 \# K_2) = F(K_1) \cdot F(K_2)$;
3. $F(L_1 \sqcup L_2) = \left((a+a^{-1})z^{-1} - 1\right) \cdot F(L_1) \cdot F(L_2)$, where $L_1 \sqcup L_2$ means the split union of links;

$8_8 = $

$\overline{10_{129}} = $

Table 2.4 *Kauffman polynomials of knots with up to 7 crossings*

3_1 $(-2a^2 - a^4) + (a^3 + a^5)z + (a^2 + a^4)z^2$

4_1 $(-a^{-2} - 1 - a^2) + (-a^{-1} - a)z + (a^{-2} + 2 + a^2)z^2 + (a^{-1} + a)z^3$

5_1 $(3a^4 + 2a^6) + (-2a^5 - a^7 + a^9)z + (-4a^4 - 3a^6 + a^8)z^2$
$\quad + (a^5 + a^7)z^3 + (a^4 + a^6)z^4$

5_2 $(-a^2 + a^4 + a^6) + (-2a^5 - 2a^7)z + (a^2 - a^4 - 2a^6)z^2$
$\quad + (a^3 + 2a^5 + a^7)z^3 + (a^4 + a^6)z^4$

6_1 $(-a^{-2} + a^2 + a^4) + (2a + 2a^3)z + (a^{-2} - 4a^2 - 3a^4)z^2$
$\quad + (a^{-1} - 2a - 3a^3)z^3 + (1 + 2a^2 + a^4)z^4 + (a + a^3)z^5$

6_2 $(2 + 2a^2 + a^4) + (-a^3 - a^5)z + (-3 - 6a^2 - 2a^4 + a^6)z^2$
$\quad + (-2a + 2a^5)z^3 + (1 + 3a^2 + 2a^4)z^4 + (a + a^3)z^5$

6_3 $(a^{-2} + 3 + a^2) + (-a^{-3} - 2a^{-1} - 2a - a^3)z + (-3a^{-2} - 6 - 3a^2)z^2$
$\quad + (a^{-3} + a^{-1} + a + a^3)z^3 + (2a^{-2} + 4 + 2a^2)z^4 + (a^{-1} + a)z^5$

7_1 $(-4a^6 - 3a^8) + (3a^7 + a^9 - a^{11} + a^{13})z + (10a^6 + 7a^8 - 2a^{10} + a^{12})z^2$
$\quad + (-4a^7 - 3a^9 + a^{11})z^3 + (-6a^6 - 5a^8 + a^{10})z^4 + (a^7 + a^9)z^5$
$\quad + (a^6 + a^8)z^6$

7_2 $(-a^2 - a^6 - a^8) + (3a^7 + 3a^9)z + (a^2 + 3a^6 + 4a^8)z^2$
$\quad + (a^3 - a^5 - 6a^7 - 4a^9)z^3 + (a^4 - 3a^6 - 4a^8)z^4 + (a^5 + 2a^7 + a^9)z^5$
$\quad + (a^6 + a^8)z^6$

7_3 $(-2a^{-8} - 2a^{-6} + a^{-4}) + (-2a^{-11} + a^{-9} + 3a^{-7})z$
$\quad + (-a^{-10} + 6a^{-8} + 4a^{-6} - 3a^{-4})z^2 + (a^{-11} - a^{-9} - 4a^{-7} - 2a^{-5})z^3$
$\quad + (a^{-10} - 3a^{-8} - 3a^{-6} + a^{-4})z^4 + (a^{-9} + 2a^{-7} + a^{-5})z^5$
$\quad + (a^{-8} + a^{-6})z^6$

7_4 $(-a^{-8} + 2a^{-4}) + (4a^{-9} + 4a^{-7})z + (2a^{-8} - 3a^{-6} - 4a^{-4} + a^{-2})z^2$
$\quad + (-4a^{-9} - 8a^{-7} - 2a^{-5} + 2a^{-3})z^3 + (-3a^{-8} + 3a^{-4})z^4$
$\quad + (a^{-9} + 3a^{-7} + 2a^{-5})z^5 + (a^{-8} + a^{-6})z^6$

7_5 $(2a^4 - a^8) + (-a^5 + a^7 + a^9 - a^{11})z + (-3a^4 + a^8 - 2a^{10})z^2$
$\quad + (-a^5 - 4a^7 - 2a^9 + a^{11})z^3 + (a^4 - a^6 + 2a^{10})z^4$
$\quad + (a^5 + 3a^7 + 2a^9)z^5 + (a^6 + a^8)z^6$

7_6 $(1 + a^2 + 2a^4 + a^6) + (a + 2a^3 - a^7)z + (-2 - 4a^2 - 4a^4 - 2a^6)z^2$
$\quad + (-4a - 6a^3 - a^5 + a^7)z^3 + (1 + a^2 + 2a^4 + 2a^6)z^4$
$\quad + (2a + 4a^3 + 2a^5)z^5 + (a^2 + a^4)z^6$

7_7 $(a^{-4} + 2a^{-2} + 2) + (2a^{-3} + 3a^{-1} + a)z + (-2a^{-4} - 6a^{-2} - 7 - 3a^2)z^2$
$\quad + (-4a^{-3} - 8a^{-1} - 3a + a^3)z^3 + (a^{-4} + 2a^{-2} + 4 + 3a^2)z^4$
$\quad + (2a^{-3} + 5a^{-1} + 3a)z^5 + (a^{-2} + 1)z^6$

Table 2.5 *Kauffman polynomials of knots with 8 crossings*

8_1 $(-a^{-2} - a^4 - a^6) + (-3a^3 - 3a^5)z + (a^{-2} + 7a^4 + 6a^6)z^2$
$+(a^{-1} - a + 5a^3 + 7a^5)z^3 + (1 - 2a^2 - 8a^4 - 5a^6)z^4$
$+(a - 4a^3 - 5a^5)z^5 + (a^2 + 2a^4 + a^6)z^6 + (a^3 + a^5)z^7$

8_2 $(-3a^2 - 3a^4 - a^6) + (a^3 + a^5 - a^7 - a^9)z$
$+(7a^2 + 12a^4 + 3a^6 - a^8 + a^{10})z^2 + (3a^3 - a^5 - 2a^7 + 2a^9)z^3$
$+(-5a^2 - 12a^4 - 5a^6 + 2a^8)z^4 + (-4a^3 - 2a^5 + 2a^7)z^5$
$+(a^2 + 3a^4 + 2a^6)z^6 + (a^3 + a^5)z^7$

8_3 $(a^{-4} - 1 + a^4) + (-4a^{-1} - 4a)z + (-3a^{-4} + a^{-2} + 8 + a^2 - 3a^4)z^2$
$+(-2a^{-3} + 8a^{-1} + 8a - 2a^3)z^3 + (a^{-4} - 2a^{-2} - 6 - 2a^2 + a^4)z^4$
$+(a^{-3} - 4a^{-1} - 4a + a^3)z^5 + (a^{-2} + 2 + a^2)z^6 + (a^{-1} + a)z^7$

8_4 $(-2a^{-2} - 2 + a^4) + (-a^{-1} + a + 2a^3)z$
$+(7a^{-2} + 10 - a^2 - 3a^4 + a^6)z^2 + (4a^{-1} - 3a - 5a^3 + 2a^5)z^3$
$+(-5a^{-2} - 11 - 3a^2 + 3a^4)z^4 + (-4a^{-1} - a + 3a^3)z^5$
$+(a^{-2} + 3 + 2a^2)z^6 + (a^{-1} + a)z^7$

8_5 $(-2a^{-6} - 5a^{-4} - 4a^{-2}) + (4a^{-7} + 7a^{-5} + 3a^{-3})z$
$+(a^{-10} - 2a^{-8} + 4a^{-6} + 15a^{-4} + 8a^{-2})z^2 + (2a^{-9} - 8a^{-7} - 10a^{-5})z^3$
$+(3a^{-8} - 7a^{-6} - 15a^{-4} - 5a^{-2})z^4 + (4a^{-7} + a^{-5} - 3a^{-3})z^5$
$+(3a^{-6} + 4a^{-4} + a^{-2})z^6 + (a^{-5} + a^{-3})z^7$

8_6 $(2 + a^2 - a^4 - a^6) + (-a - 3a^3 - a^5 + a^7)z$
$+(-3 - 2a^2 + 6a^4 + 3a^6 - 2a^8)z^2 + (-a + 5a^3 + 2a^5 - 4a^7)z^3$
$+(1 - 6a^4 - 4a^6 + a^8)z^4 + (a - 2a^3 - a^5 + 2a^7)z^5$
$+(a^2 + 3a^4 + 2a^6)z^6 + (a^3 + a^5)z^7$

8_7 $(-2a^{-4} - 4a^{-2} - 1) + (-a^{-7} + 2a^{-3} + 2a^{-1} + a)z$
$+(-2a^{-6} + 4a^{-4} + 12a^{-2} + 6)z^2 + (a^{-7} - a^{-5} - 2a^{-3} - 3a^{-1} - 3a)z^3$
$+(2a^{-6} - 3a^{-4} - 12a^{-2} - 7)z^4 + (2a^{-5} - a^{-1} + a)z^5$
$+(2a^{-4} + 4a^{-2} + 2)z^6 + (a^{-3} + a^{-1})z^7$

8_8 $(-a^{-4} - a^{-2} + 2 + a^2) + (2a^{-5} + 3a^{-3} + a^{-1} - a - a^3)z$
$+(4a^{-4} + 5a^{-2} - 1 - 2a^2)z^2 + (-3a^{-5} - 5a^{-3} - 3a^{-1} + a^3)z^3$
$+(-6a^{-4} - 9a^{-2} - 1 + 2a^2)z^4 + (a^{-5} + a^{-1} + 2a)z^5$
$+(2a^{-4} + 4a^{-2} + 2)z^6 + (a^{-3} + a^{-1})z^7$

8_9 $(-2a^{-2} - 3 - 2a^2) + (a^{-3} + a^{-1} + a + a^3)z$
$+(-2a^{-4} + 4a^{-2} + 12 + 4a^2 - 2a^4)z^2 + (-4a^{-3} - a^{-1} - a - 4a^3)z^3$
$+(a^{-4} - 4a^{-2} - 10 - 4a^2 + a^4)z^4 + (2a^{-3} + 2a^3)z^5$
$+(2a^{-2} + 4 + 2a^2)z^6 + (a^{-1} + a)z^7$

8_{10} $(-3a^{-4} - 6a^{-2} - 2) + (-a^{-7} + 2a^{-5} + 6a^{-3} + 5a^{-1} + 2a)z$
$+(-a^{-6} + 6a^{-4} + 12a^{-2} + 5)z^2 + (a^{-7} - 3a^{-5} - 9a^{-3} - 8a^{-1} - 3a)z^3$
$+(2a^{-6} - 5a^{-4} - 13a^{-2} - 6)z^4 + (3a^{-5} + 3a^{-3} + a^{-1} + a)z^5$
$+(3a^{-4} + 5a^{-2} + 2)z^6 + (a^{-3} + a^{-1})z^7$

Table 2.6 *Kauffman polynomials of knots with 8 crossings (continuation)*

8_{11} $(1 - a^2 - 2a^4 - a^6) + (a^3 + 3a^5 + 2a^7)z + (-2 + 6a^4 + 2a^6 - 2a^8)z^2$
$+(-3a - 2a^3 - 3a^5 - 4a^7)z^3 + (1 - 2a^2 - 7a^4 - 3a^6 + a^8)z^4$
$+(2a + a^3 + a^5 + 2a^7)z^5 + (2a^2 + 4a^4 + 2a^6)z^6 + (a^3 + a^5)z^7$

8_{12} $(a^{-4} + a^{-2} + 1 + a^2 + a^4) + (a^{-3} + a^3)z$
$+(-2a^{-4} - 2a^{-2} - 2a^2 - 2a^4)z^2 + (-3a^{-3} - 3a^{-1} - 3a - 3a^3)z^3$
$+(a^{-4} - a^{-2} - 4 - a^2 + a^4)z^4 + (2a^{-3} + 2a^{-1} + 2a + 2a^3)z^5$

$+(2a^{-2} + 4 + 2a^2)z^6 + (a^{-1} + a)z^7$

8_{13} $(-a^{-4} - 2a^{-2}) + (2a^{-5} + 4a^{-3} + 3a^{-1} + a)z + (5a^{-4} + 7a^{-2} - 2a^2)z^2$
$+(-3a^{-5} - 7a^{-3} - 9a^{-1} - 4a + a^3)z^3 + (-6a^{-4} - 11a^{-2} - 2 + 3a^2)z^4$
$+(a^{-5} + a^{-3} + 4a^{-1} + 4a)z^5 + (2a^{-4} + 5a^{-2} + 3)z^6 + (a^{-3} + a^{-1})z^7$

8_{14} $1 + (a + 3a^3 + 3a^5 + a^7)z + (-2 - a^2 + 3a^4 + a^6 - a^8)z^2$
$+(-3a - 6a^3 - 8a^5 - 5a^7)z^3 + (1 - a^2 - 7a^4 - 4a^6 + a^8)z^4$
$+(2a + 3a^3 + 4a^5 + 3a^7)z^5 + (2a^2 + 5a^4 + 3a^6)z^6 + (a^3 + a^5)z^7$

8_{15} $(a^4 - 3a^6 - 4a^8 - a^{10}) + (6a^7 + 8a^9 + 2a^{11})z$
$+(-2a^4 + 5a^6 + 8a^8 - a^{12})z^2 + (-2a^5 - 11a^7 - 14a^9 - 5a^{11})z^3$
$+(a^4 - 5a^6 - 10a^8 - 3a^{10} + a^{12})z^4 + (2a^5 + 5a^7 + 6a^9 + 3a^{11})z^5$
$+(3a^6 + 6a^8 + 3a^{10})z^6 + (a^7 + a^9)z^7$

8_{16} $(-2a^2 - a^4) + (a^{-1} + 3a + 4a^3 + 2a^5)z + (5 + 10a^2 + 4a^4 - a^6)z^2$
$+(-2a^{-1} - 6a - 10a^3 - 5a^5 + a^7)z^3 + (-8 - 18a^2 - 7a^4 + 3a^6)z^4$
$+(a^{-1} - a + 3a^3 + 5a^5)z^5 + (3 + 8a^2 + 5a^4)z^6 + (2a + 2a^3)z^7$

8_{17} $(-a^{-2} - 1 - a^2) + (a^{-3} + 2a^{-1} + 2a + a^3)z$
$+(-a^{-4} + 3a^{-2} + 8 + 3a^2 - a^4)z^2 + (-4a^{-3} - 6a^{-1} - 6a - 4a^3)z^3$
$+(a^{-4} - 6a^{-2} - 14 - 6a^2 + a^4)z^4 + (3a^{-3} + 2a^{-1} + 2a + 3a^3)z^5$
$+(4a^{-2} + 8 + 4a^2)z^6 + (2a^{-1} + 2a)z^7$

8_{18} $(a^{-2} + 3 + a^2) + (a^{-1} + a)z + (3a^{-2} + 6 + 3a^2)z^2$
$+(-4a^{-3} - 9a^{-1} - 9a - 4a^3)z^3 + (a^{-4} - 9a^{-2} - 20 - 9a^2 + a^4)z^4$
$+(4a^{-3} + 3a^{-1} + 3a + 4a^3)z^5 + (6a^{-2} + 12 + 6a^2)z^6 + (3a^{-1} + 3a)z^7$

8_{19} $(-a^{-10} - 5a^{-8} - 5a^{-6}) + (5a^{-9} + 5a^{-7})z + (10a^{-8} + 10a^{-6})z^2$
$+(-5a^{-9} - 5a^{-7})z^3 + (-6a^{-8} - 6a^{-6})z^4 + (a^{-9} + a^{-7})z^5$
$+(a^{-8} + a^{-6})z^6$

8_{20} $(-1 - 4a^2 - 2a^4) + (a^{-1} + 3a + 5a^3 + 3a^5)z + (2 + 6a^2 + 4a^4)z^2$
$+(-3a - 7a^3 - 4a^5)z^3 + (-4a^2 - 4a^4)z^4 + (a + 2a^3 + a^5)z^5$
$+(a^2 + a^4)z^6$

8_{21} $(-3a^2 - 3a^4 - a^6) + (2a^3 + 4a^5 + 2a^7)z + (3a^2 + 5a^4 - 2a^8)z^2$
$+(-a^3 - 6a^5 - 5a^7)z^3 + (-2a^4 - a^6 + a^8)z^4 + (a^3 + 3a^5 + 2a^7)z^5$
$+(a^4 + a^6)z^6$

4. $F(11^a_{30}) = F(11^a_{189})$;
    (these knots can be distinguished by the Conway
    and, hence, by the HOMFLY polynomial; note
    that we use the Knotscape numbering of knots
    (Hoste and Thistlethwaite 1999), while in Lick-
    orish (1997), the old Perko's notation is used).
5. $F(L^*) = a^{4lk(K,L-K)} F(L)$, where the link L^* is
    obtained from an oriented link L by reversing the
    orientation of a connected component K.

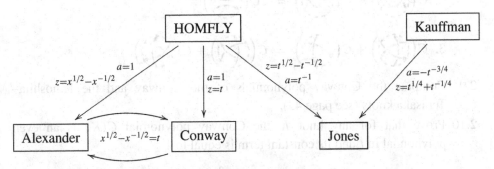

2.7.3 Comparative strength of polynomial invariants

Let us say that an invariant I_1 *dominates* an invariant I_2, if the equality $I_1(K_1) = I_1(K_2)$ for any two knots K_1 and K_2 implies the equality $I_2(K_1) = I_2(K_2)$. Denoting this relation by arrows, we have the following comparison chart:

(the absence of an arrow between the two invariants means that neither of them dominates the other).

Exercise. Find in this chapter all the facts sufficient to justify this chart.

Exercises

2.1 The *bridge number* $b(K)$ of a knot K can be defined as the minimal num-
     ber of local maxima of the projection of the knot onto a straight line, where
     the minimum is taken over all projections and over all closed curves in $\mathbb{R}^3$
     representing the knot. Show that

$$b(K_1 \# K_2) = b(K_1) + b(K_2) - 1.$$

     Knots of bridge number 2 are also called *rational knots* (see Murasugi 1996).

2.2 Prove that the Conway and the Jones polynomials of a knot are preserved
     when the knot orientation is reversed.

2.3 Compute the Conway and the Jones polynomials for the links from Sec-
     tion 1.1.3 with some orientations of their components.

2.4 A link is called *split* if it is equivalent to a link which has some components inside some ball while the other components are located outside of the ball. Prove that the Conway polynomial of a split link is trivial: $C(L) = 0$.

2.5 For a split link $L_1 \sqcup L_2$ prove that

$$J(L_1 \sqcup L_2) = (-t^{1/2} - t^{-1/2}) \cdot J(L_1) \cdot J(L_2).$$

2.6 Prove that $C(K_1 \# K_2) = C(K_1) \cdot C(K_2)$.

2.7 Prove that $J(K_1 \# K_2) = J(K_1) \cdot J(K_2)$.

2.8 (cf. Conway 1997) Check that the Conway polynomial satisfies the following relations.

1. $C\left(\vcenter{\hbox{}}\right) + C\left(\vcenter{\hbox{}}\right) = (2 + t^2)C\left(\vcenter{\hbox{}}\right)$;

2. $C\left(\vcenter{\hbox{}}\right) + C\left(\vcenter{\hbox{}}\right) = 2C\left(\vcenter{\hbox{}}\right)$;

3. $C\left(\vcenter{\hbox{}}\right) + C\left(\vcenter{\hbox{}}\right) = C\left(\vcenter{\hbox{}}\right) + C\left(\vcenter{\hbox{}}\right)$.

2.9 Compute the Conway polynomials of the Conway and the Kinoshita–Terasaka knots (see page 43).

2.10 Prove that for any knot K the Conway polynomial $C(K)$ is an even polynomial in t and its constant term is equal to 1:

$$C(K) = 1 + c_2(K)t^2 + c_4(K)t^4 + \dots$$

2.11 Let L be a link with two components K_1 and K_2. Prove that the Conway polynomial $C(L)$ is an odd polynomial in t and its lowest coefficient is equal to the linking number $lk(K_1, K_2)$:

$$C(L) = lk(K_1, K_2)t + c_3(L)t^3 + c_5(L)t^5 + \dots$$

2.12 Prove that for a link L with k components the Conway polynomial $C(L)$ is divisible by t^{k-1} and is odd or even depending on the parity of k:

$$C(L) = c_{k-1}(L)t^{k-1} + c_{k+1}(L)t^{k+1} + c_{k+3}(L)t^{k+3} + \dots$$

2.13 For a knot K, show that $C(K)\big|_{t-2i} \equiv 1$ or 5 (mod 8) depending on the parity of $c_2(K)$. The reduction of $c_2(K)$ modulo 2 is called the *Arf invariant* of K.

2.14 Show that $J(L)\big|_{t=-1} = C(L)\big|_{t=2i}$ for any link L. The absolute value of this number is called the *determinant* of the link L.

   Hint. Choose $\sqrt{t}$ in such a way that $\sqrt{-1} = -i$.

2.15 Check the following *switching formula* for the Jones polynomial.

$$J\left(\overset{\diagup}{\diagdown\!\!\!\diagup}\right) - tJ\left(\overset{\diagup}{\diagdown\!\!\!\diagup}\right) = t^{3\lambda_0}(1-t)J\left(\smile\atop\frown\right),$$

where λ_0 is the linking number of two components of the link, $\big)\big($, obtained by smoothing the crossing according to the orientation. Note that the knot in the right-hand side of the formula is unoriented. That is because such a smoothing destroys the orientation. Since the Jones polynomial does not distinguish the orientation of a knot, we may choose it arbitrarily.

2.16 Interlacing crossings formulae. Suppose K_{++} is a knot diagram with two positive crossings which are *interlaced*. That means when we trace the knot we first pass the first crossing, then the second, then again the first, and after that the second. Consider the following four knots and one link:

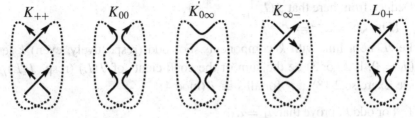

Check that the Jones polynomial satisfies the relation

$$J(K_{++}) = tJ(K_{00}) + t^{3\lambda_{0+}}\Big(J(K_{0\infty}) - tJ(K_{\infty-})\Big),$$

where λ_{0+} is the linking number of two components of the link L_{0+}.
   Check the similar relations for K_{+-} and K_{--}:

$$J(K_{+-}) = J(K_{00}) + t^{3\lambda_{0-}+1}\Big(J(K_{0\infty}) - J(K_{\infty+})\Big),$$

$$J(K_{--}) = t^{-1}J(K_{00}) + t^{3\lambda_{0-}}\Big(J(K_{0\infty}) - t^{-1}J(K_{\infty+})\Big).$$

If a knot diagram does not contain interlacing crossings, then it represents the unknot. Thus the three relations above allow one to compute the Jones polynomial for knots recursively without referring to links.

2.17 Show that the Jones polynomial satisfies the following relations.

1. $t^{-2}J\left(\overset{\diagup}{\diagdown}\right) + t^{2}J\left(\overset{\diagdown}{\diagup}\right) = (t + t^{-1})J\left(\asymp\right);$

2. $tJ\left(\overset{\diagup}{\diagdown}\right) + t^{-1}J\left(\overset{\diagdown}{\diagup}\right) = (t + t^{-1})J\left(\asymp\right);$

3. $t^2 J\left(\begin{smallmatrix} \end{smallmatrix}\right) + t^{-2} J\left(\begin{smallmatrix} \end{smallmatrix}\right) = t^{-2} J\left(\begin{smallmatrix} \end{smallmatrix}\right) + t^2 J\left(\begin{smallmatrix} \end{smallmatrix}\right).$

Compare these relations with those of Exercise 2.8 for the Conway polynomial.

2.18 Prove that for a link L with an odd number of components, $J(L)$ is a polynomial in t and t^{-1}, and for a link L with an even number of components $J(L) = t^{1/2} \cdot$ (a polynomial in t and t^{-1}).

2.19 Prove that for a link L with k components $J(L)\big|_{t=1} = (-2)^{k-1}$. In particular, $J(K)\big|_{t=1} = 1$ for a knot K.

2.20 Prove that $\dfrac{d(J(K))}{dt}\bigg|_{t=1} = 0$ for any knot K.

2.21 Evaluate the Kauffman bracket $\langle L \rangle$ at $a = e^{\pi i/3}$, $b = a^{-1}$, $c = -a^2 - a^{-2}$. Deduce from here that $J(L)\big|_{t=e^{2\pi i/3}} = 1$.
   Hint. $\sqrt{t} = a^{-2} = e^{4\pi i/3}$.

2.22 Let L be a link with k components. For odd (respectively, even) k let a_j ($j = 0, 1, 2,$ or 3) be the sum of the coefficients of $J(L)$ (resp. $J(L)/\sqrt{t}$, see Exercise 2.18) at t^s for all $s \equiv j \pmod 4$.

   1. For odd k, prove that $a_1 = a_3$.
   2. For even k, prove that $a_0 + a_1 = a_2 + a_3$.

2.23 (Lickorish 1997, theorem 10.6) Let $t = i$ with $t^{1/2} = e^{\pi i/4}$. Prove that for a knot K, $J(K)\big|_{t=i} = (-1)^{c_2(K)}$.

2.24 For the mirror reflection $\overline{L}$ of a link L prove that $J(\overline{L})$ is obtained from $J(L)$ by substituting t^{-1} for t.

2.25 For the link L^* obtained from an oriented link L by reversing the orientation of one of its components K, prove that $J(L^*) = t^{-3lk(K,L-K)}J(L)$.

2.26* Find a non-trivial knot K with $J(K) = 1$.

2.27 (Kauffman 1987b, Murasugi 1987, Thistlethwaite 1987). Prove for a reduced alternating knot diagram K (Section 1.2.3) the number of crossings is equal to $span(J(K))$, that is, to the difference beween the maximal and minimal degrees of t in the Jones polynomial $J(K)$. (This exercise is not particularly difficult, although it solves a hundred-year-old conjecture of Tait. Anyway, the reader can find a rather simple solution in Turaev 1987.)

2.28 Let L be a link with k components. Show that its HOMFLY polynomial $P(L)$ is an even function in each of the variables a and z if k is odd, and it is an odd function if k is even.

2.29 For a link L with k components, show that the lowest power of z in its HOM-FLY polynomial is z^{-k+1}. In particular the HOMFLY polynomial $P(K)$ of a knot K is a genuine polynomial in z. This means that it does not contain terms with z raised to a negative power.

2.30 For a knot K let $p_0(a) := P(K)|_{z=0}$ be the constant term of the HOMFLY polynomial. Show that its derivative at $a = 1$ equals zero.

2.31 Let L be a link with two components K_1 and K_2. Consider $P(L)$ as a Laurent polynomial in z with coefficients in Laurent polynomials in a. Let $p_{-1}(a)$ and $p_1(a)$ be the coefficients of z^{-1} and z. Check that $p_{-1}|_{a=1} = 0$, $p'_{-1}|_{a=1} = 2$, $p''_{-1}|_{a=1} = -8lk(K_1, K_2) - 2$, and $p_1|_{a=1} = lk(K_1, K_2)$.

2.32 Compute the HOMFLY polynomial of the four links shown on page 32. Note that, according to the result, the behaviour of the HOMFLY polynomial under the change of orientation of one component is rather unpredictable. (The same is true for the Conway polynomial, but not true for the Jones and the Kauffman polynomials.)

2.33 (Lickorish 1987) Prove that for an oriented link L with k components,

$$(J(L))^2\Big|_{t=-q^{-2}} = (-1)^{k-1} F(L)\Big|_{\substack{a=q^3 \\ z=q+q^{-1}}} ,$$

where $J(L)$ is the Jones polynomial and $F(L)$ is the two-variable Kauffman polynomial defined in Section 2.7.2.

2.34 Let L be a link with k components. Show that its two-variable Kauffman polynomial $F(L)$ is an even function of both variables a and z (that is, it consists of monomials $a^i z^j$ with i and j of the same parity) if k is odd, and it is an odd function (different parities of i and j) if k is even.

2.35 Prove that the Kauffman polynomial $F(K)$ of a knot K is a genuine polynomial in z.

2.36 For a knot K let $f_0(a) := F(K)|_{z=0}$ be the constant term of the Kauffman polynomial. Show that it is related to the constant term of the HOMFLY polynomial of K as $f_0(a) = p_0(\sqrt{-1} \cdot a)$.

2.37 **Quantum $\mathfrak{sl}_2$-invariant.** Let $\theta(\cdot)$ and $\theta^{fr}(\cdot)$ be the quantum invariants constructed in Sections 2.6.3 and 2.6.4 for the Lie algebra $\mathfrak{sl}_2$ and its standard two-dimensional representation.

1. Prove the following dependence of $\theta^{fr}(\cdot)$ on the first Reidemeister move

$$\theta^{fr}\left(\;\begin{matrix}\end{matrix}\;\right) = q^{3/4}\theta^{fr}\left(\;\begin{matrix}\end{matrix}\;\right).$$

2. Prove that $\theta(\cdot)$ remains unchanged under the first Reidemeister move.

3. Compute the value $\theta(4_1)$.

4. Show that the R-matrix defined in Section 2.6.2 satisfies the equation

$$q^{1/4} R - q^{-1/4} R^{-1} = (q^{1/2} - q^{-1/2})\mathrm{id}_{v \otimes v}.$$

5. Prove that $\theta^{fr}(\cdot)$ satisfies the skein relation

$$q^{1/4}\theta^{fr}\left(\underset{\dashleftarrow}{\left(\bigtimes\right)}\right) - q^{-1/4}\theta^{fr}\left(\underset{\dashleftarrow}{\left(\bigtimes\right)}\right) \quad = \quad (q^{1/2}-q^{-1/2})\theta^{fr}\left(\underset{\dashleftarrow}{\left(\big)\big(\right)}\right).$$

6. Prove that $\theta(\cdot)$ satisfies the skein relation

$$q\theta\left(\underset{\dashleftarrow}{\left(\bigtimes\right)}\right) - q^{-1}\theta\left(\underset{\dashleftarrow}{\left(\bigtimes\right)}\right) \quad = \quad (q^{1/2} - q^{-1/2})\theta\left(\underset{\dashleftarrow}{\left(\big)\big(\right)}\right).$$

7. For any link L with k components prove that

$$\theta^{fr}(L) = (-1)^k(q^{1/2}+q^{-1/2})\cdot\langle L\rangle\Big|_{a=-q^{1/4}},$$

   where $\langle\cdot\rangle$ is the Kauffman bracket defined in Section 2.4.1.

2.38 **Quantum $\mathfrak{sl}_N$ invariants.** Let V be an N-dimensional vector space of the standard representation of the Lie algebra $\mathfrak{sl}_N$ with a basis $e_1,\ldots,e_N$. Consider the operator $R : V \otimes V \to V \otimes V$ given by the formulae

$$R(e_i \otimes e_j) = \begin{cases} q^{\frac{-1}{2N}}e_j \otimes e_i & \text{if } i < j \\ q^{\frac{N-1}{2N}}e_i \otimes e_j & \text{if } i = j \\ q^{\frac{-1}{2N}}e_j \otimes e_i + \left(q^{\frac{N-1}{2N}} - q^{\frac{-N-1}{2N}}\right)e_i \otimes e_j & \text{if } i > j \end{cases}$$

which for $N = 2$ coincides with the operator from Section 2.6.2.

1. Prove that it satisfies the quantum Yang–Baxter equation

$$R_{12}R_{23}R_{12} = R_{23}R_{12}R_{23},$$

   where R_{ij} is the operator R acting on the ith and jth factors of $V \otimes V \otimes V$, that is, $R_{12} = R \otimes \mathrm{id}_V$ and $R_{23} = \mathrm{id}_V \otimes R$.

2. Show that its inverse is given by the formulae

$$R^{-1}(e_i \otimes e_j) = \begin{cases} q^{\frac{1}{2N}}e_j \otimes e_i + \left(-q^{\frac{N+1}{2N}} + q^{\frac{-N+1}{2N}}\right)e_i \otimes e_j & \text{if } i < j \\ q^{\frac{-N+1}{2N}}e_i \otimes e_j & \text{if } i = j \\ q^{\frac{1}{2N}}e_j \otimes e_i & \text{if } i > j \end{cases}$$

3. Check that $\quad q^{\frac{1}{2N}}R - q^{\frac{-1}{2N}}R^{-1} = (q^{1/2} - q^{-1/2})\mathrm{id}_{V\otimes V}.$

4. Extending the assignments of operators for maximum/minimum tangles from page 40 we set:

$$\underrightarrow{\min} : \mathbb{C} \to V^* \otimes V, \qquad \underrightarrow{\min}(1) := \sum_{k=1}^{N} q^{\frac{-N-1}{2}+k} e^k \otimes e_k \, ;$$

$$\underleftarrow{\min} : \mathbb{C} \to V \otimes V^*, \qquad \underleftarrow{\min}(1) := \sum_{k=1}^{N} e^k \otimes e_k \, ;$$

$$\underrightarrow{\max} : V \otimes V^* \to \mathbb{C}, \qquad \underrightarrow{\max}(e_i \otimes e^j) := \begin{cases} 0 & \text{if } i \neq j \\ q^{\frac{N+1}{2}-i} & \text{if } i = j \end{cases} \, ;$$

$$\underleftarrow{\max} : V^* \otimes V \to \mathbb{C}, \qquad \underleftarrow{\max}(e^i \otimes e_j) := \begin{cases} 0 & \text{if } i \neq j \\ 1 & \text{if } i = j \end{cases} \, .$$

Prove that all these operators are consistent in the sense that their appropriate combinations are consistent with the oriented Turaev moves from Section 1.6.6. Thus we get a link invariant denoted by $\theta_{\mathfrak{sl}_N}^{fr,St}$.

5. Show the $\theta_{\mathfrak{sl}_N}^{fr,St}$ satisfies the following skein relation:

$$q^{\frac{1}{2N}} \theta_{\mathfrak{sl}_N}^{fr,St}\left(\vcenter{\hbox{\includegraphics{}}}\right) - q^{-\frac{1}{2N}} \theta_{\mathfrak{sl}_N}^{fr,St}\left(\vcenter{\hbox{\includegraphics{}}}\right) = (q^{1/2} - q^{-1/2}) \theta_{\mathfrak{sl}_N}^{fr,St}\left(\vcenter{\hbox{\includegraphics{}}}\right)$$

and the following framing and initial conditions:

$$\theta_{\mathfrak{sl}_N}^{fr,St}\left(\vcenter{\hbox{\includegraphics{}}}\right) = q^{\frac{N-1/N}{2}} \theta_{\mathfrak{sl}_N}^{fr,St}\left(\vcenter{\hbox{\includegraphics{}}}\right)$$

$$\theta_{\mathfrak{sl}_N}^{fr,St}\left(\vcenter{\hbox{\includegraphics{}}}\right) = \frac{q^{N/2} - q^{-N/2}}{q^{1/2} - q^{-1/2}} \, .$$

6. The quadratic Casimir number for the standard $\mathfrak{sl}_N$ representation is equal to $N - 1/N$. Therefore, the deframing of this invariant gives

$$\theta_{\mathfrak{sl}_N}^{St} := q^{-\frac{N-1/N}{2} \cdot w} \theta_{\mathfrak{sl}_N}^{fr,St}$$

which satisfies

$$q^{N/2} \theta_{\mathfrak{sl}_N}^{St}\left(\vcenter{\hbox{\includegraphics{}}}\right) - q^{-N/2} \theta_{\mathfrak{sl}_N}^{St}\left(\vcenter{\hbox{\includegraphics{}}}\right) = (q^{1/2} - q^{-1/2}) \theta_{\mathfrak{sl}_N}^{St}\left(\vcenter{\hbox{\includegraphics{}}}\right) \, ;$$

$$\theta_{\mathfrak{sl}_N}^{St}\left(\vcenter{\hbox{\includegraphics{}}}\right) = \frac{q^{N/2} - q^{-N/2}}{q^{1/2} - q^{-1/2}} \, .$$

Check that this invariant is essentially a specialization of the HOMFLY polynomial,

$$\theta_{\mathfrak{sl}_N}^{St}(L) = \frac{q^{N/2} - q^{-N/2}}{q^{1/2} - q^{-1/2}} P(L)\Bigg|_{\substack{a=q^{N/2} \\ z=q^{1/2}-q^{-1/2}}} .$$

Prove that the set of invariants $\{\theta_{\mathfrak{sl}_N}^{St}\}$ for all values of N is equivalent to the HOMFLY polynomial. Thus $\{\theta_{\mathfrak{sl}_N}^{fr,St}\}$ may be considered as a framed version of the HOMFLY polynomial.

2.39 A different framed version of the HOMFLY polynomial is defined in Kauffman (1993, page 54): $P^{fr}(L) := a^{w(L)} P(L)$. Show that P^{fr} satisfies the following skein relation:

$$P^{fr}\left(\vcenter{\hbox{\includegraphics{}}}\right) - P^{fr}\left(\vcenter{\hbox{\includegraphics{}}}\right) \ = \ z P^{fr}\left(\vcenter{\hbox{\includegraphics{}}}\right)$$

and the following framing and initial conditions:

$$P^{fr}\left(\vcenter{\hbox{\includegraphics{}}}\right) \ = \ a P^{fr}\left(\vcenter{\hbox{\includegraphics{}}}\right), \ P^{fr}\left(\vcenter{\hbox{\includegraphics{}}}\right)$$

$$= \ a^{-1} P^{fr}\left(\vcenter{\hbox{\includegraphics{}}}\right)$$

$$P^{fr}\left(\bigcirc\right) \ = \ 1.$$

3

Finite type invariants

In this chapter we introduce the main protagonist of this book: the *finite type*, or *Vassiliev* knot invariants.

First we define the Vassiliev skein relation and extend, with its help, arbitrary knot invariants to knots with double points. A Vassiliev invariant of order at most n is then defined as a knot invariant which vanishes identically on knots with more than n double points.

After that, we introduce a combinatorial object of great importance: the *chord diagrams*. Chord diagrams serve as a means to describe the symbols (highest parts) of the Vassiliev invariants.

Then we prove that classical invariant polynomials are all, in a sense, of finite type, explain a simple method of calculating the values of Vassiliev invariants on any given knot and give a table of basis Vassiliev invariants up to degree 5.

Finally, we show how Vassiliev invariants can be defined for framed knots and for arbitrary tangles.

3.1 Definition of Vassiliev invariants

The original definition of finite type knot invariants was just an application of the general machinery developed by V. Vassiliev to study *complements of discriminants* in spaces of maps.

The discriminants in question are subspaces of maps with singularities of some kind. In particular, consider the space of all smooth maps of the circle into $\mathbb{R}^3$. Inside this space, define the discriminant as the subspace formed by maps that fail to be embeddings, such as curves with self-intersections, cusps, etc. Then the complement of this discriminant can be considered as *the space of knots*. The connected components of the space of knots are precisely the isotopy classes of knots; knot invariants are locally constant functions on the space of knots.

Vassiliev's machinery produces a spectral sequence that may (or may not, nobody knows it yet) converge to the cohomology of the space of knots. The zero-dimensional classes produced by this spectral sequence correspond to knot invariants which are now known as Vassiliev invariants.

This approach is indispensable if one wants to understand the higher cohomology of the space of knots. However, if we are only after the zero-dimensional classes, that is, knot invariants, the definitions can be greatly simplified. In this chapter we follow the easy path that requires no knowledge of algebraic topology whatsoever. For the reader who is not intimidated by spectral sequences, we outline Vassiliev's construction in Chapter 15.

3.1.1 Singular knots and the Vassiliev skein relation

A *singular knot* is a smooth map $S^1 \to \mathbb{R}^3$ that fails to be an embedding. We shall only consider singular knots with the simplest singularities, namely transversal self-intersections, or *double points*.

Definition 3.1. Let f be a map of a one-dimensional manifold to $\mathbb{R}^3$. A point $p \in \text{im}(f) \subset \mathbb{R}^3$ is a *double point* of f if $f^{-1}(p)$ consists of two points t_1 and t_2 and the two tangent vectors $f'(t_1)$ and $f'(t_2)$ are linearly independent. Geometrically, this means that in a neighbourhood of the point p the curve f has two branches with non-collinear tangents.

A double point

Remark 3.2. In fact, we gave a definition of a *simple double point*. We omit the word "simple" since these are the only double points we shall see.

Any knot invariant can be extended to knots with double points by means of the *Vassiliev skein relation*:

$$v\left(\raisebox{-0.5ex}{\includegraphics[height=2ex]{x1}}\right) = v\left(\raisebox{-0.5ex}{\includegraphics[height=2ex]{x2}}\right) - v\left(\raisebox{-0.5ex}{\includegraphics[height=2ex]{x3}}\right). \tag{3.1}$$

Here v is the knot invariant with values in some abelian group, the left-hand side is the value of v on a singular knot K (shown in a neighbourhood of a double point) and the right-hand side is the difference of the values of v on (possibly singular) knots obtained from K by replacing the double point with a positive and a negative crossing respectively. The process of applying the skein relation is also referred to as *resolving a double point*. It is clearly independent of the plane projection of the singular knot.

Using the Vassiliev skein relation recursively, we can extend any knot invariant to knots with an arbitrary number of double points. There are many ways to do this, since we can choose to resolve double points in an arbitrary order. However, the result is independent of any choice. Indeed, the calculation of the value of v on a singular knot K with n double points is in all cases reduced to the *complete resolution* of the knot K which yields an alternating sum

$$v(K) = \sum_{\varepsilon_1 = \pm 1, \ldots, \varepsilon_n = \pm 1} (-1)^{|\varepsilon|} v(K_{\varepsilon_1, \ldots, \varepsilon_n}), \qquad (3.2)$$

where $|\varepsilon|$ is the number of -1's in the sequence $\varepsilon_1, \ldots, \varepsilon_n$, and $K_{\varepsilon_1, \ldots, \varepsilon_n}$ is the knot obtained from K by a positive or negative resolution of the double points according to the sign of ε_i for the point number i.

Definition 3.3. (Vassiliev 1990). A knot invariant is said to be a *Vassiliev invariant* (or a *finite type invariant*) of order (or degree) $\leqslant n$ if its extension vanishes on all singular knots with more than n double points. A Vassiliev invariant is said to be of order (degree) n if it is of order $\leqslant n$ but not of order $\leqslant n - 1$.

In general, a Vassiliev invariant may take values in an arbitrary abelian group. In practice, however, all our invariants will take values in commutative rings and it will be convenient to make this assumption from now on.

Notation 3.4. We shall denote by $\mathscr{V}_n$ the set of Vassiliev invariants of order $\leqslant n$ with values in a ring $\mathscr{R}$. Whenever necessary, we shall write $\mathscr{V}_n^{\mathscr{R}}$ to indicate the range of the invariants explicitly. It follows from the definition that, for each n, the set $\mathscr{V}_n$ is an $\mathscr{R}$-module. Moreover, $\mathscr{V}_n \subseteq \mathscr{V}_{n+1}$, so we have an increasing filtration

$$\mathscr{V}_0 \subseteq \mathscr{V}_1 \subseteq \mathscr{V}_2 \subseteq \cdots \subseteq \mathscr{V}_n \subseteq \cdots \subseteq \mathscr{V} := \bigcup_{n=0}^{\infty} \mathscr{V}_n.$$

We shall further discuss this definition in the next section. First, let us see that there are indeed many (in fact, infinitely many) independent Vassiliev invariants.

Example 3.5. (Bar-Natan 1991a). The nth coefficient of the Conway polynomial is a Vassiliev invariant of order $\leqslant n$.

Indeed, the definition of the Conway polynomial, together with the Vassiliev skein relation, implies that

$$C\left(\!\!\begin{array}{c}\times\end{array}\!\!\right) = t\,C\left(\!\!\begin{array}{c})(\end{array}\!\!\right).$$

Applying this relation several times, we get

$$C\left(\ \right) = t^k C\left(\ \right)$$

for a singular knot with k double points. If $k \geqslant n + 1$, then the coefficient of t^n in this polynomial is zero.

3.2 Algebra of Vassiliev invariants

3.2.1 The singular knot filtration

Consider the "tautological knot invariant" $\mathcal{K} \to \mathbb{Z}\mathcal{K}$ which sends a knot to itself. Applying the Vassiliev skein relation, we extend it to knots with double points; a knot with n double points is then sent to an alternating sum of 2^n genuine knots.

Recall that we denote by $\mathbb{Z}\mathcal{K}$ the free abelian group spanned by the equivalence classes of knots with multiplication induced by the connected sum of knots. Let $\mathcal{K}_n$ be the $\mathbb{Z}$-submodule of the algebra $\mathbb{Z}\mathcal{K}$ spanned by the images of knots with n double points.

Exercise. Prove that $\mathcal{K}_n$ is an ideal of $\mathbb{Z}\mathcal{K}$.

A knot with $n + 1$ double points gives rise to a difference of two knots with n double points in $\mathbb{Z}\mathcal{K}$; hence, we have the descending *singular knot filtration*

$$\mathbb{Z}\mathcal{K} = \mathcal{K}_0 \supseteq \mathcal{K}_1 \supseteq \ldots \supseteq \mathcal{K}_n \supseteq \ldots$$

The definition of Vassiliev invariants can now be re-stated in the following terms:

Definition 3.6. Let $\mathcal{R}$ be a commutative ring. A Vassiliev invariant of order $\leqslant n$ is a linear function $\mathbb{Z}\mathcal{K} \to \mathcal{R}$ which vanishes on $\mathcal{K}_{n+1}$.

According to this definition, the module of $\mathcal{R}$-valued Vassiliev invariants of order $\leqslant n$ is naturally isomorphic to the space of linear functions $\mathbb{Z}\mathcal{K} / \mathcal{K}_{n+1} \to \mathcal{R}$. So, in a certain sense, the study of the Vassiliev invariants is equivalent to studying the filtration $\mathcal{K}_n$. In the next several chapters we shall mostly speak about invariants, rather than the filtration on the algebra of knots. Nevertheless, the latter approach, developed by Goussarov (1993), is important and we cannot skip it here altogether.

Definition 3.7. Two knots K_1 and K_2 are *n-equivalent* if they cannot be distinguished by Vassiliev invariants of degree n and smaller with values in an arbitrary abelian group. A knot that is n-equivalent to the trivial knot is called *n-trivial*.

In other words, K_1 and K_2 are n-equivalent if and only if $K_1 - K_2 \in \mathcal{K}_{n+1}$.

Definition 3.8. Let $\Gamma_n \mathcal{K}$ be the set of $(n-1)$-trivial knots. The *Goussarov filtration* on $\mathcal{K}$ is the descending filtration

$$\mathcal{K} = \Gamma_1 \mathcal{K} \supseteq \Gamma_2 \mathcal{K} \supseteq \cdots \supseteq \Gamma_n \mathcal{K} \supseteq \cdots$$

The sets $\Gamma_n \mathcal{K}$ are, in fact, abelian monoids under the connected sum of knots (this follows from the fact that each $\mathcal{K}_n$ is a subalgebra of $\mathbb{Z}\mathcal{K}$). Goussarov proved that the monoid quotient $\mathcal{K}/\Gamma_n \mathcal{K}$ is an (abelian) group. We shall consider n-equivalence in greater detail in Chapters 12 and 14.

3.2.2 Vassiliev invariants as polynomials

A useful way to think of Vassiliev invariants is as follows. Let v be an invariant of singular knots with n double points and $\nabla(v)$ be the extension of v to singular knots with $n+1$ double points using the Vassiliev skein relation. We can consider ∇ as an operator between the corresponding spaces of invariants. Now, a function $v : \mathcal{K} \to \mathcal{R}$ is a Vassiliev invariant of degree $\leqslant n$, if it satisfies the difference equation $\nabla^{n+1}(v) = 0$. This can be seen as an analogy between Vassiliev invariants as a subspace of all knot invariants and polynomials as a subspace of all smooth functions on a real line: the role of differentiation is played by the operator ∇. It is well known that continuous functions on a real line can be approximated by polynomials. The main open problem of the theory of finite type invariants is to find an analogue of this statement in the knot-theoretic context, namely, to understand to what extent an arbitrary numerical knot invariant can be approximated by Vassiliev invariants. More on this in Section 3.2.4.

3.2.3 The filtration on the algebra of Vassiliev invariants

The set of all Vassiliev invariants forms a commutative filtered algebra with respect to the usual (pointwise) multiplication of functions.

Theorem 3.9. *The product of two Vassiliev invariants of degrees $\leqslant p$ and $\leqslant q$ is a Vassiliev invariant of degree $\leqslant p + q$.*

Proof Let f and g be two invariants with values in a ring $\mathcal{R}$, of degrees p and q respectively. Consider a singular knot K with $n = p + q + 1$ double points. The complete resolution of K via the Vassiliev skein relation gives

$$(fg)(K) = \sum_{\varepsilon_1 = \pm 1, \ldots, \varepsilon_n = \pm 1} (-1)^{|\varepsilon|} f(K_{\varepsilon_1, \ldots, \varepsilon_n}) g(K_{\varepsilon_1, \ldots, \varepsilon_n})$$

in the notation of (3.2). The alternating sum on the right-hand side is taken over all points of an n-dimensional binary cube

$$Q_n = \{(\varepsilon_1, \ldots, \varepsilon_n) \mid \varepsilon_i = \pm 1\}.$$

In general, given a function v on Q_n and a subset $S \subseteq Q_n$, the *alternating sum* of v over S is defined as $\sum_{\varepsilon \in S} (-1)^{|\varepsilon|} v(\varepsilon)$.

If we set

$$f(\varepsilon_1, \ldots, \varepsilon_n) = f(K_{\varepsilon_1, \ldots, \varepsilon_n})$$

and define $g(\varepsilon_1, \ldots, \varepsilon_n)$ similarly, we can think of f and g as functions on Q_n. The fact that f is of degree p means that the alternating sum of f on each $(p+1)$-face of Q_n is zero. Similarly, on each $(q+1)$-face of Q_n the alternating sum of g vanishes. Now, the theorem is a consequence of the following lemma.

Lemma 3.10. *Let f, g be functions on Q_n, where $n = p+q+1$. If the alternating sums of f over any $(p+1)$-face, and of g over any $(q+1)$-face of Q_n are zero, so is the alternating sum of the product fg over the entire cube Q_n.*

Proof of the lemma. Use induction on n. For $n = 1$ we have $p = q = 0$ and the premises of the lemma read $f(-1) = f(1)$ and $g(-1) = g(1)$. Therefore, $(fg)(-1) = (fg)(1)$, as required.

For the general case, denote by $\mathcal{F}_n$ the space of functions $Q_n \to \mathcal{R}$. We have two operators

$$\rho_-, \rho_+ : \mathcal{F}_n \to \mathcal{F}_{n-1}$$

which take a function v to its restrictions to the $(n-1)$-dimensional faces $\varepsilon_1 = -1$ and $\varepsilon_1 = 1$ of Q_n:

$$\rho_-(v)(\varepsilon_2, \ldots, \varepsilon_n) = v(-1, \varepsilon_2, \ldots, \varepsilon_n)$$

and

$$\rho_+(v)(\varepsilon_2, \ldots, \varepsilon_n) = v(1, \varepsilon_2, \ldots, \varepsilon_n).$$

Let

$$\delta = \rho_+ - \rho_-.$$

Observe that if the alternating sum of v over any r-face of Q_n is zero, then the alternating sum of $\rho_\pm(v)$ (respectively, $\delta(v)$) over any r-face (respectively, $(r-1)$-face) of Q_{n-1} is zero.

A direct check shows that the operator δ satisfies the following Leibniz rule:

$$\delta(fg) = \rho_+(f) \cdot \delta(g) + \delta(f) \cdot \rho_-(g).$$

Applying the induction assumption to each of the two summands on the right-hand side, we see that the alternating sum of $\delta(fg)$ over the cube Q_{n-1} vanishes. By the definition of δ, this sum coincides with the alternating sum of fg over Q_n. $\square$

Remark 3.11. The existence of the filtration on the algebra of Vassiliev invariants can be thought of as a manifestation of their polynomial character. Indeed, a polynomial of degree $\leqslant n$ in one variable can be defined as a function whose $n + 1$st derivative is identically zero. Then the fact that a product of polynomials of degrees $\leqslant p$ and $\leqslant q$ has degree $\leqslant p + q$ can be proved by induction using the Leibniz formula. In our argument on Vassiliev invariants we have used the very same logic. A further discussion of the Leibniz formula for finite type invariants can be found in Willerton (1996).

3.2.4 Approximation by Vassiliev invariants

The analogy between finite type invariants and polynomials would be even more satisfying if there existed a Stone–Weierstraß-type theorem for knot invariants that would affirm that any invariant can be approximated by Vassiliev invariants. At the moment no such statement is known. In fact, understanding the strength of the class of finite type invariants is the main problem in the theory.

There are various ways of formulating this problem as a precise question. Let us say that a class $\mathcal{U}$ of knot invariants is *complete* if for any finite set of knots the invariants from $\mathcal{U}$ span the space of all functions on these knots. We say that invariants from $\mathcal{U}$ *distinguish knots* if for any two different knots K_1 and K_2 there exists $f \in \mathcal{U}$ such that $f(K_1) \neq f(K_2)$. Finally, the class $\mathcal{U}$ *detects the unknot* if any knot can be distinguished from the trivial knot by an invariant from $\mathcal{U}$. A priori, completeness is the strongest of these properties. In this terminology, the main outstanding problem in the theory of finite type invariants is to determine whether the Vassiliev invariants distinguish knots. While it is conjectured that the set of rational-valued Vassiliev knot invariants is complete, it is not even known if the class of all Vassiliev knot invariants detects the unknot.

Note that the rational-valued Vassiliev invariants are complete if and only if the intersection $\cap \mathcal{K}_n$ of all the terms of the singular knot filtration is zero. Indeed, a non-zero element of $\cap \mathcal{K}_n$ produces a universal relation among the values of the invariants on a certain set of knots. On the other hand, let $\cap \mathcal{K}_n = 0$. Then the map $\mathcal{K} \to \mathbb{Z}\mathcal{K}/\mathcal{K}_{n+1}$ is a Vassiliev invariant of order n whose values on any given set of knots become linearly independent as n grows. As for the Goussarov filtration $\Gamma_n \mathcal{K}$, the intersection of all of its terms consists of the trivial knot if and only if the Vassiliev invariants detect the unknot.

There are knot invariants, of which we shall see many examples, which are not of finite type, but, nevertheless, can be approximated by Vassiliev invariants in a certain sense. These are the *polynomial* and the *power series Vassiliev invariants*. A polynomial Vassiliev invariant is an element of the vector space

$$\mathcal{V}_\bullet = \bigoplus_{n=0}^{\infty} \mathcal{V}_n.$$

Since the product of two invariants of degrees m and n has degree at most $m + n$, the space $\mathcal{V}_\bullet$ is, in fact, a commutative graded algebra. The power series Vassiliev invariants are, by definition, the elements of its *graded completion* $\widehat{\mathcal{V}}_\bullet$ (see Definition A.2.5).

The Conway polynomial C is an example of a power series invariant. Observe that even though for any knot K the value $C(K)$ is a polynomial, the Conway polynomial C is not a polynomial invariant according to the definition of this paragraph.

Power series Vassiliev invariants are just one possible approach to defining approximation by finite type invariants. A wider class of invariants are those *dominated by Vassiliev invariants*. We say that a knot invariant u is dominated by Vassiliev invariants if $u(K_1) \neq u(K_2)$ for some knots K_1 and K_2 implies that there is a Vassiliev knot invariant v with $v(K_1) \neq v(K_2)$. Clearly, if Vassiliev invariants distinguish knots, then each knot invariant is dominated by Vassiliev invariants. At the moment, however, it is an open question whether, for instance, the signature of a knot (Rolfsen 1976) is dominated by Vassiliev invariants.

3.3 Vassiliev invariants of degrees 0, 1 and 2

Proposition 3.12. $\mathcal{V}_0 = \{const\}$, dim $\mathcal{V}_0 = 1$.

Proof Let $f \in \mathcal{V}_0$. By definition, the value of (the extension of) f on any singular knot with one double point is 0. Pick an arbitrary knot K. Any diagram of K can be turned into a diagram of the trivial knot K_0 by crossing changes done one at a time. By assumption, the jump of f at every crossing change is 0, therefore, $f(K) = f(K_0)$. Thus f is constant.   □

Proposition 3.13. $\mathcal{V}_1 = \mathcal{V}_0$.

Proof A singular knot with one double point is divided by the double point into two closed curves. An argument similar to the last proof shows that the value of v on any knot with one double point is equal to its value on the "figure infinity" singular knot and, hence, to 0:

$$v\left(\begin{array}{c}\includegraphics\end{array}\right) = v\left(\bigcirc\!\!\bigcirc\right) = 0. \tag{3.3}$$

Therefore, $\mathcal{V}_1 = \mathcal{V}_0$. □

The first non-trivial Vassiliev invariant appears in degree 2: it is the second coefficient c_2 of the Conway polynomial, also known as the *Casson invariant*.

Proposition 3.14. $\dim \mathcal{V}_2 = 2$.

Proof Let us explain why the argument of the proof of the previous two propositions does not work in this case. Take a knot with two double points and try to transform it into some fixed knot with two double points using smooth deformations and crossing changes. It is easy to see that any knot with two double points can be reduced to one of the following two basic knots:

Basic knot K_1 Basic knot K_2

– but these two knots cannot be obtained one from the other! The essential difference between them is in the order of the double points on the curve.

Let us label the double points of K_1 and K_2, say, by 1 and 2. When travelling along the first knot, K_1, the two double points are encountered in the order 1122 (or 1221, 2211, 2112 if you start from a different initial point). For the knot K_2 the sequence is 1212 (equivalent to 2121). The two sequences 1122 and 1212 are different even if cyclic permutations are allowed.

Now take an arbitrary singular knot K with two double points. If the cyclic order of these points is 1122, then we can transform the knot to K_1, passing in the process of deformation through some singular knots with three double points; if the order is 1212, we can reduce K in the same way to the second basic knot K_2.

The above argument shows that, to any $\mathcal{R}$-valued order 2 Vassiliev invariant there corresponds a function on the set of two elements $\{K_1, K_2\}$ with values in $\mathcal{R}$. We thus obtain a linear map $\mathcal{V}_2 \to \mathcal{R}^2$. The kernel of this map is equal to $\mathcal{V}_1$: indeed, the fact that a given invariant $f \in \mathcal{V}_2$ satisfies $f(K_1) = f(K_2) = 0$ means that it vanishes on *any* singular knot with 2 double points, which is by definition equivalent to saying that $f \in \mathcal{V}_1$.

On the other hand, the image of this linear map is no more than one-dimensional, since for *any* knot invariant f we have $f(K_1) = 0$. This proves that $\dim \mathcal{V}_2 \leqslant 2$. In fact, $\dim \mathcal{V}_2 = 2$, since the second coefficient c_2 of the Conway polynomial is not constant (see Table 2.1). □

3.4 Chord diagrams

Now let us give a formal definition of the combinatorial structure which is implicit in the proof of the last proposition.

Definition 3.15. A *chord diagram* of order n (or degree n) is an oriented circle with a distinguished set of n disjoint pairs of distinct points, considered up to orientation preserving diffeomorphisms of the circle. The set of all chord diagrams of order n will be denoted by $\mathbf{A}_n$.

We shall usually omit the orientation of the circle in pictures of chord diagrams, assuming that it is oriented counterclockwise.

Example 3.16.

$$\mathbf{A}_1 = \{\ominus\},$$

$$\mathbf{A}_2 = \{\text{⬭}, \text{⊗}\},$$

$$\mathbf{A}_3 = \{\ominus, \text{⬭}, \text{⊗}, \text{⊕}, \text{⊗}\}.$$

Remark 3.17. Chord diagrams that differ by a mirror reflection are, in general, different:

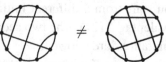

This observation reflects the fact that we are studying *oriented* knots.

3.4.1 The chord diagram of a singular knot

Chord diagrams are used to code certain information about singular knots.

Definition 3.18. The chord diagram $\sigma(K) \in \mathbf{A}_n$ of a singular knot with n double points is obtained by marking on the parametrizing circle n pairs of points whose images are the n double points of the knot.

Examples 3.19.

$$\sigma\left(\text{⬁}\right) = \text{⬭}, \qquad \sigma\left(\text{⬤}\right) = \text{⊗}.$$

Proposition 3.20. (Vassiliev 1990a). *The value of a Vassiliev invariant v of order $\leqslant n$ on a knot K with n double points depends only on the chord diagram of K:*

$$\sigma(K_1) = \sigma(K_2) \Rightarrow v(K_1) = v(K_2).$$

Proof Suppose that $\sigma(K_1) = \sigma(K_2)$. Then there is a one-to-one correspondence between the chords of both chord diagrams, and, hence, between the double points of K_1 and K_2. Place K_1, K_2 in $\mathbb{R}^3$ so that the corresponding double points coincide together with both branches of the knot in the vicinity of each double point.

Knot K_1　　　　　　　Knot K_2

Now we can deform K_1 into K_2 in such a way that some small neighbourhoods of the double points do not move. We can assume that the only new singularities created in the process of this deformation are a finite number of double points, all at distinct values of the deformation parameter. By the Vassiliev skein relation, in each of these events the value of v does not change, and this implies that $v(K_1) = v(K_2)$. □

The fact that we have just proved shows that there is a well defined map $\alpha_n : \mathcal{V}_n \to \mathcal{R}\mathbf{A}_n$ (the $\mathcal{R}$-module of $\mathcal{R}$-valued functions on the set $\mathbf{A}_n$):

$$\alpha_n(v)(D) = v(K),$$

where K is an arbitrary knot with $\sigma(K) = D$.

We want to understand the size and the structure of the space $\mathcal{V}_n$, so it would be of use to have a description of the kernel and the image of α_n.

The description of the kernel follows immediately from the definitions: $\ker \alpha_n = \mathcal{V}_{n-1}$. Therefore, we obtain an injective homomorphism

$$\overline{\alpha}_n : \mathcal{V}_n/\mathcal{V}_{n-1} \to \mathcal{R}\mathbf{A}_n. \tag{3.4}$$

The problem of describing the image of α_n is much more difficult. The answer to it will be given in Section 4.2.1.

Since there is only a finite number of diagrams of each order, we get the following

Corollary 3.21. *The module of $\mathcal{R}$-valued Vassiliev invariants of degree at most n is finitely generated over $\mathcal{R}$.*

Since the map α_n discards the order $(n-1)$ part of a Vassiliev invariant v, we can, by analogy with differential operators, call the function $\alpha_n(v)$ on chord diagrams the *symbol* of the Vassiliev invariant v:

$$\text{symb}(v) = \alpha_n(v),$$

where n is the order of v.

Example 3.22. The symbol of the Casson invariant is equal to 0 on the chord diagram with two parallel chords, and to 1 on the chord diagram with two intersecting chords.

Remark 3.23. It may be instructive to state all the above in the dual setting of the singular knot filtration. The argument in the proof of the proposition in the beginning of this section essentially says that $\mathbf{A}_n$ is the set of singular knots with n double points modulo isotopies and crossing changes. In terms of the singular knot filtration, we have shown that if two knots with n double points have the same chord diagram, then their difference lies in $\mathcal{K}_{n+1} \subset \mathbb{Z}\mathcal{K}$. Since $\mathcal{K}_n$ is spanned by the complete resolutions of knots with n double points, we have a surjective map

$$\mathbb{Z}\mathbf{A}_n \to \mathcal{K}_n / \mathcal{K}_{n+1}.$$

The kernel of this map, after tensoring with the rational numbers, is spanned by the so-called *4T* and *1T relations*, defined in the next chapter. This is the content of the Fundamental Theorem (see Section 4.2).

3.5 Invariants of framed knots

A *singular framing* on a closed curve immersed in $\mathbb{R}^3$ is a smooth normal vector field with a finite number of simple zeroes on this curve. A *singular framed knot* is a knot with simple double points in $\mathbb{R}^3$ equipped with a singular framing whose set of zeroes is disjoint from the set of double points.

Invariants of framed knots are extended to singular framed knots by means of the Vassiliev skein relation; for double points it has the same form as before, and for the zeroes of the singular framing it can be drawn as

$$v\!\left(\includegraphics{}\right) = v\!\left(\includegraphics{}\right) - v\!\left(\includegraphics{}\right).$$

An invariant of framed knots is of order $\leqslant n$ if its extension vanishes on knots with more than n singularities (double points or zeroes of the framing).

Let us denote the space of invariants of order $\leqslant n$ by $\mathcal{V}_n^{fr}$. There is a natural inclusion $i : \mathcal{V}_n \to \mathcal{V}_n^{fr}$ defined by setting $i(f)(K) = f(K')$ where K is a framed knot, and K' is the same knot without framing. It turns out that this is a proper inclusion for all $n \geqslant 1$.

Let us determine the framed Vassiliev invariants of small degree. Any invariant of degree zero is, in fact, an unframed knot invariant and, hence, is constant. Indeed, increasing the framing by one can be thought of as passing a singularity of the framing, and this does not change the value of a degree zero invariant.

Exercise. (1) Prove that dim $\mathcal{V}_1^{fr} = 2$, and that $\mathcal{V}_1^{fr}$ is spanned by the constants and the self-linking number.
(2) Find the dimension and a basis of the vector space $\mathcal{V}_2^{fr}$.

Exercise. Let v be a framed Vassiliev invariant degree n, and K an unframed knot. Let $v(K, k)$ be the value of v on K equipped with a framing with self-linking number k. Show that $v(K, k)$ is a polynomial in k of degree at most n.

3.5.1 Chord diagrams for framed knots

We have seen that chord diagrams on n chords can be thought of as singular knots with n double points modulo isotopies and crossing changes. Following the same logic, we should define a chord diagram for framed knots as an equivalence class of framed singular knots with n singularities modulo isotopies, crossing changes and additions of zeroes of the framing. In this way, the value of a degree n Vassiliev invariant on a singular framed knot with n singularities will only depend on the chord diagram of the knot.

As a combinatorial object, a framed chord diagram of degree n can be defined as a usual chord diagram of degree $n - k$ together with k dots marked on the circle. The chords correspond to the double points of a singular knot and the dots represent the zeroes of the framing.

In the sequel we shall not make any use of diagrams with dots, for the following reason. If $\mathcal{R}$ is a ring where 2 is invertible, a zero of the framing on a knot with n singularities can be replaced, modulo knots with $n + 1$ singularities, by "half of a double point":

$$v(\underset{}{\overbrace{}}) = \frac{1}{2}v(\underset{}{\overbrace{}}) - \frac{1}{2}v(\underset{}{\overbrace{}})$$

for any invariant v. In particular, if we replace a dot with a chord whose endpoints are next to each other on some diagram, the symbol of any Vassiliev invariant on this diagram is simply multiplied by 2.

On the other hand, the fact that we can use the same chord diagrams for both framed and unframed knots does not imply that the corresponding theories of Vassiliev invariants are the same. In particular, we shall see that the symbol of any invariant of unframed knots vanishes on a diagram which has a chord that has no intersections with other chords. This does not hold for an arbitrary framed invariant.

Example 3.24. The symbol of the self-linking number is the function equal to 1 on the chord diagram with one chord.

3.6 Classical knot polynomials as Vassiliev invariants

In Example 3.5, we have seen that the coefficients of the Conway polynomial are Vassiliev invariants. The Conway polynomial, taken as a whole, is not, of course, a finite type invariant, but it is an infinite linear combination of such; in other words, it is a power series Vassiliev invariant. This property holds for all classical knot polynomials – but only after a suitable substitution.

3.6.1 The Jones polynomial

Modify the Jones polynomial of a knot K substituting $t = e^h$ and then expanding it into a formal power series in h. Let $j_n(K)$ be the coefficient of h^n in this expansion.

Theorem 3.25 (Goussarov 1991; Birman and Lin 1993; Bar-Natan 1995a). *The coefficient $j_n(K)$ is a Vassiliev invariant of order $\leqslant n$.*

Proof Plugging $t = e^h = 1 + h + \ldots$ into the skein relation on page 34 we get

$$(1 - h + \ldots) \cdot J\left(\begin{smallmatrix}\end{smallmatrix}\right) - (1 + h + \ldots) \cdot J\left(\begin{smallmatrix}\end{smallmatrix}\right) = (h + \ldots) \cdot J\left(\begin{smallmatrix}\end{smallmatrix}\right).$$

We see that the difference

$$J\left(\begin{smallmatrix}\end{smallmatrix}\right) - J\left(\begin{smallmatrix}\end{smallmatrix}\right) = J\left(\begin{smallmatrix}\end{smallmatrix}\right)$$

is congruent to 0 modulo h. Therefore, the Jones polynomial of a singular knot with k double points is divisible by h^k. In particular, for $k \geqslant n + 1$ the coefficient of h^n equals zero. ∎

Below we shall give an explicit description of the symbols of the finite type invariants j_n; the similar description for the Conway polynomial is left as an exercise (see Exercise 3.16).

3.6.2 Symbol of the Jones invariant j_n

To find the symbol of $j_n(K)$, we must compute the coefficient of h^n in the Jones polynomial $J(K_n)$ of a singular knot K_n with n double points in terms of its chord diagram $\sigma(K_n)$. Since

$$J\left(\begin{array}{c}\includegraphics{}\end{array}\right) = J\left(\begin{array}{c}\includegraphics{}\end{array}\right) - J\left(\begin{array}{c}\includegraphics{}\end{array}\right)$$

$$= h\left(j_0\left(\begin{array}{c}\includegraphics{}\end{array}\right) + j_0\left(\begin{array}{c}\includegraphics{}\end{array}\right) + j_0\left(\begin{array}{c}\includegraphics{}\end{array}\right)\right) + \cdots$$

the contribution of a double point of K_n to the coefficient $j_n(K_n)$ is the sum of the values of $j_0(\cdot)$ on the three links in the parentheses above. The values of $j_0(\cdot)$ for the last two links are equal, according to Exercise 3.4 to this chapter, to $j_0(L) = (-2)^{\#(\text{components of } L)-1}$. So it does not depend on the specific way L is knotted and linked and we can freely change the under/over-crossings of L. On the level of chord diagrams these two terms mean that we just forget about the chord corresponding to this double point. The first term, $j_0\left(\begin{array}{c}\includegraphics{}\end{array}\right)$, corresponds to the smoothing of the double point according to the orientation of our knot (link). On the level of chord diagrams this corresponds to the doubling of a chord:

This leads to the following procedure of computing the value of the symbol of $j_n(D)$ on a chord diagram D. Introduce *a state s* for D as an arbitrary function on the set chords of D with values in the set $\{1, 2\}$. With each state s we associate an immersed plane curve obtained from D by resolving (either doubling or deleting) all its chords according to s:

, if $s(c) = 1$;   , if $s(c) = 2$.

Let $|s|$ denote the number of components of the curve obtained in this way. Then

$$\text{symb}(j_n)(D) = \sum_s \left(\prod_c s(c)\right) (-2)^{|s|-1},$$

where the product is taken over all n chords of D, and the sum is taken over all 2^n states for D.

For example, to compute the value of the symbol of j_3 on the chord diagram ⊞ we must consider eight states:

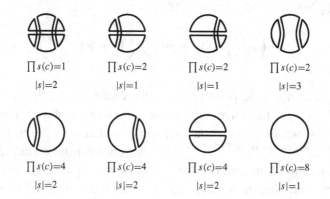

$$\prod s(c)=1 \qquad \prod s(c)=2 \qquad \prod s(c)=2 \qquad \prod s(c)=2$$
$$|s|=2 \qquad\qquad |s|=1 \qquad\qquad |s|=1 \qquad\qquad |s|=3$$

$$\prod s(c)=4 \qquad \prod s(c)=4 \qquad \prod s(c)=4 \qquad \prod s(c)=8$$
$$|s|=2 \qquad\qquad |s|=2 \qquad\qquad |s|=2 \qquad\qquad |s|=1$$

Therefore,

$$\mathrm{symb}(j_3)\left(\bigoplus\right) = -2+2+2+2(-2)^2+4(-2)+4(-2)+4(-2)+8 = -6.$$

Similarly one can compute the values of $\mathrm{symb}(j_3)$ on all chord diagrams with three chords. Here is the result:

D	⊖	◍	⊗	⊞	✳
$\mathrm{symb}(j_3)(D)$	0	0	0	-6	-12

This function on chord diagrams, as well as the whole Jones polynomial, is closely related to the Lie algebra $\mathfrak{sl}_2$ and its standard two-dimensional representation. We shall return to this subject several times (see Sections 6.1.3, 6.1.7, etc.).

3.6.3 Jones invariant j_n and the mirror reflection

According to Exercise 2.24, for the mirror reflection $\overline{K}$ of a knot K the power series expansion of $J(\overline{K})$ can be obtained from the series $J(K)$ by substituting $-h$ for h. This means that $j_{2k}(\overline{K}) = j_{2k}(K)$ and $j_{2k+1}(\overline{K}) = -j_{2k+1}(K)$.

3.6.4 Some values of the Jones invariant j_n

Table 3.1 displays the first five terms of the power series expansion of the Jones polynomial after the substitution $t = e^h$.

Table 3.1 *Taylor expansion of the modified Jones polynomial*

3_1	1	$-3h^2$	$+6h^3$	$-(29/4)h^4$	$+(13/2)h^5$	$+\dots$
4_1	1	$+3h^2$		$+(5/4)h^4$		$+\dots$
5_1	1	$-9h^2$	$+30h^3$	$-(243/4)h^4$	$+(185/2)h^5$	$+\dots$
5_2	1	$-6h^2$	$+18h^3$	$-(65/2)h^4$	$+(87/2)h^5$	$+\dots$
6_1	1	$+6h^2$	$-6h^3$	$+(17/2)h^4$	$-(13/2)h^5$	$+\dots$
6_2	1	$+3h^2$	$-6h^3$	$+(41/4)h^4$	$-(25/2)h^5$	$+\dots$
6_3	1	$-3h^2$		$-(17/4)h^4$		$+\dots$
7_1	1	$-18h^2$	$+84h^3$	$-(477/2)h^4$	$+511h^5$	$+\dots$
7_2	1	$-9h^2$	$+36h^3$	$-(351/4)h^4$	$+159h^5$	$+\dots$
7_3	1	$-15h^2$	$-66h^3$	$-(697/4)h^4$	$-(683/2)h^5$	$+\dots$
7_4	1	$-12h^2$	$-48h^3$	$-113h^4$	$-196h^5$	$+\dots$
7_5	1	$-12h^2$	$+48h^3$	$-119h^4$	$+226h^5$	$+\dots$
7_6	1	$-3h^2$	$+12h^3$	$-(89/4)h^4$	$+31h^5$	$+\dots$
7_7	1	$+3h^2$	$+6h^3$	$+(17/4)h^4$	$+(13/2)h^5$	$+\dots$
8_1	1	$+9h^2$	$-18h^3$	$+(135/4)h^4$	$-(87/2)h^5$	$+\dots$
8_2	1		$-6h^3$	$+27h^4$	$-(133/2)h^5$	$+\dots$
8_3	1	$+12h^2$		$+17h^4$		$+\dots$
8_4	1	$+9h^2$	$-6h^3$	$+(63/4)h^4$	$-(25/2)h^5$	$+\dots$
8_5	1	$+3h^2$	$+18h^3$	$+(209/4)h^4$	$+(207/2)h^5$	$+\dots$
8_6	1	$+6h^2$	$-18h^3$	$+(77/2)h^4$	$-(123/2)h^5$	$+\dots$
8_7	1	$-6h^2$	$-12h^3$	$-(47/2)h^4$	$-31h^5$	$+\dots$
8_8	1	$-6h^2$	$-6h^3$	$-(29/2)h^4$	$-(25/2)h^5$	$+\dots$
8_9	1	$+6h^2$		$+(23/2)h^4$		$+\dots$
8_{10}	1	$-9h^2$	$-18h^3$	$-(123/4)h^4$	$-(75/2)h^5$	$+\dots$
8_{11}	1	$+3h^2$	$-12h^3$	$+(125/4)h^4$	$-55h^5$	$+\dots$
8_{12}	1	$+9h^2$		$+(51/4)h^4$		$+\dots$
8_{13}	1	$-3h^2$	$-6h^3$	$-(53/4)h^4$	$-(25/2)h^5$	$+\dots$
8_{14}	1			$+6h^4$	$-18h^5$	$+\dots$
8_{15}	1	$-12h^2$	$+42h^3$	$-80h^4$	$+(187/2)h^5$	$+\dots$
8_{16}	1	$-3h^2$	$+6h^3$	$-(53/4)h^4$	$+(37/2)h^5$	$+\dots$

8_{17}	1	$+3h^2$		$+(29/4)h^4$		$+\ldots$
8_{18}	1	$-3h^2$		$+(7/4)h^4$		$+\ldots$
8_{19}	1	$-15h^2$	$-60h^3$	$-(565/4)h^4$	$-245h^5$	$+\ldots$
8_{20}	1	$-6h^2$	$+12h^3$	$-(35/2)h^4$	$+19h^5$	$+\ldots$
8_{21}	1		$-6h^3$	$+21h^4$	$-(85/2)h^5$	$+\ldots$

Example 3.26. In the following examples the h-expansion of the Jones polynomial starts with a power of h equal to the number of double points in a singular knot, in compliance with Theorem 3.25.

$$J\left(\;\right) = \underbrace{J\left(\;\right) - J\left(\;\right)}_{0} = -\underbrace{J\left(\;\right) + J\left(\;\right)}_{1}$$

$$= -1 + J(3_1) = -3h^2 + 6h^3 - \tfrac{29}{4}h^4 + \tfrac{13}{2}h^5 + \ldots .$$

Similarly,

$$J\left(\;\right) = J(\overline{3_1}) - 1 = -3h^2 - 6h^3 - \frac{29}{4}h^4 - \frac{13}{2}h^5 + \ldots .$$

Thus we have

$$J\left(\;\right) = J\left(\;\right) - J\left(\;\right) = -12h^3 - 13h^5 + \ldots .$$

3.6.5 Quantum invariants

It was proved in Birman and Lin (1993) that all quantum invariants produce Vassiliev invariants in the same way as the Jones polynomial. More precisely, let $\theta(K)$ be the quantum invariant constructed as in Section 2.6. It is a polynomial in q and q^{-1}. Now let us make substitution $q = e^h$ and consider the coefficient $\theta_n(K)$ of h^n in the Taylor expansion of $\theta(K)$.

Theorem 3.27 (Birman and Lin 1993; Bar-Natan 1995a). *The coefficient $\theta_n(K)$ is a Vassiliev invariant of order $\leqslant n$.*

The argument is similar to that of the proof of Theorem 3.25: it is based on the fact that an R-matrix R and its inverse R^{-1} are congruent modulo h.

3.6.6 The Casson invariant

The second coefficient of the Conway polynomial, or the *Casson invariant*, can be computed directly from any knot diagram by counting (with signs) pairs of crossings of a certain type.[1]

Namely, fix a based Gauss diagram G of a knot K, with an arbitrary basepoint, and consider all pairs of arrows of G that form a subdiagram of the following form:

$$(3.5)$$

The Casson invariant $a_2(K)$ is defined as the number of such pairs of arrows with $\varepsilon_1 \varepsilon_2 = 1$ minus the number of pairs of this form with $\varepsilon_1 \varepsilon_2 = -1$.

Theorem 3.28. *The Casson invariant coincides with the second coefficient of the Conway polynomial c_2.*

Proof We shall prove that the Casson invariant as defined above is a Vassiliev invariant of degree 2. It can be checked directly that it vanishes on the unknot and is equal to 1 on the left trefoil. Since the same holds for the invariant c_2 and $\dim \mathcal{V}_2 = 2$, the assertion of the theorem will follow.

First, let us verify that a_2 does not depend on the location of the basepoint on the Gauss diagram. It is enough to prove that whenever the basepoint is moved over the endpoint of one arrow, the value of a_2 remains the same.

Let c be an arrow of some Gauss diagram. For another arrow c' of the same Gauss diagram with the sign $\varepsilon(c')$, the *flow* of c' through c is equal to $\varepsilon(c')$ if c' intersects c, and is equal to 0 otherwise. The *flow to the right* through c is the sum of the flows through c of all arrows c' such that c' and c, in this order, form a positive basis of $\mathbb{R}^2$. The *flow to the left* is defined as the sum of the flows of all c' such that c', c form a negative basis. The *total flow* through the arrow c is the difference of the right and the left flows through c.

Now, let us observe that if a Gauss diagram is realizable, then the total flow through each of its arrows is equal to zero. Indeed, let us cut and re-connect the branches of the knot represented by the Gauss diagram in the vicinity of the crossing point that corresponds to the arrow c. What we get is a two-component link:

---

[1] The Casson invariant was defined in 1985 by Casson as an invariant of homology 3-spheres. The Casson invariant of a knot can be interpreted as the difference between the Casson invariants of the homology spheres obtained by surgeries on the knot with different framings, see Akbulut and McCarthy (1990).

It is easy to see that the two ways of computing the linking number of the two components A and B (see Section 2.2) are equal to the right and the left flow through c respectively. Since the linking number is an invariant, the difference of the flows is 0.

Now, let us see what happens when the basepoint is moved over an endpoint of an arrow c. If this endpoint corresponds to an overcrossing, this means that the arrow c does not appear in any subdiagram of the form (3.5) and, hence, the value of a_2 remains unchanged. If the basepoint of the diagram is moved over an undercrossing, the value of a_2 changes by the amount that is equal to the number of all subdiagrams of G involving c, counted with signs. Taking the signs into account, we see that this amount is equal to the total flow through the chord c in G, that is, zero.

Let us now verify that a_2 is invariant under the Reidemeister moves. This is clear for the move $V\Omega_1$, since an arrow with adjacent endpoints cannot participate in a subdiagram of the form (3.5).

The move $V\Omega_2$ involves two arrows; denote them by c_1 and c_2. Choose the basepoint "far" from the endpoints of c_1 and c_2, namely, in such a way that it belongs neither to the interval between the sources of c_1 and c_2, nor to the interval between the targets of these arrows. (Since a_2 does not depend on the location of the basepoint, there is no loss of generality in this choice.) Then the contribution to a_2 of any pair that contains the arrow c_1 cancels with the corresponding contribution for c_2.

The moves of type 3 involve three arrows. If we choose a basepoint far from all of these endpoints, only one of the three distinguished arrows can participate in a subdiagram of the form (3.5). It is then clear that exchanging the endpoints of the three arrows as in the move $V\Omega_3$ does not affect the value of a_2.

It remains to show that a_2 has degree 2. Consider a knot with three double points. Resolving the double point, we obtain an alternating sum of eight knots whose Gauss diagrams are the same except for the directions and signs of three arrows. Any subdiagram of the form (3.5) fails to contain at least one of these three arrows. It is therefore clear that for each instance that the Gauss diagram of one of the eight knots contains the diagram (3.5) as a subdiagram, there is another occurrence of (3.5) in another of the eight knots, counted in a_2 with the opposite sign. $\square$

Remark 3.29. This method of calculating c_2 (invented by Polyak and Viro 1994, 2001) is an example of a *Gauss diagram formula*. See Chapter 13 for details and for more examples.

3.7 Actuality tables

In general, the amount of information needed to describe a knot invariant v is infinite, since v is a function on an infinite domain: the set of isotopy classes of knots. However, Vassiliev invariants require only a finite amount of information for their description. We already mentioned the analogy between Vassiliev invariants and polynomials. A polynomial of degree n can be described, for example, using the Lagrange interpolation formula, by its values in $n + 1$ particular points. In a similar way, a given Vassiliev invariant can be described by its values on a finitely many knots. These values are organized in the *actuality table* (see Vassiliev 1990a; Birman and Lin 1993; Birman 1993).

3.7.1 Basic knots and actuality tables

To construct the actuality table we must choose a representative (*basic*) singular knot for every chord diagram. A possible choice of basic knots up to degree 3 is shown in the table.

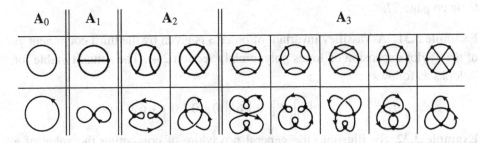

A_0	A_1	A_2		A_3			

The actuality table for a particular invariant v of order $\leqslant n$ consists of the set of its values on the set of all basic knots with at most n double points. The knowledge of this set is sufficient for calculating v for any knot.

Indeed, any knot K can be transformed into any other knot, in particular, into the basic knot with no singularities (in the table above this is the unknot), by means of crossing changes and isotopies. The difference of two knots that participate in a crossing change is a knot with a double point, hence in $\mathbb{Z}\mathcal{K}$ the knot K can be written as a sum of the basic non-singular knot and several knots with one double point. In turn, each knot with one double point can be transformed, by crossing changes and isotopies, into the basic singular knot *with the same chord diagram*, and can be written, as a result, as a sum of a basic knot with one double point and several knots with two double points. This process can be iterated until we obtain a representation of the knot K as a sum of basic knots with at most n double points and several knots with $n+1$ double points. Now, since v is of order $\leqslant n$, it vanishes on the knots with $n+1$ double points, so $v(K)$ can be written as a sum of the values of v on the basic knots with at most n singularities.

By Proposition 3.20, the values of v on the knots with precisely n double points depend only on their chord diagrams. For a smaller number of double points, the values of v in the actuality table depend not only on chord diagrams, but also on the basic knots. Of course, the values in the actuality table cannot be arbitrary. They satisfy certain relations which we shall discuss later (see Section 4.1). The simplest of these relations, however, is easy to spot from the examples: the value of any invariant on a diagram with a chord that has no intersections with other chords is zero.

Example 3.30. The second coefficient c_2 of the Conway polynomial (Section 3.1.1) is a Vassiliev invariant of order $\leqslant 2$. Here is an actuality table for it.

$$c_2: \qquad 0 \parallel 0 \parallel 0 \mid 1$$

The order of the values in this table corresponds to the order of basic knots in the table on page 77.

Example 3.31. A Vassiliev invariant of order 3 is given by the third coefficient j_3 of the Taylor expansion of Jones polynomial (Section 2.4). The actuality table for j_3 looks as follows.

$$j_3: \qquad 0 \parallel 0 \parallel 0 \mid 6 \parallel 0 \mid 0 \mid 0 \mid -6 \mid -12$$

Example 3.32. To illustrate the general procedure of computing the value of a Vassiliev invariant on a particular knot by means of actuality tables, let us compute the value of j_3 on the right-hand trefoil. The right-hand trefoil is an ordinary knot, without singular points, so we have to deform it (using crossing changes) to our basic knot without double points, that is, the unknot. This can be done by one crossing change, and by the Vassiliev skein relation we have

because $j_3(unknot) = 0$ in the actuality table. Now the knot with one double point we got is not quite the one from our basic knots. We can deform it to a basic knot changing the upper right crossing.

$$j_3\left(\text{⬡}\right) = j_3\left(\text{⬡}\right) + j_3\left(\text{⬡}\right) = j_3\left(\text{⬡}\right)$$

Here we used the fact that *any* invariant vanishes on the basic knot with a single double point. The knot with two double points on the right-hand side of the equation still differs by one crossing from the basic knot with two double points. This means that we have to do one more crossing change. Combining these equations together and using the values from the actuality table, we get the final answer:

$$j_3\left(\text{⬡}\right) = j_3\left(\text{⬡}\right) = j_3\left(\text{⬡}\right) + j_3\left(\text{⬡}\right) = 6 - 12 = -6.$$

3.7.2 The first ten Vassiliev invariants

Using actuality tables, one can find the values of the Vassiliev invariants of low degree. Table 3.2 uses a certain basis in the space of Vassiliev invariants up to degree 5. It represents an abridged version of the table compiled by T. Stanford (2002), where the values of invariants up to degree 6 are given on all knots with at most 10 crossings.

Some of the entries in Table 3.2 are different from Stanford (2002); this is due to the fact that, for some non-amphicheiral knots, Stanford uses mirror reflections of the Rolfsen's knots shown in Table 1.1.

The two signs after the knot number refer to their symmetry properties: a plus in the first position means that the knot is amphicheiral, a plus in the second position means that the knot is invertible.

3.8 Vassiliev invariants of tangles

Knots are tangles whose skeleton is a circle. A theory of Vassiliev invariants, similar to the theory for knots, can be constructed for isotopy classes of tangles with any given skeleton X.

Indeed, similar to the case of knots, one can introduce *tangles with double points*, with the only extra assumption that the double points lie in the interior of the tangle box. Then, any invariant of tangles can be extended to tangles with double points with the help of the Vassiliev skein relation. An invariant of tangles is a Vassiliev invariant of degree $\leqslant n$ if it vanishes on all tangles with more than n double points.

We stress that we define Vassiliev invariants separately for each skeleton X. Nevertheless, there are relations among invariants of tangles with different skeleta.

Table 3.2 *Vassiliev invariants of order $\leqslant 5$*

		v_0	v_2	v_3	v_{41}	v_{42}	v_2^2	v_{51}	v_{52}	v_{53}	v_2v_3
0_1	++	1	0	0	0	0	0	0	0	0	0
3_1	-+	1	1	-1	1	-3	1	-3	1	-2	-1
4_1	++	1	-1	0	-2	3	1	0	0	0	0
5_1	-+	1	3	-5	1	-6	9	-12	4	-8	-15
5_2	-+	1	2	-3	1	-5	4	-7	3	-5	-6
6_1	-+	1	-2	1	-5	5	4	4	-1	2	-2
6_2	-+	1	-1	1	-3	1	1	3	-1	1	-1
6_3	++	1	1	0	2	-2	1	0	0	0	0
7_1	-+	1	6	-14	-4	-3	36	-21	7	-14	-84
7_2	-+	1	3	-6	0	-5	9	-9	6	-7	-18
7_3	-+	1	5	11	-3	-6	25	16	-8	13	55
7_4	-+	1	4	8	-2	-8	16	10	-8	10	32
7_5	-+	1	4	-8	0	-5	16	-14	6	-9	-32
7_6	-+	1	1	-2	0	-3	1	-2	3	-2	-2
7_7	-+	1	-1	-1	-1	4	1	0	2	0	1
8_1	-+	1	-3	3	-9	5	9	12	-3	5	-9
8_2	-+	1	0	1	-3	-6	0	2	0	-3	0
8_3	++	1	-4	0	-14	8	16	0	0	0	0
8_4	-+	1	-3	1	-11	4	9	0	-2	-1	-3
8_5	-+	1	-1	-3	-5	-5	1	-5	3	2	3
8_6	-+	1	-2	3	-7	0	4	9	-3	2	-6
8_7	-+	1	2	2	4	-2	4	7	-1	3	4
8_8	-+	1	2	1	3	-4	4	2	-1	1	2
8_9	++	1	-2	0	-8	1	4	0	0	0	0
8_{10}	-+	1	3	3	3	-6	9	5	-3	3	9
8_{11}	-+	1	-1	2	-4	-2	1	8	-1	2	-2
8_{12}	++	1	-3	0	-8	8	9	0	0	0	0
8_{13}	-+	1	1	1	3	0	1	6	0	3	1
8_{14}	-+	1	0	0	-2	-3	0	-2	0	-3	0
8_{15}	-+	1	4	-7	1	-7	16	-16	5	-10	-28
8_{16}	-+	1	1	-1	3	0	1	2	2	2	-1
8_{17}	+-	1	-1	0	-4	0	1	0	0	0	0
8_{18}	++	1	1	0	0	-5	1	0	0	0	0
8_{19}	-+	1	5	10	0	-5	25	18	-6	10	50
8_{20}	-+	1	2	-2	2	-5	4	-1	3	-1	-4
8_{21}	-+	1	0	1	-1	-3	0	1	-1	-1	0

Example 3.33. Assume that the isotopy classes of tangles with the skeleta X_1 and X_2 can be multiplied. Given a tangle T with skeleton X_1 and a Vassiliev invariant v of tangles with skeleton X_1X_2, we can define an invariant of tangles on X_2 of the same order as v by composing a tangle with T and applying v.

Example 3.34. In the above example the product of tangles can be replaced by their tensor product. (Of course, the condition that X_1 and X_2 can be multiplied is no longer necessary here.)

In particular, the Vassiliev invariants of tangles whose skeleton has one component can be identified with the Vassiliev invariants of knots.

Example 3.35. Assume that X' is obtained from X by dropping one or several components. Then any Vassiliev invariant v' of tangles with skeleton X' gives rise to an invariant v of tangles on X of the same order; to compute v, drop the components of the tangle that are not in X' and apply v'.

This example immediately produces a lot of tangle invariants of finite type: namely, those coming from knots. The simplest example of a Vassiliev invariant that does not come from knots is the linking number of two components of a tangle. So far, we have defined the linking number only for pairs of closed curves. If one or both of the components are not closed, we can use the constructions above to close them up in some fixed way.

Lemma 3.36. *The linking number of two components of a tangle is a Vassiliev invariant of order 1.*

Proof Consider a two-component link with one double point. This double point can be of two types: either it is a self-intersection point of a single component, or it is an intersection of two different components. Using the Vassiliev skein relation and the formula on page 28, we see that in the first case the linking number vanishes, while in the second case it is equal to 1. It follows that for a two-component link with two double points the linking number is always zero. $\qquad\square$

Among the invariants for all classes of tangles, the string link invariants have attracted most attention. Two particular classes of string link invariants are the knot invariants (recall that string links on one strand are in one-to-one correspondence with knots) and the invariants of pure braids. We shall treat the Vassiliev invariants of pure braids in detail in Chapter 12.

Exercises

3.1 Using the actuality tables, compute the value of j_3 on the left-hand trefoil.

3.2 Choose the basic knots with four double points and construct the actuality tables for the fourth coefficients c_4 and j_4 of the Conway and Jones polynomials.

3.3 Prove that $j_0(K) = 1$ and $j_1(K) = 0$ for any knot K.

3.4 Show that the value of j_0 on a link with k components is equal to $(-2)^{k-1}$.

3.5   For a link L with two components K_1 and K_2 prove that
      $j_1(L) = -3 \cdot lk(K_1, K_2)$. In other words,

$$J(L) = -2 - 3 \cdot lk(K_1, K_2) \cdot h + j_2(L) \cdot h^2 + j_3(L) \cdot h^3 + \dots .$$

3.6   Prove that for any knot K the integer $j_3(K)$ is divisible by 6.

3.7   For a knot K, find the relation between the second coefficients $c_2(K)$ and
      $j_2(K)$ of the Conway and Jones polynomials.

3.8   Prove that $v(3_1 \# 3_1) = 2v(3_1) - v(0)$, where 0 is the trivial knot, for any
      Vassiliev invariant $v \in \mathcal{V}_3$.

3.9   Prove that for a knot K the nth derivative at 1 of the Jones polynomial

$$\left. \frac{d^n(J(K))}{dt^n} \right|_{t=1}$$

      is a Vassiliev invariant of order $\leqslant n$. Find the relation between these invariants
      and $j_1, \dots, j_n$ for small values of n.

3.10 Express the coefficients c_2, c_4, j_2, j_3, j_4, j_5 of the Conway and the modified
      Jones polynomials in terms of the basis Vassiliev invariants from Table 3.2.

3.11 Find the symbols of the Vassiliev invariants from Table 3.2.

3.12 Express the invariants of Table 3.2 via the coefficients of the Conway and the
      Jones polynomials.

3.13 Find the actuality tables for some of the Vassiliev invariants appearing in
      Table 3.2.

3.14 Explain the correlation between the first sign and the zeroes in the last four
      columns of Table 3.2.

3.15 Check that Vassiliev invariants up to order 4 are enough to distinguish, up to
      orientation, all knots with at most 8 crossings from Table 1.1.

3.16 Prove that the symbol of the coefficient c_n of the Conway polynomial can
      be calculated as follows. Double every chord of a given chord diagram D as
      in Section 3.6.2, and let $|D|$ be equal to the number of components of the
      obtained curve. Then

$$\text{symb}(c_n)(D) = \begin{cases} 1, & \text{if } |D| = 1 \\ 0, & \text{otherwise.} \end{cases}$$

3.17 Prove that if n is even, then c_n is a Vassiliev invariant of degree exactly n.

3.18 Prove that there is a well-defined extension of knot invariants to singular
      knots with a non-degenerate *triple* point according to the rule

$$f\left(\raisebox{-0.5em}{\includegraphics{triple}}\right) = f\left(\raisebox{-0.5em}{\includegraphics{resolved1}}\right) - f\left(\raisebox{-0.5em}{\includegraphics{resolved2}}\right).$$

Is it true that, according to this extension, a Vassiliev invariant of degree 2 is equal to 0 on any knot with a triple point?

Is it possible to use the same method to define an extension of knot invariants to knots with self-intersections of multiplicity higher than 3?

3.19 Following Example 3.26, find the power series expansion of the modified Jones polynomial of the singular knot .

3.20 Prove the following relation between the Casson knot invariant c_2, extended to singular knots, and the linking number of two curves. Let K be a knot with one double point. Smoothing the double point by the rule , one obtains a 2-component link L. Then $lk(L) = c_2(K)$.

3.21 Is there a prime knot K such that $j_4(K) = 0$?

3.22 **Vassiliev invariants coming from the HOMFLY polynomial.** For a link L, make a substitution $a = e^h$ in the HOMFLY polynomial $P(L)$ and take the Taylor expansion in h. The result will be a Laurent polynomial in z and a power series in h. Let $p_{k,l}(L)$ be its coefficient at $h^k z^l$.

(a) Show that for any link L the total degree $k + l$ is not negative.

(b) If l is odd, then $p_{k,l} = 0$.

(c) Prove that $p_{k,l}(L)$ is a Vassiliev invariant of order $\leqslant k + l$.

(d) Describe the symbol of $p_{k,l}(L)$.

4

Chord diagrams

A chord diagram encodes the order of double points along a singular knot. We saw
in the last chapter that a Vassiliev invariant of degree n gives rise to a function on
chord diagrams with n chords. Here we shall describe the conditions, called one-
term and four-term relations, that a function on chord diagrams should satisfy in
order to come from a Vassiliev invariant. We shall see that the vector space spanned
by chord diagrams modulo these relations has the structure of a Hopf algebra. This
Hopf algebra turns out to be dual to the graded algebra of the Vassiliev invariants.

4.1 Four- and one-term relations

Recall that $\mathscr{R}$ denotes a commutative ring and $\mathscr{V}_n$ is the space of $\mathscr{R}$-valued Vassiliev
invariants of order $\leqslant n$. Some of our results will only hold when $\mathscr{R}$ is a field of
characteristic 0; sometimes we shall take $\mathscr{R} = \mathbb{C}$. In Section 3.1.1 we constructed
a linear inclusion (the symbol of an invariant)

$$\bar{\alpha}_n : \mathscr{V}_n/\mathscr{V}_{n-1} \to \mathscr{R}\mathbf{A}_n,$$

where $\mathscr{R}\mathbf{A}_n$ is the space of $\mathscr{R}$-valued functions on the set $\mathbf{A}_n$ of chord diagrams of
order n.

To describe the image of $\bar{\alpha}_n$, we need the following definition.

Definition 4.1. A function $f \in \mathscr{R}\mathbf{A}_n$ is said to satisfy the *4-term (or 4T) relations*
if the alternating sum of the values of f is zero on the following quadruples of
diagrams:

$$f\left(\begin{array}{c}\bigcirc\end{array}\right) - f\left(\begin{array}{c}\bigcirc\end{array}\right) + f\left(\begin{array}{c}\bigcirc\end{array}\right) - f\left(\begin{array}{c}\bigcirc\end{array}\right) = 0. \qquad (4.1)$$

In this case f is also called a *(framed) weight system of order n*.

Here it is assumed that the diagrams in the pictures may have other chords with endpoints on the dotted arcs, while all the endpoints of the chords on the solid portions of the circle are explicitly shown. For example, this means that in the first and second diagrams the two bottom points are adjacent. The chords omitted from the pictures should be the same in all the four cases.

Example 4.2. Let us find all 4-term relations for chord diagrams of order 3. We must add one chord in one and the same way to all the four terms of Equation (4.1). Since there are 3 dotted arcs, there are 6 different ways to do that, in particular,

$$f\left(\bigcirc\right) - f\left(\bigcirc\right) + f\left(\bigcirc\right) - f\left(\bigcirc\right) = 0$$

and

$$f\left(\bigcirc\right) - f\left(\bigcirc\right) + f\left(\bigcirc\right) - f\left(\bigcirc\right) = 0.$$

Some of the diagrams in these equations are equal, and the relations can be simplified as $f\left(\bigcirc\right) = f\left(\bigcirc\right)$ and $f\left(\bigcirc\right) - 2f\left(\bigcirc\right) + f\left(\bigcirc\right) = 0$. The reader is invited to check that the remaining four 4-term relations (we wrote only two out of six) are either trivial or coincide with one of these two.

It is often useful to look at a 4T relation from the following point of view. We can think that one of the two chords that participate in Equation (4.1) is fixed, and the other is moving. One of the ends of the moving chord is also fixed, while the other end travels around the fixed chord, stopping at the four locations adjacent to its endpoints. The resulting four diagrams are then summed up with alternating signs. Graphically,

$$f\left(\bigcirc\right) - f\left(\bigcirc\right) + f\left(\bigcirc\right) - f\left(\bigcirc\right) = 0, \qquad (4.2)$$

where the fixed end of the moving chord is marked by ✖.

Another way of writing the 4T relation, which will be useful in Section 5.1, is to split the four terms into two pairs:

$$f\left(\bigcirc\right) - f\left(\bigcirc\right) = f\left(\bigcirc\right) - f\left(\bigcirc\right).$$

Because of the obvious symmetry, this can be completed as follows:

$$f\left(\bigcirc\right) - f\left(\bigcirc\right) = f\left(\bigcirc\right) - f\left(\bigcirc\right). \qquad (4.3)$$

Note that for each order n the choice of a specific 4-term relation depends on the following data:

- a diagram of order $n - 1$,
- a distinguished chord of this diagram ("fixed chord"), and
- a distinguished arc on the circle of this diagram (where the fixed endpoint of the "moving chord" is placed).

There are three fragments of the circle that participate in a 4-term relation, namely, those that are shown by solid lines in the equations above. If these three fragments are drawn as three vertical line segments, then the 4-term relation can be restated as follows:

$$(-1)^{\downarrow} f\left(\underset{i \;\; j \;\; k}{\vphantom{|}}\right) - (-1)^{\downarrow} f\left(\underset{i \;\; j \;\; k}{\vphantom{|}}\right)$$

$$+ (-1)^{\downarrow} f\left(\underset{i \;\; j \;\; k}{\vphantom{|}}\right) - (-1)^{\downarrow} f\left(\underset{i \;\; j \;\; k}{\vphantom{|}}\right) = 0, \quad (4.4)$$

where $\downarrow$ stands for the number of endpoints of the chords in which the orientation of the strands is directed downwards. This form of a 4T relation is called a *horizontal 4T relation*. It first appeared, in a different context, in the work by T. Kohno (1987).

Exercise. Choose some orientations of the three fragments of the circle, add the portions necessary to close it up and check that the last form of the 4-term relation carries over into the ordinary four-term relation.

Here is an example:

$$f\left(\underset{}{\vphantom{|}}\right) - f\left(\underset{}{\vphantom{|}}\right) - f\left(\underset{}{\vphantom{|}}\right) + f\left(\underset{}{\vphantom{|}}\right) = 0.$$

We shall see in the next section that the four-term relations are always satisfied by the symbols of Vassiliev invariants, both in the usual and in the framed case. For the framed knots, there are no other relations; in the unframed case, there is another set of relations, called *one-term*, or *framing independence* relations.

Definition 4.3. An *isolated chord* is a chord that does not intersect any other chord of the diagram. A function $f \in \mathcal{R}A_n$ is said to satisfy the *1-term relations* if it vanishes on every chord diagram with an isolated chord. An *unframed weight system* of order n is a weight system that satisfies the 1-term relations.

Here is an example of a 1T relation: $f\left(\vphantom{|}\right) = 0.$

Notation 4.4. We denote by $\mathscr{W}_n^{fr}$ the subspace of $\mathscr{R}\mathbf{A}_n$ consisting of all (framed) weight systems of order n and by $\mathscr{W}_n \subset \mathscr{W}_n^{fr}$ the space of all unframed weight systems of order n.

4.2 The Fundamental Theorem

4.2.1 The Fundamental Theorem of Vassiliev invariants

In Section 3.4 we showed that the symbol of an invariant gives an injective map $\overline{\alpha}_n : \mathscr{V}_n/\mathscr{V}_{n-1} \to \mathscr{R}\mathbf{A}_n$. The Fundamental Theorem on Vassiliev invariants describes its image.

Theorem 4.5 (Vassiliev–Kontsevich). *For* $\mathscr{R} = \mathbb{C}$ *the map* $\overline{\alpha}_n$ *identifies* $\mathscr{V}_n/\mathscr{V}_{n-1}$ *with the subspace of unframed weight systems* $\mathscr{W}_n \subset \mathscr{R}\mathbf{A}_n$. *In other words, the space of unframed weight systems is isomorphic to the graded space associated with the filtered space of Vassiliev invariants,*

$$\mathscr{W} = \bigoplus_{n=0}^{\infty} \mathscr{W}_n \cong \bigoplus_{n=0}^{\infty} \mathscr{V}_n/\mathscr{V}_{n+1}.$$

The theorem consists of two parts:

- (V. Vassiliev) The symbol of every Vassiliev invariant is an unframed weight system.
- (M. Kontsevich) Every unframed weight system is the symbol of a certain Vassiliev invariant.

We shall now prove the first (easy) part of the theorem. The second (difficult) part will be proved later (in Section 8.8) using the Kontsevich integral.

The first part of the theorem consists of two assertions, and we prove them one by one.

First assertion: *Any function* $f \in \mathscr{R}\mathbf{A}_n$ *coming from an invariant* $v \in \mathscr{V}_n$ *satisfies the 1-term relations.*

Proof Let K be a singular knot whose chord diagram contains an isolated chord. The double point p that corresponds to the isolated chord divides the knot into two parts: A and B.

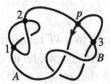

The fact that the chord is isolated means that A and B do not have common double points. There may, however, be crossings involving branches from both parts. By crossing changes, we can untangle part A from part B, thus obtaining a singular knot K' with the same chord diagram as K and with the property that the two parts lie on either side of some plane in $\mathbb{R}^3$ that passes through the double point p:

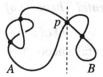

Here it is obvious that the two resolutions of the double point p give equivalent singular knots, therefore $v(K) = v(K') = v(K'_+) - v(K'_-) = 0$. □

Second assertion: *Any function $f \in \mathscr{R}\mathbf{A}_n$ coming from an invariant $v \in \mathscr{V}_n$ satisfies the 4-term relations.*

Proof We shall use the following lemma.

Lemma 4.6 (4-term relation for knots). *Any Vassiliev invariant satisfies*

$$f\left(\text{⊘}\right) - f\left(\text{⊘}\right) + f\left(\text{⊘}\right) - f\left(\text{⊘}\right) = 0.$$

Proof By the Vassiliev skein relation,

$$f\left(\text{⊘}\right) = f\left(\text{⊘}\right) - f\left(\text{⊘}\right) = a - b,$$

$$f\left(\text{⊘}\right) = f\left(\text{⊘}\right) - f\left(\text{⊘}\right) = c - d,$$

$$f\left(\text{⊘}\right) = f\left(\text{⊘}\right) - f\left(\text{⊘}\right) = c - a,$$

$$f\left(\text{⊘}\right) = f\left(\text{⊘}\right) - f\left(\text{⊘}\right) = d - b.$$

The alternating sum of these expressions is $(a-b)-(c-d)+(c-a)-(d-b) = 0$, and the lemma is proved. □

Now, denote by $D_1, \ldots, D_4$ the four diagrams in a 4T relation. In order to prove the 4-term relation for the symbols of Vassiliev invariants, let us choose for the first diagram D_1 an arbitrary singular knot K_1 such that $\sigma(K_1) = D_1$:

$$\sigma\left(\begin{array}{c} K_1 \\ \end{array} \right) = \begin{array}{c} D_1 \\ \end{array}.$$

Then the three remaining knots K_2, K_3, K_4 that participate in the 4-term relation for knots, correspond to the three remaining chord diagrams of the 4-term relation for chord diagrams, and the claim follows from the lemma.

$$\sigma\left(\begin{array}{c} K_2 \\ \end{array} \right) = \begin{array}{c} D_2 \\ \end{array}, \quad \sigma\left(\begin{array}{c} K_3 \\ \end{array} \right) = \begin{array}{c} D_3 \\ \end{array}, \quad \sigma\left(\begin{array}{c} K_4 \\ \end{array} \right) = \begin{array}{c} D_4 \\ \end{array}.$$

$\square$

4.2.2 The case of framed knots

As in the case of usual knots, for the invariants of framed knots we can define a linear map $\mathcal{V}_n^{fr}/\mathcal{V}_{n-1}^{fr} \to \mathscr{R}\mathbf{A}_n$. This map satisfies the 4T relations, but does *not* satisfy the 1T relation, since the two knots differing by a crossing change (see the proof of the first assertion on page 87) are not equivalent as framed knots (the two framings differ by 2). The Fundamental Theorem also holds, in fact, for framed knots: we have the equality

$$\mathcal{V}_n^{fr}/\mathcal{V}_{n-1}^{fr} = \mathscr{W}_n^{fr};$$

it can be proved using the Kontsevich integral for framed knots (see Section 9.1).

This explains why the 1-term relation for the Vassiliev invariants of (unframed) knots is also called the *framing independence relation*.

We see that, in a sense, the 4T relations are more fundamental than the 1T relations. Therefore, in the sequel we shall mainly study combinatorial structures involving the 4T relations only. In any case, 1T relations can be added at all times, either by considering an appropriate subspace or an appropriate quotient space (see Section 4.4.4). This is especially easy to do in terms of the primitive elements (see Section 4.6): the problem reduces to simply leaving out one primitive generator.

4.3 Bialgebras of knots and of Vassiliev knot invariants

Prerequisites on bialgebras can be found in the appendix (see Section A.2). In this section it will be assumed that $\mathscr{R} = \mathbb{F}$, a field of characteristic zero.

4.3.1 The bialgebra of knots

In Section 2.5 we noted that the algebra of knot invariants $\mathscr{I}$, as a vector space, is dual to the algebra of knots $\mathbb{F}\mathscr{K} = \mathbb{Z}\mathscr{K} \otimes \mathbb{F}$. This duality provides the algebras of knots and of knot invariants with additional structure. Indeed, the dual map to a product $V \otimes V \to V$ on a vector space V is a map $V^* \to (V \otimes V)^*$; when V is finite-dimensional it is a *coproduct* $V^* \to V^* \otimes V^*$. This observation does not apply to the algebras of knots and knot invariants directly, since they are not finite-dimensional. Nevertheless, the coproduct on the algebra of knots exists and is given by an explicit formula

$$\delta(K) = K \otimes K$$

for any knot K; by linearity this map extends to the entire space $\mathbb{F}\mathscr{K}$. Note that its dual is precisely the product in $\mathscr{I}$.

Exercise. Show that with this coproduct $\mathbb{F}\mathscr{K}$ is a bialgebra. (For this, define the counit and check the compatibility conditions for the product and the coproduct.)

The *singular knot filtration* $\mathscr{K}_n$ on $\mathbb{F}\mathscr{K}$ is obtained from the singular knot filtration on $\mathbb{Z}\mathscr{K}$ (Section 3.2.1) simply by tensoring it with the field $\mathbb{F}$.

Theorem 4.7. *The bialgebra of knots $\mathbb{F}\mathscr{K}$ considered with the singular knot filtration is a bialgebra with a decreasing filtration (Section A.2.3).*

Proof There are two assertions to prove:

1. If $x \in \mathscr{K}_m$ and $y \in \mathscr{K}_n$, then $xy \in \mathscr{K}_{m+n}$,
2. If $x \in \mathscr{K}_n$, then $\delta(x) \in \sum\limits_{p+q=n} \mathscr{K}_p \otimes \mathscr{K}_q$.

The first assertion was proved in Chapter 3.

In order to prove assertion (2), first let us introduce some additional notation.

Let K be a knot given by a plane diagram with $\geqslant n$ crossings out of which exactly n are distinguished and numbered. Consider the set $\hat{K}$ of 2^n knots that may differ from K by crossing changes at the distinguished points and the vector space $X_K \subset \mathbb{F}\mathscr{K}$ spanned by $\hat{K}$. The group $\mathbb{Z}_2^n$ acts on the set $\hat{K}$; the action of ith generator s_i consists in the flip of under/overcrossing at the distinguished point number i. We thus obtain a set of n commuting linear operators $s_i : X_K \to X_K$. Set $\sigma_i = 1 - s_i$. In these terms, a typical generator x of $\mathscr{K}_n$ can be written as $x = (\sigma_1 \circ \cdots \circ \sigma_n)(K)$. To evaluate $\delta(x)$, we must find the commutator relations between the operators δ and σ_i.

Lemma 4.8.

$$\delta \circ \sigma_i = (\sigma_i \otimes \mathrm{id} + s_i \otimes \sigma_i) \circ \delta,$$

where both the left-hand side and the right-hand side are understood as linear operators from X_K into $X_K \otimes X_K$.

Proof Just check that the values of both operators on an arbitrary element of the set $\hat{K}$ are equal. □

A successive application of the lemma yields:

$$\delta \circ \sigma_1 \circ \cdots \circ \sigma_n = \left(\prod_{i=1}^{n} (\sigma_i \otimes \mathrm{id} + s_i \otimes \sigma_i) \right) \circ \delta$$

$$= \left(\sum_{I \subset \{1,\ldots,n\}} \prod_{i \in I} \sigma_i \prod_{i \notin I} s_i \otimes \prod_{i \notin I} \sigma_i \right) \circ \delta .$$

Therefore, an element $x = (\sigma_1 \circ \cdots \circ \sigma_n)(K)$ satisfies

$$\delta(x) = \sum_{I \subset \{1,\ldots,n\}} (\prod_{i \in I} \sigma_i \prod_{i \notin I} s_i)(K) \otimes (\prod_{i \notin I} \sigma_i)(K),$$

which obviously belongs to $\sum_{p+q=n} \mathbb{Z}\mathcal{K}_p \otimes \mathbb{Z}\mathcal{K}_q$. □

4.3.2 The bialgebra of Vassiliev knot invariants

In contrast with the knot algebra, the algebra of invariants does not have a natural coproduct. The map dual to the product in $\mathbb{F}\mathcal{K}$ is given by

$$\delta(f)(K_1 \otimes K_2) = f(K_1 \# K_2)$$

for an invariant f and any pair of knots K_1 and K_2. It sends $\mathscr{I} = (\mathbb{F}\mathcal{K})^*$ to $(\mathbb{F}\mathcal{K} \otimes \mathbb{F}\mathcal{K})^*$ but its image is not contained in $\mathscr{I} \otimes \mathscr{I}$.

Exercise. Find a knot invariant whose image under δ is not in $\mathscr{I} \otimes \mathscr{I}$.

Even though the map δ is not a coproduct, it becomes one if we restrict our attention to the subalgebra $\mathscr{V}^{\mathbb{F}} \subset \mathscr{I}$ consisting of all $\mathbb{F}$-valued Vassiliev invariants.

Proposition 4.9. *The algebra of $\mathbb{F}$-valued Vassiliev knot invariants $\mathscr{V}^{\mathbb{F}}$ is a bialgebra with an increasing filtration (page 471).*

Indeed, the algebra of $\mathscr{V}^{\mathbb{F}}$ is dual *as a filtered bialgebra* to the bialgebra of knots with the singular knot filtration. The filtrations on $\mathscr{V}^{\mathbb{F}}$ and $\mathbb{F}\mathcal{K}$ are of finite

92   *Chord diagrams*

type by Corollary 3.21 and, hence, the proposition follows from Theorem 4.7 and Proposition A.26.

4.3.3 Primitive and group-like elements

Let us now find all the *primitive* and the *group-like* elements in the algebras $\mathbb{F}\mathcal{K}$ and $\mathcal{V}^{\mathbb{F}}$ (see definitions in Section A.2.2). As for the algebra of knots $\mathbb{F}\mathcal{K}$, both structures are quite poor: it follows from the definitions that $\mathcal{P}(\mathbb{F}\mathcal{K}) = 0$, while $\mathcal{G}(\mathbb{F}\mathcal{K})$ consists of only one element: the trivial knot. (Non-trivial knots are semigroup-like, but not group-like!)

The case of the algebra of Vassiliev invariants is more interesting. As a consequence of Proposition A.28 we obtain a description of primitive and group-like Vassiliev knot invariants: these are nothing but the *additive* and the *multiplicative* invariants, respectively, that is, the invariants satisfying the relations

$$f(K_1 \# K_2) = f(K_1) + f(K_2),$$
$$f(K_1 \# K_2) = f(K_1) f(K_2),$$

respectively, for any two knots K_1 and K_2.

As in the case of the knot algebra, the group-like elements of $\mathcal{V}^{\mathbb{F}}$ are scarce:

Exercise. Show that the only group-like Vassiliev invariant is the constant 1.

In contrast, we shall see that primitive Vassiliev invariants abound.

4.3.4 The case of power series Vassiliev invariants

The bialgebra structure of the Vassiliev invariants extends naturally to the power series Vassiliev invariants term by term. In this framework, there are many more group-like invariants.

Example 4.10. According to Exercise 2.6, the Conway polynomial is a group-like power series Vassiliev invariant. Taking its logarithm one obtains a primitive power series Vassiliev invariant. For example, the coefficient c_2 (the Casson invariant) is primitive.

Exercise. Find a finite linear combination of coefficients j_n of the modified Jones polynomial that gives a primitive Vassiliev invariant.

4.4 Bialgebra of chord diagrams

4.4.1 The vector space of chord diagrams

A dual way to define the weight systems is to introduce the 1- and 4-term relations directly in the vector space spanned by chord diagrams.

Definition 4.11. The space $\mathscr{A}_n$ of chord diagrams of order n is the vector space generated by the set $\mathbf{A}_n$ (all diagrams of order n) modulo the subspace spanned by all 4-term linear combinations

$$\bigcirc - \bigotimes + \bigcirc - \bigcirc.$$

The space $\mathscr{A}_n'$ of *unframed* chord diagrams of order n is the quotient of $\mathscr{A}_n$ by the subspace spanned by all diagrams with an isolated chord.

In these terms, the space of framed weight systems $\mathscr{W}_n^{fr}$ is dual to the space of framed chord diagrams $\mathscr{A}_n$, and the space of unframed weight systems $\mathscr{W}_n$, to that of unframed chord diagrams $\mathscr{A}_n'$:

$$\mathscr{W}_n = \mathrm{Hom}(\mathscr{A}_n', \mathscr{R}),$$
$$\mathscr{W}_n^{fr} = \mathrm{Hom}(\mathscr{A}_n, \mathscr{R}).$$

Below, we list the dimensions and some bases of the spaces $\mathscr{A}_n$ for $n = 1, 2$ and 3:

$$\mathscr{A}_1 = \langle \bigcirc \rangle, \dim \mathscr{A}_1 = 1.$$

$$\mathscr{A}_2 = \langle \bigcirc, \bigotimes \rangle, \dim \mathscr{A}_2 = 2,$$ since the only 4-term relation involving chord diagrams of order 2 is trivial.

$$\mathscr{A}_3 = \langle \bigcirc, \bigotimes, \bigoplus \rangle, \dim \mathscr{A}_3 = 3,$$ since $\mathbf{A}_3$ consists of 5 elements, and there are two independent 4-term relations (see page 85):

$$\bigcirc = \bigcirc \quad \text{and} \quad \bigotimes - 2\bigoplus + \bigotimes = 0.$$

Taking into account the 1-term relations, we get the following result for the spaces of unframed chord diagrams of small orders:

$$\mathscr{A}_1' = 0, \dim \mathscr{A}_1' = 0.$$

$$\mathscr{A}_2' = \langle \bigotimes \rangle, \dim \mathscr{A}_2' = 1.$$

$$\mathscr{A}_3' = \langle \bigoplus \rangle, \dim \mathscr{A}_3' = 1.$$

The result of similar calculations for order 4 diagrams is presented in Table 4.1. In this case $\dim \mathscr{A}_4 = 6$; the set $\{d_3^4, d_6^4, d_7^4, d_{15}^4, d_{17}^4, d_{18}^4\}$ is used in the table as a basis. The table is obtained by running Bar-Natan's computer program, available

Table 4.1 *Chord diagrams of order 4*

CD	Code and expansion	CD	Code and expansion
	$d_1^4 = [12341234]$ $= d_3^4 + 2d_6^4 - d_7^4 - 2d_{15}^4 + d_{17}^4$		$d_2^4 = [12314324]$ $= d_3^4 - d_6^4 + d_7^4$
	$d_3^4 = [12314234]$ $= d_3^4$		$d_4^4 = [12134243]$ $= d_6^4 - d_7^4 + d_{15}^4$
	$d_5^4 = [12134234]$ $= 2d_6^4 - d_7^4$		$d_6^4 = [12132434]$ $= d_6^4$
	$d_7^4 = [12123434]$ $= d_7^4$		$d_8^4 = [11234432]$ $= d_{18}^4$
	$d_9^4 = [11234342]$ $= d_{17}^4$		$d_{10}^4 = [11234423]$ $= d_{17}^4$
	$d_{11}^4 = [11234324]$ $= d_{15}^4$		$d_{12}^4 = [11234243]$ $= d_{15}^4$
	$d_{13}^4 = [11234234]$ $= 2d_{15}^4 - d_{17}^4$		$d_{14}^4 = [11232443]$ $= d_{17}^4$
	$d_{15}^4 = [11232434]$ $= d_{15}^4$		$d_{16}^4 = [11223443]$ $= d_{18}^4$
	$d_{17}^4 = [11223434]$ $= d_{17}^4$		$d_{18}^4 = [11223344]$ $= d_{18}^4$

at Bar-Natan (1996a). The numerical notation for chord diagrams like [12314324] is easy to understand: one writes the numbers on the circle in the positive direction and connects equal numbers by chords. Of all possible codes, we choose the lexicographically minimal one.

4.4.2 *Multiplication of chord diagrams*

Now we are ready to define the structure of an algebra in the vector space $\mathscr{A} = \bigoplus_{k \geqslant 0} \mathscr{A}_k$ of chord diagrams.

Definition 4.12. The *product* of two chord diagrams D_1 and D_2 is defined by cutting and glueing the two circles as shown:

$$\text{⊕}\cdot\text{⊕} = \text{⊕—⊗} = \text{⊕}$$

This map is then extended by linearity to

$$\mu : \mathscr{A}_m \otimes \mathscr{A}_n \to \mathscr{A}_{m+n}.$$

Note that the product of diagrams depends on the choice of the points where the diagrams are cut: in the example above we could equally well cut the circles in other places and get a different result: ⊕.

Lemma 4.13. *The product is well-defined modulo 4T relations.*

Proof We shall show that the product of two diagrams is well-defined; it follows immediately that this is also true for linear combinations of diagrams. It is enough to prove that if one of the two diagrams, say D_2, is turned inside the product diagram by one "click" with respect to D_1, then the result is the same modulo 4T relations.

Note that such rotation is equivalent to the following transformation. Pick a chord in D_2 with endpoints a and b such that a is adjacent to D_1. Then, fixing the endpoint b, move a through the diagram D_1. In this process we obtain $2n + 1$ diagrams $P_0, P_1, ..., P_{2n}$, where n is the order of D_1, and we must prove that $P_0 \equiv P_{2n}$ mod 4T. Now, it is not hard to see that the difference $P_0 - P_{2n}$ is, in fact, equal to the sum of all n four-term relations which are obtained by fixing the endpoint b and all chords of D_1, one by one. For example, if we consider the two products shown above and use the following notation:

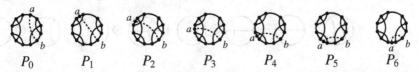

$$P_0 \qquad P_1 \qquad P_2 \qquad P_3 \qquad P_4 \qquad P_5 \qquad P_6$$

then we must take the sum of the three linear combinations

$$P_0 - P_1 + P_2 - P_3,$$
$$P_1 - P_2 + P_4 - P_5,$$
$$P_3 - P_4 + P_5 - P_6,$$

and the result is exactly $P_0 - P_6$. □

Exercise. Show that the multiplication of chord diagrams corresponds to the connected sum operation on knots in the following sense: if K_1 and K_2 are two singular knots and D_1 and D_2 are their chord diagrams, there exists a singular knot, equal to $K_1 \# K_2$ as an element of $\mathbb{Z}\mathcal{K}$, whose diagram is $D_1 \cdot D_2$.

In view of this exercise, the product of chord diagrams D_1 and D_2 is sometimes referred to as their *connected sum* and denoted by $D_1 \# D_2$.

4.4.3 Comultiplication of chord diagrams

The *coproduct* in the algebra $\mathcal{A}$

$$\delta : \mathcal{A}_n \to \bigoplus_{k+l=n} \mathcal{A}_k \otimes \mathcal{A}_l$$

is defined as follows. For a diagram $D \in \mathcal{A}_n$ we put

$$\delta(D) := \sum_{J \subseteq [D]} D_J \otimes D_{\bar{J}},$$

the summation taken over all subsets J of the set of chords of D. Here D_J is the diagram consisting of the chords that belong to J and $\bar{J} = [D] \setminus J$ is the complementary subset of chords. To the entire space $\mathcal{A}$ the operator δ is extended by linearity.

If D is a diagram of order n, the total number of summands in the right-hand side of the definition is 2^n.

Example 4.14.

Lemma 4.15. *The coproduct δ is well-defined modulo 4T relations.*

Proof Let $D_1 - D_2 + D_3 - D_4 = 0$ be a 4T relation. We must show that the sum $\delta(D_1) - \delta(D_2) + \delta(D_3) - \delta(D_4)$ can be written as a combination of 4T relations. Recall that a specific four-term relation is determined by the choice of a moving chord m and a fixed chord a. Now, take one and the same splitting $A \cup B$ of the set of chords in the diagrams D_i, the same for each i, and denote by A_i, B_i the resulting chord diagrams giving the contributions $A_i \otimes B_i$ to $\delta(D_i)$, $i = 1, 2, 3, 4$. Suppose that the moving chord m belongs to the subset A. Then $B_1 = B_2 = B_3 = B_4$ and $A_1 \otimes B_1 - A_2 \otimes B_2 + A_3 \otimes B_3 - A_4 \otimes B_4 = (A_1 - A_2 + A_3 - A_4) \otimes B_1$. If the fixed chord a belongs to A, then the $A_1 - A_2 + A_3 - A_4$ is a four-term combination; otherwise it is easy to see that $A_1 = A_2$ and $A_3 = A_4$ for an appropriate numbering. The case when $m \in B$ is treated similarly. $\qquad \square$

The *unit* and the *counit* in $\mathscr{A}$ are defined as follows:

$$\iota : \mathscr{R} \to \mathscr{A} \quad , \quad \iota(x) = x\bigcirc ,$$

$$\varepsilon : \mathscr{A} \to \mathscr{R} \quad , \quad \varepsilon\left(x\bigcirc + \ldots\right) = x.$$

Exercise. Check the axioms of a bialgebra for $\mathscr{A}$ and verify that it is commutative, cocommutative and connected.

4.4.4 Deframing the chord diagrams

The space of unframed chord diagrams $\mathscr{A}$ was defined as the quotient of the space $\mathscr{A}$ by the subspace spanned by all diagrams with an isolated chord. In terms of the multiplication in $\mathscr{A}$, this subspace can be described as the ideal of $\mathscr{A}$ generated by Θ, the chord diagram with one chord, so that we can write:

$$\mathscr{A}' = \mathscr{A}/(\Theta).$$

It turns out that there is a simple explicit formula for a linear operator $p : \mathscr{A} \to \mathscr{A}$ whose kernel is the ideal (Θ). Namely, define $p_n : \mathscr{A}_n \to \mathscr{A}_n$ by

$$p_n(D) := \sum_{J \subseteq [D]} (-\Theta)^{n-|J|} \cdot D_J ,$$

where, as earlier, $[D]$ stands for the set of chords in the diagram D and D_J means the subdiagram of D with only the chords from J left. The sum of p_n over all n is the operator $p : \mathscr{A} \to \mathscr{A}$.

Exercise. Check that

1. p is a homomorphism of algebras.
2. $p(\Theta) = 0$ and hence p takes the entire ideal (Θ) into 0.
3. p is a projector, that is, $p^2 = p$.
4. the kernel of p is exactly (Θ).

We see, therefore, that the quotient map $\bar{p} : \mathscr{A}/(\Theta) \to \mathscr{A}$ is the isomorphism of $\mathscr{A}'$ onto its image and we have a direct decomposition $\mathscr{A} = \bar{p}(\mathscr{A}') \oplus (\Theta)$. Note that the first summand here is different from the subspace spanned merely by all diagrams without isolated chords!

For example, $p(\mathscr{A}_3)$ is spanned by the two vectors

$$p\left(\vcenter{\hbox{\includegraphics{a}}}\right) = \vcenter{\hbox{\includegraphics{b}}} - 2\vcenter{\hbox{\includegraphics{c}}} + \vcenter{\hbox{\includegraphics{d}}},$$

$$p\left(\vcenter{\hbox{\includegraphics{e}}}\right) = \vcenter{\hbox{\includegraphics{f}}} - 3\vcenter{\hbox{\includegraphics{g}}} + 2\vcenter{\hbox{\includegraphics{h}}} = 2p\left(\vcenter{\hbox{\includegraphics{i}}}\right),$$

while the subspace generated by the elements $\vcenter{\hbox{\includegraphics{j}}}$ and $\vcenter{\hbox{\includegraphics{k}}}$ is two-dimensional and has a nonzero intersection with the ideal (Θ).

4.5 Bialgebra of weight systems

4.5.1 Multiplication and comultiplication of weight systems

According to Section 4.4.1 the vector space $\mathscr{W}^{fr}$ is dual to the space $\mathscr{A}$. Since now $\mathscr{A}$ is equipped with the structure of a Hopf algebra, the general construction of Section A.2.7 supplies the space $\mathscr{W}^{fr}$ with the same structure. In particular, weight systems can be multiplied: $(w_1 \cdot w_2)(D) := (w_1 \otimes w_2)(\delta(D))$ and comultiplied: $(\delta(w))(D_1 \otimes D_2) := w(D_1 \cdot D_2)$. The *unit* of $\mathscr{W}^{fr}$ is the weight system $\mathbf{I}_0$ which takes value 1 on the chord diagram without chords and vanishes elsewhere. The *counit* sends a weight system to its value on the chord diagram without chords.

For example, if w_1 is a weight system which takes value a on the chord diagram

and zero value on all other chord diagrams, and w_2 takes value b on

and vanishes elsewhere, then

$$(w_1 \cdot w_2)\left(\vcenter{\hbox{\includegraphics{l}}}\right) = (w_1 \otimes w_2)\left(\delta\left(\vcenter{\hbox{\includegraphics{m}}}\right)\right) = 2w_1\left(\vcenter{\hbox{\includegraphics{n}}}\right) \cdot w_2\left(\vcenter{\hbox{\includegraphics{o}}}\right) = 2ab.$$

Proposition 4.16. *The symbol* symb $: \mathscr{V}^{fr} \to \mathscr{W}^{fr}$ *commutes with multiplication and comultiplication.*

Proof Analyzing the proof of Theorem 3.9 in Section 3.2.3 one can conclude that for any two Vassiliev invariants of orders $\leqslant p$ and $\leqslant q$, the symbol of their product is equal to the product of their symbols. This implies that the map symb respects the multiplication. Now we prove that $\mathrm{symb}(\delta(v)) = \delta(\mathrm{symb}(v))$ for a Vassiliev invariant v of order $\leqslant n$. Let us apply both parts of this equality to the tensor product of two chord diagrams D_1 and D_2 with the number of chords p and q respectively where $p + q = n$. We have

$$\mathrm{symb}(\delta(v))\left(D_1 \otimes D_2\right) = \delta(v)\left(K^{D_1} \otimes K^{D_2}\right) = v\left(K^{D_1}\#K^{D_2}\right),$$

where the singular knots K^{D_1} and K^{D_2} represent chord diagrams D_1 and D_2. But the singular knot $K^{D_1}\#K^{D_2}$ represents the chord diagram $D_1 \cdot D_2$. Since the total number of chords in $D_1 \cdot D_2$ is equal to n, the value of v on the corresponding singular knot would be equal to the value of its symbol on the chord diagram:

$$v\left(K^{D_1}\#K^{D_2}\right) = \mathrm{symb}(v)\left(D_1 \cdot D_2\right) = \delta(\mathrm{symb}(v))\left(D_1 \otimes D_2\right).$$

$\square$

Remark 4.17. The map symb : $\mathcal{V}^{fr} \rightarrow \mathcal{W}^{fr}$ is not a bialgebra homomorphism because it does not respect the addition. Indeed, the sum of two invariants $v_1 + v_2$ of different orders p and q with, say, $p > q$ has order p. That means $\mathrm{symb}(v_1 + v_2) = \mathrm{symb}(v_1) \neq \mathrm{symb}(v_1) + \mathrm{symb}(v_2)$.

However, we can extend the map symb to power series Vassiliev invariants by sending the invariant $\prod v_i \in \widehat{\mathcal{V}}_{\bullet}^{fr}$ to the element $\sum \mathrm{symb}(v_i)$ of the graded completion $\widehat{\mathcal{W}}^{fr}$. Then the above proposition implies that the map symb : $\widehat{\mathcal{V}}_{\bullet}^{fr} \rightarrow \widehat{\mathcal{W}}^{fr}$ is a graded bialgebra homomorphism.

4.5.2 Multiplicative weight systems

We call a weight system w *multiplicative* if for any two chord diagrams D_1 and D_2 we have

$$w(D_1 \cdot D_2) = w(D_1)w(D_2).$$

This is the same as to say that w is a semigroup-like element in the bialgebra of weight systems (see Section A.2.2). Note that a multiplicative weight system always takes value 1 on the chord diagram with no chords. The unit $\mathbf{I}_0$ is the only group-like element of the bialgebra $\mathcal{W}^{fr}$ (compare with the exercise in Section 4.3.3). However, the graded completion $\widehat{\mathcal{W}}^{fr}$ contains many interesting group-like elements. The fact that the symbol commutes with the multiplication and comultiplication has the following:

Corollary 4.18. *Suppose that*

$$v = \prod_{n=0}^{\infty} v_n \in \widehat{\mathcal{V}}_{\bullet}^{fr}$$

is multiplicative. Then its symbol is also multiplicative.

Indeed, any homomorphism of bialgebras sends group-like elements to group-like elements.

4.5.3 The exponential of a weight system

A weight system that belongs to a homogeneous component $\mathcal{W}_n^{fr}$ of the space $\mathcal{W}^{fr}$ is said to be *homogeneous of degree n*. Let $w \in \widehat{\mathcal{W}}^{fr}$ be an element with homogeneous components $w_i \in \mathcal{W}_i^{fr}$ such that $w_0 = 0$. Then the exponential of w can be defined as the Taylor series

$$\exp(w) = \sum_{k=0}^{\infty} \frac{w^k}{k!}.$$

This formula makes sense because only a finite number of operations is required for the evaluation of each homogeneous component of this sum. One can easily check that the weight systems $\exp(w)$ and $\exp(-w)$ are inverse to each other:

$$\exp(w) \cdot \exp(-w) = \mathbf{I}_0.$$

By definition, a *primitive* weight system w satisfies

$$w(D_1 \cdot D_2) = \mathbf{I}_0(D_1) \cdot w(D_2) + w(D_1) \cdot \mathbf{I}_0(D_2).$$

(In particular, a primitive weight system is always zero on a product of two non-trivial diagrams $D_1 \cdot D_2$.) The exponential $\exp(w)$ of a primitive weight system w is multiplicative (group-like). Note that it always belongs to the completion $\widehat{\mathcal{W}}^{fr}$, even if w belongs to $\mathcal{W}^{fr}$.

A simple example of a homogeneous weight system of degree n is provided by the function on the set of chord diagrams which is equal to 1 on any diagram of degree n and to 0 on chord diagrams of all other degrees. This function clearly satisfies the four-term relations. Let us denote this weight system by $\mathbf{I}_n$.

Lemma 4.19. $\mathbf{I}_n \cdot \mathbf{I}_m = \binom{m+n}{n} \mathbf{I}_{n+m}$.

This directly follows from the definition of the multiplication for weight systems.

Corollary 4.20. (i) $\mathbf{I}_1^n/n! = \mathbf{I}_n$;

  (ii)   *If we set* $\mathbf{I} = \sum_{n=0}^{\infty} \mathbf{I}_n$ *(that is,* $\mathbf{I}$ *is the weight system that is equal to 1 on every chord diagram), then*

$$\exp(\mathbf{I}_1) = \mathbf{I}.$$

As we already mentioned, $\mathbf{I}$ is not an element of $\mathscr{W}^{fr} = \oplus_n \mathscr{W}_n^{fr}$ but of the graded completion $\widehat{\mathscr{W}}^{fr}$. Note that $\mathbf{I}$ is not the unit of $\widehat{\mathscr{W}}^{fr}$. Its unit, as well as the unit of $\mathscr{W}$, is represented by the element $\mathbf{I}_0$.

4.5.4 Deframing the weight systems

Since $\mathscr{A}' = \mathscr{A}/(\Theta)$ is a quotient of $\mathscr{A}$, the corresponding dual spaces are embedded one into another, $\mathscr{W} \subset \mathscr{W}^{fr}$. The elements of $\mathscr{W}$ take zero values on all chord diagrams with an isolated chord. In Section 4.1 they were called *unframed weight systems*. The deframing procedure for chord diagrams (Section 4.4.4) leads to a deframing procedure for weight systems. By duality, the projector $p : \mathscr{A} \to \mathscr{A}$ gives rise to a projector $p^* : \mathscr{W}^{fr} \to \mathscr{W}^{fr}$ whose value on an element $w \in \mathscr{W}_n^{fr}$ is defined by

$$w'(D) = p^*(w)(D) := w(p(D)) = \sum_{J \subseteq [D]} w\left((-\Theta)^{n-|J|} \cdot D_J\right).$$

Obviously, $w'(D) = 0$ for any w and any chord diagram D with an isolated chord. Hence the operator $p^* : w \mapsto w'$ is a projection of the space $\widehat{\mathscr{W}}^{fr}$ onto its subspace $\widehat{\mathscr{W}}$ consisting of unframed weight systems.

The deframing operator looks especially nice for multiplicative weight systems.

Exercise. Prove that for any number $\theta \in \mathbb{F}$ the exponent $e^{\theta \mathbf{I}_1} \in \widehat{\mathscr{W}}$ is a multiplicative weight system.

Lemma 4.21. *Let* $\theta = w(\Theta)$ *for a multiplicative weight system* w. *Then the deframing of* w *is* $w' = e^{-\theta \mathbf{I}_1} \cdot w$.

We leave the proof of this lemma to the reader as an exercise. The lemma, together with the previous exercise, implies that the deframing of a multiplicative weight system is again multiplicative.

4.6 Primitive elements in $\mathscr{A}$

The algebra of chord diagrams $\mathscr{A}$ is commutative, cocommutative and connected. Therefore, by the Milnor–Moore Theorem (Theorem A.32), any element of $\mathscr{A}$ is uniquely represented as a polynomial in basis primitive elements. Let us denote the

nth homogeneous component of the primitive subspace by $\mathscr{P}_n = \mathscr{A}_n \cap \mathscr{P}(\mathscr{A})$ and find an explicit description of $\mathscr{P}_n$ for small n.

<u>dim = 1</u>. $\mathscr{P}_1 = \mathscr{A}_1$ is one-dimensional and spanned by .

<u>dim = 2</u>. Since

$$\delta\left(\vcenter{\hbox{}}\right) = \bigcirc \otimes \otimes + 2\, \ominus \otimes \ominus + \otimes \otimes \bigcirc,$$

$$\delta\left(\vcenter{\hbox{}}\right) = \bigcirc \otimes \textcircled{||} + 2\, \ominus \otimes \ominus + \textcircled{||} \otimes \bigcirc,$$

the element $\otimes - \textcircled{||}$ is primitive. It constitutes a basis of $\mathscr{P}_2$.

<u>dim = 3</u>. The coproducts of the three basis elements of $\mathscr{A}_3$ are

$$\delta\left(\vcenter{\hbox{}}\right) = \bigcirc \otimes \boxplus + 2\, \ominus \otimes \otimes + \ominus \otimes \textcircled{||} + \cdots,$$

$$\delta\left(\vcenter{\hbox{}}\right) = \bigcirc \otimes \boxtimes + \ominus \otimes \otimes + 2\, \ominus \otimes \textcircled{||} + \cdots,$$

$$\delta\left(\vcenter{\hbox{}}\right) = \bigcirc \otimes \ominus + 3\, \ominus \otimes \textcircled{||} + \cdots,$$

(Here the dots stand for the terms symmetric to the terms that are shown explicitly.) Looking at these expressions, it is easy to check that the element

$$\boxplus - 2\, \boxtimes + \ominus$$

is the only, up to multiplication by a scalar, primitive element of $\mathscr{A}_3$.

The exact dimensions of $\mathscr{P}_n$ are currently (in 2011) known up to $n = 12$ (the last three values, corresponding to $n = 10, 11, 12$, were found by J. Kneissler (1997)):

n	1	2	3	4	5	6	7	8	9	10	11	12
dim $\mathscr{P}_n$	1	1	1	2	3	5	8	12	18	27	39	55

We shall discuss the sizes of the spaces $\mathscr{P}_n$, $\mathscr{A}_n$ and $\mathscr{V}_n$ in more detail later (see Sections 5.5 and 14.5).

If the dimensions of $\mathscr{P}_n$ were known for all n, then the dimensions of $\mathscr{A}_n$ would also be known.

Example 4.22. Let us find the dimensions of $\mathscr{A}_n$, $n \leqslant 5$, assuming that we know the values of dim $\mathscr{P}_n$ for $n = 1, 2, 3, 4, 5$, which are equal to $1, 1, 1, 2, 3$, respectively. Let p_i be the basis element of $\mathscr{P}_i$, $i = 1, 2, 3$ and denote the bases of $\mathscr{P}_4$ and $\mathscr{P}_5$ as p_{41}, p_{42} and p_{51}, p_{52}, p_{53}, respectively. Nontrivial monomials up to degree 5 that can be made out of these basis elements are:

Degree 2 monomials (1): p_1^2.
Degree 3 monomials (2): p_1^3, $p_1 p_2$.
Degree 4 monomials (4): p_1^4, $p_1^2 p_2$, $p_1 p_3$, p_2^2.
Degree 5 monomials (7): p_1^5, $p_1^3 p_2$, $p_1^2 p_3$, $p_1 p_2^2$, $p_1 p_{41}$, $p_1 p_{42}$, $p_2 p_3$.

A basis of each $\mathscr{A}_n$ can be made of the primitive elements and their products of the corresponding degree. For $n = 0, 1, 2, 3, 4, 5$ we get: dim $\mathscr{A}_0 = 1$, dim $\mathscr{A}_1 = 1$, dim $\mathscr{A}_2 = 1 + 1 = 2$, dim $\mathscr{A}_3 = 1 + 2 = 3$, dim $\mathscr{A}_4 = 2 + 4 = 6$, dim $\mathscr{A}_5 = 3 + 7 = 10$.

The partial sums of this sequence give the dimensions of the spaces of framed Vassiliev invariants: dim $\mathscr{V}_0^{fr} = 1$, dim $\mathscr{V}_1^{fr} = 2$, dim $\mathscr{V}_2^{fr} = 4$, dim $\mathscr{V}_3^{fr} = 7$, dim $\mathscr{V}_4^{fr} = 13$, dim $\mathscr{V}_5^{fr} = 23$.

Exercise. Let p_n be the sequence of dimensions of primitive spaces in a Hopf algebra and a_n the sequence of dimensions of the entire algebra. Prove the relation

$$1 + a_1 t + a_2 t^2 + \cdots = \frac{1}{(1-t)^{p_1}(1-t^2)^{p_2}(1-t^3)^{p_3}\cdots}.$$

Note that primitive elements of $\mathscr{A}$ are represented by rather complicated linear combinations of chord diagrams. A more concise and clear representation can be obtained via connected closed diagrams, to be introduced in the next chapter (Section 5.5).

4.7 Linear chord diagrams

The arguments of this chapter, applied to *long* knots (see Section 1.7.2), lead us naturally to considering the space of *linear* chord diagrams, that is, diagrams on an oriented line:

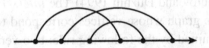

subject to the 4-term relations:

Let us temporarily denote the space of linear chord diagrams with n chords modulo the 4-term relations by $(\mathscr{A}_n)^{long}$. The space $(\mathscr{A})^{long}$ of such chord diagrams of all degrees modulo the 4T relations is a bialgebra; the product in $(\mathscr{A})^{long}$ can be defined simply by concatenating the oriented lines.

If the line is closed into a circle, linear 4-term relations become circular (that is, usual) 4-term relations; thus, we have a linear map $(\mathscr{A}_n)^{long} \to \mathscr{A}_n$. This map is evidently onto, as one can find a preimage of any circular chord diagram by cutting the circle at an arbitrary point. This preimage, in general, depends on the place where the circle is cut, so it may appear that this map has a non-trivial kernel. For example, the linear diagram shown above closes up to the same diagram as the one drawn below:

Remarkably, *modulo 4-term relations*, all the preimages of any circular chord diagram are equal in $(\mathscr{A}_3)^{long}$ (in particular, the two diagrams in the above pictures give the same element of $(\mathscr{A}_3)^{long}$). This fact is proved by exactly the same argument as the statement that the product of chord diagrams is well-defined (Lemma 4.13); we leave it to the reader as an exercise.

Summarizing, we have:

Proposition 4.23. *Closing up the line into the circle gives rise to a vector space isomorphism $(\mathscr{A})^{long} \to \mathscr{A}$. This isomorphism is compatible with the multiplication and comultiplication and thus defines an isomorphism of bialgebras.*

A similar statement holds for diagrams modulo 4T and 1T relations. Further, one can consider chord diagrams (and 4T relations) with chords attached to an arbitrary one-dimensional oriented manifold – see Section 5.10.

4.8 Intersection graphs

4.8.1 The intersection graph of a chord diagram

Definition 4.24. (Chmutov and Duzhin 1994) The *intersection graph* $\Gamma(D)$ of a chord diagram D is the graph whose vertices correspond to the chords of D and whose edges are determined by the following rule: two vertices are connected by an edge if and only if the corresponding chords intersect, and multiple edges are not allowed. (Two chords, a and b, are said to intersect if their endpoints a_1, a_2 and b_1, b_2 appear in the interlacing order a_1, b_1, a_2, b_2 along the circle.)

For example,

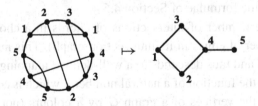

The intersection graphs of chord diagrams are also called *circle graphs* or *alternance graphs*.

Note that not every graph can be represented as the intersection graph of a chord diagram. For example, the following graphs are not intersection graphs:

Exercise. Prove that all graphs with no more than five vertices are intersection graphs.

On the other hand, distinct diagrams may have coinciding intersection graphs. For example, there are three different diagrams

with the same intersection graph ●—●—●—●—●.

A complete characterization of those graphs that can be realized as intersection graphs was given by A. Bouchet (1994).

With each chord diagram D we can associate an oriented surface Σ_D by attaching a disc to the circle of D and thickening the chords of D. Then the chords determine a basis in $H_1(\Sigma_D, \mathbb{Z}_2)$ as in the picture below. The intersection matrix for this basis coincides with the adjacency matrix of Γ_D. Using the terminology of singularity theory we may say that the intersection graph Γ_D is the *Dynkin diagram* of the intersection form in $H_1(\Sigma_D, \mathbb{Z}_2)$ constructed for the basis of $H_1(\Sigma_D, \mathbb{Z}_2)$.

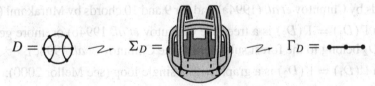

4.8.2 Some weight systems

Intersections graphs are useful, for one thing, because they provide a simple way to define some weight systems. We shall describe two framed weight systems which depend only on the intersection graph. The reader is invited to find their deframings, using the formulae of Section 4.5.4.

(1) Let ν be the number of intersections of chords in a chord diagram (or, if you like, the number of edges in its intersection graph). This number satisfies the four-term relations and thus descends to a well-defined mapping $\nu : \mathscr{A} \mapsto \mathbb{Z}$.

(2) Let $\chi(G)$ be the function of a natural number n which is equal to the number of ways to colour the vertices of a graph G by n colours (not necessarily using all the colours) so that the endpoints of any edge are coloured differently. It is easy to see (Harary 1969) that $\chi(G)$ is a polynomial in n called the *chromatic polynomial* of G. If D is a chord diagram, then keeping the notation $\chi(D)$ for the chromatic polynomial of $\Gamma(D)$, one can prove that this function satisfies the 4T relations and therefore produces a weight system $\chi : \mathscr{A} \to \mathbb{Z}[n]$ (this follows from the deletion–contraction relation for the chromatic polynomial and the relation in Definition 14.11).

Exercise. Prove that the primitivization (see Exercise 4.10 at the end of this chapter) of the chord diagram with the complete intersection graph provides one non-zero primitive element of $\mathscr{A}$ in each degree, thus giving the first non-trivial lower estimate on the dimensions of the spaces $\mathscr{A}$.

4.8.3 Intersection graph conjecture

Intersection graphs contain a good deal of information about chord diagrams. In (Chmutov *et al.* 1994a) the following conjecture was stated.

Intersection graph conjecture. *If D_1 and D_2 are two chord diagrams whose intersection graphs are equal, $\Gamma(D_1) = \Gamma(D_2)$, then $D_1 = D_2$ as elements of $\mathscr{A}$ (that is, modulo four-term relations).*

Although wrong in general (see Section 11.1.3), this assertion is true in some particular situations:

1. for all diagrams D_1, D_2 with up to 10 chords (a direct computer check up to 8 chords by Chmutov *et al.* (1994a) and for 9 and 10 chords by Murakami (2000));
2. when $\Gamma(D_1) = \Gamma(D_2)$ is a tree (see Chmutov *et al.* 1994b) or, more generally, D_1, D_2 belong to the forest subalgebra (see Chmutov *et al.* 1994c);
3. when $\Gamma(D_1) = \Gamma(D_2)$ is a graph with a single loop (see Mellor 2000);

4. for weight systems w coming from standard representations of Lie algebras $\mathfrak{gl}_N$ or $\mathfrak{so}_N$. This means that $\Gamma(D_1) = \Gamma(D_2)$ implies $w(D_1) = w(D_2)$; see Proposition 6.10 and Exercise 6.15;

5. for the universal $\mathfrak{sl}_2$ weight system and the weight system coming from the standard representation of the Lie superalgebra $\mathfrak{gl}(1|1)$ (see Chmutov and Lando 2007).

In fact, the intersection graph conjecture can be refined to the following theorem which covers items (4) and (5) above.

Theorem 4.25 (Chmutov and Lando 2007). *The symbol of a Vassiliev invariant that does not distinguish mutant knots depends on the intersection graph only.*

We postpone the discussion of mutant knots, the proof of this theorem and its converse to Section 11.1.

4.8.4 Chord diagrams representing a given graph

To describe all chord diagrams representing a given intersection graph, we need the notion of a *share* (Chmutov et al. 1994b, Chmutov and Lando 2007). Informally, a *share* of a chord diagram is a subset of chords whose endpoints are separated into at most two parts by the endpoints of the complementary chords. More formally,

Definition 4.26. A *share* is a part of a chord diagram consisting of two arcs of the outer circle with the following property: each chord one of whose ends belongs to these arcs has both ends on these arcs.

Here are some examples:

A share   Not a share   Two shares

The complement of a share also is a share. The whole chord diagram is its own share whose complement contains no chords.

Definition 4.27. A *mutation of a chord diagram* is another chord diagram obtained by a flip of a share.

For example, three mutations of the share in the first chord diagram above produce the following chord diagrams:

Obviously, mutations preserve the intersection graphs of chord diagrams.

Theorem 4.28. *Two chord diagrams have the same intersection graph if and only if they are related by a sequence of mutations.*

This theorem is contained implicitly in papers (Bouchet 1987, Gabor *et al.* 1989) where chord diagrams are written as *double occurrence words*.

Proof of the theorem.   The proof uses Cunningham's theory of graph decompositions (Cunningham 1982).

A *split* of a (simple) graph Γ is a disjoint bipartition $\{V_1, V_2\}$ of its set of vertices $V(\Gamma)$ such that each part contains at least 2 vertices, and with the property that there are subsets $W_1 \subseteq V_1$, $W_2 \subseteq V_2$ such that all the edges of Γ connecting V_1 with V_2 form the complete bipartite graph $K(W_1, W_2)$ with the parts W_1 and W_2. Thus for a split $\{V_1, V_2\}$ the whole graph Γ can be represented as a union of the induced subgraphs $\Gamma(V_1)$ and $\Gamma(V_2)$ linked by a complete bipartite graph.

Another way to think about splits, which is sometimes more convenient and which we shall use in the pictures below, is as follows. Consider two graphs, Γ_1 and Γ_2, each with a distinguished vertex $v_1 \in V(\Gamma_1)$ and $v_2 \in V(\Gamma_2)$, respectively, called *markers*. Construct the new graph

$$\Gamma = \Gamma_1 \boxtimes_{(v_1,v_2)} \Gamma_2$$

whose set of vertices is $V(\Gamma) = \{V(\Gamma_1) - v_1\} \cup \{V(\Gamma_2) - v_2\}$, and whose set of edges is

$$E(\Gamma) = \{(v_1', v_1'') \in E(\Gamma_1) : v_1' \neq v_1 \neq v_1''\} \cup \{(v_2', v_2'') \in E(\Gamma_2) : v_2' \neq v_2 \neq v_2''\}$$

$$\cup \{(v_1', v_2') : (v_1', v_1) \in E(\Gamma_1) \text{ and } (v_2, v_2') \in E(\Gamma_2)\}.$$

Representation of Γ as $\Gamma_1 \boxtimes_{(v_1,v_2)} \Gamma_2$ is called a *decomposition* of Γ, the graphs Γ_1 and Γ_2 are called the *components* of the decomposition. The partition $\{V(\Gamma_1) - v_1, V(\Gamma_2) - v_2\}$ is a split of Γ. Graphs Γ_1 and Γ_2 might be decomposed further, giving a finer decomposition of the initial graph Γ. Graphically, we represent a decomposition by pictures of its components where the corresponding markers are connected by a dashed edge.

A *prime* graph is a graph with at least three vertices admitting no splits. A decomposition of a graph is said to be *canonical* if the following conditions are satisfied:

(i) each component is either a prime graph, or a complete graph K_n, or a star S_n, which is the tree with a vertex, the *centre*, adjacent to n other vertices;

(ii) no two components that are complete graphs are neighbours, that is, their markers are not connected by a dashed edge;

(iii) the markers of two components that are star graphs connected by a dashed edge are either both centres or both not centres of their components.

W. H. Cunningham proved (1982, theorem 3) that each graph with at least six vertices possesses a unique canonical decomposition.

Let us illustrate the notions introduced above by an example of canonical decomposition of an intersection graph. We number the chords and the corresponding vertices in our graphs, so that the unnumbered vertices are the markers of the components.

A chord diagram

The intersection graph

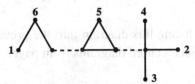

The canonical decomposition

The key observation in the proof of the theorem is that components of the canonical decomposition of any intersection graph admit a unique representation by chord diagrams. For a complete graph and star components, this is obvious. For a prime component, this was proved by A. Bouchet (1987, statement 4.4) (see also (Gabor *et al.* 1989, section 6) for an algorithm finding such a representation for a prime graph).

Now, in order to describe all chord diagrams with a given intersection graph, we start with a component of its canonical decomposition. There is only one way to realize the component by a chord diagram. We draw the chord corresponding to the marker as a dashed chord and call it the *marked chord*. This chord indicates the places where we must cut the circle removing the marked chord together with small arcs containing its endpoints. As a result we obtain a chord diagram on two arcs. Repeating the same procedure with the next component of the canonical decomposition, we get another chord diagram on two arcs. We have to glue the arcs of these two diagrams together in the alternating order. There are four possibilities to do this, and they differ by mutations of the share corresponding to one of the two components. This completes the proof of the theorem. □

To illustrate the last stage of the proof, consider our standard example and take the star 2-3-4 component first and then the triangle component. We get

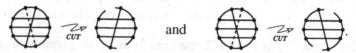

Because of the symmetry, the four ways of glueing these diagrams produce only two distinct chord diagrams with a marked chord:

Repeating the same procedure with the marked chord for the last 1-6 component of the canonical decomposition, we get

Glueing this diagram into the previous two in all possible ways, we get the four mutant chord diagrams from pages 107–108.

4.8.5 2-term relations and the genus of a diagram

A 2-term (or *endpoint sliding*) relation for chord diagrams has the form

The 4-term relations are evidently a consequence of the 2-term relations; therefore, any function on chord diagrams that satisfies 2-term relations is a weight system. An example of such a weight system is the *genus of a chord diagram* defined as follows.

Replace the outer circle of the chord diagram and all its chords by narrow untwisted bands – this yields an orientable surface with boundary. Attaching a disk to each boundary component gives a closed orientable surface. Its genus is by definition the genus of the chord diagram. The genus can be calculated from the number of boundary components using Euler characteristic. Indeed, the Euler characteristic of the surface with boundary obtained by the above procedure from a chord diagram of degree n is equal to $-n$. If this surface has c boundary components and genus g, then we have $-n = 2 - 2g - c$ while $g = 1 + (n - c)/2$. For example, the two chord diagrams of degree 2 have genera 0 and 1, because the number of connected components of the boundary is 4 and 2, respectively, as one can see in the following picture:

The genus of a chord diagram satisfies 2-term relations, since sliding an endpoint of a chord along another adjacent chord does not change the topological type of the corresponding surface with boundary.

An interesting way to compute the genus from the intersection graph of the chord diagram was found by Moran (1984). Moran's theorem states that the genus of a chord diagram is half the rank of the adjacency matrix over $\mathbb{Z}_2$ of the intersection graph. This theorem can be proved transforming a given chord diagram to the canonical form using the following two exercises.

Exercise. Let D_1 and D_2 be two chord diagrams differing by a 2-term relation. Check that the corresponding adjacency matrices over $\mathbb{Z}_2$ are conjugate (one is obtained from the other by adding the ith column to the jth column and the ith row to the jth row).

Exercise. A *caravan of* m_1 *"one-humped camels" and* m_2 *"two-humped camels"* is the product of m_1 diagrams with one chord and m_2 diagrams with 2 crossing chords:

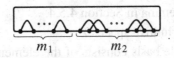

Show that any chord diagram is equivalent, modulo 2-term relations, to a caravan. Show that the caravans form a basis in the vector space of chord diagrams modulo 2-term relations.

The algebra generated by caravans is thus a quotient algebra of the algebra of chord diagrams.

Remark 4.29. The last exercise is, essentially, equivalent to the classical topological classification of compact oriented surfaces with boundary by the genus, m_2, and the number of boundary components, $m_1 + 1$.

Exercises

4.1 A *short* chord is a chord whose endpoints are adjacent, that is, one of the arcs that it bounds contains no endpoints of other chords. In particular, short chords are isolated. Prove that the linear span of all diagrams with a short chord and all four-term relations contains all diagrams with an isolated chord. This means that the restricted one-term relations (only for diagrams with a short chord) imply general one-term relations provided that the four-term relations hold.

4.2 Find the number of different chord diagrams of order n with n isolated chords. Prove that all of them are equal to each other modulo the four-term relations.

4.3 Using Table 4.1, find the space of unframed weight systems $\mathcal{W}_4$.

Answer. The basis weight systems are:

1	0	-1	-1	0	-1	-2
0	1	1	2	0	1	3
0	0	0	0	1	1	1

The table shows that the three diagrams 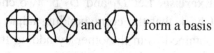 form a basis in the space $\mathcal{A}_4'$.

4.4 *Is it true that any chord diagram of order 13 is equivalent to its mirror image modulo 4-term relations?

4.5 Prove that the deframing operator $'$ (Section 4.5.4) is a homomorphism of algebras: $(w_1 \cdot w_2)' = w_1' \cdot w_2'$.

4.6 Give a proof of the lemma in Section 4.5.4.

4.7 Find a basis in the primitive space $\mathcal{P}_4$.

Answer. A possible basis consists of the elements $d_6^4 - d_7^4$ and $d_2^4 - 2d_7^4$ from Table 4.1.

4.8 Prove that for any primitive element p of degree > 1, $w(p) = w'(p)$ where w' is the deframing of a weight system w.

4.9 Prove that the symbol of a primitive Vassiliev invariant is a primitive weight system.

4.10 Prove that the projection onto the space of the primitive elements (see exercise on page 479) in the algebra $\mathcal{A}$ can be given by the following explicit formula:

$$\pi(D) = D - 1! \sum D_1 D_2 + 2! \sum D_1 D_2 D_3 - \dots,$$

where the sums are taken over all unordered splittings of the set of chords of D into 2, 3, etc. nonempty subsets.

4.11 Let Θ be the chord diagram with a single chord. By a direct computation, check that $\exp(\alpha\Theta) := \sum_{n=0}^{\infty} \frac{\alpha^n \Theta^n}{n!} \in \hat{\mathcal{A}}$ is a group-like element in the completed Hopf algebra of chord diagrams.

4.12 (a) Prove that no chord diagram is equal to 0 modulo 4-term relations.
(b) Let D be a chord diagram without isolated chords. Prove that $D \neq 0$ modulo 1- and 4-term relations.

4.13 Let $c(D)$ be the number of chord intersections in a chord diagram D. Check that c is a weight system. Find its deframing c'.

4.14 **The generalized 4-term relations.**
(a) Prove the following relation:

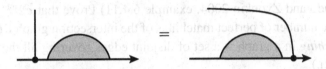

Here the horizontal line is a fragment of the circle of the diagram, while the grey region denotes an arbitrary conglomeration of chords.

(b) Prove the following relation:

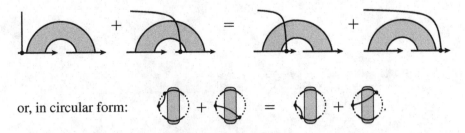

or, in circular form:

4.15 Using the generalized 4-term relation, prove the following identity:

4.16 Prove Proposition 4.23 in Section 4.7.

4.17 Check that for the chord diagram below, the intersection graph and its canonical decomposition are as shown:

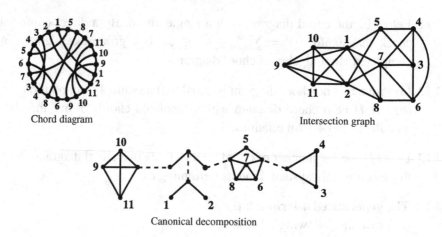

Chord diagram   Intersection graph

Canonical decomposition

4.18 (Lando and Zvonkin 2004, example 6.4.11) Prove that $e^{\mathrm{symb}(c_2)}(D)$ is equal to the number of perfect matchings of the intersection graph $\Gamma(D)$. (A *perfect matching* in a graph is a set of disjoint edges covering all the vertices of the graph.)

5

Jacobi diagrams

In the previous chapter we saw that the study of Vassiliev knot invariants, at least complex-valued, is largely reduced to the study of the algebra of chord diagrams. Here we introduce two different types of diagrams representing elements of this algebra, namely *closed Jacobi diagrams* and *open Jacobi diagrams*. These diagrams provide better understanding of the primitive space $\mathscr{P}\mathscr{A}$ and bridge the way to the applications of the Lie algebras in the theory of Vassiliev invariants; see Chapter 6.

The name *Jacobi diagrams* is justified by a close resemblance of the basic relations imposed on Jacobi diagrams (STU and IHX) to the Jacobi identity for Lie algebras.

5.1 Closed Jacobi diagrams

Definition 5.1. A *closed Jacobi diagram* (or, simply, a *closed diagram*) is a connected trivalent graph with a distinguished simple oriented cycle, called *Wilson loop*,[1] and a fixed cyclic order of half-edges at each vertex not on the Wilson loop. Half the number of the vertices of a closed diagram is called the *degree*, or *order*, of the diagram. This number is always an integer.

Remark 5.2. Some authors (see, for instance, Habegger and Masbaum 2000) also include the cyclic order of half-edges at the vertices on the Wilson loop into the structure of a closed Jacobi diagram; this leads to the same theory.

Remark 5.3. A Jacobi diagram is allowed to have multiple edges and hanging loops, that is, edges with both ends at the same vertex. It is the possible presence of hanging loops that requires introducing the cyclic order on half-edges rather than edges.

---

[1] This terminology, introduced by Bar-Natan, makes an allusion to field theory where a Wilson loop is an observable that assigns to a connection (field potential) its holonomy along a fixed closed curve.

Example 5.4. Here is a closed diagram of degree 4:

The orientation of the Wilson loop and the cyclic orders of half-edges at the internal vertices are indicated by arrows. In the pictures below, we shall always draw the diagram inside its Wilson loop, which will be assumed to be oriented counterclockwise unless explicitly specified otherwise. Inner vertices will also be assumed to be oriented counterclockwise. (This convention is referred to as the *blackboard orientation*.) Note that the intersection of two edges in the centre of the diagram above is not actually a vertex.

Chord diagrams are closed Jacobi diagrams, all of whose vertices lie on the Wilson loop.

Other terms used for closed Jacobi diagrams in the literature include *Chinese character diagrams* (Bar-Natan 1995a), *circle diagrams* (Kneissler 1997), *round diagrams* (Willerton 2000) and *Feynman diagrams* (Kricker *et al.* 1997).

Definition 5.5. The vector space of closed diagrams $\mathscr{C}_n$ is the space spanned by all closed diagrams of degree n modulo the *STU relations*:

$$\underset{S}{\bigvee} \; = \; \underset{T}{\bigcup\bigvee} \; - \; \underset{U}{\bigvee} \; .$$

The three diagrams S, T and U must be identical outside the shown fragment. We write $\mathscr{C}$ for the direct sum of the spaces $\mathscr{C}_n$ for all $n \geqslant 0$.

The two diagrams T and U are referred to as the *resolutions* of the diagram S. The choice of the plus and minus signs in front of the two resolutions in the right-hand side of the STU relation depends on the orientation for the Wilson loop and on the cyclic order of the three edges meeting at the internal vertex of the S-term. Should we reverse one of them, say, the orientation of the Wilson loop, the signs of the T- and U-terms change. Indeed,

$$\bigvee \; = \; \bigvee \; \overset{STU}{=} \; \bigcup\bigvee \; - \; \bigvee \; = \; \bigvee \; - \; \bigcup\bigvee \; .$$

This remark will be important in Section 5.5.3 where we discuss the problem of detecting knot orientation. One may think of the choice of the direction for the

Wilson loop in an STU relation as a choice of the cyclic order "forward-sideways-backwards" at the vertex lying on the Wilson loop. In these terms, the signs in the STU relation depend on the cyclic orders at both vertices of the S-term, the relation above may be thought of as a consequence of the antisymmetry relation AS for the vertex on the Wilson loop, and the STU relation itself can be regarded as a particular case of the IHX relation (see Section 5.2).

Exercise. There exist two different closed diagrams of order 1: , ,

one of which vanishes due to the STU relation:

$$\text{} = \text{} - \text{} = 0.$$

There are ten closed diagrams of degree 2:

The last six diagrams are zero. This is easy to deduce from the STU relations, but the most convenient way of seeing it is by using the AS relations which follow from the STU relations (see Section 5.2.1 below).

Furthermore, there are at least two relations among the first four diagrams:

$$\text{} = \text{} - \text{} ;$$

$$\text{} = \text{} - \text{} = 2\,\text{}.$$

It follows that $\dim \mathscr{C}_2 \leqslant 2$. Note that the first of the above equalities gives a concise representation, , for the basis primitive element of degree 2.

Exercise. Using the STU relations, rewrite the basis primitive element of order 3 in a concise way.

Answer.

$$\bigoplus - 2\,\bigotimes + \bigominus = \bigotimes.$$

We have already mentioned that chord diagrams are a particular case of closed diagrams. Using the STU relations, one can rewrite any closed diagram as a linear combination of chord diagrams. (Examples were given just above.)

A vertex of a closed diagram that lies on the Wilson loop is called *external*; otherwise it is called *internal*. External vertices are also called *legs*. There is an increasing filtration on the space $\mathscr{C}_n$ by subspaces $\mathscr{C}_n^m$ spanned by diagrams with at most m external vertices:

$$\mathscr{C}_n^1 \subset \mathscr{C}_n^2 \subset \dots \subset \mathscr{C}_n^{2n}.$$

Exercise. Prove that $\mathscr{C}_n^1 = 0$.

Hint. In a diagram with only two legs, one of the legs can go all around the circle and change places with the second leg.

5.2 IHX and AS relations

5.2.1 The definition of the IHX and the AS relations

Lemma 5.6. *The* STU *relations imply the* 4T *relations for chord diagrams.*

Proof Indeed, writing the four-term relation in the form

$$\bigcirc - \bigcirc = \bigcirc - \bigcirc$$

and applying the STU relations to both parts of this equation, we get the same closed diagrams. □

Definition 5.7. An AS (=*antisymmetry*) *relation* is:

$$\curlyvee = -\,\curlyvee.$$

In other words, a diagram changes sign when the cyclic order of three edges at a trivalent vertex is reversed.

Definition 5.8. An IHX *relation* is:

$$\text{I} = \text{H} - \text{X}.$$

As usual, the unfinished fragments of the pictures denote graphs that are identical (and arbitrary) everywhere but in this explicitly shown fragment.

Exercise. Check that the three terms of the IHX relation "have equal rights." For example, an H turned 90 degrees looks like an I; write an IHX relation starting from that I and check that it is the same as the initial one. Also, a portion of an X looks like an H; write down an IHX relation with that H and check that it is again the same. The IHX relation is in a sense unique; this is discussed in Exercise 5.15.

Lemma 5.9. *The STU relations imply the AS relations for the internal vertices of a closed Jacobi diagram.*

Proof Induction on the distance (in edges) of the vertex in question from the Wilson loop.

Induction base. If the vertex is adjacent to an external vertex, then the assertion follows by one application of the STU relation:

Induction step. Take two closed diagrams f_1 and f_2 that differ only by a cyclic order of half-edges at one internal vertex v. Apply STU relations to both diagrams *in the same way* so that v gets closer to the Wilson loop. □

Lemma 5.10. *The STU relations imply the IHX relations for the internal edges of a closed diagram.*

Proof The argument is similar to the one used in the previous proof. We take an IHX relation somewhere inside a closed diagram and, applying the same sequence of STU moves to each of the three diagrams, move the IHX fragment closer to the Wilson loop. The proof of the induction base is shown in these pictures:

Therefore,

$$\text{(diagram)} = \text{(diagram)} + \text{(diagram)} .$$

□

5.2.2 *Other forms of the* IHX *relation*

The IHX relation can be drawn in several forms, for example:

- (rotationally symmetric form)

$$\text{(diagram)} + \text{(diagram)} + \text{(diagram)} = 0.$$

- (Jacobi form)

$$\text{(diagram)} = \text{(diagram)} + \text{(diagram)} .$$

- (Kirchhoff form)

$$\text{(diagram)} = \text{(diagram)} + \text{(diagram)} .$$

Exercise. By turning your head and pulling the strings of the diagrams, check that all these forms are equivalent.

The Jacobi form of the IHX relation can be interpreted as follows. Suppose that to the upper 3 endpoints of each diagram we assign 3 elements of a Lie algebra, x, y and z, while every trivalent vertex, traversed downwards, takes the pair of "incoming" elements into their commutator:

$$\underset{[x,\,y]}{\overset{x \qquad y}{\bigvee}} .$$

Then the IHX relation means that

$$[x, [y, z]] = [[x, y], z] + [y, [x, z]],$$

which is the classical Jacobi identity. This observation, properly developed, leads to the construction of Lie algebra weight systems – see Chapter 6.

The Kirchhoff presentation is reminiscent of the Kirchhoff's law in electrotechnics. Let us view the portion —— of the given graph as a piece of electrical circuit, and the variable vertex as an "electron" e with a "tail" whose endpoint is fixed. Suppose that the electron moves towards a node of the circuit:

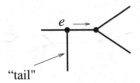

Then the IHX relation expresses the well-known Kirchhoff rule: *the sum of currents entering a node is equal to the sum of currents going out of it.* This electrotechnical analogy is very useful, for instance, in the proof of the generalized IHX relation:

Lemma 5.11. *(Kirchhoff law, or generalized IHX relation). The following identity holds:*

$$\ldots = \sum_{i=1}^{k} \ldots \quad,$$

where the grey box is an arbitrary subgraph which has only 3-valent vertices.

Proof Fix a horizontal line in the plane and consider an immersion of the given graph into the plane with smooth edges, generic with respect to the projection onto this line. More precisely, we assume that (1) the projections of all vertices onto the horizontal line are distinct, (2) when restricted to an arbitrary edge, the projection has only non-degenerate critical points, and (3) the images of all critical points are distinct and different from the images of vertices.

Bifurcation points are the images of vertices and critical points of the projection. Imagine a vertical line that moves from left to right; for every position of this line take the sum of all diagrams obtained by attaching the loose end to one of the intersection points. This sum does not depend on the position of the vertical line, because it does not change when the line crosses one bifurcation point.

Indeed, bifurcation points fall into six categories:

(1) —<   (2) >—   (3) >   (4) <   (5) >   (6) < .

In the first two cases the assertion follows from the IHX relation, in cases 3 and 4 – from the AS relation. Cases 5 and 6 by a deformation of the immersion are reduced to a combination of the previous cases (also, they can be dealt with by one application of the IHX relation in the symmetric form).   □

Example 5.12.

Remark 5.13. The difference between inputs and outputs in the Kirchhoff law is purely notational. We may bend the left-hand leg to the right and move the corresponding term to the right-hand side of the equation, changing its sign because of the antisymmetry relation, and thus obtain:

Or we may prefer to split the legs into two arbitrary subsets, putting one part on the left and another on the right. Then:

5.2.3 A corollary of the AS relation

A simple corollary of the antisymmetry relation in the space $\mathscr{C}$ is that any diagram D containing a hanging loop ⚬ is equal to zero. Indeed, there is an automorphism of the diagram that changes the two half-edges of the small circle and thus takes D to $-D$, which implies that $D = -D$ and $D = 0$. This observation also applies to the case when the small circle has other vertices on it and contains a subdiagram, symmetric with respect to the vertical axis. In fact, the assertion is true even if the diagram inside the circle is not symmetric at all. This is a generalization of the last exercise of Section 5.1 on page 118, but cannot be proved by the same argument. In Section 5.6.2 we shall prove a similar statement about *open* Jacobi diagrams; that proof also applies here.

5.3 Isomorphism $\mathscr{A} \simeq \mathscr{C}$

Let $\mathbf{A}_n$ be the set of chord diagrams of order n and $\mathbf{C}_n$ the set of closed diagrams of the same order. We have a natural inclusion $\lambda : \mathbf{A}_n \to \mathbf{C}_n$.

Theorem 5.14. *The inclusion λ gives rise to an isomorphism of vector spaces* $\lambda : \mathscr{A}_n \to \mathscr{C}_n$.

Proof We must check:

(A) that λ leads to a well-defined linear map from $\mathscr{A}_n$ to $\mathscr{C}_n$;

(B) that this map is a linear isomorphism.

Part (A) is easy. Indeed, $\mathscr{A}_n = \langle \mathbf{A}_n \rangle / \langle 4T \rangle$, $\mathscr{C}_n = \langle \mathbf{C}_n \rangle / \langle STU \rangle$, where angular brackets denote linear span. Lemma 5.6 implies that $\lambda(\langle 4T \rangle) \subseteq \langle STU \rangle$, therefore the map of the quotient spaces is well-defined.

(B) We shall construct a linear map $\rho : \mathscr{C}_n \to \mathscr{A}_n$ and prove that it is inverse to λ.

As we mentioned before, any closed diagram by the iterative use of STU relations can be transformed into a combination of chord diagrams. This gives rise to a map $\rho : \mathbf{C}_n \to \langle \mathbf{A}_n \rangle$ which is, however, multivalued, since the result may depend on the specific sequence of relations used. Here is an example of such a situation (the place where the STU relation is applied is marked by an asterisk):

However, the combination $\rho(C)$ is well-defined as an element of $\mathscr{A}_n$, that is, modulo the 4T relations. The proof of this fact proceeds by induction on the number k of internal vertices in the diagram C.

If $k = 1$, then the diagram C consists of one tripod and several chords and may look something like this:

There are three ways to resolve the internal triple point by an STU relation, and the fact that the results are the same in $\mathscr{A}_n$ is exactly the definition of the 4T relation.

Suppose that ρ is well-defined on closed diagrams with $< k$ internal vertices. Pick a diagram in $\mathbf{C}_n^{2n-k}$. The process of eliminating the triple points starts with a pair of neighbouring external vertices. Let us prove, modulo the inductive hypothesis, that if we change the order of these two points, the final result will remain the same.

There are three cases to consider: the two chosen points on the Wilson loop are (1) adjacent to a common internal vertex, (2) adjacent to neighbouring internal vertices, (3) adjacent to non-neighbouring internal vertices. The proof for the cases (1) and (2) is shown in the pictures that follow.

(1)

The position of an isolated chord does not matter, because, as we know, the multiplication in $\mathscr{A}$ is well-defined.

(2)

After the first resolution, we can choose the sequence of further resolutions arbitrarily, by the inductive hypothesis.

Exercise. Give a similar proof for the case (3).

We thus have a well-defined linear map $\rho : \mathscr{C}_n \to \mathscr{A}_n$. The fact that it is a two-sided inverse to λ is clear. □

5.4 Product and coproduct in $\mathscr{C}$

Now we shall define a bialgebra structure in the space $\mathscr{C}$.

Definition 5.15. The product of two closed diagrams is defined in the same way as for chord diagrams: the two Wilson loops are cut at arbitrary places and then glued together into one loop, in agreement with the orientations:

$$\langle\!\!\!\langle \cdot \rangle\!\!\!\rangle \cdot \langle\!\!\!\langle \rangle\!\!\!\rangle = \langle\!\!\!\langle \rangle\!\!\!\rangle .$$

Proposition 5.16. *This multiplication is well-defined, that is, it does not depend on the place of cuts.*

Proof The isomorphism $\mathscr{A} \cong \mathscr{C}$ constructed in Section 5.3 identifies the product in $\mathscr{A}$ with the above product in $\mathscr{C}$.

Since the multiplication is well-defined in $\mathscr{A}$, it is also well-defined in $\mathscr{C}$. □

To define the *coproduct* in the space $\mathscr{C}$, we need the following definition:

Definition 5.17. The *internal graph* of a closed diagram is the graph obtained by stripping off the Wilson loop. A closed diagram is said to be *connected* if its internal graph is connected. The *connected components* of a closed diagram are defined as the connected components of its internal graph.

In the sense of this definition, any chord diagram of order n consists of n connected components – the maximal possible number.

Now, the construction of the coproduct proceeds in the same way as for chord diagrams.

Definition 5.18. Let D be a closed diagram and $[D]$ the set of its connected components. For any subset $J \subseteq [D]$ denote by D_J the closed diagram with only those components that belong to J and by $D_{\overline{J}}$ the "complementary" diagram ($\overline{J} := [D] \setminus J$). We set

$$\delta(D) := \sum_{J \subseteq [D]} D_J \otimes D_{\overline{J}}.$$

Example 5.19.

$$\delta\left(\raisebox{-0.5em}{\includegraphics{}}\right) = 1 \otimes \raisebox{-0.5em}{\includegraphics{}} + \raisebox{-0.5em}{\includegraphics{}} \otimes \raisebox{-0.5em}{\includegraphics{}} + \raisebox{-0.5em}{\includegraphics{}} \otimes \raisebox{-0.5em}{\includegraphics{}} + \raisebox{-0.5em}{\includegraphics{}} \otimes 1.$$

We know that the algebra $\mathscr{C}$, as a vector space, is spanned by chord diagrams. For chord diagrams, algebraic operations defined in $\mathscr{A}$ and $\mathscr{C}$ tautologically coincide. It follows that the coproduct in $\mathscr{C}$ is compatible with its product and that the isomorphisms λ, ρ are, in fact, isomorphisms of bialgebras.

5.5 Primitive subspace of $\mathscr{C}$

5.5.1 Description of the primitive elements in $\mathscr{C}$

By definition, connected closed diagrams are primitive with respect to the coproduct δ. It may sound surprising that the converse is also true:

Theorem 5.20 (Bar-Natan 1995a). *The primitive space $\mathscr{P}$ of the bialgebra $\mathscr{C}$ coincides with the linear span of connected closed diagrams.*

Note the contrast of this straightforward characterization of the primitive space in $\mathscr{C}$ with the case of chord diagrams.

Proof If the primitive space $\mathscr{P}$ were bigger than the span of connected closed diagrams, then, according to the Milnor–Moore Theorem (Theorem A.32), it would contain an element that cannot be represented as a polynomial in connected closed diagrams. Therefore, to prove the theorem it is enough to show that every closed diagram is a polynomial in connected diagrams. This can be done by induction on the number of legs of a closed diagram C. Suppose that the diagram C consists of several connected components (see Definition 5.17). The STU relation tells us that we can freely interchange the legs of C modulo closed diagrams with fewer legs. Using such permutations, we can separate the connected components of C. This means that modulo closed diagrams with fewer legs C is equal to the product of its connected components. $\square$

5.5.2 Filtration of $\mathscr{P}_n$ by the number of legs

The primitive space $\mathscr{P}_n$ cannot be graded by the number of legs k, because the STU relation is not homogeneous with respect to k. However, it can be *filtered*:

$$0 = \mathscr{P}_n^1 \subseteq \mathscr{P}_n^2 \subseteq \mathscr{P}_n^3 \subseteq \cdots \subseteq \mathscr{P}_n^{n+1} = \mathscr{P}_n,$$

where $\mathscr{P}_n^k$ is the subspace of $\mathscr{P}_n$ spanned by connected closed diagrams with *at most k* legs.

The connectedness of a closed diagram with $2n$ vertices implies that the number of its legs cannot be bigger than $n + 1$. That is why the filtration ends at the term $\mathscr{P}_n^{n+1}$.

The following facts about the filtration are known.

- (Chmutov and Varchenko 1997) The filtration stabilizes even sooner. Namely, $\mathscr{P}_n^n = \mathscr{P}_n$ for even n, and $\mathscr{P}_n^{n-1} = \mathscr{P}_n$ for odd n. Moreover, for even n the quotient space $\mathscr{P}_n^n / \mathscr{P}_n^{n-1}$ has dimension one and is generated by the *wheel* $\overline{w}_n$ with n spokes:

$$\overline{w}_n = \qquad \qquad$$

n spokes

  This fact is related to the Melvin-Morton conjecture (see Section 14.1 and Exercise 5.13).

- (Dasbach 1996) The quotient space $\mathscr{P}_n^{n-1} / \mathscr{P}_n^{n-2}$ has dimension $[n/6] + 1$ for odd n, and 0 for even n.

- (Dasbach 1998) For even n

$$\dim(\mathscr{P}_n^{n-2} / \mathscr{P}_n^{n-3}) = \left[\frac{(n-2)^2 + 12(n-2)}{48} \right] + 1.$$

- For small degrees the dimensions of the quotient spaces $\mathscr{P}_n^k / \mathscr{P}_n^{k-1}$ were calculated by J. Kneissler (1997) (empty entries in the table are zeroes):

n\k	1	2	3	4	5	6	7	8	9	10	11	12	dim $\mathscr{P}_n$
1	1												1
2	1												1
3	1												1
4	1	1											2
5	2	1											3
6	2	2	1										5
7	3	3	2										8
8	4	4	3	1									12
9	5	6	5	2									18
10	6	8	8	4	1								27
11	8	10	11	8	2								39
12	9	13	15	12	5	1							55

5.5.3 Detecting the knot orientation

One may notice that in the table above, all entries with odd k vanish. This means that any connected closed diagram with an odd number of legs is equal to a suitable linear combination of diagrams with fewer legs. This observation is closely related to the problem of distinguishing knot orientation by Vassiliev invariants. The existence of the universal Vassiliev invariant given by the Kontsevich integral reduces the problem of detecting the knot orientation to a purely combinatorial problem. Denote by τ the operation of reversing the orientation of the Wilson loop of a chord diagram; its action is equivalent to a mirror reflection of the diagram as a planar picture. This operation descends to $\mathscr{A}'$; we call an element of $\mathscr{A}'$ *symmetric*, if τ acts on it as identity. Then, Vassiliev invariants do not distinguish the orientation of knots if and only if all chord diagrams are symmetric: $D = \tau(D)$ for all $D \in \mathscr{A}'$. The following theorem translates this fact into the language of primitive subspaces.

Theorem 5.21. *Vassiliev invariants do not distinguish the orientation of knots if and only if $\mathscr{P}_n^k = \mathscr{P}_n^{k-1}$ for any odd k and arbitrary n.*

To prove the theorem we need to reformulate the question whether $D = \tau(D)$ in terms of closed diagrams. Reversing the orientation of the Wilson loop on closed diagrams should be done with some caution; see the discussion on page 116. The correct way of doing it is carrying the operation τ from chord diagrams to closed diagrams by the isomorphism $\lambda : \mathscr{A} \to \mathscr{C}$; then we have the following assertion:

Lemma 5.22. *Let $P = \left(\begin{array}{c} P' \end{array} \right)$ be a closed diagram with k external vertices.*

Then

$$\tau(P) = (-1)^k \left(\begin{array}{c} P' \end{array} \right).$$

Proof Represent P as a linear combination of chord diagrams using STU relations, and then reverse the orientation of the Wilson loop of all chord diagrams obtained. After that, convert the resulting linear combination back to a closed diagram. Each application of the STU relation multiplies the result by -1 because of the reversed Wilson loop (see page 116). In total, we have to perform the STU relation $2n - k$ times, where n is the degree of P. Therefore, the result gets multiplied by $(-1)^{2n-k} = (-1)^k$. $\square$

In the particular case $k = 1$ the lemma asserts that $\mathscr{P}_n^1 = 0$ for all n – this fact appeared earlier as the last exercise in Section 5.1.

The operation $\tau : \mathscr{C} \to \mathscr{C}$ is, in fact, an algebra automorphism, $\tau(C_1 \cdot C_2) = \tau(C_1) \cdot \tau(C_2)$. Therefore, to check the equality $\tau = \mathrm{id}_{\mathscr{C}}$ it is enough to check it on the primitive subspace, that is, determine whether $P = \tau(P)$ for every connected closed diagram P.

Corollary of the Lemma. *Let $P \in \mathscr{P}^k = \bigoplus\limits_{n=1}^{\infty} \mathscr{P}_n^k$ be a connected closed diagram with k legs. Then $\tau(P) \equiv (-1)^k P \mod \mathscr{P}^{k-1}$.*

Proof of the Corollary. Rotating the Wilson loop in 3-space by 180° about the vertical axis, we get:

$$\tau(P) = (-1)^k \left(\begin{array}{c} P' \\ \cdots \end{array} \right) = (-1)^k \left(\begin{array}{c} P' \\ \cdots \end{array} \right).$$

The STU relations allow us to permute the legs modulo diagrams with fewer number of legs. Applying this procedure to the last diagram, we can straighten out all legs and get $(-1)^k P$. $\qquad\qquad\square$

Proof of the Theorem. Suppose that the Vassiliev invariants do not distinguish the orientation of knots. Then $\tau(P) = P$ for every connected closed diagram P. In particular, for a diagram P with an odd number of legs k we have $P \equiv -P$ mod $\mathscr{P}^{k-1}$. Hence, $2P \equiv 0 \mod \mathscr{P}^{k-1}$, which means that P is equal to a linear combination of diagrams with fewer legs, and therefore $\dim(\mathscr{P}_n^k / \mathscr{P}_n^{k-1}) = 0$.

Conversely, suppose that Vassiliev invariants do distinguish the orientation. Then there is a connected closed diagram P such that $\tau(P) \neq P$. Choose such P with the smallest possible number of legs k. Let us show that k cannot be even. Consider $X = P - \tau(P) \neq 0$. Since τ is an involution, $\tau(X) = -X$. But, in the case of even k, the non-zero element X has fewer legs than k, and $\tau(X) = -X \neq X$, so k cannot be minimal. Therefore, the minimal such k is odd, and $\dim(\mathscr{P}_n^k / \mathscr{P}_n^{k-1}) \neq 0$. $\qquad\qquad\square$

Exercise. Check that, for invariants of fixed degree, the theorem can be specialized as follows. *Vassiliev invariants of degree $\leqslant n$ do not distinguish the orientation of knots if and only if $\mathscr{P}_m^k = \mathscr{P}_m^{k-1}$ for any odd k and arbitrary $m \leqslant n$.*

Exercise. Similarly to the filtration in the primitive space $\mathscr{P}$, one can introduce the leg filtration in the whole space $\mathscr{C}$. Prove the following version of the above theorem: *Vassiliev invariants of degree n do not distinguish the orientation of knots if and only if $\mathscr{C}_n^k = \mathscr{C}_n^{k-1}$ for any odd k and arbitrary n.*

5.6 Open Jacobi diagrams

The subject of this section is the combinatorial bialgebra $\mathscr{B}$ which is isomorphic to the bialgebras $\mathscr{A}$ and $\mathscr{C}$ as a vector space and as a coalgebra, but has a different natural multiplication. This leads to the remarkable fact that in the vector space $\mathscr{A} \simeq \mathscr{C} \simeq \mathscr{B}$ there are two multiplications both compatible with one and the same coproduct.

5.6.1 The definition of open Jacobi diagrams

Definition 5.23. An *open Jacobi diagram* is a graph with 1- and 3-valent vertices, cyclic order of (half-)edges at every 3-valent vertex and with at least one 1-valent vertex in every connected component.

An open diagram is not required to be connected. It may have loops and multiple edges. We shall see later that, modulo the natural relations any diagram with a loop vanishes. However, it is important to include the diagrams with loops in the definition, because the loops may appear during natural operations on open diagrams, and it is exactly because of this fact that we introduce the cyclic order on half-edges, not on whole edges.

The total number of vertices of an open diagram is even. Half of this number is called the *degree* (or *order*) of an open diagram. We denote the set of all open diagrams of degree n by $\mathbf{B}_n$. The univalent vertices will sometimes be referred to as *legs*.

In the literature, open diagrams are also referred to as *1-3-valent diagrams*, *Jacobi diagrams*, *web diagrams* and *Chinese characters*.

Definition 5.24. An isomorphism between two open diagrams is a one-to-one correspondence between their respective sets of vertices and half-edges that preserves the vertex-edge adjacency and the cyclic order of half-edges at every vertex.

Example 5.25. Below is the complete list of open diagrams of degree 1 and 2, up to isomorphism just introduced.

Most of the elements listed above will be of no importance to us, as they are killed by the following definition.

Definition 5.26. The space of open diagrams of degree n is the quotient space

$$\mathscr{B}_n := \langle \mathbf{B}_n \rangle / \langle \mathrm{AS},\ \mathrm{IHX} \rangle,$$

where $\langle \mathbf{B}_n \rangle$ is the vector space formally generated by all open diagrams of degree n and $\langle \mathrm{AS, \ IHX} \rangle$ stands for the subspace spanned by all AS and IHX relations (see Section 5.2). By definition, $\mathscr{B}_0$ is one-dimensional, spanned by the empty diagram, and $\mathscr{B} := \bigoplus_{n=0}^{\infty} \mathscr{B}_n$.

5.6.2 Criteria for the vanishing of a diagram

Just as in the case of closed diagrams (Section 5.2.3), the AS relation immediately implies that any open diagram with a loop (⟳) vanishes in $\mathscr{B}$. Let us give a most general statement of this observation – valid, in fact, both for open and for closed Jacobi diagrams.

Definition 5.27. An *anti-automorphism* of a Jacobi diagram $b \in \mathbf{B}_n$ is a graph automorphism of b such that the cyclic order of half-edges is reversed in an odd number of vertices.

Lemma 5.28. *If a diagram $b \in \mathbf{B}_n$ admits an anti-automorphism, then $b = 0$ in the vector space $\mathscr{B}$.*

Proof Indeed, it follows from the definitions that in this case $b = -b$. □

Example 5.29.

$$= \ 0.$$

Exercise. Show that $\dim \mathscr{B}_1 = 1$, $\dim \mathscr{B}_2 = 2$.

The relations AS and IHX imply the generalized IHX relation, or Kirchhoff law (Lemma 5.11) and many other interesting identities among the elements of the space $\mathscr{B}$. Some of them are proved in Section 7.2.2 in the context of the algebra Γ. Here is one more assertion that makes sense only in $\mathscr{B}$, as its formulation refers to univalent vertices (legs).

Lemma 5.30. *If $b \in \mathscr{B}$ is a diagram with an odd number of legs, all of which are attached to one and the same edge, then $b = 0$ modulo AS and IHX relations.*

Example 5.31.

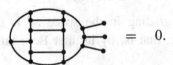

$$= \ 0.$$

Note that in this example the diagram does not have an anti-automorphism, so the previous lemma does not apply.

Proof   Any diagram satisfying the premises of the lemma can be put into the form on the left of the next picture. Then by the generalized IHX relation it is equal to the diagram on the right which obviously possesses an anti-automorphism and therefore is equal to zero:

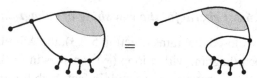

where the grey region is an arbitrary subdiagram.   □

In particular, any diagram with exactly one leg vanishes in $\mathcal{B}$. This is an exact counterpart of the corresponding property of closed diagrams (see the last exercise of Section 5.1 on page 118); both facts are, furthermore, equivalent to each other in view of the isomorphism $\mathcal{C} \cong \mathcal{B}$ that we shall speak about later (in Section 5.7).

Conjecture 5.32.   *Any diagram with an odd number of legs is 0 in $\mathcal{B}$.*

This important conjecture is equivalent to the conjecture that Vassiliev invariants do not distinguish the orientation of knots (see Section 5.8.4).

5.6.3 Gradings on $\mathcal{B}$

Relations AS and IHX, unlike STU, preserve the separation of vertices into 1- and 3-valent. Therefore, the space $\mathcal{B}$ has a much finer grading than $\mathcal{A}$. Apart from the main grading by half the number of vertices, indicated by the subscript in $\mathcal{B}$, it also has a grading by the number of univalent vertices

$$\mathcal{B} = \bigoplus_{n} \bigoplus_{k} \mathcal{B}_n^k,$$

indicated by the superscript in $\mathcal{B}$, so that $\mathcal{B}_n^k$ is the subspace spanned by all diagrams with k legs and $2n$ vertices in total.

For disconnected diagrams the second grading can, in turn, be refined to a multigrading by the number of legs in each connected component of the diagram:

$$\mathcal{B} = \bigoplus_{n} \bigoplus_{k_1 \leqslant \ldots \leqslant k_m} \mathcal{B}_n^{k_1,\ldots,k_m}.$$

Yet another important grading in the space $\mathcal{B}$ is the grading by the number of *loops* in a diagram, that is, by its first Betti number. In fact, we have a decomposition:

$$\mathcal{B} = \bigoplus_n \bigoplus_k \bigoplus_l {}^l\mathcal{B}_n^k,$$

where l can also be replaced by m (the number of connected components) because of the relation $l + k = n + m$, which can be proved by a simple argument involving the Euler characteristic.

The abundance of gradings makes the work with the space $\mathcal{B}$ more convenient than with $\mathcal{C}$, although both are isomorphic, as we shall soon see.

5.6.4 The bialgebra structure on $\mathcal{B}$

Both the product and the coproduct in the vector space $\mathcal{B}$ are defined in a rather straightforward way. We first define the product and coproduct on diagrams, then extend the operations by linearity to the free vector space spanned by the diagrams, and then note that they are compatible with the AS and IHX relations and thus descend to the quotient space $\mathcal{B}$.

Definition 5.33. The product of two open diagrams is their disjoint union.

Example 5.34.

Definition 5.35. Let D be an open diagram and $[D]$ be the set of its connected components. For a subset $J \subseteq [D]$, denote by D_J the union of the components that belong to J and by $D_{\bar{J}}$, the union of the components that do not belong to J. We set

$$\delta(D) := \sum_{J \subseteq [D]} D_J \otimes D_{\bar{J}}.$$

Example 5.36.

As the relations in $\mathcal{B}$ do not intermingle different connected components of a diagram, the product of an AS or IHX combination of diagrams by an arbitrary open diagram belongs to the linear span of the relations of the same types. Also, the coproduct of any AS or IHX relation vanishes modulo these relations. Therefore, we have well-defined algebraic operations in the space $\mathcal{B}$, and they are evidently compatible with each other. The space $\mathcal{B}$ thus becomes a graded bialgebra.

5.7 Linear isomorphism $\mathscr{B} \simeq \mathscr{C}$

In this section we construct a linear isomorphism between vector spaces $\mathscr{B}_n$ and $\mathscr{C}_n$. The question whether it preserves multiplication will be discussed later (Section 5.8). Our exposition follows Bar-Natan (1995a), with some details omitted, but some examples added.

5.7.1 The symmetrization map

To convert an open diagram into a closed diagram, we join all of its univalent vertices by a Wilson loop. Fix k distinct points on the circle. For an open diagram with k legs $D \in \mathbf{B}_n^k$ there are $k!$ ways of glueing its legs to the Wilson loop at these k points, and we set $\chi(D)$ to be equal to the arithmetic mean of all the resulting closed diagrams. Thus we get the *symmetrization map*

$$\chi : \mathbf{B} \to \mathscr{C}.$$

For example,

Scrutinizing these pictures, one can see that 16 out of 24 summands are equivalent to the first diagram, while the remaining 8 are equivalent to the second one. Therefore,

Exercise. Express this element via chord diagrams, using the isomorphism $\mathscr{C} \simeq \mathscr{A}$.

Answer:

Theorem 5.37. *The symmetrization map* $\chi : \mathbf{B} \to \mathscr{C}$ *descends to a linear map* $\chi : \mathscr{B} \to \mathscr{C}$, *which is a graded isomorphism between the vector spaces $\mathscr{B}$ and $\mathscr{C}$.*

The theorem consists of two parts:

- Easy part: χ is well-defined.
- Difficult part: χ is bijective.

The proof of bijectivity of χ is difficult because not every closed diagram can be obtained by a symmetrization of an open diagram. For example, the diagram is not a symmetrization of any open diagram, even though it looks very much symmetric. Notice that symmetrizing the internal graph of this diagram we get 0.

Easy part of the theorem. To prove the easy part, we must show that the AS and IHX combinations of open diagrams go to 0 in the space $\mathscr{C}$. This follows from the fact that the STU relations imply both IHX and AS (see page 119).

Difficult part of the theorem. To prove the difficult part, we construct a linear map τ from $\mathscr{C}$ to $\mathscr{B}$, inverse to χ. This will be done inductively by the number of legs of the diagrams. We shall write τ_k for the restriction of τ to the subspace spanned by diagrams with at most k legs.

There is only one way to attach the only leg of an open diagram to the Wilson loop. Therefore, we can define τ_1 on a closed diagram C with one leg as the internal graph of C. (In fact, both open and closed diagrams with one leg are all zero in $\mathscr{B}$ and $\mathscr{C}$ respectively; see the last exercise of Section 5.1 and Lemma 5.30.) For diagrams with two legs the situation is similar. Every closed diagram with two legs is a symmetrization of an open diagram, since there is only one cyclic order on the set of two elements. For example, is the symmetrization of the diagram . Therefore, for a closed diagram C with two legs we can define $\tau_2(C)$ to be the internal graph of C.

In what follows, we shall often speak of the action of the symmetric group S_k on closed diagrams with k legs. This action preserves the internal graph of a closed diagram and permutes the points where the legs of the internal graph are attached to the Wilson loop. Strictly speaking, to define this action we need the legs of the diagrams to be numbered. We shall always assume that such numbering is chosen; the particular form of this numbering will be irrelevant.

The difference of a closed diagram D and the same diagram whose legs are permuted by some permutation σ, is equivalent, modulo STU relations, to a combination of diagrams with a smaller number of external vertices. For every given D and σ we fix such a linear combination.

Assuming that the map τ is defined for closed diagrams having less than k legs, we define it for a diagram D with exactly k legs by the formula:

$$\tau_k(D) = \widetilde{D} + \frac{1}{k!} \sum_{\sigma \in S_k} \tau_{k-1}(D - \sigma(D)),\tag{5.1}$$

where $\widetilde{D}$ is the internal graph of D, and $D - \sigma(D)$ is represented as a combination of diagrams with less than k legs according to the choice above.

For example, we know that $\tau\!\left(\text{◯} \right) = \text{•—◯—•}$, and we want to find $\tau\!\left(\text{⊘} \right)$. By the above formula, we have:

$$\tau_3\!\left(\text{⊘}\right) = \text{Y} + \tfrac{1}{6}\left(\tau_2\!\left(\text{⊘}-\text{⊘}\right) + \tau_2\!\left(\text{⊘}-\text{⊘}\right)\right.$$

$$+\tau_2\!\left(\text{⊘}-\text{⊘}\right) + \tau_2\!\left(\text{⊘}-\text{⊘}\right)$$

$$\left.+\tau_2\!\left(\text{⊘}-\text{⊘}\right) + \tau_2\!\left(\text{⊘}-\text{⊘}\right)\right)$$

$$= \tfrac{1}{2}\tau_2\!\left(\text{◯}\right) = \tfrac{1}{2}\,\text{•—◯—•}.$$

We have to prove the following assertions:

(i) The value $\tau_{k-1}(D - \sigma(D))$ in the formula (5.1) does not depend on the presentation of $D - \sigma(D)$ as a combination of diagrams with a smaller number of external vertices.

(ii) The map τ respects STU relations.

(iii) $\chi \circ \tau = \mathrm{id}_{\mathscr{C}}$ and τ is surjective.

The first two assertions imply that τ is well-defined and the third means that τ is an isomorphism. The rest of the section is dedicated to the proof of these statements.

In the vector space spanned by all closed diagrams (with no relations imposed), let $\mathscr{D}^k$ be the subspace spanned by all diagrams with at most k external vertices. We have a chain of inclusions

$$\mathscr{D}^0 \subset \mathscr{D}^1 \subset \mathscr{D}^2 \subset \dots.$$

We denote by $\mathscr{I}^k$ the subspace in $\mathscr{D}^k$ spanned by all STU, IHX and antisymmetry relations that do not involve diagrams with more than k external vertices.

5.7.2 *Action of permutations on closed diagrams*

The action of the symmetric group S_k on closed diagrams with k legs can be represented graphically as the "composition" of a closed diagram with the diagram of the permutation:

$$k = 4 \; ; \qquad \sigma = (4132) = \text{⊠} \; ; \qquad D = \text{⬭} \; ;$$

$$\sigma D = \text{⬚} = \text{⬭}.$$

Lemma 5.38. *Let* $D \in \mathscr{D}^k$.

- *Modulo* $\mathscr{I}^k$, *the difference* $D - \sigma D$ *belongs to* $\mathscr{D}^{k-1}$.
- *Any choice* U_σ *of a presentation of* σ *as a product of transpositions determines in a natural way an element* $\Gamma_D(U_\sigma) \in \mathscr{D}^{k-1}$ *such that*

$$\Gamma_D(U_\sigma) \equiv D - \sigma D \quad \text{mod } \mathscr{I}^k.$$

- *Furthermore, if* U_σ *and* U'_σ *are two such presentations, then* $\Gamma_D(U_\sigma)$ *is equal to* $\Gamma_D(U'_\sigma)$ *modulo* $\mathscr{I}^{k-1}$.

This is lemma 5.5 from Bar-Natan (1995a). Rather than giving the details of the proof (which can be found in Bar-Natan 1995a), we illustrate it on a concrete example.

Take the permutation $\sigma = (4132)$ and let D be the diagram considered above. Choose two presentations of σ as a product of transpositions:

$$U_\sigma = (34)(23)(34)(12) = \text{⊠} \; ; \qquad U'_\sigma = (23)(34)(23)(12) = \text{⊠} \; .$$

(Here the product is drawn in such a way that reading it from the left to the right corresponds to going from the bottom to the top.)

For each of these products we represent $D - \sigma D$ as a sum:

$$\begin{aligned} D - \sigma D &= (D - {}_{(12)}D) + ({}_{(12)}D - {}_{(34)(12)}D) + ({}_{(34)(12)}D - {}_{(23)(34)(12)}D) \\ &\quad + ({}_{(23)(34)(12)}D - {}_{(34)(23)(34)(12)}D) \end{aligned}$$

and

$$\begin{aligned} D - \sigma D &= (D - {}_{(12)}D) + ({}_{(12)}D - {}_{(23)(12)}D) + ({}_{(23)(12)}D - {}_{(34)(23)(12)}D) \\ &\quad + ({}_{(34)(23)(12)}D - {}_{(23)(34)(23)(12)}D). \end{aligned}$$

Here, the two terms in every pair of parentheses differ only by a transposition of two neighbouring legs, so their difference is the right-hand side of an STU relation.

Modulo the subspace $\mathscr{I}^4$ each difference can be replaced by the corresponding left-hand side of the STU relation, which is a diagram in $\mathscr{D}^3$. We get

$$\Gamma_D(U_\sigma) = \left(\text{\small ⬡}\right) + \left(\text{\small ⬡}\right) + \left(\text{\small ⬡}\right) + \left(\text{\small ⬡}\right),$$

$$\Gamma_D(U_\sigma') = \left(\text{\small ⬡}\right) + \left(\text{\small ⬡}\right) + \left(\text{\small ⬡}\right) + \left(\text{\small ⬡}\right).$$

Now the difference $\Gamma_D(U_\sigma) - \Gamma_D(U_\sigma')$ equals

$$\left(\left(\text{\small ⬡}\right) - \left(\text{\small ⬡}\right)\right) + \left(\left(\text{\small ⬡}\right) - \left(\text{\small ⬡}\right)\right) + \left(\left(\text{\small ⬡}\right) - \left(\text{\small ⬡}\right)\right).$$

Using the STU relation in $\mathscr{I}^3$ we can represent it in the form

$$\Gamma_D(U_\sigma) - \Gamma_D(U_\sigma') = \left(\text{\small ⬡}\right) + \left(\text{\small ⬡}\right) - \left(\text{\small ⬡}\right) = 0$$

which is zero because of the IHX relation.

5.7.3 *Proof of assertions (i) and (ii) on page 136*

Let us assume that the map τ, defined by the formula (5.1), is (a) well-defined on $\mathscr{D}^{k-1}$ and (b) vanishes on $\mathscr{I}^{k-1}$.

Define $\tau'(D)$ to be equal to $\tau(D)$ if $D \in \mathscr{D}^{k-1}$, and if $D \in \mathscr{D}^k - \mathscr{D}^{k-1}$ set

$$\tau'(D) = \tilde{D} + \frac{1}{k!} \sum_{\sigma \in S_k} \tau(\Gamma_D(U_\sigma)).$$

Lemma 5.38 means that for any given $D \in \mathscr{D}^k$ with exactly k external vertices $\tau(\Gamma_D(U_\sigma))$ does not depend on a specific presentation U_σ of the permutation σ as a product of transpositions. Therefore, τ' gives a well-defined map $\mathscr{D}^k \to \mathscr{B}$.

Let us now show that τ' vanishes on $\mathscr{I}^k$. It is obvious that τ' vanishes on the IHX and antisymmetry relations since these relations hold in $\mathscr{B}$. So we only need to check the STU relation which relates a diagram D^{k-1} with $k-1$ external vertices and the corresponding two diagrams D^k and $U_i D^k$ with k external vertices, where

U_i is a transposition $U_i = (i, i + 1)$. Let us apply τ' to the right-hand side of the STU relation:

$$\tau'(D^k - U_i D^k) = \widetilde{D^k} + \tfrac{1}{k!} \sum_{\sigma \in S_k} \tau(\Gamma_{D^k}(U_\sigma))$$
$$- \widetilde{U_i D^k} - \tfrac{1}{k!} \sum_{\sigma' \in S_k} \tau(\Gamma_{U_i D^k}(U_{\sigma'})).$$

Note that $\widetilde{D^k} = \widetilde{U_i D^k}$. Reparametrizing the first sum, we get

$$\tau'(D^k - U_i D^k) = \frac{1}{k!} \sum_{\sigma \in S_k} \tau(\Gamma_{D^k}(U_\sigma U_i) - \Gamma_{U_i D^k}(U_\sigma)).$$

Using the obvious identity $\Gamma_D(U_\sigma U_i) = \Gamma_D(U_i) + \Gamma_{U_i D^k}(U_\sigma)$ and the fact that $D^{k-1} = \Gamma_D(U_i)$, we now obtain

$$\tau'(D^k - U_i D^k) = \frac{1}{k!} \sum_{\sigma \in S_k} \tau(D^{k-1}) = \tau(D^{k-1}) = \tau'(D^{k-1}),$$

which means that τ' vanishes on the STU relation, and, hence, on the whole of $\mathscr{I}^k$.

Now, it follows from the second part of Lemma 5.38 that $\tau' = \tau$ on $\mathscr{D}^k$. In particular, this means that τ is well-defined on $\mathscr{D}^k$ and vanishes on $\mathscr{I}^k$. By induction, this implies the assertions (i) and (ii).

5.7.4 Proof of assertion (iii) on page 136

Assume that $\chi \circ \tau$ is the identity for diagrams with at most $k-1$ legs. Take $D \in \mathscr{D}^k$ representing an element of $\mathscr{C}$. Then

$$(\chi \circ \tau)(D) = \chi\left(\widetilde{D} + \tfrac{1}{k!} \sum_{\sigma \in S_k} \tau(\Gamma_D(U_\sigma))\right)$$
$$= \tfrac{1}{k!} \sum_{\sigma \in S_k} (\sigma D + (\chi \circ \tau)(\Gamma_D(U_\sigma))).$$

Since $\Gamma_D(U_\sigma)$ is a combination of diagrams with at most $k-1$ legs, by the induction hypothesis $\chi \circ \tau(\Gamma_D(U_\sigma)) = \Gamma_D(U_\sigma)$ and, hence,

$$(\chi \circ \tau)(D) = \frac{1}{k!} \sum_{\sigma \in S_k} (\sigma D + \Gamma_D(U_\sigma)) = \frac{1}{k!} \sum_{\sigma \in S_k} (\sigma D + D - \sigma D) = D.$$

The surjectivity of τ is clear from the definition, so we have established that χ is a linear isomorphism between $\mathscr{B}$ and $\mathscr{C}$. $\qquad\square$

5.8 More on the relation between $\mathscr{B}$ and $\mathscr{C}$

5.8.1 Symmetrization and products

It is easy to check that the isomorphism χ is compatible with the coproduct in the algebras $\mathscr{B}$ and $\mathscr{C}$. (**Exercise**: pick a decomposable diagram $b \in \mathscr{B}$ and check that $\delta_C(\chi(b))$ and $\chi(\delta_B(b))$ coincide.) However, χ is *not* compatible with the product. For example,

$$\chi(\bullet\!\!-\!\!\bullet) = \bigominus.$$

The square of the element $\bullet\!\!-\!\!\bullet$ in $\mathscr{B}$ is $\,\vcenter{\hbox{\rule{0pt}{1pt}}}\!\!=\!\!\vcenter{\hbox{\rule{0pt}{1pt}}}\,$. However, the corresponding element of $\mathscr{C}$

$$\chi(\,\vcenter{\hbox{}}\!\!=\!\!\vcenter{\hbox{}}\,) = \frac{1}{3}\,\bigotimes + \frac{2}{3}\,\bigoplus$$

is not equal to the square of $\bigominus$.

We can, of course, carry the natural multiplication of the algebra $\mathscr{B}$ to the algebra $\mathscr{C}$ with the help of the isomorphism χ, thus obtaining a bialgebra with two different products, both compatible with one and the same coproduct.

5.8.2 Primitive elements

By definition, any connected diagram $p \in \mathscr{B}$ is primitive. Similar to the case of the closed diagrams (page 126) we have:

Theorem 5.39. *The primitive space of the bialgebra $\mathscr{B}$ is spanned by connected open diagrams.*

Proof The same argument as in the case of the closed diagrams, with the simplification that in the present case we do not have to prove that every element of $\mathscr{B}$ has a polynomial expression in terms of connected diagrams: this holds by definition. $\qquad\square$

Although the isomorphism χ does not respect the multiplication, the two algebras $\mathscr{B}$ and $\mathscr{C}$ are isomorphic. This is clear from what we know about their structure: by the Milnor–Moore theorem both algebras are commutative polynomial algebras over the corresponding primitive subspaces. But the primitive subspaces coincide, since χ preserves the coproduct! An explicit algebra isomorphism between $\mathscr{B}$ and $\mathscr{C}$ will be the subject of Section 11.3.

Situations of this kind appear in the theory of Lie algebras. Namely, the bialgebra of invariants in the symmetric algebra of a Lie algebra $\mathfrak{g}$ has a natural map into the centre of the universal enveloping algebra of $\mathfrak{g}$. This map, which is very similar in spirit to the symmetrization map χ, is an isomorphism of coalgebras, but does not respect the multiplication. In fact, this analogy is anything but superficial. It turns out that the algebra $\mathscr{C}$ is isomorphic to the centre of the universal enveloping algebra for a certain Casimir Lie algebra in a certain tensor category. For further details see Hinich and Vaintrob (2002).

5.8.3 Unframed version of $\mathscr{B}$

The unframed version of the algebras $\mathscr{A}$ and $\mathscr{C}$ are obtained by taking the quotient by the ideal generated by the diagram with one chord Θ. Although the product in $\mathscr{B}$ is different, it is easy to see that multiplication in $\mathscr{C}$ by Θ corresponds to multiplication in $\mathscr{B}$ by the *strut s*: the diagram of degree 1 consisting of two univalent vertices and one edge. Therefore, the unframed version of the algebra $\mathscr{B}$ is its quotient by the ideal generated by s and we have: $\mathscr{B}' := \mathscr{B}/(s) \cong \mathscr{C}/(\Theta) =: \mathscr{C}'$.

5.8.4 Grading in $\mathscr{PB}$ and filtration in $\mathscr{PC}$

The space of primitive elements $\mathscr{PB}$ is carried by χ isomorphically onto $\mathscr{PC}$. The space $\mathscr{PC} = \mathscr{P}$ is filtered (see Section 5.5.2), the space $\mathscr{PB}$ is graded (see Section 5.6.3). It turns out that χ intertwines the grading on $\mathscr{B}$ with the filtration on $\mathscr{C}$; as a corollary, the filtration on $\mathscr{PC}$ comes from a grading. Indeed, the definition of χ and the construction of the inverse mapping τ imply two facts:

$$\chi(\mathscr{PB}^i) \subset \mathscr{P}^i \subset \mathscr{P}^k, \text{ if } i < k,$$

$$\tau(\mathscr{P}^k) \subset \bigoplus_{i=1}^{k} \mathscr{PB}^k.$$

Therefore, we have an isomorphism

$$\tau : \mathscr{P}_n^k \longrightarrow \mathscr{PB}_n^1 \oplus \mathscr{PB}_n^2 \oplus \ldots \oplus \mathscr{PB}_n^{k-1} \oplus \mathscr{PB}_n^k,$$

and, hence, an isomorphism $\mathscr{P}_n^k / \mathscr{P}_n^{k-1} \cong \mathscr{PB}_n^k$.

Using this fact, we can give an elegant reformulation of the theorem about detecting the orientation of knots (Section 5.5.3):

Corollary 5.40. *Vassiliev invariants do distinguish the orientation of knots if and only if $\mathscr{PB}_n^k \neq 0$ for an odd k and some n.*

Let us clarify that by saying that Vassiliev invariants do distinguish the orientation of knots we mean that there exists a knot K non-equivalent to its inverse K^* and a Vassiliev invariant f such that $f(K) \neq f(K^*)$.

Exercise. Check that in the previous statement the letter $\mathscr{P}$ can be dropped: Vassiliev invariants do distinguish the orientation of knots if and only if $\mathscr{B}_n^k \neq 0$ for an odd k and some n.

The relation between $\mathscr{C}$ and $\mathscr{B}$ in this respect can also be stated in the form of a commutative diagram:

$$
\begin{array}{ccc}
\mathscr{B} & \xrightarrow{\ \chi\ } & \mathscr{C} \\
{\scriptstyle \tau_B}\Big\downarrow & & \Big\downarrow{\scriptstyle \tau_C} \\
\mathscr{B} & \xrightarrow[\ \chi\ ]{} & \mathscr{C}
\end{array}
$$

where χ is the symmetrization isomorphism, τ_C is the orientation reversing map in $\mathscr{C}$ defined by Lemma 5.22, while τ_B on an individual diagram from $\mathscr{B}$ acts as multiplication by $(-1)^k$ where k is the number of legs. The commutativity of this diagram is a consequence of the corollary to the above mentioned lemma (see page 129).

5.9 The three algebras in small degrees

Here is a comparative table which displays some linear bases of the algebras $\mathscr{A}$, $\mathscr{C}$ and $\mathscr{B}$ in small degrees.

n	$\mathscr{A}$	$\mathscr{C}$	$\mathscr{B}$
0			$\varnothing$
1			
2			
3			
4			

In every order up to 4, for each of the three algebras, this table displays a basis of the corresponding homogeneous component. Starting from order 2, decomposable elements (products of elements of smaller degree) appear on the left, while the new indecomposable elements appear on the right. The bases of $\mathscr{C}$ and $\mathscr{B}$ are chosen to consist of primitive elements and their products. We remind that the difference between the $\mathscr{A}$ and $\mathscr{C}$ columns is notational rather than anything else, since chord diagrams are a special case of closed Jacobi diagrams, the latter can be considered as linear combinations of the former, and the two algebras are in any case isomorphic.

5.10 Jacobi diagrams for tangles

In order to define chord diagrams and, more generally, closed Jacobi diagrams, for arbitrary tangles it suffices to make only minor adjustments to the definitions. Namely, one simply replaces the Wilson loop with an arbitrary oriented manifold of dimension one (the *skeleton* of the Jacobi diagram). In the 4-term relations the points of attachment of chords are allowed to belong to different components of the skeleton, while the STU relations remain the same.

The Vassiliev invariants for tangles with a given skeleton can be described with the help of chord diagrams or closed diagrams with the same skeleton; in fact the Vassiliev–Kontsevich Theorem is valid for tangles and not only for knots.

Open Jacobi diagrams can also be defined for arbitrary tangles. If we consider tangles whose skeleton is not connected, the legs of corresponding open diagrams have to be labeled by the connected components of the skeleton. Moreover, for such tangles there are mixed spaces of diagrams, some of whose legs are attached to the skeleton, while others are "hanging free." Defining spaces of open and mixed diagrams for tangles is a more delicate matter than generalizing chord diagrams: here new relations, called *link relations*, may appear in addition to the STU, IHX and AS relations.

5.10.1 Jacobi diagrams for tangles

Definition 5.41. Let X be a tangle skeleton (see Section 1.6). A *tangle closed Jacobi diagram D with skeleton X* is a unitrivalent graph with a distinguished oriented subgraph identified with X, a fixed cyclic order of half-edges at each vertex not on X, and such that:

- it has no univalent vertices other than the boundary points of X;
- each connected component of D contains at least one connected component of X.

A tangle Jacobi diagram whose all vertices belong to the skeleton is called a *tangle chord diagram*. As with usual closed Jacobi diagrams, half the number of the vertices of a closed diagram is called the *degree*, or *order*, of the diagram.

Example 5.42. A tangle diagram whose skeleton consists of a line segment and a circle:

The vector space of tangle closed Jacobi diagrams with skeleton X modulo the STU relations is denoted by $\mathscr{C}(X)$, or by $\mathscr{C}(x_1, \ldots, x_n)$ where the x_i are the connected components of X. The space $\mathscr{C}_n(X)$ is the subspace of $\mathscr{C}(X)$ spanned by diagrams of degree n. It is clear that for any X the space $\mathscr{C}_n(X)$ is spanned by chord diagrams with n chords.

Two tangle diagrams are considered to be equivalent if there is a graph isomorphism between them which preserves the skeleton and the cyclic order of half-edges at the trivalent vertices outside the skeleton.

Weight systems of degree n for tangles with skeleton X can now be defined as linear functions on $\mathscr{C}_n(X)$. The Fundamental Theorem (Section 4.2) extends to the present case:

Theorem 5.43. *Each tangle weight system of degree n is a symbol of some degree n Vassiliev invariant of framed tangles.*

In fact, we shall prove this more general version of the Fundamental Theorem in Chapter 8 and deduce the corresponding statement for knots as a corollary.

Now, assume that X is a union of connected components x_i and y_j and suppose that the y_j have no boundary.

Definition 5.44. A *mixed tangle Jacobi diagram* is a unitrivalent graph with a distinguished oriented subgraph (the *skeleton*) identified with $\cup x_i$, with all univalent vertices, except those on the skeleton, labelled by elements of the set $\{y_j\}$ and a fixed cyclic order of edges at each trivalent vertex not on the skeleton, and such that each connected component either contains at least one of the x_i, or at least one univalent vertex. A *leg* of a mixed diagram is a univalent vertex that does not belong to the skeleton.

Here is an example of a mixed Jacobi diagram:

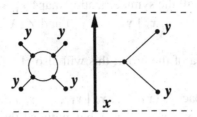

Mixed Jacobi diagrams, apart from the usual STU, IHX and antisymmetry relations, are subject to a new kind of relations, called *link relations* (Bar-Natan *et al.* 2002). To obtain a link relation, take a mixed diagram, choose one of its legs and one label y. For each y-labeled vertex, attach the chosen leg to the edge, adjacent to this vertex, sum all the results and set this sum to be equal to 0. The attaching is done according to the cyclic order as illustrated by the following picture:

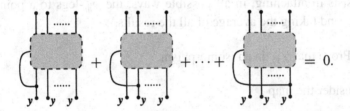

Here the shaded parts of all diagrams coincide, the skeleton is omitted from the pictures and the unlabelled legs are assumed to have labels distinct from y.

Note that when the skeleton is empty and y is the only label (that is, we are speaking about the usual open Jacobi diagrams), the link relations are an immediate consequence from the Kirchhoff law.

Now, define the vector space $\mathscr{C}(x_1, \ldots, x_n \mid y_1, \ldots, y_m)$ to be spanned by all mixed diagrams with the skeleton $\cup x_i$ and labels y_j, modulo the STU, IHX, antisymmetry and link relations.

Both closed and open diagrams are particular cases of this construction. In particular, $\mathscr{C}(x_1, \ldots, x_n \mid \emptyset) = \mathscr{C}(X)$ and $\mathscr{C}(\emptyset \mid y) = \mathscr{B}$. The latter equality justifies the notation $\mathscr{B}(y_1, \ldots, y_m)$ or just $\mathscr{B}(m)$ for the space of m-coloured open Jacobi diagrams $\mathscr{C}(\emptyset \mid y_1, \ldots, y_m)$.

Given a diagram D in $\mathscr{C}(x_1, \ldots, x_n \mid y_1, \ldots, y_m)$ we can perform "symmetrization of D with respect to the label y_m" by taking the average of all possible ways of attaching the y_m-legs of D to a circle with the label y_m. This way we get the map

$$\chi_{y_m} : \mathscr{C}(x_1, \ldots, x_n \mid y_1, \ldots, y_m) \to \mathscr{C}(x_1, \ldots, x_n, y_m \mid y_1, \ldots, y_{m-1}).$$

Theorem 5.45. *The symmetrization map χ_{y_m} is an isomorphism of vector spaces.*

In particular, applying all the symmetrization maps χ_{y_i} we get the isomorphism between the spaces $\mathscr{C}(x_1, \ldots, x_n \mid y_1, \ldots, y_m)$ and $\mathscr{C}(X \cup Y)$, where $X = \cup x_i$ and $Y = \cup y_j$.

Let us indicate the idea of the proof; this will also clarify the origin of the link relations.

Consider the vector space $\mathscr{C}(x_1, \ldots, x_n \mid y_1, \ldots, y_m^*)$ defined in the same way as the space $\mathscr{C}(x_1, \ldots, x_n \mid y_1, \ldots, y_m)$ but without the link relations on the y_m-legs. Also, define the space $\mathscr{C}(x_1, \ldots, x_n, y_m^* \mid y_1, \ldots, y_{m-1})$ in the same way as $\mathscr{C}(x_1, \ldots, x_n, y_m \mid y_1, \ldots, y_{m-1})$ but with an additional feature that all diagrams have a marked point on the component y_m.

Then we have the symmetrization map

$$\chi_{y_m^*} : \mathscr{C}(x_1, \ldots, x_n \mid y_1, \ldots, y_m^*) \rightarrow \mathscr{C}(x_1, \ldots, x_n, y_m^* \mid y_1, \ldots, y_{m-1})$$

which consists in attaching, in all possible ways, the y_m-legs to a pointed circle labelled y_m, and taking the average of all the results.

Exercise. Prove that $\chi_{y_m^*}$ is an isomorphism.

Now, consider the map

$$\mathscr{C}(x_1, \ldots, x_n, y_m^* \mid y_1, \ldots, y_{m-1}) \rightarrow \mathscr{C}(x_1, \ldots, x_n, y_m \mid y_1, \ldots, y_{m-1})$$

that simply forgets the marked point on the circle y_m. The kernel of this map is spanned by differences of diagrams of the form

(The diagrams above illustrate the particular case of four legs attached to the component y_m.) By the STU relations the above is equal to the following "attached link relation":

Exercise. Show that the symmetrization map $\chi_{y_m^*}$ identifies the subspace of link relations in $\mathscr{C}(x_1, \ldots, x_n \mid y_1, \ldots, y_m^*)$ with the subspace of $\mathscr{C}(x_1, \ldots, x_n, y_m^* \mid y_1, \ldots, y_{m-1})$ spanned by all "attached link relations."

5.10.2 Pairings on diagram spaces

There are several kinds of pairings on diagram spaces. The first pairing is induced by the product on tangles; it generalizes the multiplication in the algebra $\mathscr{C}$. This pairing exists between the vector spaces $\mathscr{C}(X_1)$ and $\mathscr{C}(X_2)$ such that the bottom part of X_1 coincides with the top part of X_2 and these manifolds can be concatenated into an oriented 1-manifold $X_1 \circ X_2$. In this case we have the bilinear map

$$\mathscr{C}(X_1) \otimes \mathscr{C}(X_2) \to \mathscr{C}(X_1 \circ X_2),$$

obtained by putting one diagram on top of another.

If X is a collection of n intervals, with one top and one bottom point on each of them, $X \circ X$ is the same thing as X and in this case we have an algebra structure on $\mathscr{C}(X)$. This is the algebra of closed Jacobi diagrams for string links on n strands. When $n = 1$, we, of course, come back to the algebra $\mathscr{C}$.

Remark 5.46. While $\mathscr{C}(X)$ is not necessarily an algebra, it is always a coalgebra with the coproduct defined in the same way as for the usual closed Jacobi diagrams:

$$\delta(D) := \sum_{J \subseteq [D]} D_J \otimes D_{\bar{J}},$$

where $[D]$ is the set of connected components of the internal graph of D.

The second multiplication is the *tensor product* of tangle diagrams. It is induced by the tensor product of tangles, and consists in placing the diagrams side by side.

There is yet another pairing on diagram spaces, which is sometimes called "inner product." For diagrams $C \in \mathscr{C}(x \mid y)$ and $D \in \mathscr{B}(y)$ define the diagram $\langle C, D \rangle_y \in \mathscr{C}(x)$ as the sum of all ways of glueing all the y-legs of C to the y-legs of D. If the numbers of y-legs of C and D are not equal, we set $\langle C, D \rangle_y$ to be zero. It may happen that in the process of glueing we get closed circles with no vertices on them (this happens if C and D contain intervals with both ends labelled by y). We set such diagrams containing circles to be equal to zero.

Lemma 5.47. *The inner product*

$$\langle \, , \, \rangle_y : \mathscr{C}(x \mid y) \otimes \mathscr{B}(y) \to \mathscr{C}(x)$$

is well-defined.

Proof We need to show that the class of the resulting diagram in $\mathcal{C}(x)$ does not change if we modify the second argument of $\langle\ ,\ \rangle_y$ by IHX or antisymmetry relations, and the first argument – by STU or link relations. This is clear for the first three kinds of relations. For link relations it follows from the Kirchhoff rule and the antisymmetry relation. For example, we have

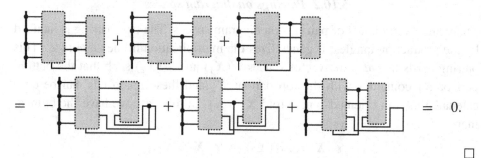

$$= 0.$$

$\square$

The definition of the inner product can be extended. For example, if two diagrams C, D have the same number of y_1-legs and the same number of y_2-legs, they can be glued together along the y_1-legs and then along the y_2-legs. The sum of the results of all such glueings is denoted by $\langle C, D \rangle_{y_1,y_2}$. This construction, clearly, can be generalized further.

5.10.3 Actions of $\mathcal{C}$ and $\mathcal{B}$ on tangle diagrams

While the coalgebra $\mathcal{C}(X)$, in general, does not have a product, it carries an algebraic structure that generalizes the product in $\mathcal{C}$. Namely, for each component x of X, there is an action of $\mathcal{C}(x)$ on $\mathcal{C}(X)$, defined as the connected sum along the component x. We denote this action by #, as if it were the usual connected sum. More generally, the spaces of mixed tangle diagrams $\mathcal{C}(x_1, \ldots, x_n \mid y_1, \ldots, y_m)$ are two-sided modules over $\mathcal{C}(x_i)$ and $\mathcal{B}(y_j)$. The algebra $\mathcal{C}(x_i)$ acts, as before, by the connected sum on the component x_i, while the action of $\mathcal{B}(y_j)$ consists in taking the disjoint union with diagrams in $\mathcal{B}(y_j)$. We shall denote the action of $\mathcal{B}(y_j)$ by $\cup$.

We cannot expect the relation of the module structures on the space of mixed diagrams with the symmetrization map to be straightforward, since the symmetrization map from $\mathcal{B}$ to $\mathcal{C}$ fails to be multiplicative. We shall clarify this remark in Section 11.3.7.

Exercise. Prove that the above actions are well-defined. In particular, prove that the action of $\mathcal{C}(x_i)$ does not depend on the location where the diagram is inserted into the corresponding component of the tangle diagram, and show that the action of $\mathcal{B}(y_j)$ respects the link relations.

5.10.4 Sliding property

There is one important corollary of the IHX relation (Kirchhoff law), called *sliding property* (Bar-Natan *et al.* 2003), which holds in the general context of tangle Jacobi diagrams. To formulate it, we need to define the operation $\Delta_x^{(n)} : \mathscr{C}(x \cup Y) \rightarrow \mathscr{C}(x_1 \cup \cdots \cup x_n \cup Y)$. By definition, $\Delta_x^{(n)}(D)$ is the lift of D to the nth disconnected cover of the component x, that is, for each x-leg of the diagram D we take the sum over all ways to attach it to x_i for any $i = 1, \ldots, n$ (the sum consists of n^k terms, if k is the number of vertices of D belonging to x). Example:

$$\Delta_x^{(2)}\left(\begin{array}{c}x\\ \vdash\!\bigcirc\end{array}\right) = \overset{x_1\ x_2}{\Vert\!\bigcirc} + \overset{x_1\ x_2}{\Vert\!\bigcirc} + \overset{x_1\ x_2}{\Vert\!\bigcirc} + \overset{x_1\ x_2}{\vert\!\vdash\!\bigcirc}.$$

Proposition 5.48. *(Sliding relation) Suppose that $D \in \mathscr{C}(x \cup Y)$; let $D_1 = \Delta_x^{(n)}(D)$. Then for any diagram $D_2 \in \mathscr{C}(x_1 \cup \cdots \cup x_n)$ we have $D_1 D_2 = D_2 D_1$. In pictures:*

Proof Indeed, take the leg in D_1 which is closest to D_2 and consider the sum of all diagrams on $x_1 \cup \cdots \cup x_n \cup Y$ where this leg is attached to x_i, $i = 1, \ldots, n$, while all the other legs are fixed. By the Kirchhoff law, this sum is equal to the similar sum where the chosen leg has jumped over D_2. In this way, all the legs jump over D_2 one by one, and the commutativity follows. □

5.10.5 Closing a component of a Jacobi diagram

Recall that long knots can be closed up to produce usual knots. This closure induces a bijection of the corresponding isotopy classes and an isomorphism of the corresponding diagram spaces.

This fact can be generalized to tangles whose skeleton consists of one interval and several circles.

Theorem 5.49. *Let X be a tangle skeleton with only one interval component, and X' be a skeleton obtained by closing this component into a circle. The induced map*

$$\mathscr{C}(X) \to \mathscr{C}(X')$$

is an isomorphism of vector spaces.

The proof of this theorem consists in applying the Kirchhoff's law and we leave it to the reader.

We should point out that closing one component of a skeleton with more than one interval component does not produce an isomorphism of the corresponding diagram spaces. Indeed, let us denote by $\mathscr{A}(2)$ the space of closed diagrams for string links on 2 strands. A direct calculation shows that the two diagrams of order 2 below on the left are different in $\mathscr{A}(2)$, while their images under closing one strand of the skeleton are obviously equal:

$$\bowtie \neq \mathbb{H} \qquad\qquad \bowtie\!\!\bigcirc = \vdash\!\!\bigcirc$$

The above statements about tangle diagrams, of course, are not arbitrary, but reflect the following topological fact that we state as an exercise:

Exercise. Define the map of closing one component on isotopy classes of tangles with a given skeleton and show that it is bijective if and only if it is applied to tangles whose skeleton has only one interval component.

5.11 Horizontal chord diagrams

5.11.1 The algebra of horizontal diagrams

There is yet another diagram algebra which will be of great importance in what follows, namely, the algebra $\mathscr{A}^h(n)$ of *horizontal chord diagrams* on n strands.

A horizontal chord diagram on n strands is a tangle diagram whose skeleton consists of n vertical intervals (all oriented, say, upwards) and all of whose chords are horizontal. Two such diagrams are considered to be equivalent if one can be deformed into the other through horizontal diagrams.

A product of two horizontal diagrams is clearly a horizontal diagram; by definition, the algebra $\mathscr{A}^h(n)$ is generated by the equivalence classes of all such diagrams modulo the horizonal 4T relations 4.4 (see Section 4.1). We denote by $\mathbf{1}_n$ the empty diagram in $\mathscr{A}^h(n)$ which is the multiplicative unit.

Each horizontal chord diagram is equivalent to a diagram whose chords are all situated on different levels, that is, to a product of diagrams of degree 1. Set

$$u_{jk} = \Big| \cdots \underset{j \quad\;\; k}{\overset{}{\prod}} \cdots \Big| \cdots \Big|, \qquad 1 \leqslant j < k \leqslant n,$$

and for $1 \leqslant k < j \leqslant n$ set $u_{jk} = u_{kj}$. Then $\mathscr{A}^h(n)$ is generated by the u_{jk} subject to the following relations (*infinitesimal pure braid relations*, first appeared in Kohno 1987)

$$[u_{jk}, u_{jl} + u_{kl}] = 0, \quad \text{if } j, k, l \text{ are different,}$$
$$[u_{jk}, u_{lm}] = 0, \quad \text{if } j, k, l, m \text{ are different.}$$

Indeed, the first relation is just the horizontal 4T relation. The second relation is similar to the far commutativity relation in braids. The products of the u_{jk} up to this relation are precisely the equivalence classes of horizontal diagrams.

The algebra $\mathscr{A}^h(2)$ is simply the free commutative algebra on one generator u_{12}.

Proposition 5.50. $\mathscr{A}^h(3)$ *is a direct product of the free algebra on two generators* u_{12} *and* u_{23}*, and the free commutative algebra on one generator*

$$u = u_{12} + u_{23} + u_{13}.$$

In particular, $\mathscr{A}^h(3)$ is highly noncommutative.

Proof Choose u_{12}, u_{23} and u as the set of generators for $\mathscr{A}^h(3)$. In terms of these generators, all the relations in $\mathscr{A}^h(3)$ can be written as

$$[u_{12}, u] = 0, \quad \text{and} \quad [u_{23}, u] = 0.$$

$\square$

For $n > 3$ the multiplicative structure of the algebra $\mathscr{A}^h(n)$ is rather more involved, even though it admits a simple description as a vector space. We shall treat this subject in more detail in Chapter 12, as the algebra $\mathscr{A}^h(n)$ plays the same role in the theory of finite type invariants for pure braids as the algebra $\mathscr{A}$ in the theory of the Vassiliev knot invariants.

We end this section with one property of $\mathscr{A}^h(n)$ which will be useful in Chapter 10.

Lemma 5.51. *Let* $J, K \subseteq \{1, \dots, n\}$ *be two non-empty subsets with* $J \cap K = \emptyset$*. Then the element* $\sum_{j \in J, k \in K} u_{jk}$ *commutes in* $\mathscr{A}^h(n)$ *with any generator* u_{pq} *with* p *and* q *either both in* J *or both in* K*.*

Proof It is clearly sufficient to prove the lemma for the case when K consists of one element, say k, and both p and q are in J. Now, any u_{jk} commutes with u_{pq} if j is different from both p and q. But $u_{pk} + u_{qk}$ commutes with u_{pq} by the horizontal 4T relation, and this proves the lemma. $\square$

5.11.2 Horizontal diagrams and string link diagrams

Denote by $\mathscr{A}(n)$ the algebra of closed diagrams for string links. Horizontal diagrams are examples of string link diagrams and the horizontal 4T relations are a particular case of the usual 4T relations, and, hence, there is an algebra homomorphism

$$\mathscr{A}^h(n) \to \mathscr{A}(n).$$

This homomorphism is injective, but this is a surprisingly nontrivial fact; see Bar-Natan (1996c); Habegger (2000). We shall give a proof of this in Chapter 12; see page 370.

Exercise.

(a) Prove that the chord diagram consisting of one chord connecting the two components of the skeleton belongs to the centre of the algebra $\mathscr{A}(2)$.

(b) Prove that any chord diagram consisting of two intersecting chords belongs to the centre of the algebra $\mathscr{A}(2)$.

(c) Prove that Lemma 5.51 is also valid for $\mathscr{A}(n)$. Namely, show that the element $\sum_{j\in J, k\in K} u_{jk}$ commutes in $\mathscr{A}(n)$ with any chord diagram whose chords have either both ends on the strands in J or on the strands in K.

Exercises

5.1   Prove that $\left(\vcenter{\hbox{}}\right) = \dfrac{1}{4}\,\vcenter{\hbox{}}$.

5.2   Let $a_1 = \vcenter{\hbox{}}$, $a_2 = \vcenter{\hbox{}}$, $a_3 = \vcenter{\hbox{}}$, $a_4 = \vcenter{\hbox{}}$, $a_5 = \vcenter{\hbox{}}$.

(a) Find a relation between a_1 and a_2.

(b) Represent the sum $a_3 + a_4 - 2a_5$ as a connected closed diagram.

(c) Prove the linear independence of a_3 and a_4 in $\mathscr{C}$.

5.3   Express the primitive elements $\vcenter{\hbox{}}$, $\vcenter{\hbox{}}$ and $\vcenter{\hbox{}}$ of degrees 3 and 4 as linear combinations of chord diagrams.

5.4   Prove the following identities in the algebra $\mathscr{C}$:

$$\vcenter{\hbox{}} = \vcenter{\hbox{}} - \frac{1}{2}\,\vcenter{\hbox{}}\;;$$

$$\vcenter{\hbox{}} = \vcenter{\hbox{}} - \vcenter{\hbox{}} + \frac{1}{4}\,\vcenter{\hbox{}}\;;$$

$$\bigotimes = \bigominus - \frac{3}{2}\bigoplus + \frac{3}{4}\overline{\text{oo}} - \frac{1}{8}\text{ooo} \; ;$$

$$\bigotimes = \bigominus - \frac{3}{2}\bigoplus + \frac{1}{2}\overline{\text{oo}} + \frac{1}{4}\bigoplus - \frac{1}{8}\text{ooo} \; ;$$

$$\bigoplus = \bigominus - 2\bigoplus + \overline{\text{oo}} + \frac{1}{2}\bigoplus - \frac{1}{2}\text{ooo} + \bigotimes .$$

5.5 Show that the symbols of the coefficients of the Conway polynomial (Section 2.3) take the following values on the basis primitive diagrams of degree 3 and 4.

$$\mathrm{symb}(c_3)\left(\overline{\text{oo}}\right) = 0,$$

$$\mathrm{symb}(c_4)\left(\text{ooo}\right) = 0, \qquad \mathrm{symb}(c_4)\left(\bigotimes\right) = -2.$$

5.6 Show that the symbols of the coefficients of the Jones polynomial (Section 3.6.2) take the following values on the basis primitive diagrams of degrees 3 and 4.

$$\mathrm{symb}(j_3)\left(\overline{\text{oo}}\right) = -24,$$

$$\mathrm{symb}(j_4)\left(\text{ooo}\right) = 96, \qquad \mathrm{symb}(j_4)\left(\bigotimes\right) = 18.$$

5.7 (Chmutov and Varchenko 1997) Let $\bar{t}_n \in \mathscr{P}_{n+1}$ be the closed diagram shown on the right. Prove the following identity:

$$\bar{t}_n = \overline{\text{(n legs)}}$$

$$\bar{t}_n = \frac{1}{2^n}\,\overline{\text{o}\cdots\text{o}}$$

n bubbles

Deduce that $\bar{t}_n \in \mathscr{P}_{n+1}^2$.

5.8 Express $\bar{t}_n$ as a linear combination of chord diagrams. In particular, show that the intersection graph of every chord diagram that occurs in this expression is a forest.

5.9 (Chmutov and Varchenko 1997) Prove the following identity in the space $\mathscr{C}$ of closed diagrams:

$$\text{(dodecagon diagram)} = \frac{3}{4}\,\text{(diagram)} - \frac{1}{12}\,\text{(diagram)} - \frac{1}{48}\,\text{(diagram)}.$$

Hint. Turn the internal pentagon of the left-hand side of the identity in the 3-space by 180° about the vertical axis. The result will represent the same graph with the cyclic orders at all five vertices of the pentagon changed to the opposite:

$$\text{(diagram)} = (-1)^5\,\text{(diagram)} = -\,\text{(diagram)} + (\text{terms with at most 4 legs}).$$

The last equality follows from the STU relations which allow us to rearrange the legs modulo diagrams with a smaller number of legs. To finish the solution, the reader must figure out the terms in the parentheses.

5.10 Prove the linear independence of the three elements in the right-hand side of the last equality, using Lie algebra invariants defined in Chapter 6.

5.11 (Chmutov and Varchenko 1997) Prove that the primitive space in the algebra $\mathscr{C}$ is generated by the closed diagrams whose internal graph is a tree.

5.12 (Chmutov and Varchenko 1997) With each permutation σ of n objects, associate a closed diagram P_σ acting as in Section 5.7.2 by the permutation on the lower legs of a closed diagram $P_{(12\ldots n)} = \overline{t_n}$ from Exercise 5.7. Here are some examples:

$$P_{(2143)} = \text{(diagram)}\,; \qquad P_{(4123)} = \text{(diagram)}\,; \qquad P_{(4132)} = \text{(diagram)}.$$

Prove that the diagrams P_σ span the vector space $\mathscr{P}_{n+1}$.

5.13 (Chmutov and Varchenko 1997) Prove that

- $\mathscr{P}_n^n = \mathscr{P}_n$ for even n, and $\mathscr{P}_n^{n-1} = \mathscr{P}_n$ for odd n;
- for even n the quotient space $\mathscr{P}_n^n / \mathscr{P}_n^{n-1}$ has dimension one and generated by the wheel $\overline{w}_n$.

5.14 Let $b_1 =$, $b_2 =$, $b_3 =$, $b_4 =$.

Which of these diagrams are zero in $\mathscr{B}$, that is, vanish modulo AS and IHX relations?

5.15 Prove that the algebra generated by all open diagrams modulo the AS and the modified IHX equation $I = aH - bX$, where a and b are arbitrary complex numbers, is isomorphic (equal) to $\mathscr{B}$ if and only if $a = b = 1$, in all other cases it is a free polynomial algebra on one generator.

5.16 • Indicate an explicit form of the isomorphisms $\mathscr{A} \cong \mathscr{C} \cong \mathscr{B}$ in the bases given in Section 5.9.
• Compile the multiplication table for $\mathscr{B}_m \times \mathscr{B}_n \to \mathscr{B}_{m+n}, m+n \leqslant 4$, for the second product in $\mathscr{B}$ (the one pulled back from $\mathscr{C}$ along the isomorphism $\mathscr{C} \cong \mathscr{B}$).
• Find some bases of the spaces $\mathscr{A}_n, \mathscr{C}_n, \mathscr{B}_n$ for $n = 5$.

5.17 (*J. Kneissler*). Let $\mathscr{B}_n^u$ be the space of open diagrams of degree n with u univalent vertices. Denote by $\omega_{i_1 i_2 \ldots i_k}$ the element of $\mathscr{B}_{i_1 + \cdots + i_k + k - 1}^{i_1 + \cdots + i_k}$ represented by a *caterpillar* diagram consisting of k body segments with $i_1, \ldots, i_k$ "legs", respectively. Using the AS and IHX relations, prove that $\omega_{i_1 i_2 \ldots i_k}$ is well-defined, that is, for inner segments it makes no difference on which side of the body they are drawn. For example,

$$\omega_{0321} = \quad \text{} \quad = \quad \text{} \quad .$$

5.18* (*J. Kneissler*) Is it true that any caterpillar diagram in the algebra $\mathscr{B}$ can be expressed via caterpillar diagrams with even indices $i_1, \ldots, i_k$? Is it true that the primitive space $\mathscr{P}(\mathscr{B})$ (that is, the space spanned by connected open diagrams) is generated by caterpillar diagrams?

5.19 Prove the equivalence of the two claims:
• all chord diagrams are symmetric modulo one- and four-term relations.
• all chord diagrams are symmetric modulo only four-term relations.

5.20 Similar to symmetric chord diagrams (page 128), we can speak of *anti-symmetric* diagrams: an element D of $\mathscr{A}'$ or $\mathscr{A}$ is anti-symmetric if $\tau(D) = -D$. Prove that under the isomorphism $\chi^{-1} : \mathscr{A} \to \mathscr{B}$:
• the image of a symmetric chord diagram is a linear combination of open diagrams with an even number of legs,
• the image of an anti-symmetric chord diagram in is a linear combination of open diagrams with an odd number of legs.

5.21* (The simplest unsolved case of Conjecture 5.32.) Is it true that an open diagram with three univalent vertices is always equal to 0 as an element of the algebra $\mathscr{B}$?

5.22 Prove that the diagram is equal to 0 in the space $\mathscr{B}(2)$.

5.23 Let u_{ij} be the diagram in $\mathscr{A}(3)$ with one chord connecting the ith and the jth component of the skeleton. Prove that for any k the combination $u_{12}^k + u_{23}^k + u_{13}^k$ belongs to the centre of $\mathscr{A}(3)$.

5.24 Let $D^3 = $ $\in \mathscr{C}(X, y)$ and $D^4 = $ $\in \mathscr{C}(X, y)$ be tangle diagrams with exactly three and four y-legs respectively. Show that

$$\mathscr{C}(X|y) \ni \chi_y^{-1}(D^3) = \text{\includegraphics{}} + \tfrac{1}{2}\text{\includegraphics{}} = \text{\includegraphics{}} + \tfrac{1}{2}\text{\includegraphics{}} = \text{\includegraphics{}} - \tfrac{1}{2}\text{\includegraphics{}} \; ;$$

$$\mathscr{C}(X|y) \ni \chi_y^{-1}(D^4) = \text{\includegraphics{}} + \tfrac{1}{2}\text{\includegraphics{}} + \tfrac{1}{2}\text{\includegraphics{}} + \tfrac{1}{8}\text{\includegraphics{}} + \tfrac{5}{24}\text{\includegraphics{}} .$$

Hint. Follow the proof of Theorem 5.37 and then use link relations.

6

Lie algebra weight systems

Given a Lie algebra $\mathfrak{g}$ equipped with a non-degenerate invariant bilinear form, one can construct a weight system with values in the centre of the universal enveloping algebra $U(\mathfrak{g})$. In a similar fashion one can define a map from the space $\mathcal{B}$ into the ad-invariant part of the symmetric algebra $S(\mathfrak{g})$. These constructions are due to M. Kontsevich (1993), with basic ideas already appearing in Penrose (1971). If, in addition, we have a finite dimensional representation of the Lie algebra, then taking the trace of the corresponding operator we get a numeric weight system. It turns out that these weight systems are the symbols of the quantum group invariants (Section 3.6.5). The construction of weight systems based on representations first appeared in D. Bar-Natan's paper (1991a). The reader is invited to consult the Appendix for basics on Lie algebras and their universal envelopes.

A useful tool to compute Lie algebra weight systems is Bar-Natan's computer program called `main.c` and available online at (Bar-Natan 1996a). The tables in this chapter were partially obtained using that program.

There is another construction of weight systems, also invented by Kontsevich: the weight systems coming from *marked surfaces*. As proved in Bar-Natan (1995a), this construction gives the same set of weight systems as the classical Lie algebras, and we shall not speak about it here.

6.1 Lie algebra weight systems for the algebra $\mathcal{A}$

6.1.1 Universal Lie algebra weight systems

Kontsevich's construction proceeds as follows. Let $\mathfrak{g}$ be a metrized Lie algebra over $\mathbb{R}$ or $\mathbb{C}$, that is, a Lie algebra with an ad-invariant non-degenerate bilinear form $\langle \cdot, \cdot \rangle$ (see A.1.1). Choose a basis $e_1, \ldots, e_m$ of $\mathfrak{g}$ and let $e_1^*, \ldots, e_m^*$ be the dual basis with respect to the form $\langle \cdot, \cdot \rangle$.

Given a chord diagram D with n chords, we first choose a base point on its Wilson loop, away from the chords of D. This gives a linear order on the endpoints of the chords, increasing in the positive direction of the Wilson loop. Assign to each chord a an *index*, that is, an integer-valued variable, i_a. The values of i_a will range from 1 to m, the dimension of the Lie algebra. Mark the first endpoint of the chord with the symbol e_{i_a} and the second endpoint with $e_{i_a}^*$.

Now, write the product of all the e_{i_a} and all the $e_{i_a}^*$, in the order in which they appear on the Wilson loop of D, and take the sum of the m^n elements of the universal enveloping algebra $U(\mathfrak{g})$ obtained by substituting all possible values of the indices i_a into this product. Denote by $\varphi_{\mathfrak{g}}(D)$ the resulting element of $U(\mathfrak{g})$.

For example,

$$\varphi_{\mathfrak{g}}\left(\bigominus\right) = \sum_{i=1}^{m} e_i e_i^* =: c$$

is the *quadratic Casimir element* associated with the chosen invariant form. The next theorem shows, in particular, that the Casimir element does not depend on the choice of the basis in $\mathfrak{g}$. Another example: if

$$D = \bigoplus{}^{i\ j}_{k}$$

then

$$\varphi_{\mathfrak{g}}(D) = \sum_{i=1}^{m} \sum_{j=1}^{m} \sum_{k=1}^{m} e_i e_j e_k e_i^* e_k^* e_j^*.$$

Theorem 6.1. *The above construction has the following properties:*

1. *The element $\varphi_{\mathfrak{g}}(D)$ does not depend on the choice of the base point on the diagram;*
2. *It does not depend on the choice of the basis $\{e_i\}$ of the Lie algebra;*
3. *It belongs to the ad-invariant subspace*

$$U(\mathfrak{g})^{\mathfrak{g}} = \{x \in U(\mathfrak{g}) \mid xy = yx \text{ for all } y \in \mathfrak{g}\}$$

   of the universal enveloping algebra $U(\mathfrak{g})$ (that is, to the centre $ZU(\mathfrak{g})$);
4. *The function $D \mapsto \varphi_{\mathfrak{g}}(D)$ satisfies 4-term relations;*
5. *The resulting map $\varphi_{\mathfrak{g}} : \mathscr{A}^{fr} \to ZU(\mathfrak{g})$ is a homomorphism of algebras.*

Proof (1) Introducing a base point means that a circular chord diagram is replaced by a linear chord diagram (see Section 4.7). Modulo 4-term relations, this map is an isomorphism, and, hence, the assertion follows from (4).

(2) An exercise in linear algebra: take two different bases $\{e_i\}$ and $\{f_j\}$ of $\mathfrak{g}$ and reduce the expression for $\varphi_{\mathfrak{g}}(D)$ in one basis to the expression in the other using the transition matrix between the two bases. Technically, it is enough to do this exercise only for $m = \dim \mathfrak{g} = 2$, since the group of transition matrices $GL(m)$ is generated by linear transformations in the two-dimensional coordinate planes. This also follows from the invariant construction of this weight system in Section 6.1.2 which does not use any basis.

(3) It is enough to prove that $\varphi_{\mathfrak{g}}(D)$ commutes with any basis element e_r. By property (2), we can choose the basis to be orthonormal with respect to the ad-invariant form $\langle \cdot, \cdot \rangle$, so that $e_i^* = e_i$ for all i. Now, the commutator of e_r and $\varphi_{\mathfrak{g}}(D)$ can be expanded into a sum of $2n$ expressions, similar to $\varphi_{\mathfrak{g}}(D)$, only with one of the e_i replaced by its commutator with e_r. Due to the antisymmetry of the structure constants c_{ijk} (Lemma A.2), these expressions cancel in pairs that correspond to the ends of each chord.

To take a concrete example,

$$[e_r, \sum_{ij} e_i e_j e_i e_j]$$

$$= \sum_{ij} [e_r, e_i] e_j e_i e_j + \sum_{ij} e_i [e_r, e_j] e_i e_j + \sum_{ij} e_i e_j [e_r, e_i] e_j + \sum_{ij} e_i e_j e_i [e_r, e_j]$$

$$= \sum_{ijk} c_{rik} e_k e_j e_i e_j + \sum_{ijk} c_{rjk} e_i e_k e_i e_j + \sum_{ijk} c_{rik} e_i e_j e_k e_j + \sum_{ijk} c_{rjk} e_i e_j e_i e_k$$

$$= \sum_{ijk} c_{rik} e_k e_j e_i e_j + \sum_{ijk} c_{rjk} e_i e_k e_i e_j + \sum_{ijk} c_{rki} e_k e_j e_i e_j + \sum_{ijk} c_{rkj} e_i e_k e_i e_j.$$

Here the first and the second sums cancel with the third and the fourth sums, respectively.

(4) We still assume that the basis $\{e_i\}$ is $\langle \cdot, \cdot \rangle$–orthonormal. Then one of the pairwise differences of the chord diagrams that constitute the 4-term relation in Equation (4.3) is sent by $\varphi_{\mathfrak{g}}$ to

$$\sum c_{ijk} \ldots e_i \ldots e_j \ldots e_k \ldots ,$$

while the other goes to

$$\sum c_{ijk} \ldots e_j \ldots e_k \ldots e_i \cdots = \sum c_{kij} \ldots e_i \ldots e_j \ldots e_k \ldots .$$

By the cyclic symmetry of the structure constants c_{ijk} in an orthonormal basis (again see Lemma A.2), these two expressions are equal.

(5) Using property (1), we can place the base point in the product diagram $D_1 \cdot D_2$ between D_1 and D_2. Then the identity $\varphi_{\mathfrak{g}}(D_1 \cdot D_2) = \varphi_{\mathfrak{g}}(D_1) \varphi_{\mathfrak{g}}(D_2)$ becomes evident. $\qquad \square$

Remark 6.2. If D is a chord diagram with n chords, then

$$\varphi_{\mathfrak{g}}(D) = c^n + \{\text{terms of degree less than } 2n \text{ in } U(\mathfrak{g})\},$$

where c is the quadratic Casimir element as on page 158. Indeed, we can permute the endpoints of chords on the circle without changing the highest term of $\varphi_{\mathfrak{g}}(D)$ since all the additional summands arising as commutators have degrees smaller than $2n$. Therefore, the highest degree term of $\varphi_{\mathfrak{g}}(D)$ does not depend on D. Finally, if D is a diagram with n isolated chords, that is, the nth power of the diagram with one chord, then $\varphi_{\mathfrak{g}}(D) = c^n$.

The centre $ZU(\mathfrak{g})$ of the universal enveloping algebra is precisely the $\mathfrak{g}$-invariant subspace $U(\mathfrak{g})^{\mathfrak{g}} \subset U(\mathfrak{g})$, where the action of $\mathfrak{g}$ on $U(\mathfrak{g})$ consists in taking the commutator. According to the Harish–Chandra theorem (see Humphreys 1980), for a semi-simple Lie algebra $\mathfrak{g}$, the centre $ZU(\mathfrak{g})$ is isomorphic to the algebra of polynomials in certain variables $c_1 = c, c_2, \ldots, c_r$, where $r = rank(\mathfrak{g})$.

6.1.2 A basis-free description of Lie algebra weight systems

The construction of Lie algebra weight systems can be described without referring to any particular basis.

A based chord diagram D with n chords gives a permutation σ_D of the set $\{1, 2, \ldots, 2n\}$ as follows. As we have noted before, the endpoints of chords of a based chord diagram are ordered, so we can order the chords of D by their first endpoint. Let us number the chords from 1 to n, and their endpoints from 1 to $2n$, in the increasing order. Then, for $1 \leqslant i \leqslant n$ the permutation σ_D sends $2i - 1$ to the (number of the) first endpoint of the ith chord, and $2i$ to the second endpoint of the same chord. In the terminology of Section 5.7.2 the permutation σ_D sends the diagram with n consecutive isolated chords into D. For instance:

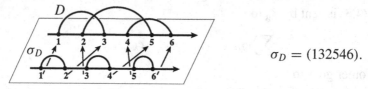

$$\sigma_D = (132546).$$

The bilinear form $\langle \cdot, \cdot \rangle$ on $\mathfrak{g}$ is a tensor in $\mathfrak{g}^* \otimes \mathfrak{g}^*$. The algebra $\mathfrak{g}$ is metrized, so we can identify $\mathfrak{g}^*$ with $\mathfrak{g}$ and think of $\langle \cdot, \cdot \rangle$ as an element of $\mathfrak{g} \otimes \mathfrak{g}$. The permutation σ_D acts on $\mathfrak{g}^{\otimes 2n}$ by interchanging the factors. The value of the universal Lie algebra weight system $\varphi_{\mathfrak{g}}(D)$ is then the image of the nth tensor power $\langle \cdot, \cdot \rangle^{\otimes n}$ under the map

$$\mathfrak{g}^{\otimes 2n} \xrightarrow{\sigma_D} \mathfrak{g}^{\otimes 2n} \to U(\mathfrak{g}),$$

where the second map is the natural projection of the tensor algebra on $\mathfrak{g}$ to its universal enveloping algebra.

6.1.3 The universal $\mathfrak{sl}_2$ weight system

Consider the Lie algebra $\mathfrak{sl}_2$ of 2×2 matrices with zero trace. It is a three-dimensional Lie algebra spanned by the matrices

$$H = \begin{pmatrix} 1 & 0 \\ 0 & -1 \end{pmatrix}, \qquad E = \begin{pmatrix} 0 & 1 \\ 0 & 0 \end{pmatrix}, \qquad F = \begin{pmatrix} 0 & 0 \\ 1 & 0 \end{pmatrix}$$

with the commutators

$$[H,E] = 2E, \qquad [H,F] = -2F, \qquad [E,F] = H.$$

We shall use the symmetric bilinear form $\langle x, y \rangle = \text{Tr}(xy)$:

$$\langle H,H \rangle = 2, \; \langle H,E \rangle = 0, \; \langle H,F \rangle = 0, \; \langle E,E \rangle = 0, \; \langle E,F \rangle = 1, \; \langle F,F \rangle = 0.$$

One can easily check that it is ad-invariant and non-degenerate. The corresponding dual basis is

$$H^* = \frac{1}{2}H, \qquad E^* = F, \qquad F^* = E,$$

and, hence, the Casimir element is $c = \frac{1}{2}HH + EF + FE$.

The centre $ZU(\mathfrak{sl}_2)$ is isomorphic to the algebra of polynomials in a single variable c. The value $\varphi_{\mathfrak{sl}_2}(D)$ is thus a polynomial in c. In this section, following Chmutov and Varchenko (1997), we explain a combinatorial procedure to compute this polynomial for a given chord diagram D.

The algebra $\mathfrak{sl}_2$ is simple, hence, any invariant form is equal to $\lambda\langle\cdot, \cdot\rangle$ for some constant λ. The corresponding Casimir element c_λ, as an element of the universal enveloping algebra, is related to $c = c_1$ by the formula $c_\lambda = \frac{c}{\lambda}$. Therefore, the weight system

$$\varphi_{\mathfrak{sl}_2}(D) = c^n + a_{n-1}c^{n-1} + a_{n-2}c^{n-2} + \cdots + a_2c^2 + a_1c$$

and the weight system corresponding to $\lambda\langle\cdot, \cdot\rangle$

$$\varphi_{\mathfrak{sl}_2,\lambda}(D) = c_\lambda^n + a_{n-1,\lambda}c_\lambda^{n-1} + a_{n-2,\lambda}c_\lambda^{n-2} + \cdots + a_{2,\lambda}c_\lambda^2 + a_{1,\lambda}c_\lambda$$

are related by the formula $\varphi_{\mathfrak{sl}_2,\lambda}(D) = \frac{1}{\lambda^n} \cdot \varphi_{\mathfrak{sl}_2}(D)|_{c=\lambda\cdot c_\lambda}$, or

$$a_{n-1} = \lambda a_{n-1,\lambda}, \quad a_{n-2} = \lambda^2 a_{n-2,\lambda}, \quad \ldots \quad a_2 = \lambda^{n-2}a_{2,\lambda}, \quad a_1 = \lambda^{n-1}a_{1,\lambda}.$$

Theorem 6.3. *Let $\varphi_{\mathfrak{sl}_2}$ be the weight system associated with $\mathfrak{sl}_2$, with the invariant form $\langle \cdot, \cdot \rangle$. Take a chord diagram D and choose a chord a of D. Then*

$$\varphi_{\mathfrak{sl}_2}(D) = (c - 2k)\varphi_{\mathfrak{sl}_2}(D_a) + 2 \sum_{1 \leqslant i < j \leqslant k} \left(\varphi_{\mathfrak{sl}_2}(D_{i,j}^{\shortparallel}) - \varphi_{\mathfrak{sl}_2}(D_{i,j}^{\times}) \right),$$

where:

- *k is the number of chords that intersect the chord a;*
- *D_a is the chord diagram obtained from D by deleting the chord a;*
- *$D_{i,j}^{\shortparallel}$ and $D_{i,j}^{\times}$ are the chord diagrams obtained from D_a in the following way. Draw the diagram D so that the chord a is vertical. Consider an arbitrary pair of chords a_i and a_j different from a and such that each of them intersects a. Denote by p_i and p_j the endpoints of a_i and a_j that lie to the left of a and by p_i^*, p_j^* the endpoints of a_i and a_j that lie to the right. There are three ways to connect the four points p_i, p_i^*, p_j, p_j^* by two chords. D_a is the diagram where these two chords are $(p_i, p_i^*), (p_j, p_j^*)$, the diagram $D_{i,j}^{\shortparallel}$ has the chords $(p_i, p_j), (p_i^*, p_j^*)$ and $D_{i,j}^{\times}$ has the chords $(p_i, p_j^*), (p_i^*, p_j)$. All other chords are the same in all the diagrams:*

The theorem allows one to compute $\varphi_{\mathfrak{sl}_2}(D)$ recursively, as each of the three diagrams D_a, $D_{i,j}^{\shortparallel}$ and $D_{i,j}^{\times}$ has one chord less than D.

Examples 6.4.

1. $\varphi_{\mathfrak{sl}_2}\left(\bigotimes \right) = (c - 2)c$. In this case, $k = 1$ and the sum in the right-hand side is zero, since there are no pairs (i, j).

2.
$$\varphi_{\mathfrak{sl}_2}\left(\bigoplus \right) = (c - 4)\varphi_{\mathfrak{sl}_2}\left(\bigominus \right) + 2\varphi_{\mathfrak{sl}_2}\left(\big)\big(\right) - 2\varphi_{\mathfrak{sl}_2}\left(\bigotimes \right)$$
$$= (c - 4)c^2 + 2c^2 - 2(c - 2)c = (c - 2)^2 c.$$

3.
$$\varphi_{\mathfrak{sl}_2}\left(\bigotimes \right) = (c - 4)\varphi_{\mathfrak{sl}_2}\left(\bigotimes \right) + 2\varphi_{\mathfrak{sl}_2}\left(\big)\big(\right) - 2\varphi_{\mathfrak{sl}_2}\left(\bigominus \right)$$
$$= (c - 4)(c - 2)c + 2c^2 - 2c^2 = (c - 4)(c - 2)c.$$

Remark 6.5. Choosing the invariant form $\lambda \langle \cdot, \cdot \rangle$, we obtain a modified relation

$$\varphi_{\mathfrak{sl}_2,\lambda}(D) = \left(c_\lambda - \frac{2k}{\lambda}\right)\varphi_{\mathfrak{sl}_2,\lambda}(D_a) + \frac{2}{\lambda}\sum_{1\leqslant i<j\leqslant k}\left(\varphi_{\mathfrak{sl}_2,\lambda}(D_{i,j}^{||}) - \varphi_{\mathfrak{sl}_2,\lambda}(D_{i,j}^{\times})\right).$$

If $k = 1$, the second summand vanishes. In particular, for the Killing form ($\lambda = 4$) and $k = 1$ we have

$$\varphi_\mathfrak{g}(D) = (c - 1/2)\varphi_\mathfrak{g}(D_a).$$

It is interesting that the last formula is valid for any simple Lie algebra $\mathfrak{g}$ with the Killing form and any chord a which intersects precisely one other chord. See Exercise 6.8 for a generalization of this fact in the case $\mathfrak{g} = \mathfrak{sl}_2$.

Exercise. Deduce the theorem of this section from the following lemma by induction (in case of difficulty, see the proof in Chmutov and Varchenko 1997).

Lemma 6.6 (6-term relations for the universal $\mathfrak{sl}_2$ weight system). *Let* $\varphi_{\mathfrak{sl}_2}$ *be the weight system associated with* $\mathfrak{sl}_2$ *and the invariant form* $\langle \cdot, \cdot \rangle$. *Then*

These relations also provide a recursive way to compute $\varphi_{\mathfrak{sl}_2}(D)$ as the two chord diagrams on the right-hand side have one chord less than the diagrams on the left-hand side, and the last three diagrams on the left-hand side are simpler than the first one since they have less intersections between their chords. See Section 6.2.3 for a proof of this lemma.

6.1.4 Weight systems associated with representations

The construction of Bar-Natan, in comparison with that of Kontsevich, uses one additional ingredient: a representation of a Lie algebra (see A.1.1).

A linear representation $T : \mathfrak{g} \to \mathrm{End}(V)$ extends to a homomorphism of associative algebras $U(T) : U(\mathfrak{g}) \to \mathrm{End}(V)$. The composition of the following three maps (with the last map being the trace)

$$\mathcal{A} \xrightarrow{\varphi_{\mathfrak{g}}} U(\mathfrak{g}) \xrightarrow{U(T)} \mathrm{End}(V) \xrightarrow{\mathrm{Tr}} \mathbb{C}$$

by definition gives the *weight system associated with the representation*

$$\varphi_{\mathfrak{g}}^{T} = \mathrm{Tr} \circ U(T) \circ \varphi_{\mathfrak{g}}$$

(by abuse of notation, we shall sometimes write $\varphi_{\mathfrak{g}}^{V}$ instead of $\varphi_{\mathfrak{g}}^{T}$).

The map $\varphi_{\mathfrak{g}}^{T}$ is not in general multiplicative (the reader may check this for the diagram Θ and its square in the standard representation of the algebra $\mathfrak{gl}_{N}$, see Section 6.1.6). However, if the representation T is irreducible, then, according to the Schur Lemma (Humphreys 1980), every element of the centre $ZU(\mathfrak{g})$ is represented (via $U(T)$) by a scalar operator $\mu \cdot \mathrm{id}_{V}$. Therefore, its trace equals $\varphi_{\mathfrak{g}}^{T}(D) = \mu \dim V$. The number $\mu = \frac{\varphi_{\mathfrak{g}}^{T}(D)}{\dim V}$, as a function of the chord diagram D, is a weight system which is clearly multiplicative.

6.1.5 Algebra $\mathfrak{sl}_2$ with the standard representation

Consider the standard two-dimensional representation St of $\mathfrak{sl}_{2}$. Then the Casimir element is represented by the matrix

$$c = \frac{1}{2}HH + EF + FE = \begin{pmatrix} 3/2 & 0 \\ 0 & 3/2 \end{pmatrix} = \frac{3}{2} \cdot \mathrm{id}_{2}.$$

In degree 3 we have the following weight systems:

D					
$\varphi_{\mathfrak{sl}_2}(D)$	c^3	c^3	$c^2(c-2)$	$c(c-2)^2$	$c(c-2)(c-4)$
$\varphi_{\mathfrak{sl}_2}^{St}(D)$	$27/4$	$27/4$	$-9/4$	$3/4$	$15/4$
$\varphi_{\mathfrak{sl}_2}^{\prime St}(D)$	0	0	0	12	24

Here the last row represents the unframed weight system obtained from $\varphi_{\mathfrak{sl}_2}^{St}$ by the deframing procedure from Section 4.5.4. A comparison of this computation with the one from Section 3.6.2 shows that on these elements $\mathrm{symb}(j_3) = -\frac{1}{2}\varphi_{\mathfrak{sl}_2}^{\prime St}$. See Exercises 6.12 and 6.13 at the end of the chapter for more information about these weight systems.

6.1.6 Algebra $\mathfrak{gl}_N$ with the standard representation

Consider the Lie algebra $\mathfrak{g} = \mathfrak{gl}_N$ of all $N \times N$ matrices and its standard representation St. Fix the trace of the product of matrices as the preferred ad-invariant form: $\langle x, y \rangle = \mathrm{Tr}(xy)$.

The algebra $\mathfrak{gl}_N$ is linearly spanned by matrices e_{ij} with 1 on the intersection of ith row with jth column and zero elsewhere. We have $\langle e_{ij}, e_{kl} \rangle = \delta_i^l \delta_j^k$, where δ is the Kronecker delta. Therefore, the duality between $\mathfrak{gl}_N$ and $(\mathfrak{gl}_N)^*$ defined by $\langle \cdot, \cdot \rangle$ is given by the formula $e_{ij}^* = e_{ji}$.

Exercise. Prove that the form $\langle \cdot, \cdot \rangle$ equals $2(N-1)$ times the Killing form. (*Hint:* It is enough to compute the trace of just one operator $(\mathrm{ad}e_{11})^2$.)

One can verify that $[e_{ij}, e_{kl}] \neq 0$ only in the following cases:

- $[e_{ij}, e_{jk}] = e_{ik}$, if $i \neq k$,
- $[e_{ij}, e_{ki}] = -e_{kj}$, if $j \neq k$,
- $[e_{ij}, e_{ji}] = e_{ii} - e_{jj}$, if $i \neq j$.

This gives the following formula for the Lie bracket as a tensor in $\mathfrak{gl}_N^* \otimes \mathfrak{gl}_N^* \otimes \mathfrak{gl}_N$:

$$[\cdot, \cdot] = \sum_{i,j,k=1}^{N} (e_{ij}^* \otimes e_{jk}^* \otimes e_{ik} - e_{ij}^* \otimes e_{ki}^* \otimes e_{kj}).$$

When transferred to $\mathfrak{gl}_N \otimes \mathfrak{gl}_N \otimes \mathfrak{gl}_N$ by duality, this tensor takes the form

$$J = \sum_{i,j,k=1}^{N} (e_{ji} \otimes e_{kj} \otimes e_{ik} - e_{ji} \otimes e_{ik} \otimes e_{kj}).$$

This formula will be used later in Section 6.2.

D. Bar-Natan found the following elegant way of computing the weight system $\varphi_{\mathfrak{gl}_N}^{St}$.

Theorem 6.7 (Bar-Natan 1991a). *Denote by $s(D)$ the number of connected components of the curve obtained by doubling all chords of a chord diagram D.*

Then $\varphi_{\mathfrak{gl}_N}^{St}(D) = N^{s(D)}$.

Remark 6.8. By definition, the number $s(D)$ equals $c - 1$, where c is the number of boundary components of the surface described in Section 4.8.5.

Example 6.9. For $D = $  we obtain the picture . Here $s(D) = 2$, hence $\varphi_{\mathfrak{gl}_N}^{St}(D) = N^2$.

Proof We take the matrices e_{ij} as the chosen basis of $\mathfrak{gl}_N$. The values of the index variables associated with the chords are pairs (ij); each chord has one end labelled by a matrix e_{ij} and the other end by $e_{ji} = e_{ij}^*$.

Now, consider the curve γ obtained by doubling the chords. Given a chord whose ends are labelled by e_{ij} and e_{ji}, we can label the two copies of this chord in γ, as well as the four pieces of the Wilson loop adjacent to its endpoints, by the indices i and j as follows:

To compute the value of the weight system $\varphi_{\mathfrak{gl}_N}^{St}(D)$, we must sum up the products $\ldots e_{ij}e_{kl}\ldots$. Since we are dealing with the standard representation of $\mathfrak{gl}_N$, the product should be understood as genuine matrix multiplication, rather than the formal product in the universal enveloping algebra. Since $e_{ij} \cdot e_{kl} = \delta_{jk} \cdot e_{il}$, we get a non-zero summand only if $j = k$. This means that the labels of the chords must follow the pattern:

Therefore, all the labels on one and the same connected component of the curve γ are equal. If we take the whole product of matrices along the circle, we get the operator e_{ii} whose trace is 1. Now, we must sum up the traces of all such operators over all possible labellings. The number of labellings is equal to the number of values the indices $i, j, l, \ldots$ take on the connected components of the curve γ. Each component gives exactly N possibilities, so the total number is $N^{s(D)}$. □

Proposition 6.10. *The weight system $\varphi_{\mathfrak{gl}_N}^{St}(D)$ depends only on the intersection graph of D.*

Proof The value $\varphi_{\mathfrak{gl}_N}^{St}(D)$ is defined by the number $s(D) = c - 1$ (where c has the meaning given on page 111), therefore it is a function of the genus of the diagram D. In Section 4.8.5 we proved that the genus depends only on the intersection graph. □

6.1.7 Algebra $\mathfrak{sl}_N$ with the standard representation

Here we describe the weight system $\varphi^{St}_{\mathfrak{sl}_N}(D)$ associated with the Lie algebra $\mathfrak{sl}_N$, its standard representation by $N \times N$ matrices with zero trace and the invariant form $\langle x, y \rangle = \mathrm{Tr}(xy)$. In the same spirit as in Section 3.6.2, introduce *a state σ* for a chord diagram D as an arbitrary function on the set $[D]$ of chords of D with values in the set $\{1, -\frac{1}{N}\}$. With each state σ we associate an immersed plane curve obtained from D by resolutions of all its chords according to s:

, if $\sigma(a) = 1$;   , if $\sigma(a) = -\frac{1}{N}$.

Let $|\sigma|$ denote the number of components of the curve obtained in this way.

Theorem 6.11. $\varphi^{St}_{\mathfrak{sl}_N}(D) = \sum_{\sigma} \left(\prod_{a} \sigma(a) \right) N^{|\sigma|}$, *where the product is taken over all n chords of D, and the sum is taken over all 2^n states for D.*

One can prove this theorem in the same way as we did for $\mathfrak{gl}_N$ by picking an appropriate basis for the vector space $\mathfrak{sl}_N$ and then working with the product of matrices (see Exercise 6.11). However, we prefer to prove it in a different way, via several reformulations, using the algebra structure of weight systems which is dual to the coalgebra structure of chord diagrams (Section 4.5).

Reformulation 1. *For a subset $J \subseteq [D]$ (the empty set and the whole $[D]$ are allowed) of chords of D, denote by $|J|$ the cardinality of J. Write D_J for the chord diagram formed by the chords from J, and denote by $s(D_J)$ the number of connected components of the curve obtained by doubling all the chords of D_J. Then*

$$\varphi^{St}_{\mathfrak{sl}_N}(D) = \sum_{J \subseteq [D]} (-1)^{n-|J|} N^{s(D_J)-n+|J|}.$$

This assertion is obviously equivalent to the Theorem: for every state s, the subset J consists of all chords c with value $s(c) = 1$.

Consider the weight system $e^{-\mathbf{I}_1/N}$ from Section 4.5.4, which is equal to the constant $\frac{1}{(-N)^n}$ on any chord diagram with n chords.

Reformulation 2.

$$\varphi^{St}_{\mathfrak{sl}_N} = e^{-\frac{\mathbf{I}_1}{N}} \cdot \varphi^{St}_{\mathfrak{gl}_N}.$$

Indeed, by the definition of the product of weight systems (Section 4.5),

$$\left(e^{-\frac{\mathbf{I}_1}{N}} \cdot \varphi^{St}_{\mathfrak{gl}_N}\right)(D) = \left(e^{-\frac{\mathbf{I}_1}{N}} \otimes \varphi^{St}_{\mathfrak{gl}_N}\right)(\delta(D)),$$

where $\delta(D)$ is the coproduct (Section 4.4) of the chord diagram D. It splits D into two complementary parts $D_{\bar{J}}$ and D_J: $\delta(D) = \sum_{J \subseteq [D]} D_{\bar{J}} \otimes D_J$. The weight system $\varphi^{St}_{\mathfrak{gl}_N}(D_J)$ gives $N^{s(D_J)}$. The remaining part is given by $e^{-\mathbf{I}_1/N}(D_{\bar{J}})$.

Reformulation 3.

$$\varphi^{St}_{\mathfrak{gl}_N} = e^{\frac{\mathbf{I}_1}{N}} \cdot \varphi^{St}_{\mathfrak{sl}_N}.$$

The equivalence of this and the foregoing formulae follows from the fact that the weight systems $e^{\mathbf{I}_1/N}$ and $e^{-\mathbf{I}_1/N}$ are inverse to each other as elements of the completed algebra of weight systems.

Proof We shall prove the theorem in Reformulation 3. The Lie algebra $\mathfrak{gl}_N$ is a direct sum of $\mathfrak{sl}_N$ and the trivial one-dimensional Lie algebra generated by the identity matrix id_N. Its dual is $\mathrm{id}^*_N = \frac{1}{N}\mathrm{id}_N$. We can choose a basis for the vector space $\mathfrak{gl}_N$ consisting of the basis for $\mathfrak{sl}_N$ and the unit matrix id_N. To every chord we must assign either a pair of dual basis elements of $\mathfrak{sl}_N$, or the pair $(\mathrm{id}_N, \frac{1}{N}\mathrm{id}_N)$, which is equivalent to forgetting the chord and multiplying the obtained diagram by $\frac{1}{N}$. This means precisely that we are applying the weight system $e^{\mathbf{I}_1/N}$ to the chord subdiagram $D_{\bar{J}}$ formed by the forgotten chords, and the weight system $\varphi^{St}_{\mathfrak{sl}_N}$ to the chord subdiagram D_J formed by the remaining chords. $\qquad\square$

6.1.8 Algebra $\mathfrak{so}_N$ with the standard representation

In this case *a state* σ for D is a function on the set $[D]$ of chords of D with values in the set $\{1/2, -1/2\}$. The rule for the resolution of a chord according to its state is

, if $\sigma(a) = \frac{1}{2}$;   , if $\sigma(a) = -\frac{1}{2}$.

As before, $|\sigma|$ denotes the number of components of the obtained curve.

Theorem 6.12 (Bar-Natan 1991a, 1995a). *For the invariant form* $\langle x, y \rangle = Tr(xy)$,

$$\varphi^{St}_{\mathfrak{so}_N}(D) = \sum_\sigma \left(\prod_a \sigma(a)\right) N^{|\sigma|},$$

where the product is taken over all n chords of D, and the sum is taken over all 2^n states for D.

We leave the proof of this theorem to the reader as an exercise (Exercise 6.14 at the end of the chapter, to be precise). The outline of the proof is the same as in the case of $\mathfrak{gl}_N$.

Here is the table of values of $\varphi_{\mathfrak{so}_N}^{St}(D)$ for some basis elements of $\mathscr{A}$ of small degree:

D				
$\varphi_{\mathfrak{so}_N}^{St}(D)$	$\frac{1}{2}(N^2-N)$	$\frac{1}{4}(N^2-N)$	$\frac{1}{8}(N^2-N)$	$\frac{1}{8}N(-N^2+4N-3)$

D			
$\varphi_{\mathfrak{so}_N}^{St}(D)$	$\frac{1}{16}N(3N^2-8N+5)$	$\frac{1}{16}(2N^3-5N^2+3N)$	$\frac{1}{16}N(N^3-4N^2+6N-3)$

Exercises 6.15–6.19 contain additional information about this weight system.

6.1.9 Algebra $\mathfrak{sp}_{2N}$ with the standard representation

It turns out that

$$\varphi_{\mathfrak{sp}_{2N}}^{St}(D) = (-1)^{n+1}\varphi_{\mathfrak{so}_{-2N}}(D),$$

where the last notation means the formal substitution of $-2N$ instead of the variable N in the polynomial $\varphi_{\mathfrak{so}_N}(D)$, and n, as usual, is the degree of D. This implies that the weight system $\varphi_{\mathfrak{sp}_{2N}}^{St}$ does not provide any new knot invariant. Some details about it can be found in Bar-Natan (1991a, 1995a).

It would be interesting to find a combinatorial description of the weight systems for the exceptional simple Lie algebras E_6, E_7, E_8, F_4, G_2.

6.2 Lie algebra weight systems for the algebra $\mathscr{C}$

6.2.1 The construction

Since every closed diagram is a linear combination of chord diagrams, the weight system $\varphi_{\mathfrak{g}}$ can be treated as a function on $\mathscr{C}$ with values in $U(\mathfrak{g})$. It turns out that $\varphi_{\mathfrak{g}}$ can be evaluated on any closed diagram directly, often in a more convenient way.

The STU relation (page 116), which defines the algebra $\mathscr{C}$, gives us a hint how to do it. Namely, if we assign elements e_i, e_j to the endpoints of chords of the T- and U-diagrams from the STU relations,

$$T \; \underset{e_i \quad e_j}{\overset{e_i^* \quad e_j^*}{\bigvee}} \; - \; U \; \underset{e_j \quad e_i}{\overset{e_i^* \quad e_j^*}{\bigvee}} \; = \; S \; \underset{[e_i,e_j]}{\overset{e_i^* \quad e_j^*}{\bigvee}},$$

then it is natural to assign the commutator $[e_i, e_j]$ to the trivalent vertex on the Wilson loop of the S-diagram.

Strictly speaking, $[e_i, e_j]$ may not be a basis vector. A diagram with an endpoint marked by a linear combination of the basis vectors should be understood as a corresponding linear combination of diagrams marked by basis vectors. For this reason it will be more convenient to use the description of $\varphi_{\mathfrak{g}}$ given in Section 6.1.2, which does not depend on the choice of a basis. The formal construction goes as follows.

Let $C \in \mathbf{C}_n$ be a closed Jacobi diagram with a base point and $V = \{v_1, \ldots, v_m\}$ be the set of its external vertices ordered according to the orientation of the Wilson loop. We shall construct a tensor $T_{\mathfrak{g}}(C) \in \mathfrak{g}^{\otimes m}$ whose ith tensor factor $\mathfrak{g}$ corresponds to the element v_i of the set V. The weight system $\varphi_{\mathfrak{g}}$ evaluated on C is the image of $T_{\mathfrak{g}}(C)$ in $U(\mathfrak{g})$ under the natural projection.

In order to construct the tensor $T_{\mathfrak{g}}(C)$, consider the internal graph of C and cut all the edges connecting the trivalent vertices of C. This splits the internal graph of C into a union of elementary pieces of two types: chords and tripods, the latter consisting of one trivalent vertex and three legs with a fixed cyclic order. Here is an example:

To each leg of a chord or of a tripod we associate a copy of $\mathfrak{g}$, marked by this leg. Just as in Section 6.1.2, to each chord we can assign the tensor $\langle \cdot , \cdot \rangle$ considered as an element of $\mathfrak{g} \otimes \mathfrak{g}$, where the copies of $\mathfrak{g}$ in the tensor product are labelled by the ends of the chord. Similarly, to a tripod we associate the tensor $-J \in \mathfrak{g} \otimes \mathfrak{g} \otimes \mathfrak{g}$ defined as follows. The Lie bracket $[\cdot , \cdot]$ is an element of $\mathfrak{g}^* \otimes \mathfrak{g}^* \otimes \mathfrak{g}$. Identifying $\mathfrak{g}^*$ and $\mathfrak{g}$ by means of $\langle \cdot , \cdot \rangle$ we see that it corresponds to a tensor in $\mathfrak{g} \otimes \mathfrak{g} \otimes \mathfrak{g}$ which we denote by J. The order on the three copies of $\mathfrak{g}$ should be consistent with the cyclic order of legs in the tripod.

Now, take the tensor product $\widetilde{T}_{\mathfrak{g}}(C)$ of all the tensors assigned to the elementary pieces of the internal graph of C, with an arbitrary order of the factors. It is an element of the vector space $\mathfrak{g}^{\otimes(m+2k)}$ which has one copy of $\mathfrak{g}$ for each external vertex v_i of C and two copies of $\mathfrak{g}$ for each of the k edges where the internal graph of C has been cut. The form $\langle \cdot , \cdot \rangle$, considered now as a bilinear map of $\mathfrak{g} \otimes \mathfrak{g}$ to the ground field, induces a map

$$\mathfrak{g}^{\otimes(m+2k)} \longrightarrow \mathfrak{g}^{\otimes m}$$

by contracting a tensor over all pairs of coinciding labels. Apply this contraction to $\widetilde{T}_{\mathfrak{g}}(C)$; the result is a tensor in $\mathfrak{g}^{\otimes m}$ where the factors are indexed by the v_i, but possibly in a wrong order. Finally, re-arranging the factors in $\mathfrak{g}^{\otimes m}$ according to the cyclic order of vertices on the Wilson loop, we obtain the tensor $T_{\mathfrak{g}}(C)$ we were looking for.

Remark 6.13. Note that we associate the tensor $-J$, not J, to each tripod. This is not a matter of choice, but a reflection of our convention for the default cyclic order at the 3-valent vertices and the signs in the STU relation.

Remark 6.14. The construction of the tensor $T_{\mathfrak{g}}(C)$ consists of two steps: taking the product of the tensors that correspond to the elementary pieces of the internal graph of C, and contracting these tensors on the coinciding labels. These two steps can be performed "locally." For instance, let

$$C = \bigotimes.$$

The internal graph of C consists of three tripods. To obtain $T_{\mathfrak{g}}(C)$ we first take the tensor product of two copies of $-J$ and contract the resulting tensor on the coinciding labels, thus obtaining a tensor in $\mathfrak{g}^{\otimes 4}$. Graphically, this could be illustrated by glueing together the two tripods into a graph with four univalent vertices. Next, this graph is glued to the remaining tripod; this means taking product of the corresponding tensors and contracting it on a pair of labels:

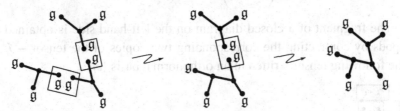

The result is, of course, the same as if we took first the tensor product of all three copies of $-J$ and then performed all the contractions.

The only choice involved in the construction of $T_{\mathfrak{g}}(C)$ is the order of the factors in the tensor product $\mathfrak{g}^{\otimes 3}$ that corresponds to a tripod. The following exercise shows that this order does not matter as long as it is consistent with the cyclic order of legs:

Exercise. Use the properties of $[\cdot,\cdot]$ and $\langle\cdot,\cdot\rangle$ to prove that the tensor J is skew-symmetric under the permutations of the three tensor factors (for the solution see Lemma A.2; note that we have already used this fact earlier in Section 6.1.1, in the proof of Theorem 6.1).

This shows that $T_{\mathfrak{g}}(C)$ is well-defined. Moreover, it produces a weight system: the definition of the commutator in the universal enveloping algebra implies that the element $\varphi_{\mathfrak{g}}(C)$, which is the image of $T_{\mathfrak{g}}(C)$ in $U(\mathfrak{g})$, satisfies the STU relation. If C is a chord diagram, this definition of $\varphi_{\mathfrak{g}}(C)$ coincides with the definition given in Section 6.1.2.

Since the STU relation implies both the AS and the IHX relations, $\varphi_{\mathfrak{g}}$ satisfies these relations too. Moreover, it is easy to see that the AS and the IHX relations are already satisfied for the function $C \mapsto T_{\mathfrak{g}}(C)$:

- The AS relation follows from the fact that the tensor J changes sign under odd permutations of the three factors in $\mathfrak{g} \otimes \mathfrak{g} \otimes \mathfrak{g}$.
- The IHX relation is a corollary of the Jacobi identity in $\mathfrak{g}$.

6.2.2 Example: inserting a bubble

Let us show how the construction of $T_{\mathfrak{g}}$ works on an example and prove the following lemma that relates the tensor corresponding to a "bubble" with the quadratic Casimir tensor.

Lemma 6.15. *For the Killing form $\langle \cdot, \cdot \rangle^K$ as the preferred invariant form, the tensor $T_{\mathfrak{g}}$ does not change if a bubble is inserted into an internal edge of a diagram:*

$$T_{\mathfrak{g}}\left(\overset{\cdots}{\underset{\cdots}{\multimap\!\!\circ\!\!\multimap}} \right) = T_{\mathfrak{g}}\left(\overset{\cdots}{\underset{\cdots}{\longmapsto}} \right).$$

Proof The fragment of a closed diagram on the left-hand side is obtained from two tripods by contracting the corresponding two copies of the tensor $-J$. This gives the following tensor written in an orthonormal basis $\{e_i\}$:

$$\leadsto \sum_{i,l} \sum_{k,j,k',j'} c_{ijk} c_{lk'j'} \langle e_k, e_{k'} \rangle^K \langle e_j, e_{j'} \rangle^K e_i \otimes e_l$$

$$= \sum_{i,l} \left(\sum_{j,k} c_{ijk} c_{lkj} \right) e_i \otimes e_l,$$

where c_{ijk} are the structure constants: $J = \sum_{i,j,k=1}^{d} c_{ijk} e_i \otimes e_j \otimes e_k$.

To compute the coefficient $\left(\sum_{j,k} c_{ijk} c_{lkj} \right)$ let us find the value of the Killing form

$$\langle e_i, e_l \rangle^K = \mathrm{Tr}(\mathrm{ad}_{e_i} \mathrm{ad}_{e_l}).$$

Since

$$\mathrm{ad}_{e_i}(e_s) = \sum_k c_{isk}e_k \quad \text{and} \quad \mathrm{ad}_{e_l}(e_t) = \sum_k c_{ltk}e_k \,,$$

the (j, r)-entry of the matrix of the product $\mathrm{ad}_{e_i}\,\mathrm{ad}_{e_l}$ will be $\sum_k c_{ikj}c_{lrk}$. Therefore,

$$\langle e_i, e_l \rangle^K = \sum_{k,j} c_{ikj}c_{ljk} = \sum_{j,k} c_{ijk}c_{lkj} \,.$$

Orthonormality of the basis $\{e_i\}$ implies that

$$\sum_{j,k} c_{ijk}c_{lkj} = \delta_{i,l}.$$

This means that the tensor corresponding to the fragment on the left-hand side in the statement of the lemma equals

$$\sum_i e_i \otimes e_i \,,$$

which is the quadratic Casimir tensor from the right-hand side. $\square$

Remark 6.16. If in the above lemma we use the bilinear form $\mu\langle \cdot, \cdot \rangle^K$ instead of the Killing form, the rule changes as follows:

$$T_{\mathfrak{g}}\left(\vcenter{\hbox{\includegraphics{fig1}}}\right) = \frac{1}{\mu}T_{\mathfrak{g}}\left(\vcenter{\hbox{\includegraphics{fig2}}}\right).$$

6.2.3 The universal $\mathfrak{sl}_2$ weight system for $\mathscr{C}$

Theorem 6.17 (Chmutov and Varchenko 1997). *For the invariant form $\langle x, y \rangle = \mathrm{Tr}(xy)$ the tensor $T_{\mathfrak{sl}_2}$ satisfies the following skein relation:*

$$T_{\mathfrak{sl}_2}\left(\vcenter{\hbox{\includegraphics{fig3}}}\right) = 2T_{\mathfrak{sl}_2}\left(\vcenter{\hbox{\includegraphics{fig4}}}\right) - 2T_{\mathfrak{sl}_2}\left(\vcenter{\hbox{\includegraphics{fig5}}}\right).$$

If the chosen invariant form is $\lambda\langle \cdot, \cdot \rangle$, then the coefficient 2 in this equation is replaced by $\frac{2}{\lambda}$.

Proof For the algebra $\mathfrak{sl}_2$ the Casimir tensor and the Lie bracket tensor are

$$C = \frac{1}{2}H \otimes H + E \otimes F + F \otimes E \,;$$

$$-J = -H \otimes F \otimes E + F \otimes H \otimes E + H \otimes E \otimes F - E \otimes H \otimes F - F \otimes E \otimes H + E \otimes F \otimes H.$$

Then the tensor corresponding to the elementary pieces on the right-hand side is equal to (we enumerate the vertices according to the tensor factors)

$$
T_{\mathfrak{sl}_2}\left(\begin{array}{c}1\qquad 4\\ \\ 2\qquad 3\end{array}\right) = \begin{aligned}&-H\otimes F\otimes H\otimes E+H\otimes F\otimes E\otimes H+F\otimes H\otimes H\otimes E-F\otimes H\otimes E\otimes H\\ &-H\otimes E\otimes H\otimes F+H\otimes E\otimes F\otimes H+E\otimes H\otimes H\otimes F-E\otimes H\otimes F\otimes H\\ &+2F\otimes E\otimes F\otimes E-2F\otimes E\otimes E\otimes F-2E\otimes F\otimes F\otimes E+2E\otimes F\otimes E\otimes F\end{aligned}
$$

$$
\begin{aligned}
= \ &2\Big(\tfrac{1}{4}H\otimes H\otimes H\otimes H+\tfrac{1}{2}H\otimes E\otimes F\otimes H+\tfrac{1}{2}H\otimes F\otimes E\otimes H+\tfrac{1}{2}E\otimes H\otimes H\otimes F\\
&+E\otimes E\otimes F\otimes F+E\otimes F\otimes E\otimes F+\tfrac{1}{2}F\otimes H\otimes H\otimes E+F\otimes E\otimes F\otimes E+F\otimes F\otimes E\otimes E\Big)\\
&-2\Big(\tfrac{1}{4}H\otimes H\otimes H\otimes H+\tfrac{1}{2}H\otimes E\otimes H\otimes F+\tfrac{1}{2}H\otimes F\otimes H\otimes E+\tfrac{1}{2}E\otimes H\otimes F\otimes H\\
&+E\otimes E\otimes F\otimes F+E\otimes F\otimes F\otimes E+\tfrac{1}{2}F\otimes H\otimes E\otimes H+F\otimes E\otimes E\otimes F+F\otimes F\otimes E\otimes E\Big)
\end{aligned}
$$

$$
= 2T_{\mathfrak{sl}_2}\left(\begin{array}{c}1\qquad 4\\ \\ 2\qquad 3\end{array}\right) - 2T_{\mathfrak{sl}_2}\left(\begin{array}{c}1\qquad 4\\ \\ 2\qquad 3\end{array}\right).
$$

$\square$

Remark 6.18. While transforming a closed diagram according to this theorem, a closed circle different from the Wilson loop may occur (see the example below). In this situation the circle should be replaced by the numeric factor $3 = \dim \mathfrak{sl}_2$, which is the trace of the identity operator in the adjoint representation of $\mathfrak{sl}_2$.

Remark 6.19. In the context of weight systems this relation was first noted in Chmutov and Varchenko (1997); afterwards, it was rediscovered several times. In a more general context of graphical notation for tensors it appeared already in Penrose (1971). In a certain sense, this relation goes back to Euler and Lagrange because it is an exact counterpart of the classical "*bac − cab*" rule,

$$
\mathbf{a} \times (\mathbf{b} \times \mathbf{c}) = \mathbf{b}(\mathbf{a} \cdot \mathbf{c}) - \mathbf{c}(\mathbf{a} \cdot \mathbf{b}),
$$

for the ordinary cross product of vectors in 3-space.

Example 6.20.

$$
\varphi_{\mathfrak{sl}_2}\left(\bigotimes\right) = 2\varphi_{\mathfrak{sl}_2}\left(\bigotimes\right) - 2\varphi_{\mathfrak{sl}_2}\left(\bigotimes\right) = 4\varphi_{\mathfrak{sl}_2}\left(\bigotimes\right)
$$

$$
-4\varphi_{\mathfrak{sl}_2}\left(\bigotimes\right) - 4\varphi_{\mathfrak{sl}_2}\left(\bigotimes\right) + 4\varphi_{\mathfrak{sl}_2}\left(\bigotimes\right)
$$

$$
= 12c^2 - 4c^2 - 4c^2 + 4c^2 = 8c^2.
$$

The next corollary implies the 6-term relation from Section 6.1.3.

Corollary 6.21.

$$\varphi_{\mathfrak{sl}_2}\left(\,\begin{array}{c}\text{(image)}\end{array}\,\right) = 2\varphi_{\mathfrak{sl}_2}\left(\begin{array}{c}\text{(image)}\end{array} - \begin{array}{c}\text{(image)}\end{array}\right); \qquad \varphi_{\mathfrak{sl}_2}\left(\begin{array}{c}\text{(image)}\end{array}\right) = 2\varphi_{\mathfrak{sl}_2}\left(\begin{array}{c}\text{(image)}\end{array} - \begin{array}{c}\text{(image)}\end{array}\right);$$

$$\varphi_{\mathfrak{sl}_2}\left(\begin{array}{c}\text{(image)}\end{array}\right) = 2\varphi_{\mathfrak{sl}_2}\left(\begin{array}{c}\text{(image)}\end{array} - \begin{array}{c}\text{(image)}\end{array}\right); \qquad \varphi_{\mathfrak{sl}_2}\left(\begin{array}{c}\text{(image)}\end{array}\right) = 2\varphi_{\mathfrak{sl}_2}\left(\begin{array}{c}\text{(image)}\end{array} - \begin{array}{c}\text{(image)}\end{array}\right).$$

6.2.4 The universal $\mathfrak{gl}_N$ weight system for $\mathscr{C}$

Let us apply the general procedure of the beginning of this section to the Lie algebra $\mathfrak{gl}_N$ equipped with the bilinear form $\langle e_{ij}, e_{kl}\rangle = \delta_{il}\delta_{jk}$ so that $e_{ij}^* = e_{ji}$. The corresponding universal weight system $\varphi_{\mathfrak{gl}_N}$ can be calculated with the help of a graphical calculus similar to that invented by R. Penrose (1971). (A modification of this calculus is used in Bar-Natan (1995a) to treat the standard representation of $\mathfrak{gl}_N$; see Section 6.2.5 below.)

According to the general procedure, in order to construct $T_{\mathfrak{gl}_N}$ we first erase the Wilson loop of the diagram, then place a copy of the tensor

$$-J = \sum_{i,j,k=1}^{N} (e_{ij} \otimes e_{jk} \otimes e_{ki} - e_{ij} \otimes e_{ki} \otimes e_{jk})$$

into each trivalent vertex and, finally, make contractions along all edges. Any interval component (that is, chord) of the internal graph of the diagram is replaced simply by a copy of the bilinear form understood as the element $\sum e_{ij} \otimes e_{ji}$. The cyclic order of the endpoints is remembered. The universal weight system $\varphi_{\mathfrak{gl}_N}$ is the image of $T_{\mathfrak{gl}_N}$ in the universal enveloping algebra $U(\mathfrak{gl}_N)$; in order to obtain it we simply omit the symbol of the tensor product in the above expressions:

$$-J = \sum_{i,j,k=1}^{N} (e_{ij}e_{jk}e_{ki} - e_{ij}e_{ki}e_{jk}).$$

Now, the formula for $-J$ can be visualized as a "resolution" of a trivalent vertex:

$$-J = \begin{array}{c}\text{(image)}\end{array} - \begin{array}{c}\text{(image)}\end{array}.$$

One should imagine a basis element e_{ij} attached to any of the three pairs of adjacent endpoints marked i and j, with i on the incoming line and j on the outgoing line. More generally, one may encode tensors by pictures as follows: specify k pairs of points, each point connected to some other point with an arrow, so that each pair consists of one arrowhead and one arrowtail. Each of the k arrows carries an index and each of the k pairs carries the generator e_{ij}, where i is the index of the incoming arrow and j is the index of the outgoing arrow. The tensor that corresponds to such a picture is obtained by fixing an order on the set of pairs (for closed Jacobi diagrams the order is defined below), taking the product of the k elements e_{ij} that correspond to the pairs, in the corresponding order, and then taking the sum over all the possible values of all the indices.

Choose one of the two pictures as above for each trivalent vertex (this may be thought of as "resolving" the trivalent vertex in a positive or negative way). The contraction along the edges means that we must glue together the small pictures. This is done in the following manner. For any edge connecting two trivalent vertices, the contraction along it always gives zero except for the case when we have $\langle e_{ij}, e_{ji} \rangle = 1$. Graphically, this means that we must connect the endpoints of the tripods and write one and the same letter on each connected component of the resulting curve. Note that the orientations on the small pieces of curves (that come from the cyclic order of the edges at every vertex) always agree for any set of resolutions, so that we get a set of oriented curves. We shall, further, add small intervals at each univalent vertex (now doubled) thus obtaining one connected oriented curve for every connected component of the initial diagram. To convert this curve into an element of the universal enveloping algebra, we write, at every univalent vertex, the element e_{ij} where the subscripts i and j are written in the order induced by the orientation on the curve. Then we take the product of all such elements in the order coming from the cyclic order of univalent vertices on the Wilson loop. As we know, the result is invariant under cyclic permutations of the factors. Finally, we sum up these results over all the resolutions of the triple points and all the indices from 1 to N.

Example 6.22. Let us compute the value of φ on the diagram

We have:

$$\longmapsto \sum_{i,j,k,l=1}^{N} (e_{ij}e_{jk}e_{kl}e_{li} - e_{ij}e_{jk}e_{li}e_{kl} - e_{ij}e_{ki}e_{jl}e_{lk} + e_{ij}e_{ki}e_{lk}e_{jl}).$$

As we know, $\varphi_{\mathfrak{gl}_N}$ of any diagram always belongs to the centre of $U(\mathfrak{gl}_N)$, so it can be written as a polynomial in N commuting variables $c_1, \ldots, c_N$ (the generalized Casimir elements, see, for instance, Zhelobenko 1973):

$$c_s = \sum_{i_1,\ldots,i_j=1}^{N} e_{i_1 i_2} e_{i_2 i_3} \cdots e_{i_{s-1} i_s} e_{i_s i_1}.$$

In the graphical notation

In particular,

$$c_1 = \bigcap = \sum_{i=1}^{N} e_{ii}$$

is the unit matrix (note that it is *not* the unit of the algebra $U(\mathfrak{g})$),

$$c_2 = \boxed{} = \sum_{i,j=1}^{N} e_{ij}e_{ji}$$

is the quadratic Casimir element. It is convenient to extend the list $c_1, \ldots, c_N$ of our variables by setting $c_0 = N$; the graphical notation for c_0 will be a circle. This is especially useful when speaking about the direct limit $\mathfrak{gl}_\infty = \lim_{N \to \infty} \mathfrak{gl}_N$.

For instance, the first term in the expansion of $\varphi(C)$ in the previous example is nothing but c_4; the whole alternating sum, after some transformations, turns out to be equal to $c_0^2(c_2 - c_1^2)$. Expressing the values of φ on closed Jacobi diagrams via the generators c_i is, in general, a nontrivial operation; a much clearer description exists for the analog of the map φ defined for the algebra of open diagrams; see Section 6.3.2.

Remark 6.23. If the resulting picture contains curves which have no univalent vertices, then, in the corresponding element of $U(\mathfrak{gl}_N)$ every such curve is replaced by the numerical factor N. This happens because every such curve leads to a sum where one of the indices does not appear among the subscripts of the product $e_{i_1 j_1} \ldots e_{i_s j_s}$, but the summation over this index still must be done. The proof is similar to that of the general lemma in Section 6.2.2, where a different bilinear form is used. If C stands for the diagram Θ with an inserted bubble, we obtain

$$\varphi(C) = \sum_{i,j,k=1}^{N} e_{ij} e_{ji} = N \sum_{i,j=1}^{N} e_{ij} e_{ji}.$$

6.2.5 *Algebra* $\mathfrak{gl}_N$ *with the standard representation*

The procedure for the closed diagrams repeats what we did with chord diagrams in Section 6.1.6. For a closed diagram $C \in \mathcal{C}_n$ with the set IV of t internal trivalent vertices, we double each internal edge and count the number of components of the resulting curve as before. The only problem here is how to connect the lines near an internal vertex. This can be decided by means of a state function $s : IV \to \{-1, 1\}$.

Theorem 6.24 (Bar-Natan 1995a). *Let* $\varphi_{\mathfrak{gl}_N}^{St}$ *be the weight system associated with the standard representation of the Lie algebra* $\mathfrak{gl}_N$ *with the invariant form* $\langle x, y \rangle = \mathrm{Tr}(xy)$.

For a closed diagram C and a state $s : IV \to \{-1, 1\}$ double every internal edge and connect the lines together in a neighbourhood of a vertex $v \in IV$ according to the state s:

and replace each external vertex as follows .

Let $|s|$ denote the number of components of the curve obtained in this way. Then

$$\varphi_{\mathfrak{gl}_N}^{St}(C) = \sum_{s} \left(\prod_{v} s(v) \right) N^{|s|},$$

where the product is taken over all t internal vertices of C, and the sum is taken over all the 2^t states for C.

A straightforward way to prove this theorem is to use the STU relation and Theorem 6.7. We leave the details to the reader.

Example 6.25. Let us compute the value $\varphi_{\mathfrak{gl}_N}^{St}$. There are four resolutions

of the triple points:

$$\prod s(v)=1 \qquad \prod s(v)=-1 \qquad \prod s(v)=-1 \qquad \prod s(v)=1$$
$$|s|=4 \qquad\qquad |s|=2 \qquad\qquad |s|=2 \qquad\qquad |s|=2$$

Therefore, $\varphi_{\mathfrak{gl}_N}^{St}$ $= N^4 - N^2$.

Other properties of the weight system $\varphi_{\mathfrak{gl}_N}^{St}$ are formulated in Exercises 6.27–6.31.

6.2.6 Algebra $\mathfrak{so}_N$ with standard representation

Here, a *state* for $C \in \mathbf{C}_n$ will be a function $s : IE \to \{-1, 1\}$ on the set IE of internal edges (those which are not on the Wilson loop). The value of a state indicates the way of doubling the corresponding edge:

, if $s(e) = 1$;   , if $s(e) = -1$.

In the neighbourhoods of trivalent and external vertices, we connect the lines in the standard fashion as before. For example, if the values of the state on three edges e_1, e_2, e_3 meeting at a vertex v are $s(e_1) = -1$, $s(e_2) = 1$, and $s(e_3) = -1$, then we resolve it as follows:

As before, $|s|$ denotes the number of components of the curve obtained in this way.

Theorem 6.26 (Bar-Natan 1995a). *Let $\varphi_{\mathfrak{so}_N}^{St}$ be the weight system associated with the standard representation of the Lie algebra $\mathfrak{so}_N$ with the invariant form $\langle x, y \rangle = Tr(xy)$. Then*

$$\varphi_{\mathfrak{so}_N}^{St}(C) = 2^{-\deg C} \sum_s \left(\prod_e s(e) \right) N^{|s|},$$

where the product is taken over all internal edges of C and the sum is taken over all the states $s : IE(C) \to \{1, -1\}$.

Proof First let us note that $\deg C = \#(IE) - \#(IV)$, where $\#(IV)$ and $\#(IE)$ denote the numbers of internal vertices and edges respectively. We prove the theorem by induction on $\#(IV)$.

If $\#(IV) = 0$ then C is a chord diagram. In this case the theorem coincides with Theorem 6.12.

If $\#(IV) \neq 0$ we can use the STU relation to decrease the number of internal vertices. Thus it remains to prove that the formula for $\varphi^{St}_{\mathfrak{so}_N}$ satisfies the STU relation. For this we split the 8 resolutions of the S diagram corresponding to the various values of s on the three edges of S into two groups which can be deformed to the corresponding resolutions of the T and the U diagrams:

$$ S\ \,\rightsquigarrow\ \left(\ -\ -\ +\ \right) $$

$$ -\left(\ -\ -\ +\ \right); $$

$$ T\ \,\rightsquigarrow\ \ -\ -\ +\ ; $$

$$ U\ \,\rightsquigarrow\ \ -\ -\ +\ . $$

$\square$

Example 6.27.

$$ \varphi^{St}_{\mathfrak{so}_N}\!\left(\ \right) = \frac{1}{4}(N^3 - 3N^2 + 2N) = \frac{1}{4}N(N-1)(N-2). $$

6.2.7 A small table of values of φ

The following table shows the values of φ on the generators of the algebra $\mathscr{C}$ of degrees $\leqslant 4$:

$$ t_1 = \ ,\quad t_2 = \ ,\quad t_3 = \ ,\quad t_4 = \ ,\quad w_4 = \ $$

for the simple Lie algebras A_1, A_2, A_3, A_4, B_2, B_3, C_3, D_4, G_2, computed by A. Kaishev (2000).

	t_1	t_2	t_3	t_4	w_4
A_1	c	$2c$	$4c$	$8c$	$8c^2$
A_2	c	$3c$	$9c$	$27c$	$9c^2 + 9c$
A_3	c	$4c$	$16c$	$64c$	e
A_4	c	$5c$	$25c$	$125c$	e
B_2	c	$3/2c$	$9/4c$	$27/8c$	d
B_3	c	$5/2c$	$25/4c$	$125/8c$	d
C_3	c	$4c$	$16c$	$64c$	d
D_4	c	$3c$	$9c$	$27c$	$3c^2 + 15c$
G_2	c	$2c$	$4c$	$8c$	$5/2c^2 + 11/3c$

Here c is the quadratic Casimir element of the corresponding enveloping algebra $U(\mathfrak{g})$, while d and e are the following (by degree) independent generators of $ZU(\mathfrak{g})$. Note that in this table all d's and e's have degree 4 and are defined modulo elements of smaller degrees. The exact expressions for d and e can be found in Kaishev (2000). In the table, the following ad-invariant forms are used: $\mathrm{Tr}(xy)$ for the algebras of A series, $\frac{1}{2}\mathrm{Tr}(xy)$ for the algebras of B, C, D series, and $\frac{1}{6}\mathrm{Tr}(xy)$ for the algebra G_2.

A look at the table shows that the mapping φ for almost all simple Lie algebras has a nontrivial kernel. In fact, $\varphi_{\mathfrak{g}}(t_1 t_3 - t_2^2) = 0$.

Exercise. Find a metrized Lie algebra $\mathfrak{g}$ such that the mapping $\varphi_{\mathfrak{g}}$ has a nontrivial cokernel.

6.3 Lie algebra weight systems for the algebra $\mathscr{B}$

The construction of the Lie algebra weight systems for open Jacobi diagrams is very similar to the procedure for closed diagrams. For a metrized Lie algebra $\mathfrak{g}$ we construct a weight system $\rho_{\mathfrak{g}} : \mathscr{B} \to S(\mathfrak{g})$, defined on the space of open diagrams $\mathscr{B}$ and taking values in the symmetric algebra of the vector space $\mathfrak{g}$ (in fact, even in its $\mathfrak{g}$-invariant subspace $S(\mathfrak{g})^{\mathfrak{g}}$).

Let $O \in \mathbf{B}$ be an open diagram. Choose an order on the set of its univalent vertices; then O can be treated as the internal graph of some closed diagram C_O. Following the recipe of Section 6.2, construct a tensor $T_{\mathfrak{g}}(C_O) \in \mathfrak{g}^{\otimes m}$, where m is the number of legs of the diagram O. Now we define $\rho_{\mathfrak{g}}(O)$ as the image of the tensor $T_{\mathfrak{g}}(C_O)$ in $S^m(\mathfrak{g})$ under the natural projection of the tensor algebra on $\mathfrak{g}$ onto $S(\mathfrak{g})$.

The choice of an order on the legs of O is of no importance. Indeed, it amounts to choosing an order on the tensor factors in the space $\mathfrak{g}^{\otimes m}$ to which the tensor $T_{\mathfrak{g}}(C_O)$ belongs. Since the algebra $S(\mathfrak{g})$ is commutative, the image of $T_{\mathfrak{g}}(C_O)$ is always the same.

6.3.1 The formal PBW theorem

The relation between the Lie algebra weight systems for the open diagrams and for the closed diagrams is expressed by the following theorem.

Theorem 6.28. *For any metrized Lie algebra $\mathfrak{g}$ the diagram*

$$
\begin{array}{ccc}
\mathscr{B} & \xrightarrow{\ \rho_{\mathfrak{g}}\ } & S(\mathfrak{g}) \\
{\scriptstyle \chi}\downarrow & & \downarrow{\scriptstyle \beta_{\mathfrak{g}}} \\
\mathscr{C} & \xrightarrow[\ \varphi_{\mathfrak{g}}\ ]{} & U(\mathfrak{g})
\end{array}
$$

where $\beta_{\mathfrak{g}}$ is the Poincaré–Birkhoff–Witt isomorphism, commutes.

Proof The assertion becomes evident as soon as one recalls the definitions of all the ingredients of the diagram: the isomorphism χ between the vector spaces $\mathscr{C}$ and $\mathscr{B}$ described in Section 5.7, the weight systems $\varphi_{\mathfrak{g}}$ and $\rho_{\mathfrak{g}}$, defined in Sections 6.1 and 6.3, and $\beta_{\mathfrak{g}}$, the Poincaré–Birkhoff–Witt isomorphism taking an element $x_1 x_2 ... x_n$ into the arithmetic mean of $x_{i_1} x_{i_2} ... x_{i_n}$ over all permutations $(i_1, i_2, ..., i_n)$ of the set $\{1, 2, ..., n\}$. Its restriction to the invariant subspace $S(\mathfrak{g})^{\mathfrak{g}}$ is a vector space isomorphism with the centre of $U(\mathfrak{g})$. $\square$

Example 6.29. Let $\mathfrak{g}$ be the Lie algebra $\mathfrak{so}_3$. It has a basis $\{a, b, c\}$ which is orthonormal with respect to the Killing form $\langle \cdot, \cdot \rangle^K$ and with the commutators $[a, b] = c$, $[b, c] = a$, $[c, a] = b$. As a metrized Lie algebra $\mathfrak{so}_3$ is isomorphic to the Euclidean 3-space with the cross product as a Lie bracket. The tensor that we put in every trivalent vertex in this case is

$$-J = -a \wedge b \wedge c$$
$$= -a \otimes b \otimes c - b \otimes c \otimes a - c \otimes a \otimes b + b \otimes a \otimes c + c \otimes b \otimes a + a \otimes c \otimes b.$$

Since the basis is orthonormal, the only way to get a non-zero element in the process of contraction along the edges is to choose the same basis element on either end of each edge. On the other hand, the formula for J shows that in every vertex we must choose a summand with different basis elements corresponding to the three edges. This leads to the following algorithm for computing the tensor $T_{\mathfrak{so}_3}(O)$ for a given diagram O: one must list all 3-colourings of the edges of the graph by three colours a, b, c such that the three colours at every vertex are always different, then sum up the tensor products of the elements written on the legs, each taken with the sign $(-1)^s$, where s is the number of *negative* vertices (that is, vertices where the colours, read counterclockwise, come in the negative order a, c, b).

For example, consider the diagram (the *Pont-Neuf diagram with parameters* $(1, 3, 0)$ in the terminology of O. Dasbach (2000); see also page 431 below):

$$O = \quad$$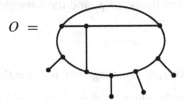

It has 18 edge 3-colourings, which can be obtained from the following three by permutations of (a, b, c):

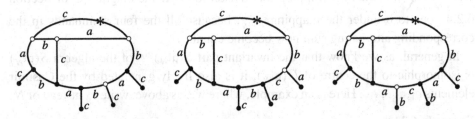

In these pictures, negative vertices are marked by small empty circles. Writing the tensors in the counterclockwise order starting from the marked point, we get:

$$2(a \otimes a \otimes a \otimes a + b \otimes b \otimes b \otimes b + c \otimes c \otimes c \otimes c)$$
$$+ a \otimes b \otimes b \otimes a + a \otimes c \otimes c \otimes a + b \otimes a \otimes a \otimes b$$
$$+ b \otimes c \otimes c \otimes b + c \otimes a \otimes a \otimes c + c \otimes b \otimes b \otimes c$$
$$+ a \otimes a \otimes b \otimes b + a \otimes a \otimes c \otimes c + b \otimes b \otimes a \otimes a$$
$$+ b \otimes b \otimes c \otimes c + c \otimes c \otimes a \otimes a + c \otimes c \otimes b \otimes b.$$

Projecting onto the symmetric algebra, we get:

$$\rho_{\mathfrak{so}_3}(O) = 2(a^2 + b^2 + c^2)^2.$$

This example shows that the weight system defined by the Lie algebra $\mathfrak{so}_3$, is closely related to the 4-colour theorem; see Bar-Natan (1997b) for details.

Example 6.30. For an arbitrary metrized Lie algebra $\mathfrak{g}$, let us calculate $\rho_{\mathfrak{g}}(w_n)$ where $w_n \in \mathscr{B}$ is the *wheel with n spokes*:

$$w_n := \quad$$
$$n \text{ spokes}$$

Note that n must be even; otherwise by Lemma 5.28 $w_n = 0$.

Dividing the wheel into n tripods, contracting the resulting tensors of rank 3 and projecting the result to $S(\mathfrak{g})$ we get

$$c_{j_1 i_1 j_2} \cdots c_{j_n i_n j_1} \cdot e_{i_1} \cdots e_{i_n} = \mathrm{Tr}\,(\mathrm{ad}\, e_{i_1} \cdots \mathrm{ad}\, e_{i_n}) \cdot e_{i_1} \cdots e_{i_n},$$

where $\{e_i\}$ is an orthonormal basis for $\mathfrak{g}$, and the summation by repeating indices is implied.

6.3.2 The universal $\mathfrak{gl}_N$ weight system for the algebra $\mathcal{B}$

The $\mathfrak{gl}_N$ weight system for the algebra $\mathcal{B}$ of open Jacobi diagrams is computed in exactly the same way as for the closed diagrams (see Section 6.2.4), only now we treat the variables e_{ij} as commuting elements of $S(\mathfrak{gl}_N)$. For instance, the diagram $B = $ >—< obtained by stripping the Wilson loop off the diagram C of Section 6.2.4 goes to 0 under the mapping $\rho_{\mathfrak{gl}_N}$, because all the four summands in the corresponding alternating sum now become equal.

In general, as we know that the invariant part $S(\mathfrak{gl}_N)^{\mathfrak{gl}_N}$ of the algebra $S(\mathfrak{gl}_N)$ is isomorphic to the centre of $U(\mathfrak{gl}_N)$, it is also freely generated by the Casimir elements $c_1, \ldots, c_N$. Here is an example, where we, as above, write c_0 instead of N:

Example 6.31.

$$\rho_{\mathfrak{gl}_N}\left(\ \right) = \quad - \quad - \quad + \quad$$

$$= \quad - \quad - \quad + \quad = 2(c_0 c_2 - c_1^2).$$

6.3.3 Invariants of string links and the algebra of necklaces

Recall that the algebra $\mathscr{A}(n)$ of closed diagrams for string links on n strands (see Section 5.11.2) has a $\mathscr{B}$-analog, denoted by $\mathscr{B}(n)$ and called the *algebra of coloured open Jacobi diagrams*; see page 145. In this section we shall describe the weight system generalizing $\rho_{\mathfrak{gl}_N} : \mathscr{B} \to S(\mathfrak{gl}_N)$ to a mapping

$$\rho_{\mathfrak{gl}_N}^{(n)} : \mathscr{B}(n) \to S(\mathfrak{gl}_N)^{\otimes n}.$$

A diagram in $\mathscr{B}(n)$ is an open Jacobi diagram with univalent vertices marked by numbers between 1 and n (or coloured by n colours). The vector space spanned by these elements modulo AS and IHX relations is what we call $\mathscr{B}(n)$. The colour-respecting averaging map $\chi_n : \mathscr{B}(n) \to \mathscr{A}(n)$, defined similarly to the simplest case $\chi : \mathscr{B} \to \mathscr{A}$ (see Section 5.7), is a linear isomorphism (see Bar-Natan 1995a).

Given a coloured open Jacobi diagram, we consider positive and negative resolutions of all its t trivalent vertices and get the alternating sum of 2^t pictures as on page 176 with the univalent legs marked additionally by the colours. For each resolution, mark the connected components by different variables i, j etc., then add small arcs near the univalent vertices and obtain a set of oriented closed curves. To each small arc (which was a univalent vertex before) there corresponds a pair of indices, say i and j. Write e_{ij} in the tensor factor of $S(\mathfrak{gl}_N)^{\otimes n}$ whose number is the number of that univalent vertex, and where i and j go in the order consistent with the orientation on the curve. Then take the sum over all subscripts from 1 to N.

To make this explanation clearer, let us illustrate it on a concrete example. Take the coloured diagram

$$D \quad = \quad 1\text{—} \underset{}{\overset{3}{\boxed{}}} \text{—} 2$$

with the blackboard (counterclockwise) cyclic order of edges meeting at trivalent vertices. Resolving all the trivalent vertices positively, we get the following collection of directed curves:

which, according to the above procedure, after filling in the gaps at univalent vertices, transcribes as the following element of $S(\mathfrak{gl}_N)^{\otimes 3}$:

$$\sum_{i,j,k,l,m,p=1}^{N} e_{lm}e_{jk} \otimes e_{ml}e_{ij} \otimes e_{ki} = N \cdot \sum_{i,j,k=1}^{N} e_{jk} \otimes e_{ij} \otimes e_{ki} \cdot \sum_{l,m=1}^{N} e_{lm} \otimes e_{ml} \otimes 1.$$

We see that the whole expression is the product of three elements corresponding to the three connected components of the closed curve. In particular, the factor N corresponds to the circle without univalent vertices and can be represented alternatively as multiplication by $\sum_{p=1}^{N} 1 \otimes 1 \otimes 1$.

As the choice of notations for the summation indices does not matter, we can write the obtained formula schematically as the product of three *necklaces*:

An n-coloured necklace is an arrangement of several *beads*, numbered between 1 to n, along an oriented circle (the default orientation is counterclockwise). A necklace can be uniquely denoted by a letter, say x, with a subscript consisting of the sequence of bead numbers chosen to be lexicographically smallest among all its cyclic shifts. Any n-coloured necklace corresponds to an element of the tensor power of $S(\mathfrak{gl}_N)$ according to the following rule. Mark each arc of the circle between two beads by a different integer variable i, j, etc. To each bead we assign the element e_{ij}, where i is the variable written on the incoming arc and j, on the outgoing arc. Then compose the tensor product of all these e_{ij}'s putting each into the tensor factor of $S(\mathfrak{gl}_N)^{\otimes n}$ whose number is the number of the bead under consideration, and take the sum of these expressions where each integer variable runs from 1 to N.

Examples (for $n = 3$):

$$x_{123} := \quad \begin{array}{c} {}^{2}\!\bigcirc^{\,1}_{\,3} \end{array} \quad \mapsto \quad \sum_{i,j,k=1}^{N} e_{ij} \otimes e_{jk} \otimes e_{ki}$$

$$x_{132} := \quad \begin{array}{c} {}^{3}\!\bigcirc^{\,1}_{\,2} \end{array} \quad \mapsto \quad \sum_{i,j,k=1}^{N} e_{jk} \otimes e_{ij} \otimes e_{ki}$$

$$x_{12123} := \quad \begin{array}{c} {}^{1}\!\bigcirc^{\,2}_{\,3}\!{}_{2} \end{array} \quad \mapsto \quad \sum_{i,j,k,l,m=1}^{N} e_{ij} e_{kl} \otimes e_{jk} e_{lm} \otimes e_{mi}.$$

(All the circles are oriented counterclockwise.)

We will call such elements of $S(\mathfrak{gl}_N)^{\otimes n}$ the *necklace elements*. By a theorem of S. Donkin (1993), the $\mathfrak{gl}_N$-invariant subspace of the algebra $S(\mathfrak{gl}_N)^{\otimes n}$ is generated by the necklace elements, and the algebraic relations between them may exist for small values of N, but disappear as $N \to \infty$, so that the invariant subspace of the direct limit $S(\mathfrak{gl}_\infty)^{\otimes n}$ is isomorphic to the free polynomial algebra generated by n-coloured necklaces.

Summing up, we can formulate the algorithm of finding the image of any given diagram in $S(\mathfrak{gl}_\infty)^{\otimes n}$ immediately in terms of necklaces. For a given coloured $\mathscr{B}$-diagram, take the alternating sum over all resolutions of the triple points. For each resolution convert the obtained picture into a collection of oriented closed curves, put the numbers $(1,\dots,n)$ of the univalent vertices on the places where they were before closing and thus get a product of necklaces. For instance:

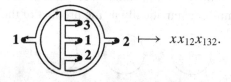

$$1 \!\!\overset{\text{}}{\underset{\text{}}{\Longleftarrow}}\!\! \begin{array}{c} 3 \\ 1 \\ 2 \end{array} \!\!\Longrightarrow\!\! 2 \quad \longmapsto \quad xx_{12}x_{132}.$$

Exercise. Find a direct proof, without appealing to Lie algebras, that the described mapping ρ into the necklace algebra provides a weight system, that is, satisfies the

AS and IHX relations. *Hint*: see Dasbach (2000), where this is done for the case $n = 1$.

One application of (unicoloured) necklace weight system is the lower bound on the dimensions of the spaces $\mathcal{V}_n$ for knots; see Section 14.5.4.

Another application – of the 2-coloured necklace weight system – is the proof that there exists a degree 7 Vassiliev invariant that is capable of detecting the change of orientation in two-component string links; see Duzhin and Karev (2007). This fact follows from the computation

$$\rho\left(\begin{array}{c} \end{array} \right) = x(x_{1121222} - x_{1122212}) + 3x_2(x_{112212} - x_{112122})$$

which implies that the depicted diagram is non-zero in $\mathcal{B}(2)$.

6.4 Lie superalgebra weight systems

The construction of Lie algebra weight systems works for algebraic structures more general than Lie algebras (Vaintrob 1994; Figueroa-O'Farrill *et al.* 1997; Hinich and Vaintrob 2002), namely for the analogs of metrized Lie algebras in categories more general than the category of vector spaces. An example of such a category is that of super vector spaces; Lie algebras in this category are called *Lie superalgebras*. The definition and basic properties of Lie superalgebras are discussed in Section A.1.8; we refer the reader to Kac (1977a, 1977b) for more details.

6.4.1 Weight systems for Lie superalgebras

Recall the construction of the Lie algebra weight systems for the closed diagrams as described in Sections 6.1.2 and 6.2. It consists of several steps. First, the internal graph of the diagram is cut into tripods and chords. Then to each tripod we assign a tensor in $\mathfrak{g}^{\otimes 3}$ coming from the Lie bracket, and to each chord – a tensor in $\mathfrak{g}^{\otimes 2}$ coming from the invariant form. Next, we take the tensor product $\widetilde{T}_\mathfrak{g}$ of all these tensors and perform contractions on the pairs of indices corresponding to the points where the diagram was cut. Finally, we re-arrange the factors in the tensor product and this gives the tensor $T_\mathfrak{g}$ whose image in $U(\mathfrak{g})$ is the weight system we were after.

If $\mathfrak{g}$ is a metrized Lie superalgebra, the very same construction works with only one modification: re-arranging the factors in the final step should be done with certain care. Instead of simply permuting the factors in the tensor product, one should use a certain representation of the symmetric group S_m on m letters that exists on the mth tensor power of any super vector space.

This representation is defined as follows. Let

$$S : \mathfrak{g} \otimes \mathfrak{g} \to \mathfrak{g} \otimes \mathfrak{g}$$

be the linear map that sends $u \otimes v$ to $(-1)^{p(u)p(v)} v \otimes u$, where u, v are homogeneous (that is, purely even or purely odd) elements of $\mathfrak{g}$ and $p(x)$ stands for the parity of x. The map S is an involution; in other words, it defines a representation of the symmetric group S_2 on the vector space $\mathfrak{g}^{\otimes 2}$. More generally, the representation of S_m on $\mathfrak{g}^{\otimes m}$ is defined by sending the transposition $(i, i+1)$ to $\mathrm{id}^{\otimes i-1} \otimes S \otimes \mathrm{id}^{m-i-1}$. If the odd part of $\mathfrak{g}$ is zero, this representation simply permutes the factors in the tensor product.

We shall use the same notation $\varphi_{\mathfrak{g}}$ for the resulting weight system.

Example 6.32. Let $\mathfrak{g}$ be a metrized Lie superalgebra with the orthonormal bases $e_1, \ldots, e_m$ and $f_1, \ldots, f_r$ for the even and the odd parts, respectively. Denote by D the diagram ⊗. Then

$$\varphi_{\mathfrak{g}}(D) = \sum_{i=1}^{m} \sum_{j=1}^{m} e_i e_j e_i e_j - \sum_{i=1}^{m} \sum_{j=1}^{r} (e_i f_j e_i f_j + f_j e_i f_j e_i) + \sum_{i=1}^{r} \sum_{j=1}^{r} f_i f_j f_i f_j.$$

Exercise. Write down the expression for $\varphi_{\mathfrak{g}}\left(\bigoplus\right)$. This exercise is resolved in Figueroa-O'Farrill *et al.* (1997) (example 2 in section 1.3) though with a different base point and a not necessarily orthonormal basis.

Exercise. Show that $\varphi_{\mathfrak{g}}$ is a well-defined weight system with values in the (super) centre of $U(\mathfrak{g})$. In particular, prove that $\varphi_{\mathfrak{g}}$ satisfies the 4T relation.

6.4.2 The $\mathfrak{gl}(1|1)$ weight system

The simplest nontrivial example of a Lie superalgebra is the space $\mathfrak{gl}(1|1)$ of endomorphisms of the super vector space of dimension $1 + 1$. The universal weight system for $\mathfrak{gl}(1|1)$ can be calculated with the help of a recursive formula similar to the formula for $\mathfrak{sl}_2$ (see Section 6.1.3).

The (super) centre of $U(\mathfrak{gl}(1|1))$ is a polynomial algebra in two generators c and h, where c is the quadratic Casimir element and $h \in \mathfrak{gl}(1|1)$ is the identity matrix.

Theorem 6.33 (Figueroa-O'Farrill *et al.* 1997). *Let $\varphi_{\mathfrak{gl}(1|1)}$ be the weight system associated with $\mathfrak{gl}(1|1)$ with the invariant form $\langle x, y \rangle = \mathrm{sTr}(xy)$.*

Take a chord diagram D and choose a chord a of D. Then

$$\varphi_{\mathfrak{gl}(1|1)}(D) = c\varphi_{\mathfrak{gl}(1|1)}(D_a) + h^2 \sum_{1 \leqslant i \leqslant k} \varphi_{\mathfrak{gl}(1|1)}(D_i)$$

$$- h^2 \sum_{1 \leqslant i < j \leqslant k} \left(\varphi_{\mathfrak{gl}(1|1)}(D_{i,j}^{+-}) + \varphi_{\mathfrak{gl}(1|1)}(D_{i,j}^{-+}) - \varphi_{\mathfrak{gl}(1|1)}(D_{i,j}^{l}) - \varphi_{\mathfrak{gl}(1|1)}(D_{i,j}^{r}) \right),$$

where:

- *k is the number of chords that intersect the chord a;*
- *D_a is the chord diagram obtained from D by deleting the chord a;*
- *for each chord a_i that intersects a, the diagram D_i is obtained from D by deleting the chords a and a_i;*
- *$D_{i,j}^{+-}$, $D_{i,j}^{-+}$, $D_{i,j}^{l}$ and $D_{i,j}^{r}$ are the chord diagrams obtained from D_a in the following way. Draw the diagram D so that the chord a is vertical. Consider an arbitrary pair of chords a_i and a_j different from a and such that each of them intersects a. Denote by p_i and p_j the endpoints of a_i and a_j that lie to the left of a and by p_i^*, p_j^* the endpoints of a_i and a_j that lie to the right. Delete from D the chords a, a_i and a_j and insert one new chord: (p_i, p_j^*) for $D_{i,j}^{+-}$, (p_j, p_i^*) for $D_{i,j}^{-+}$, (p_i, p_j) for $D_{i,j}^{l}$ and (p_i^*, p_j^*) for $D_{i,j}^{r}$:*

In particular, $\varphi_{\mathfrak{gl}(1|1)}(D)$ is a polynomial in c and h^2.

We refer to Figueroa-O'Farrill *et al.* (1997) for the proof.

6.4.3 Invariants not coming from Lie algebras

Lie algebra weight systems produce infinite series of examples of Vassiliev invariants. J. Kneissler has shown in Kneissler (1997) that all invariants up to order 12 come from Lie algebras. However, in general, this is not the case. P. Vogel (1997) has used the family of Lie superalgebras $D(1, 2, \alpha)$ depending on the parameter α; he showed that these algebras produce invariants which cannot be expressed as combinations of invariants coming from Lie algebras. (J. Lieberum (1999) gave an example of an order 17 closed diagram detected by $D(1, 2, \alpha)$ but not by semi-simple Lie algebra weight systems.) Moreover, there exist Vassiliev invariants that do no come from Lie (super) algebras (Vogel 1997; Lieberum 1999).

The main technical tool for proving these results is the algebra Λ constructed by Vogel. In the next chapter we shall consider the *algebra of 3-graphs* closely related to Vogel's algebra Λ.

Exercises

6.1 Let $(\mathfrak{g}_1, \langle \cdot, \cdot \rangle_1)$ and $(\mathfrak{g}_2, \langle \cdot, \cdot \rangle_2)$ be two metrized Lie algebras. Then their direct sum $\mathfrak{g}_1 \oplus \mathfrak{g}_2$ is also a metrized Lie algebra with respect to the form $\langle \cdot, \cdot \rangle_1 \oplus \langle \cdot, \cdot \rangle_2$. Prove that $\varphi_{\mathfrak{g}_1 \oplus \mathfrak{g}_2} = \varphi_{\mathfrak{g}_1} \cdot \varphi_{\mathfrak{g}_2}$.

The general aim of Exercises (6.2)–(6.8) is to compare the behaviour of $\varphi_{\mathfrak{sl}_2}(D)$ with that of the chromatic polynomial of a graph (see page 106). In these exercises we use the form $\langle x, y \rangle = 2\mathrm{Tr}(xy)$ as the invariant form.

6.2 (Chmutov and Lando 2007). Prove that $\varphi_{\mathfrak{sl}_2}(D)$ depends only on the intersection graph $\Gamma(D)$ of the chord diagram D.

6.3 Prove that the polynomial $\varphi_{\mathfrak{sl}_2}(D)$ has alternating coefficients.

6.4 Show that for any chord diagram D the polynomial $\varphi_{\mathfrak{sl}_2}(D)$ is divisible by c.

6.5 *Prove that the sequence of coefficients of the polynomial $\varphi_{\mathfrak{sl}_2}(D)$ is unimodal (that is, its absolute values form a sequence with only one maximum).

6.6 Let D be a chord diagram with n chords for which $\Gamma(D)$ is a tree. Prove that $\varphi_{\mathfrak{sl}_2}(D) = c(c-1)^{n-1}$.

6.7 Prove that the highest three terms of the polynomial $\varphi_{\mathfrak{sl}_2}(D)$ are

$$\varphi_{\mathfrak{sl}_2}(D) = c^n - e \cdot c^{n-1} + (e(e-1)/2 - t + 2q) \cdot c^{n-2} - \dots,$$

where e is the number of intersections of chords of D; t and q are the numbers of *triangles* and *quadrangles* of D respectively. A triangle is a subset of three chords of D with all pairwise intersections. A quadrangle of D is an unordered subset of four chords a_1, a_2, a_3, a_4 which form a cycle of length four. This means that, after a suitable relabeling, a_1 intersects a_2 and a_4, a_2 intersects a_3 and a_1, a_3 intersects a_4 and a_2, a_4 intersects a_1 and a_3 and any other intersections are allowed. For example,

$$e\left(\includegraphics{}\right) = 6, \qquad t\left(\includegraphics{}\right) = 4, \qquad q\left(\includegraphics{}\right) = 1.$$

6.8 (Vaintrob 1997). Define *vertex multiplication* of chord diagrams as follows:

$$\oplus \vee \otimes \; := \; \text{[diagram]} \; = \; \oplus.$$

Of course, the result depends of the choice of vertices where multiplication is performed. Prove that for any choice

$$\varphi_{\mathfrak{sl}_2}(D_1 \vee D_2) = \frac{\varphi_{\mathfrak{sl}_2}(D_1) \cdot \varphi_{\mathfrak{sl}_2}(D_2)}{c}.$$

6.9 (Bar-Natan and Garoufalidis 1996) Let c_n be the coefficient of t^n in the Conway polynomial and D a chord diagram of degree n. Prove that $\mathrm{symb}(c_n)(D)$ is equal, modulo 2, to the determinant of the adjacency matrix for the intersection graph $\Gamma(D)$.

6.10 Let D_n be the chord diagram with n chords whose intersection graph is a circle, $n \geqslant 3$. Prove that $\varphi_{\mathfrak{gl}_N}^{St}(D_n) = \varphi_{\mathfrak{gl}_N}^{St}(D_{n-2})$. Deduce that $\varphi_{\mathfrak{gl}_N}^{St}(D_n) = N^2$ for odd n and $\varphi_{\mathfrak{gl}_N}^{St}(D_n) = N^3$ for even n.

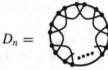

$$D_n =$$

n chords

6.11 Work out a proof of the theorem from Section 6.1.7 about the $\mathfrak{sl}_N$ weight system with standard representation, similar to the one given in Section 6.1.6. Use the basis of the vector space $\mathfrak{sl}_N$ consisting of the matrices e_{ij} for $i \neq j$ and the matrices $e_{ii} - e_{i+1,i+1}$.

6.12 Prove that $\varphi_{\mathfrak{sl}_N}^{'St} \equiv \varphi_{\mathfrak{gl}_N}^{'St}$.

Hint. $\varphi_{\mathfrak{sl}_N}^{'St} = e^{-\frac{N^2-1}{N}\mathbf{I}_1} \cdot \varphi_{\mathfrak{sl}_N}^{St} = e^{-N\mathbf{I}_1} \cdot \varphi_{\mathfrak{gl}_N}^{St} = \varphi_{\mathfrak{gl}_N}^{'St}$.

6.13 Compare the symbol of the coefficient j_n of the modified Jones polynomial (Section 3.6.2) with the weight system coming from $\mathfrak{sl}_2$, and prove that

$$\mathrm{symb}(j_n) = \frac{(-1)^n}{2} \varphi_{\mathfrak{sl}_2}^{'St}.$$

Hint. Compare the formula for $\varphi_{\mathfrak{sl}_2}^{'St}$ from the previous problem and the formula for $\mathrm{symb}(j_n)$ from Section 3.6.2, and prove that

$$(|s| - 1) \equiv \#\{\text{chords } c \text{ such that } s(c) = 1\} \mod 2.$$

6.14 Work out a proof of Theorem 6.12 about the $\mathfrak{so}_N$ weight system in standard representation. Use the basis of $\mathfrak{so}_N$ formed by the matrices $e_{ij} - e_{ji}$ for $i < j$. (In case of difficulty consult Bar-Natan 1991a, 1995a.)

6.15 Work out a proof, similar to the proof of Proposition 6.10, that $\varphi_{\mathfrak{so}_N}^{St}(D)$ depends only on the intersection graph of D.

6.16 (Mellor 2003). For any subset $J \subseteq [D]$, let M_J denote the *marked adjacency matrix* of the intersection graph of D over the field $\mathbb{F}_2$, that is the adjacency matrix M with each diagonal element corresponding to an element of J replaced by 1. Prove that

$$\varphi_{\mathfrak{so}_N}^{St}(D) = \frac{N^{n+1}}{2^n} \sum_{J \subseteq [D]} (-1)^{|J|} N^{-\text{rank}(M_J)},$$

where the rank is computed as the rank of a matrix over $\mathbb{F}_2$. This gives another proof of the fact that $\varphi_{\mathfrak{so}_N}^{St}(D)$ depends only on the intersection graph $\Gamma(D)$.

6.17 Show that $N = 0$ and $N = 1$ are roots of the polynomial $\varphi_{\mathfrak{so}_N}^{St}(D)$ for any chord diagram D.

6.18 Let D be a chord diagram with n chords, such that the intersection graph $\Gamma(D)$ is a tree. Show that $\varphi_{\mathfrak{so}_N}^{St}(D) = \frac{1}{2^n} N(N-1)$.

6.19 Let D_n be the chord diagram from Exercise 6.10. Prove that

(a) $\varphi_{\mathfrak{so}_N}^{St}(D_n) = \frac{1}{2} \left(\varphi_{\mathfrak{so}_N}^{St}(D_{n-2}) - \varphi_{\mathfrak{so}_N}^{St}(D_{n-1}) \right)$;

(b) $\varphi_{\mathfrak{so}_N}^{St}(D_n) = \frac{1}{(-2)^n} N(N-1)(a_{n-1}N - a_n)$, where the recurrent sequence a_n is defined by $a_0 = 0$, $a_1 = 1$, $a_n = a_{n-1} + 2a_{n-2}$.

6.20 Compute the values of $\varphi_{\mathfrak{sl}_2}$ on the closed diagrams and , and show that these two diagrams are linearly independent.

   Answer: $16c^2$, $64c$.

6.21 Let $\bar{t}_n \in \mathcal{C}_{n+1}$ be a closed diagram with n legs as shown in the figure.
Show that $\varphi_{\mathfrak{sl}_2}(\bar{t}_n) = 2^n c$.

$\bar{t}_n := $

n legs

6.22 Let $\bar{w}_n \in \mathcal{C}_n$ be a wheel with n spokes.
Show that

$$\varphi_{\mathfrak{sl}_2}(\bar{w}_2) = 4c, \quad \varphi_{\mathfrak{sl}_2}(\bar{w}_3) = 4c, \text{ and}$$

$$\varphi_{\mathfrak{sl}_2}(\bar{w}_n) = 2c \cdot \varphi_{\mathfrak{sl}_2}(\bar{w}_{n-2}) + 2\varphi_{\mathfrak{sl}_2}(\bar{w}_{n-1}) - 2^{n-1}c.$$

$\bar{w}_n := $

n spokes

6.23 Let $w_{2n} \in \mathcal{B}_{2n}$ be a wheel with $2n$ spokes and $(\bullet\!\!-\!\!\bullet)^n \in \mathcal{B}_n$ be the nth power of the element $\bullet\!\!-\!\!\bullet$ in the algebra $\mathcal{B}$.
Show that for the tensor $T_{\mathfrak{sl}_2}$ as in 6.2.3 the following equality holds: $T_{\mathfrak{sl}_2}(w_{2n}) = 2^{n+1} T_{\mathfrak{sl}_2}((\bullet\!\!-\!\!\bullet)^n)$. Therefore, $\rho_{\mathfrak{sl}_2}(w_{2n}) = 2^{n+1} \rho_{\mathfrak{sl}_2}((\bullet\!\!-\!\!\bullet)^n)$.

$w_{2n} := $

$2n$ spokes

$(\bullet\!\!-\!\!\bullet)^n := $ $\Big\}$ n segments

6.24 Let $p \in \mathcal{P}_n^k \subset \mathcal{C}_n$ be a primitive element of degree $n > 1$ with at most k external vertices. Show that $\varphi_{\mathfrak{sl}_2}(p)$ is a polynomial in c of degree $\leq k/2$.

   Hint. Use Theorem 6.17 and the calculation of $\varphi_{\mathfrak{sl}_2}(\bar{t}_3)$ from Exercise 6.21.

6.25 Let $\varphi'_{\mathfrak{sl}_2}$ be the deframing of the weight system $\varphi_{\mathfrak{sl}_2}$ according to the procedure of Section 4.5.4. Show that for any element $D \in \mathscr{A}'_n$, the value $\varphi'_{\mathfrak{sl}_2}(D)$ is a polynomial in c of degree $\leqslant [n/2]$.

Hint. Use the previous exercise, Exercise 4.8, and Section 5.5.2.

6.26 Denote by V_k the $(k+1)$-dimensional irreducible representation of $\mathfrak{sl}_2$ (see Example A.4 on page 459). Let $\varphi'^{V_k}_{\mathfrak{sl}_2}$ be the corresponding weight system. Show that for any element $D \in \mathscr{A}_n$ of degree n, $\varphi'^{V_k}_{\mathfrak{sl}_2}(D)/k$ is a polynomial in k of degree at most n.

Hint. The Casimir number (see page 460) in this case is $\frac{k^2-1}{2}$.

6.27 Let $D \in \mathscr{C}_n$ $(n > 1)$ be a connected closed diagram. Prove that $\varphi^{St}_{\mathfrak{gl}_N}(D) = \varphi^{St}_{\mathfrak{sl}_N}(D)$.

Hint. For the Lie algebra $\mathfrak{gl}_N$ the tensor $J \in \mathfrak{gl}_N^{\otimes 3}$ lies in the subspace $\mathfrak{sl}_N^{\otimes 3}$.

6.28 Consider a closed diagram $D \in \mathscr{C}_n$ and a $\mathfrak{gl}_N$-state s for it (see page 178). Construct a surface $\Sigma_s(D)$ by attaching a disk to the Wilson loop, replacing each edge by a narrow band and glueing the bands together at the trivalent vertices with a twist if $s = -1$, and without a twist if $s = 1$. Here is an example:

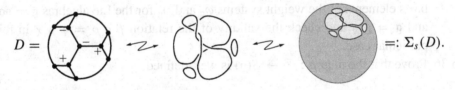

(a) Show that the surface $\Sigma_s(D)$ is orientable.

(b) Compute the Euler characteristic of $\Sigma_s(D)$ in terms of D, and show that it depends only on the degree n of D.

(c) Prove that $\varphi^{St}_{\mathfrak{gl}_N}(D)$ is an odd polynomial for even n, and it is an even polynomial for odd n.

6.29 Show that $N = 0$, $N = -1$, and $N = 1$ are roots of the polynomial $\varphi^{St}_{\mathfrak{gl}_N}(D)$ for any closed diagram $D \in \mathscr{C}_n$ $(n > 1)$.

6.30 Compute $\varphi^{St}_{\mathfrak{gl}_N}(\overline{t_n})$, where $\overline{t_n}$ is the closed diagram from Exercise 6.21.

Answer. For $n \geqslant 1$, $\varphi^{St}_{\mathfrak{gl}_N}(\overline{t_n}) = N^n(N^2 - 1)$.

6.31 For the closed diagram $\overline{w_n}$ as in Exercise 6.22, prove that $\varphi^{St}_{\mathfrak{gl}_N}(\overline{w_n}) = N^2(N^{n-1} - 1)$ for odd n, and $\varphi^{St}_{\mathfrak{gl}_N}(\overline{w_n}) = N(N^n + N^2 - 2)$ for even n.

Hint. Prove the recurrent formula $\varphi^{St}_{\mathfrak{gl}_N}(\overline{w_n}) = N^{n-1}(N^2 - 1) + \varphi^{St}_{\mathfrak{gl}_N}(\overline{w_{n-2}})$ for $n \geqslant 3$.

6.32 Extend the definition of the weight system $\mathrm{symb}(c_n)$ of the coefficient c_n of the Conway polynomial to $\mathscr{C}_n$, and prove that

$$\mathrm{symb}(c_n)(D) = \sum_s \left(\prod_v s(v) \right) \delta_{1,|s|} \,,$$

where the states s are precisely the same as in the theorem of Section 6.2.5 for the weight system $\varphi^{St}_{\mathfrak{gl}_N}$. In other words, prove that $\mathrm{symb}(c_n)(D)$ is equal to the coefficient of N in the polynomial $\varphi^{St}_{\mathfrak{gl}_N}(D)$. In particular, show that $\mathrm{symb}(c_n)(\overline{w_n}) = -2$ for even n, and $\mathrm{symb}(c_n)(\overline{w_n}) = 0$ for odd n.

6.33 (a) Let $D \in \mathscr{C}$ be a closed diagram with at least one internal trivalent vertex. Prove that $N = 2$ is a root of the polynomial $\varphi^{St}_{\mathfrak{so}_N}(D)$.

(b) Deduce that $\varphi^{St}_{\mathfrak{so}_2}(D) = 0$ for any primitive closed diagram D.

Hint. Consider the eight states that differ only on three edges meeting at an internal vertex (see page 180). Show that the sum over these eight states, $\sum \mathrm{sign}(s) 2^{|s|}$, equals zero.

6.34 Prove that $\varphi^{St}_{\mathfrak{so}_N}(\overline{t_n}) = \frac{N-2}{2} \varphi^{St}_{\mathfrak{so}_N}(\overline{t_{n-1}})$ for $n > 1$, where $\overline{t_n}$ is as in Exercise 6.21.

In particular, $\varphi^{St}_{\mathfrak{so}_N}(\overline{t_n}) = \frac{(N-2)^n}{2^{n+1}} N(N-1)$.

6.35 Using some bases in $\mathscr{C}_2$ and $\mathscr{B}_2$, find the matrix of the isomorphism χ, then calculate (express as polynomials in the standard generators) the values on the basis elements of the weight systems $\varphi_{\mathfrak{g}}$ and $\rho_{\mathfrak{g}}$ for the Lie algebras $\mathfrak{g} = \mathfrak{so}_3$ and $\mathfrak{g} = \mathfrak{gl}_N$ and check the validity of the relation $\beta \circ \rho = \varphi \circ \chi$ in this particular case.

6.36 Prove that the map $\rho : \mathscr{B} \to S(\mathfrak{g})$ is well-defined.

7

Algebra of 3-graphs

The *algebra of 3-graphs* Γ, introduced in Duzhin *et al.* (1998), is related to the diagram algebras $\mathscr{C}$ and $\mathscr{B}$. The difference between 3-graphs and closed diagrams is that 3-graphs do not have a distinguished cycle (Wilson loop); neither do they have univalent vertices, which distinguishes them from open diagrams. Strictly speaking, there are two different algebra structures on the space of 3-graphs, given by the edge (Section 7.2) and the vertex (Section 7.3) products. The space Γ is closely related to the Vassiliev invariants in several ways:

- The vector space Γ is isomorphic to the subspace $\mathscr{P}^2$ of the primitive space $\mathscr{P} \subset \mathscr{C}$ spanned by the connected diagrams with two legs (Section 7.4.1).
- The algebra Γ acts on the primitive space $\mathscr{P}$ in two ways, via the edge, and via the vertex products (see Sections 7.4.1 and 7.4.2). These actions behave nicely with respect to Lie algebra weight systems (see Chapter 6); as a consequence, the algebra Γ is as good a tool for the proof of existence of non-Lie-algebraic weight systems as the algebra Λ in Vogel's original approach (Section 7.6.4).
- The vector space Γ describes the combinatorics of finite type invariants of integral homology 3-spheres in much the same way as the space of chord diagrams describes the combinatorics of Vassiliev knot invariants. This topic, however, lies outside of the scope of our book and we refer an interested reader to Ohtsuki (2002).

Unlike $\mathscr{C}$ and $\mathscr{B}$, the algebra Γ does not have any natural coproduct.

7.1 The space of 3-graphs

A *3-graph* is a connected 3-valent graph with a fixed cyclic order of half-edges at each vertex. Two 3-graphs are *isomorphic* if there exists a graph isomorphism between them that preserves the cyclic order of half-edges at every vertex. The *degree*, or *order*, of a 3-graph is defined as half the number of its vertices. It will be convenient to consider a circle with no vertices on it as a 3-graph of degree 0 (even though, strictly speaking, it is not a graph).

Example 7.1. Up to an isomorphism, there are three different 3-graphs of degree 1:

Remark 7.2. Graphs with a cyclic order of half-edges at each vertex are often called *ribbon graphs* (see Lando and Zvonkin 2004), as every such graph can be represented as an orientable surface with boundary obtained by "thickening" the graph:

To be more precise, given a graph, we replace each of its vertices and each of its edges by an oriented disk (imagine that the disks for the vertices are "round" while the disks for the edges are "oblong"). The disks are glued together along segments of their boundary in agreement with the orientation and with the prescribed cyclic order at each vertex; the cyclic order at a vertex is taken in the positive direction of the vertex-disk boundary.

Definition 7.3. The *space of 3-graphs* Γ_n is the $\mathbb{Q}$-vector space spanned by all 3-graphs of degree n modulo the AS and IHX relations (see Section 5.2).

In particular, the space Γ_0 is one-dimensional and spanned by the circle.

Exercise. Check that the 3-graph on the right is equal to zero as an element of the space Γ_3.

7.2 Edge multiplication

7.2.1 The definition of the edge multiplication

In the graded space

$$\Gamma = \Gamma_0 \oplus \Gamma_1 \oplus \Gamma_2 \oplus \Gamma_3 \oplus \ldots$$

there is a natural structure of a commutative algebra.

Let G_1 and G_2 be two 3-graphs. Choose arbitrarily an edge in G_1 and an edge in G_2. Cut each of these two edges in the middle and re-connect them in any other way so as to get a 3-graph.

The resulting 3-graph is called the *edge product* of G_1 and G_2.

The edge product of 3-graphs can be thought of as the connected sum of G_1 and G_2 along the chosen edges, or as the result of insertion of one graph, say G_1, into an edge of G_2.

Remark 7.4. The product of two connected graphs may yield a disconnected graph, for example:

This happens, however, only in the case when each of the two graphs becomes disconnected after cutting the chosen edge, and in this case both graphs are 0 modulo AS and IHX relations (see part (b) of Lemma 7.11, page 199).

Theorem 7.5. *The edge product of 3-graphs, viewed as an element of the space Γ, is well-defined.*

Note that, as soon as this assertion is proved, one immediately sees that the edge product is commutative.

The claim that the product is well-defined consists of two parts. First, we need to prove that modulo the AS and the IHX relations the product does not depend on the *choice* of the two edges of G_1 and G_2 which are cut and re-connected. Second, we must show that the product does not depend on the *way* they are re-connected (clearly, the two loose ends of G_1 can be glued to the two loose ends of G_2 in two different ways). These two facts are established in the following two lemmas.

Lemma 7.6. *Modulo the AS and the IHX relations, a subgraph with two legs can be carried through a vertex:*

Proof This statement is a particular case ($k = 1$) of the Kirchhoff law (see page 121). □

The above lemma shows that given an insertion of a 3-graph G_1 into an edge of G_2, there exists an equivalent insertion of G_1 into any adjacent edge. Since G_2 is connected, it only remains to show that the two possible insertions of G_1 into an edge of G_2 give the same result.

Lemma 7.7. *The two different ways to re-connect two 3-graphs produce the same element of the space Γ:*

$$
\left(\begin{array}{c} G_1 \\ G_2 \end{array} \right) = \mathord{\text{⧖}}\begin{array}{c} G_1 \\ G_2 \end{array}.
$$

Proof At a vertex of G_1 which lies next to the subgraph G_2 in the product, one can, by the previous lemma, perform the following manoeuvres:

$$
G_1 \;\vert_{G_2} \;=\; G_1 \;\vert_{G_2} \;=\; G_1 \;\vert_{G_2} \;=\; G_1 \;\vert_{G_2}.
$$

Therefore,

$$
\left(\begin{array}{c} G_1 \\ G_2 \end{array} \right) = \left(\begin{array}{c} G_1 \\ G_2 \end{array} \right) = \mathord{\text{⧖}}\begin{array}{c} G_1 \\ G_2 \end{array}.
$$

The lemma is proved, and the edge multiplication of 3-graphs is thus well-defined. □

The edge product of 3-graphs extends by linearity to the whole space Γ.

Corollary 7.8. *The edge product in Γ is well-defined and associative.*

This follows from the fact that a linear combination of either AS or IHX type relations, when multiplied by an arbitrary graph, is a combination of the same type. The associativity is obvious.

7.2.2 *Some identities*

There are two natural operations defined on the space Γ: the *insertion of a bubble into an edge*:

$$
\text{———} \quad \rightsquigarrow \quad \text{—◇—} \;,
$$

and the *insertion of a triangle into a vertex*:

Inserting a bubble into an edge of a 3-graph is the same thing as multiplying this graph by $\beta = \ominus \in \Gamma_1$. In particular, this operation is well-defined and does not depend on the edge where the bubble is created. Inserting a triangle into a vertex can be expressed in a similar fashion via the *vertex multiplication* discussed below in Section 7.3. The following lemma implies that inserting a triangle into a vertex is a well-defined operation:

Lemma 7.9. *A triangle is equal to one-half of a bubble:*

$$\triangle = \frac{1}{2} \quad = \frac{1}{2} \quad = \frac{1}{2} \quad.$$

Proof

$$\triangle = \quad + \quad = \quad - $$

$$= \quad - $$

□

Remark 7.10. It was proved by Pierre Vogel (2006) that the operator of bubble insertion has nontrivial kernel. He exhibited an element of degree 15 which is killed by inserting a bubble.

The second lemma describes two classes of 3-graphs which are equal to 0 in the algebra Γ, that is, modulo the AS and IHX relations.

Lemma 7.11.

(a) *A graph with a loop is 0 in Γ.*

$$= 0$$

(b) *More generally, if the edge connectivity of the graph γ is 1, that is, if it becomes disconnected after removal of an edge, then $\gamma = 0$ in Γ.*

$$\gamma = \quad = 0$$

Proof (a) A graph with a loop is zero because of the antisymmetry relation. Indeed, changing the cyclic order at the vertex of the loop produces a graph which is, on one hand, isomorphic to the initial graph, and on the other hand, differs from it by a sign.

(b) Such a graph can be represented as a product of two graphs, one of which is a graph with a loop that vanishes according to (a):

$$\gamma = \boxed{G_1} \!\!\!\!\!\!\!\!>\!\!\!\!-\!\!\!\!<\!\!\!\!\!\boxed{G_2} = \boxed{G_1}\!\!\!\!\!\!>\!\!\!\!-\!\!\bigcirc \times \big(\boxed{G_2} = 0. \qquad \square$$

7.2.3 The Zoo

Table 7.1 shows the dimensions d_n and displays the bases of the vector spaces Γ_n for $n \leqslant 11$, obtained by computer calculations.

Note that the column for d_n coincides with the column for $k = 2$ in the table of primitive spaces on page 127. This will be proved in Section 7.4.1.

One can see from the table that the multiplicative generators of the algebra Γ up to degree 11 can be chosen as follows (here β stands for "bubble," ω_i for "wheels," δ for "dodecahedron"):

1	4	6	7	8	9	10		11
β	ω_4	ω_6	ω_7	ω_8	ω_9	ω_{10}	δ	ω_{11}

The reader may have noticed that the table of additive generators does not contain the elements ω_4^2 of degree 8 and $\omega_4\omega_6$ of degree 10. This is due to the following relations found by A. Kaishev (2000) in the algebra Γ:

$$\omega_4^2 = \frac{5}{384}\beta^8 - \frac{5}{12}\beta^4\omega_4 + \frac{5}{2}\beta^2\omega_6 - \frac{3}{2}\beta\omega_7,$$

$$\omega_4\omega_6 = \frac{305}{27648}\beta^{10} - \frac{293}{864}\beta^6\omega_4 + \frac{145}{72}\beta^4\omega_6 - \frac{31}{12}\beta^3\omega_7 + 2\beta^2\omega_8 - \frac{3}{4}\beta\omega_9.$$

In fact, as we shall see in Section 7.3, it is true in general that the product of an arbitrary pair of homogeneous elements of Γ of positive degree belongs to the ideal generated by β.

Since there are nontrivial relations between the generators, the algebra of 3-graphs, in contrast to the algebras $\mathcal{A}$, $\mathcal{B}$ and $\mathcal{C}$, is commutative but not free commutative and, hence, does not possess the structure of a Hopf algebra.

Table 7.1 *Additive generators of the algebra of 3-graphs* Γ

n	d_n	additive generators
1	1	
2	1	
3	1	
4	2	
5	2	
6	3	
7	4	
8	5	
9	6	
10	8	
11	9	

7.3 Vertex multiplication

7.3.1 The definition of the vertex multiplication

Apart from the edge product, the space

$$\Gamma_{\geqslant 1} = \Gamma_1 \oplus \Gamma_2 \oplus \Gamma_3 \oplus \Gamma_4 \oplus \dots$$

spanned by all the 3-graphs of non-zero degree has another commutative and associative product.

Let G_1 and G_2 be two 3-graphs of positive degree. Choose arbitrarily a vertex in G_1 and a vertex in G_2. Cut out each of these two vertices and attach the three loose ends that appear on G_1 to the three loose ends on G_2. There are six possible ways of doing this. Take the alternating average of all of them, assigning the negative sign to those three cases where the cyclic order on the loose ends of G_1 agrees with that for G_2, and the positive sign to the other three cases. This alternating average is called the *vertex product* of G_1 and G_2.

Pictorially, if the graphs G_1, G_2 are drawn as $G_1 = $, $G_2 = $, then, in order to draw their vertex product we have to merge them, inserting a permutation of the three strands in the middle. Then we take the result with the sign of the permutation and average it over all six permutations:

$$
\text{(vertex product)} = \frac{1}{6}\left[\; - \; - \; - \; + \; + \; \right].
$$

As an example, let us compute the vertex product with the theta graph:

$$
\beta \vee \; = \; \Theta \vee \; = \frac{1}{6}\left[\; - \; - \; - \; + \; + \; \right] = \; ,
$$

since all the summands in the brackets (taken with their signs) are equal to each other due to the AS relation. Therefore, β will be the unit for the vertex product on $\Gamma_{\geq 1}$.

In order to simplify the notation, we shall use diagrams with shaded disks, understanding them as alternating linear combinations of six graphs as above. For example:

$$
\text{(shaded vertex product)} \; = \; \; = \; .
$$

Theorem 7.12. *The vertex product in $\Gamma_{\geq 1}$ is well-defined, commutative and associative.*

Proof It is sufficient to prove that the AS and the IHX relations imply the following equality:

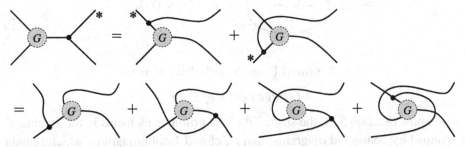

$$X_1 = \quad\quad\quad G \quad\quad\quad = \quad\quad\quad G \quad\quad\quad = X_2.$$

where G denotes an arbitrary subgraph with three legs (and each picture is the alternating sum of six diagrams).

By the Kirchhoff law we have:

(the stars indicate the place where the tail of the "moving electron" is fixed in Kirchhoff's relation). Now, in the last line the first and the fourth diagrams are equal to X_2, while the sum of the second and the third diagrams is equal to $-X_1$ (again, by an application of Kirchhoff's rule). We thus have $2X_1 = 2X_2$ and therefore $X_1 = X_2$.

Commutativity and associativity are obvious.   □

Remark 7.13. Unlike the edge product, which respects the grading on Γ, the vertex multiplication is an operation of degree -1:

$$\Gamma_n \vee \Gamma_m \subset \Gamma_{n+m-1}.$$

7.3.2 Relation between the two products in Γ

Proposition 7.14. *The edge product $\cdot$ in the algebra of 3-graphs Γ is related to the vertex product $\vee$ on $\Gamma_{\geqslant 1}$ as follows:*

$$G_1 \cdot G_2 = \beta \cdot (G_1 \vee G_2).$$

Proof Choose a vertex in each of the given graphs G_1 and G_2 and call its complement G_1' and G_2', respectively:

$$G_1 = \quad\quad G_1' \quad\quad = \quad\quad G_1' \quad\quad , \quad\quad\quad G_2 = \quad\quad G_2' \quad\quad ,$$

where, as explained above, the shaded region indicates the alternating average over the six permutations of the three legs.

Then, since the vertex product is well-defined, we have:

$$G_1 \cdot G_2 \quad = \quad \text{[diagram]} \quad = \quad \text{[diagram]}$$

$$= \quad \beta \cdot \text{[diagram]} \quad = \quad \beta \cdot (G_1 \vee G_2). \qquad \square$$

7.4 Action of Γ on the primitive space $\mathscr{P}$

7.4.1 Edge action of Γ on $\mathscr{P}$

As we know (Section 5.5), the space $\mathscr{P}$ of the primitive elements in the algebra $\mathscr{C}$ is spanned by connected diagrams, that is, closed Jacobi diagrams which remain connected after the Wilson loop is stripped off. It is natural to define the *edge action* of Γ on a primitive diagram $D \in \mathscr{P}$ simply by taking the edge product of a graph $G \in \Gamma$ with D as if D were a 3-graph, using an internal edge of D. The resulting graph $G \cdot D$ is again a closed diagram; moreover, it lies in $\mathscr{P}$. Since D is connected, the lemmas in Section 7.2.1 imply that $G \cdot D$ does not depend on the choice of the edge in G and of the internal edge in D. Therefore, we get a well-defined action of Γ on $\mathscr{P}$, which is clearly compatible with the gradings:

$$\Gamma_n \cdot \mathscr{P}_m \subset \mathscr{P}_{n+m}.$$

Proposition 7.15. *The vector space Γ is isomorphic as a graded vector space (with the grading shifted by one) to the subspace $\mathscr{P}^2 \subset \mathscr{P}$ of primitive closed diagrams spanned by connected diagrams with 2 legs: $\Gamma_n \cong \mathscr{P}^2_{n+1}$ for all $n \geqslant 0$.*

Proof The isomorphism $\Gamma \to \mathscr{P}^2$ is given by the edge action of 3-graphs on the element $\Theta \in \mathscr{P}^2$ represented by the chord diagram with a single chord, $G \mapsto G \cdot \Theta$. The inverse map is equally simple. For a connected closed diagram D with two legs strip off the Wilson loop and glue together the two loose ends of the resulting diagram, obtaining a 3-graph of degree one less than D. Obviously, this map is well-defined and inverse to the edge action on Θ. $\qquad \square$

7.4.2 Vertex action of Γ on $\mathscr{P}$

In order to perform the vertex multiplication, we need at least one vertex in each of the factors. Therefore, we shall define an action of the algebra $\Gamma_{\geqslant 1}$ (with the vertex product) on the space $\mathscr{P}_{>1}$ of primitive elements of degree strictly greater than 1.

The action $G \vee D$ of G on D is the alternated average over all six ways of insert-ing G, with one vertex removed, into D with one internal vertex taken out. Again, since D is connected, the same argument as in Section 7.3.1 works to show that this action is well-defined. Note that the vertex action decreases the total grading by 1 and preserves the number of legs:

$$\Gamma_n \vee \mathscr{P}_m^k \subset \mathscr{P}_{n+m-1}^k.$$

The simplest element of $\mathscr{P}$ on which $\Gamma_{\geqslant 1}$ acts in this way is the "Mercedes-Benz diagram":

$$\bar{t}_1 = \bigotimes.$$

Lemma 7.16.

(a) *The map $\Gamma_{\geqslant 1} \to \mathscr{P}$ defined as $G \mapsto G \vee \bar{t}_1$ is injective.*
(b) *For all $G \in \Gamma_{\geqslant 1}$ we have*

$$G \vee \bar{t}_1 = \frac{1}{2} G \cdot \Theta.$$

Proof   Indeed, $\bar{t}_1 = \frac{1}{2}\beta \cdot \Theta$. Therefore,

$$G \vee \bar{t}_1 = \frac{1}{2}(G \vee \beta) \cdot \Theta = \frac{1}{2} G \cdot \Theta.$$

Since the map $G \mapsto G \cdot \Theta$ is an isomorphism (Section 7.4.1), the map $G \mapsto G \vee \bar{t}_1$ is also an isomorphism $\Gamma_{\geqslant 1} \cong \mathscr{P}_{\geqslant 1}^2$. $\qquad\square$

7.4.3 A product on the primitive space $\mathscr{P}$

In principle, the space of primitive elements $\mathscr{P}$ of the algebra $\mathscr{C}$ does not possess any a priori defined multiplicative structure. Primitive elements only generate the algebra $\mathscr{C}$ much in the same way as the variables $x_1, \dots, x_n$ generate the polyno-mial algebra $\mathbb{R}[x_1, \dots, x_n]$. However, the link between the space $\mathscr{P}$ and the algebra of 3-graphs Γ allows to introduce a (non-commutative) multiplication in $\mathscr{P}$.

There is a projection $\pi : \mathscr{P}_n \to \Gamma_n$, which consists in introducing a cyclic order on the half-edges at the vertices of the Wilson loop according to the rule "forward–sideways–backwards" and then forgetting the fact that the Wilson loop was distinguished. The edge action $\Gamma \times \mathscr{P} \to \mathscr{P}$ then gives rise to an operation $* : \mathscr{P} \times \mathscr{P} \to \mathscr{P}$ defined by the rule

$$p * q = \pi(p) \cdot q,$$

where $\pi : \mathscr{P} \to \Gamma$ is the homomorphism of forgetting the Wilson loop defined above.

The operation $*$ is associative, but, in general, non-commutative:

$$\ominus \ * \ \oslash \ = \ \oslash \ , \quad \text{but} \quad \oslash \ * \ \ominus \ = \ \oslash .$$

These two elements of the space $\mathscr{P}$ are different; they can be distinguished, for instance, by the $\mathfrak{sl}_2$-invariant (see Exercise 6.20 at the end of Chapter 6). However, π projects these two elements into the same element $\beta \cdot \omega_4 \in \Gamma_5$.

7.5 Lie algebra weight systems for the algebra Γ

A *weight system for 3-graphs* is a function on 3-graphs that satisfies the anti-symmetry and the IHX relations. Lie algebras (and super Lie algebras) give rise to weight systems for 3-graphs in the very same fashion as for the algebra $\mathscr{C}$ (Section 6.2), using the structure tensor J. Since 3-graphs have no univalent vertices, these weight systems take values in the ground field (here assumed to be $\mathbb{C}$). For a graph $G \in \Gamma$ we put

$$\varphi_{\mathfrak{g}}(G) := T_{\mathfrak{g}}(G) \in \mathfrak{g}^0 \cong \mathbb{C}.$$

In particular, the weight system $\varphi_{\mathfrak{g}}$ evaluated on the circle (the 3-graph without vertices, which is the unit in Γ) gives the dimension of the Lie algebra.

Remark 7.17. This circle should not be confused with the circles appearing in the state sum formulae for $\varphi_{\mathfrak{sl}_N}$ and $\varphi_{\mathfrak{so}_N}$ from Sections 6.2.5 and 6.2.6. The contribution of these circles to the values of $\varphi_{\mathfrak{sl}_N}$ and $\varphi_{\mathfrak{so}_N}$ is equal to N, while the dimensions of the corresponding Lie algebras are $N^2 - 1$ and $\frac{1}{2}N(N - 1)$, respectively.

7.5.1 Changing the bilinear form

From the construction of $\varphi_{\mathfrak{g}}$ it is easy to see that the function $\varphi_{\mathfrak{g},\lambda}$ corresponding to the form $\lambda\langle \cdot, \cdot \rangle$ is a multiple of $\varphi_{\mathfrak{g}}$:

$$\varphi_{\mathfrak{g},\lambda}(G) = \lambda^{-n}\varphi_{\mathfrak{g}}(G)$$

for $G \in \Gamma_n$.

7.5.2 Multiplicativity with respect to the edge product in Γ

Proposition 7.18. *For a simple Lie algebra $\mathfrak{g}$ and any choice of an ad-invariant non-degenerate symmetric bilinear form $\langle \cdot, \cdot \rangle$ the function $\frac{1}{\dim \mathfrak{g}}\varphi_{\mathfrak{g}} : \Gamma \to \mathbb{C}$ is multiplicative with respect to the edge product in Γ.*

Proof This is a consequence of the fact that, up to a constant, the quadratic Casimir tensor of a simple Lie algebra is the only ad-invariant, symmetric, non-degenerate tensor in $\mathfrak{g} \otimes \mathfrak{g}$.

Consider two graphs $G_1, G_2 \in \Gamma$ and choose an orthonormal basis e_i for the Lie algebra $\mathfrak{g}$. Cut an arbitrary edge of the graph G and consider the tensor that corresponds to the resulting graph G'_1 with two univalent vertices. This tensor is a scalar multiple of the quadratic Casimir tensor $c \in \mathfrak{g} \otimes \mathfrak{g}$:

$$a \cdot c = a \sum_{i=1}^{\dim \mathfrak{g}} e_i \otimes e_i.$$

Now, $\varphi_{\mathfrak{g},K}(G_1)$ is obtained by contracting these two tensor factors. This gives $\varphi_{\mathfrak{g},K}(G_1) = a \dim \mathfrak{g}$, and $a = \frac{1}{\dim \mathfrak{g}} \varphi_{\mathfrak{g}}(G_1)$. Similarly, for the graph G_2 we get the tensor $\frac{1}{\dim \mathfrak{g}} \varphi_{\mathfrak{g}}(G_2) \cdot c$. Now, if we join together one pair of univalent vertices of the graphs G'_1 and G'_2 (where G'_2 is obtained from G_2 by cutting an edge), the partial contraction of the element $c^{\otimes 2} \in \mathfrak{g}^{\otimes 4}$ will give

$$\frac{1}{(\dim \mathfrak{g})^2} \varphi_{\mathfrak{g}}(G_1) \varphi_{\mathfrak{g}}(G_2) \cdot c \in \mathfrak{g} \otimes \mathfrak{g}.$$

But, on the other hand, this tensor equals

$$\frac{1}{\dim \mathfrak{g}} \varphi_{\mathfrak{g}}(G_1 \cdot G_2) \cdot c \in \mathfrak{g} \otimes \mathfrak{g}.$$

This shows that $\frac{1}{\dim \mathfrak{g}} \varphi_{\mathfrak{g}}$ is multiplicative. $\qquad\square$

7.5.3 Compatibility with the edge action of Γ on $\mathscr{C}$

Recall the definition of the edge action of 3-graphs on closed diagrams (see Section 7.4.1). We choose an edge in $G \in \Gamma$ and an internal edge in $D \in \mathscr{C}$, and then take the connected sum of G and D along the chosen edges. In fact, this action depends on the choice of the connected component of the internal graph of D containing the chosen edge. It is well defined only on the primitive subspace $\mathscr{P} \subset \mathscr{C}$. In spite of this indeterminacy we have the following lemma.

Lemma 7.19. *For any choice of the glueing edges,* $\varphi_{\mathfrak{g}}(G \cdot D) = \frac{\varphi_{\mathfrak{g}}(G)}{\dim \mathfrak{g}} \varphi_{\mathfrak{g}}(D)$.

Proof Indeed, in order to compute $\varphi_{\mathfrak{g}}(D)$ we assemble the tensor $T_{\mathfrak{g}}(D)$ from tensors that correspond to tripods and chords. The legs of these elementary pieces are glued together by contraction with the quadratic Casimir tensor c, which corresponds to the metric on the Lie algebra. By the previous argument, to compute the tensor $T_{\mathfrak{g}}(G \cdot D)$ one must use for the chosen edge the tensor $\frac{1}{\dim \mathfrak{g}} \varphi_{\mathfrak{g}}(G) \cdot c$

instead of c. This gives the coefficient $\frac{1}{\dim \mathfrak{g}}\varphi_{\mathfrak{g}}(G)$ in the expression for $\varphi_{\mathfrak{g}}(G \cdot D)$ as compared with $\varphi_{\mathfrak{g}}(D)$. $\square$

One particular case of the edge action of Γ is especially interesting: when the graph G varies, while D is fixed and equal to Θ, the chord diagram with only one chord. In this case the action is an isomorphism of the vector space Γ with the subspace $\mathscr{P}^2$ of the primitive space $\mathscr{P}$ generated by connected closed diagrams with two legs (Section 7.4.1).

Corollary 7.20. *For the weight systems associated with a simple Lie algebra* $\mathfrak{g}$ *and the Killing form* $\langle \cdot, \cdot \rangle^K$:

$$\varphi_{\mathfrak{g},K}(G) = \varphi_{\mathfrak{g},K}^{ad}(G \cdot \Theta) ,$$

where $\varphi_{\mathfrak{g},K}^{ad}$ *is the weight system corresponding to the adjoint representation of* $\mathfrak{g}$.

Proof Indeed, according to the lemma, for the universal enveloping algebra invariants we have

$$\varphi_{\mathfrak{g},K}(G \cdot \Theta) = \frac{1}{\dim \mathfrak{g}}\varphi_{\mathfrak{g}}(G)\varphi_{\mathfrak{g},K}(\Theta) = \frac{1}{\dim \mathfrak{g}}\varphi_{\mathfrak{g}}(G) \sum_{i=1}^{\dim \mathfrak{g}} e_i e_i ,$$

where $\{e_i\}$ is a basis orthonormal with respect to the Killing form. Now, to compute $\varphi_{\mathfrak{g},K}^{ad}(G \cdot \Theta)$, we take the trace of the product of operators in the adjoint representation:

$$\varphi_{\mathfrak{g},K}^{ad}(G \cdot \Theta) = \frac{1}{\dim \mathfrak{g}}\varphi_{\mathfrak{g}}(G) \sum_{i=1}^{\dim \mathfrak{g}} \mathrm{Tr}(\mathrm{ad}_{e_i}\,\mathrm{ad}_{e_i}) = \varphi_{\mathfrak{g}}(G)$$

by the definition of the Killing form. $\square$

7.5.4 Multiplicativity with respect to the vertex product in Γ

Proposition 7.21. *Let* $w : \Gamma \to \mathbb{C}$ *be an edge-multiplicative weight system, and* $w(\beta) \neq 0$. *Then* $\frac{1}{w(\beta)}w : \Gamma \to \mathbb{C}$ *is multiplicative with respect to the vertex product. In particular, for a simple Lie algebra* $\mathfrak{g}$, $\frac{1}{\varphi_{\mathfrak{g}}(\beta)}\varphi_{\mathfrak{g}}$ *is vertex-multiplicative.*

Proof According to 7.3.2 the edge product is related to the vertex product as $G_1 \cdot G_2 = \beta \cdot (G_1 \vee G_2)$. Therefore,

$$w(G_1) \cdot w(G_2) = w(\beta \cdot (G_1 \vee G_2)) = w(\beta) \cdot w(G_1 \vee G_2).$$

This means that the weight system $\frac{1}{w(\beta)}w : \Gamma \to \mathbb{C}$ is multiplicative with respect to the vertex product. $\square$

Corollary 7.22. *The weight systems* $\frac{1}{2N(N^2-1)}\varphi_{\mathfrak{sl}_N}$, $\frac{2}{N(N-1)(N-2)}\varphi_{\mathfrak{so}_N}$: Γ → ℂ *associated with the ad-invariant form* $\langle x, y \rangle = \mathrm{Tr}(xy)$ *are multiplicative with respect to the vertex product in* Γ.

This follows from a direct computation for the "bubble":

$$\varphi_{\mathfrak{sl}_N}(\beta) = 2N(N^2 - 1), \text{ and } \varphi_{\mathfrak{so}_N}(\beta) = \frac{1}{2}N(N-1)(N-2).$$

7.5.5 Compatibility with the vertex action of Γ on 𝒞

The vertex action $G \vee D$ of a 3-graph $G \in$ Γ on a closed diagram $D \in$ 𝒞 with at least one vertex (see Section 7.4.2) is defined as the alternating sum of the six ways to glue the graph G to the closed diagram D along chosen internal vertices in D and G. Again, this action is well-defined only on the primitive space $\mathscr{P}_{>1}$.

Lemma 7.23. *Let* 𝔤 *be a simple Lie algebra. Then for any choice of the glueing vertices in G and D:*

$$\varphi_{\mathfrak{g}}(G \vee D) = \frac{\varphi_{\mathfrak{g}}(G)}{\varphi_{\mathfrak{g}}(\beta)}\varphi_{\mathfrak{g}}(D).$$

Proof Using the edge action (Section 7.5.3) and its relation to the vertex action we can write

$$\frac{\varphi_{\mathfrak{g}}(G)}{\dim \mathfrak{g}}\varphi_{\mathfrak{g}}(D) = \varphi_{\mathfrak{g}}(G \cdot D) = \varphi_{\mathfrak{g}}(\beta \cdot (G_1 \vee D)) = \frac{\varphi_{\mathfrak{g}}(\beta)}{\dim \mathfrak{g}}\varphi_{\mathfrak{g}}(G_1 \vee D),$$

which is what we need.   □

7.5.6 $\mathfrak{sl}_N$- and $\mathfrak{so}_N$-polynomials

The $\mathfrak{sl}_N$- and $\mathfrak{so}_N$-*polynomials* are the weight systems $\varphi_{\mathfrak{sl}_N}$ (with respect to the bilinear form $\langle x, y \rangle = \mathrm{Tr}(xy)$), and $\varphi_{\mathfrak{so}_N}$ (with the bilinear form $\langle x, y \rangle = \frac{1}{2}\mathrm{Tr}(xy)$). In the case of $\mathfrak{so}_N$ the choice of *half* the trace as the bilinear form is more convenient since it gives polynomials in N with integral coefficients. In particular, for this form $\varphi_{\mathfrak{so}_N}(\beta) = N(N-1)(N-2)$, and in the state sum formula from Theorem 6.26 the coefficient in front of the sum equals 1.

The polynomial $\varphi_{\mathfrak{sl}_N}(G) (= \varphi_{\mathfrak{gl}_N}(G))$ is divisible by $2N(N^2-1)$ (Exercise 7.10 at the end of this chapter) and the quotient is a multiplicative function with respect to the vertex product. We call this quotient the *reduced $\mathfrak{sl}$-polynomial* and denote it by $\widetilde{\mathfrak{sl}}(G)$. Dividing the $\mathfrak{so}$-polynomial $\varphi_{\mathfrak{so}_N}(G)$ by $N(N-1)(N-2)$ (see Exercise 7.11), we obtain the *reduced $\mathfrak{so}$-polynomial* $\widetilde{\mathfrak{so}}(G)$, which is also multiplicative with respect to the vertex product.

A. Kaishev (2000) computed the values of $\widetilde{\mathfrak{sl}}$-, and $\widetilde{\mathfrak{so}}$-polynomials on the generators of Γ of small degrees (for $\widetilde{\mathfrak{so}}$-polynomial the substitution $M = N - 2$ is used in the table):

deg		$\widetilde{\mathfrak{sl}}$-polynomial	$\widetilde{\mathfrak{so}}$-polynomial
1	β	1	1
4	ω_4	N^3+12N	$M^3-3M^2+30M-24$
6	ω_6	N^5+32N^3+48N	$M^5-5M^4+80M^3-184M^2+408M-288$
7	ω_7	$N^6+64N^4+64N^2$	$M^6-6M^5+154M^4-408M^3+664M^2-384$
8	ω_8	$N^7+128N^5+128N^3$ $+192N$	$M^7-7M^6+294M^5-844M^4+1608M^3-2128M^2$ $+4576M-3456$
9	ω_9	$N^8+256N^6+256N^4$ $+256N^2$	$M^8-8M^7+564M^6-1688M^5+3552M^4-5600M^3$ $-5600M^3+6336M^2+6144M-9216$
10	ω_{10}	$N^9+512N^7+512N^5$ $+512N^3+768N$	$M^9-9M^8+1092M^7-3328M^6+7440M^5-13216M^4$ $+18048M^3-17920M^2+55680M-47616$
10	δ	$N^9+11N^7+114N^5$ $-116N^3$	$M^9-9M^8+44M^7-94M^6+627M^5+519M^4$ $-2474M^3-10916M^2+30072M-17760$
11	ω_{11}	$N^{10}+1024N^8+1024N^6$ $+1024N^4+1024N^2$	$M^{10}-10M^9+2134M^8-6536M^7+15120M^6$ $-29120M^5+45504M^4-55040M^3+48768M^2$ $+145408M-165888$

There are recognizable patterns in this table. For example, we see that

$$\widetilde{\mathfrak{sl}}(\omega_n) = N^{n-1} + 2^{n-1}(N^{n-3} + \cdots + N^2), \text{ for odd } n > 5;$$
$$\widetilde{\mathfrak{sl}}(\omega_n) = N^{n-1} + 2^{n-1}(N^{n-3} + \cdots + N^3) + 2^{n-2}3N, \text{ for even } n \geqslant 4.$$

It would be interesting to know if these observations are particular cases of some general statements.

7.6 Vogel's algebra Λ

7.6.1 Fixed diagrams

Diagrams with 1- and 3-valent vertices can be considered with different additional structures on the set of univalent vertices. If there is no structure, then we get the notion of an open Jacobi diagram; open diagrams are considered modulo AS and IHX relations. If the legs are attached to a circle or a line, then we obtain closed

Jacobi diagrams; for the closed diagrams, AS, IHX and STU relations are used. Connected diagrams with a linear order (numbering) on the set of legs and a cyclic order on the half-edges at each 3-valent vertex, considered modulo AS and IHX, but without STU relations, will be referred to as *fixed diagrams*. The set $\mathbf{X}$ of all fixed diagrams has two gradings: by the number of legs (denoted by a superscript) and by half the total number of vertices (denoted by a subscript).

Definition 7.24. The $\mathbb{Q}$-vector space spanned by fixed diagrams with k legs modulo the usual AS and IHX relations

$$\mathscr{X}_n^k = \langle \mathbf{X}_n^k \rangle / \langle AS, IHX \rangle,$$

is called the *space of fixed diagrams* of degree n with k legs.

We shall write $\mathscr{X}^k$ for the direct sum $\oplus_n \mathscr{X}_n^k$.

Remark 7.25. The spaces $\mathscr{X}^k$ for different values of k are related by various operations. For example, one may think about the diagram $\mathbb{W}$ as of a linear operator from $\mathscr{X}^4$ to $\mathscr{X}^3$. Namely, it acts on an element G of $\mathscr{X}^4$ as follows:

Exercise.

$$\mathbb{W} : \boxed{G}_{123\ 4} \longmapsto \boxed{G}_{123}.$$

(a) Prove the following relation:

$$\mathbb{W} + \mathbb{W} + \mathbb{W} = 0,$$

    among the three linear operators from $\mathscr{X}^4$ to $\mathscr{X}^3$.

(b) Prove that

$$\big)\big)\big\vert = \big\vert\big(\big($$

    as linear maps from $\mathscr{X}^3$ to $\mathscr{X}^4$.

The space of open diagrams $\mathscr{B}^k$ studied in Chapter 5 is the quotient of $\mathscr{X}^k$ by the action of the symmetric group S_k which permutes the legs of a fixed diagram. The quotient map $\mathscr{X} \to \mathscr{B}$ has a nontrivial kernel; for example, a tripod, which is nonzero in $\mathscr{X}^3$, becomes zero in $\mathscr{B}$:

$$0 \neq \underset{1\ \ 2}{\overset{3}{\curlyvee}} \longmapsto \curlyvee = 0.$$

7.6.2 The algebra Λ

The algebra Λ is the subspace of $\mathscr{X}^3$ that consists of all elements antisymmetric with respect to permutations of their legs. The product in Λ is similar to the vertex product in Γ. Given a connected fixed diagram and an element of Λ, we remove an arbitrary vertex in the diagram and insert the element of Λ instead – in compliance with the cyclic order at the vertex. This operation extends to a well-defined product on Λ, and this fact is proved in the same way as for the vertex multiplication in Γ. Since antisymmetry is presupposed, we do not need to take the alternated average over the six ways of insertion, as in Γ – all the six summands will be equal to each other.

Example 7.26.

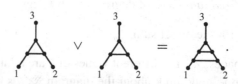

Conjecturally, the antisymmetry requirement in this definition is superfluous:

Conjecture 7.27. $\Lambda = \mathscr{X}^3$, *that is, any fixed diagram with three legs is antisymmetric with respect to leg permutations.*

Remark 7.28. Note the sign difference in the definitions of the product in Λ and the vertex product in Γ: in Γ when two graphs are glued together in compliance with the cyclic order of half-edges, the corresponding term is counted with a *negative* sign.

Remark 7.29. Vogel (1997) defines the spaces $\mathscr{X}^k$ and the algebra Λ over the integers, rather than over $\mathbb{Q}$. In this approach, the equality (b) of the exercise on page 211 no longer follows from the AS and IHX relations. It has to be postulated separately as one of the equations defining Λ in $\mathscr{X}^3$ in order to make the product in Λ well-defined.

The product in the algebra Λ naturally generalizes to the action of Λ on different spaces generated by 1- and 3-valent diagrams, such as the space of connected open diagrams $\mathscr{PB}$ and the space of 3-graphs Γ. The same argument as above shows that these actions are well-defined.

7.6.3 Relation between Λ and Γ

Recall that the space $\Gamma_{\geqslant 1}$ of all 3-graphs of degree at least one is an algebra with respect to the vertex product.

Proposition 7.30. *The algebra* Λ *is isomorphic to* $(\Gamma_{\geq 1}, \vee)$.

Proof   There are two mutually inverse maps between Λ and $\Gamma_{\geq 1}$. The map from Λ to $\Gamma_{\geq 1}$ glues the three legs of an element of Λ to a new vertex so that the cyclic order of edges at this vertex is opposite to the order of legs:

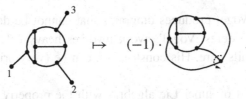

In order to define a map from $\Gamma_{\geq 1}$ to Λ, we choose an arbitrary vertex of a 3-graph, delete it and antisymmetrize:

$$\bigoplus \;\mapsto\; \frac{1}{6}\Big[-\Big(\textstyle\bigoplus\Big)+\Big(\textstyle\bigoplus\Big)+\Big(\textstyle\bigoplus\Big)+\Big(\textstyle\bigoplus\Big)-\Big(\textstyle\bigoplus\Big)-\Big(\textstyle\bigoplus\Big)\Big].$$

It is easy to see that this is indeed a well-defined map (Hint: use part (b) of the exercise on page 211).

It is evident from the definitions that both maps are inverse to each other and send products to products.   □

Remark 7.31. If Conjecture 7.27 is true for $k = 3$, then all the six terms (together with their signs) in the definition of the map $\Gamma_{\geq 1} \to$ Λ are equal to each other. This means that there is no need to antisymmetrize. What we do is remove one vertex (with a small neighbourhood) and number the three legs obtained according to their cyclic ordering at the deleted vertex. This would also simplify the definition of the vertex product in Section 7.3 as in this case

$$\overset{(G)}{\underset{}{\bigvee\!\!\bigvee}} = \overset{(G)}{\underset{}{\bigvee\!\!\bigvee}},$$

and we simply insert one graph in a vertex of another.

Conjecture 7.32 (Vogel 1997). *The algebra* Λ *is generated by the elements* t *and* x_k *with odd* $k = 3, 5, \ldots$:

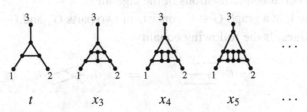

| t | x_3 | x_4 | x_5 | $\cdots$ |

7.6.4 *Weight systems not coming from Lie algebras*

In order to construct a weight system which would not be a combination of Lie algebra weight systems, it is sufficient to find a non-zero element in $\mathscr{C}$ on which all the Lie algebra weight systems vanish. The same is true, of course, for super Lie algebras.

In Vogel (1997), Vogel produces diagrams that cannot be detected by (super) Lie algebra weight systems. Vogel's work involves heavy calculations of which we shall give no details here. His construction can be (very briefly) described as follows.

First, he gives a list of super Lie algebras with the property that whenever all the weight systems for the algebras from this list vanish on an element of Λ, this element cannot be detected by any (super) Lie algebra weight system. This list includes a certain super Lie algebra $D(2, 1, \alpha)$; this algebra detects an element of Λ which all other algebras from the list do not detect. Making this element act by the vertex action on the "Mercedes-Benz" closed diagram $\bar{t}_1$, he obtains a closed diagram which is nonzero because of Lemma 7.16 but which cannot be detected by any super Lie algebra weight system. We refer to Vogel (1997) and Lieberum (1999) for the details.

Exercises

7.1 Find an explicit chain of IHX and AS relations that proves the following equality in the algebra Γ of 3-graphs:

7.2 Let $\tau_2 : \mathscr{X}^2 \to \mathscr{X}^2$ be the transposition of legs in a fixed diagram. Prove that τ_2 is the identity. *Hint:* (1) prove that a "hole" can be dragged through a trivalent vertex (2) to change the numbering of the two legs, use manoeuvres like in Lemma 7.7 with $G_2 = \emptyset$.

7.3 * Let Γ be the algebra of 3-graphs.
- Is it true that Γ is generated by plane graphs?
- Find generators and relations of the algebra Γ.
- Suppose that a graph $G \in \Gamma$ consists of two parts G_1 and G_2 connected by three edges. Is the following equality:

true?

7.4 * Is it true that the algebra of primitive elements $\mathscr{P}$ has no divisors of zero with respect to the product $*$?

7.5 Let $\mathscr{X}^k$ be the space of 1- and 3-valent graphs with k numbered legs. Consider the transposition of two legs of an element of $\mathscr{X}^k$.

  - Give an example of a nonzero element of $\mathscr{X}^k$ with even k which is changed under such a transposition.
  - * Is it true that any such transposition changes the sign of the element if k is odd? (The first nontrivial case is when $k = 3$: this is Conjecture 7.27.)

7.6 * Let Λ be Vogel's algebra, that is, the subspace of $\mathscr{X}^3$ consisting of all antisymmetric elements.

  - Is it true that $\Lambda = \mathscr{X}^3$ (this is again Conjecture 7.27)?
  - Is it true that Λ is generated by the elements t and x_k (this is Conjecture 7.32; see also Exercises 7.7 and 7.8)?

7.7 Let $t, x_3, x_4, x_5, \ldots$ be the elements of the space $\mathscr{X}^3$ described above.

  - Prove that the x_i belong to Vogel's algebra Λ, that is, that they are antisymmetric with respect to permutations of legs.
  - Prove the relation $x_4 = -\frac{4}{3}t \vee x_3 - \frac{1}{3}t^{\vee 4}$.
  - Prove that x_k with an arbitrary even k can be expressed via $t, x_3, x_5, \ldots$

7.8 Prove that the dodecahedron

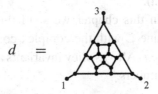

$$d \quad = \quad$$

belongs to Λ, and express it as a vertex polynomial in t, x_3, x_5, x_7, x_9.

7.9 * The group S_3 acts in the space of fixed diagrams with three legs $\mathscr{X}^3$, splitting it into three subspaces:

  - symmetric, which is isomorphic to $\mathscr{B}^3$ (open diagrams with three legs),
  - totally antisymmetric, which is Vogel's Λ by definition, and
  - some subspace Q, corresponding to a two-dimensional irreducible representation of S_3.

  Is it true that $Q = 0$?

7.10 Show that $N = 0$, $N = -1$, and $N = 1$ are roots of the polynomial $\varphi_{\mathfrak{gl}_N}(G)$ for any 3-graph $G \in \Gamma_n$ $(n > 1)$.

7.11 Show that $N = 0$, $N = 1$ and $N = 2$ are roots of the polynomial $\varphi_{\mathfrak{so}_N}(G)$ for any 3-graph $G \in \Gamma_n$ $(n > 0)$.

8

The Kontsevich integral

The Kontsevich integral appeared in the paper by M. Kontsevich (1993) as a tool to prove the Fundamental Theorem of the theory of Vassiliev invariants (Section 4.2.1). Any Vassiliev knot invariant with coefficients in a field of characteristic 0 can be factored through the universal invariant defined by the Kontsevich integral.

Detailed (and different) expositions of the construction and properties of the Kontsevich integral can be found in Bar-Natan (1995a); Chmutov and Duzhin (2001); and Lescop (1999). Other important references are Cartier (1993); Le and Murakami (1995a, 1996a).

About the notation: in this chapter we shall think of $\mathbb{R}^3$ as the product of a (horizontal) complex plane $\mathbb{C}$ with the complex coordinate z and a (vertical) real line $\mathbb{R}$ with the coordinate t. All Vassiliev invariants are always thought of as having values in the complex numbers.

8.1 First examples

We start with two examples where the Kontsevich integral appears in a simplified form and with a clear geometric meaning.

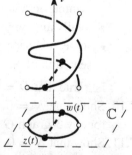

8.1.1 The braiding number of a 2-braid

A braid on two strands has a complete invariant: the number of full twists that one strand makes around the other.

Let us consider the horizontal coordinates of points on the strands, $z(t)$ and $w(t)$, as functions of the vertical coordinate t, $0 \leqslant t \leqslant 1$, then the number of full twists can be computed by the integral formula

$$\frac{1}{2\pi i} \int_0^1 \frac{dz - dw}{z - w}.$$

Note that the number of full twists is not necessarily an integer; however, the number of *half-twists* always is.

8.1.2 Kontsevich type formula for the linking number

The Gauss integral formula for the linking number of two spatial curves $lk(K, L)$ (discussed in Section 2.2.2) involves integration over a torus (namely, the product of the two curves). Here we shall give a different integral formula for the same invariant, with the integration over an interval, rather than a torus. This formula generalizes the expression for the braiding number of a braid on two strands and, as we shall later see, gives the first term of the Kontsevich integral of a two-component link.

Definition 8.1. A link in $\mathbb{R}^3$ is a *Morse* link if the function t (the vertical coordinate) on it has only non-degenerate critical points. A Morse link is a *strict* Morse link if the critical values of the vertical coordinate are all distinct. Similarly, one speaks of *Morse tangles* and *strict Morse tangles*.

Theorem 8.2. *Suppose that two disjoint connected curves K, L are embedded into $\mathbb{R}^3$ as a strict Morse link.*

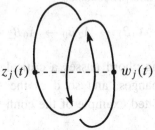

$$z_j(t) \dashleftarrow \dashrightarrow w_j(t)$$

Then

$$lk(K, L) = \frac{1}{2\pi i} \int \sum_j (-1)^{\downarrow_j} \frac{d(z_j(t) - w_j(t))}{z_j(t) - w_j(t)},$$

where the index j enumerates all possible choices of a pair of strands on the link as functions $z_j(t)$, $w_j(t)$ corresponding to K and L, respectively, and the integer $\downarrow_j$ is the number of strands in the pair which are oriented downwards.

Remark 8.3. In fact, the condition that the link in question is a strict Morse link can be relaxed. One may consider piecewise linear links with no horizontal segments, or smooth links whose vertical coordinate function has no flattening points (those where all the derivatives vanish).

Proof The proof consists of three steps which – in a more elaborate setting – will also appear in the full construction of the Kontsevich integral.

Step 1. *The value of the sum in the right-hand side is an integer.* Note that for a strict Morse link with two components K and L, the *configuration space* of all horizontal chords joining K and L is a closed one-dimensional manifold, that is, a disjoint union of several circles.

For example, assume that two adjacent critical values m and M (with $m < M$) of the vertical coordinate correspond to a minimum on the component K and a maximum on the component L respectively:

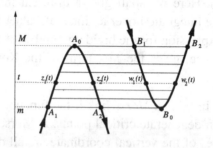

The space of all horizontal chords that join the shown parts of K and L consists of four intervals which join together to form a circle. The motion along this circle starts, say, at a chord $A_1 B_0$ and proceeds as

$$A_1 B_0 \to A_0 B_1 \to A_2 B_0 \to A_0 B_2 \to A_1 B_0.$$

Note that when the moving chord passes a critical level (either m or M), the direction of its motion changes, and so does the sign $(-1)^{\downarrow j}$. (Exercise 8.1 deals with a more complicated example of the configuration space of horizontal chords.)

It is now clear that our integral formula counts the number of complete turns made by the horizontal chord while running through the whole configuration space of chords with one end $(z_j(t), t)$ on K and the other end $(w_j(t), t)$ on L. This is, clearly, an integer.

Step 2. *The value of the right-hand side remains unchanged under a continuous horizontal deformation of the link.* (By a horizontal deformation we mean a deformation of a link which moves every point in a horizontal plane $t = $ const.) The assertion is evident, since the integral changes continuously while always remaining an integer. Note that this is true even if we allow self-intersections within each of the components; this does not influence the integral because $z_j(t)$ and $w_j(t)$ lie on the different components.

Step 3. *Reduction to the combinatorial formula for the linking number* (Section 2.2). Choose a vertical plane in $\mathbb{R}^3$ and represent the link by a generic projection to that plane. By a horizontal deformation, we can flatten the link so that it lies in the plane completely, save for the small fragments around the diagram crossings between K and L (as we noted above, *self-intersections* of each component are allowed). Now, the rotation of the horizontal chord for each crossing is by $\pm\pi$, and the signs are in agreement with the number of strands oriented downwards. The reader is invited to draw the two different possible crossings, then, for each picture, consider the four possibilities for the orientations of the strands and make sure that the sign of the half-turn of the moving horizontal chord always agrees with the factor $(-1)^{\downarrow j}$. (Note that the integral in the theorem is computed over t, so that each specific term computes the angle of rotation of the chord as it moves from bottom to top.) $\qquad\qquad\square$

The Kontsevich integral can be regarded as a generalization of this formula. Here we kept track of one horizontal chord moving along the two curves. The full Kontsevich integral keeps track of how *finite sets* of horizontal chords on the knot (or a tangle) rotate when moved in the vertical direction. This is the somewhat naïve approach that we use in the next section. Later, in Section 10.1, we shall adopt a more sophisticated point of view, interpreting the Kontsevich integral as the monodromy of the Knizhnik–Zamolodchikov connection in the complement of the union of diagonals in $\mathbb{C}^n$.

8.2 The construction

8.2.1 The construction of the Kontsevich integral

Let us recall some notation and terminology of the preceding section. For points of $\mathbb{R}^3$ we use coordinates (z, t) with z complex and t real; the planes $t = $ const are thought of being horizontal. Having chosen the coordinates, we can speak of *strict Morse knots*, namely, knots with the property that the coordinate t restricted to the knot has only non-degenerate critical points with distinct critical values.

We define the Kontsevich integral for strict Morse knots. Its values belong to the graded completion $\widehat{\mathscr{A}'}$ of the algebra of chord diagrams with 1-term relations $\mathscr{A}' = \mathscr{A}/(\Theta)$. (By definition, the elements of a graded algebra are finite linear combinations of homogeneous elements. The *graded completion* consists of all infinite combinations of such elements.)

Definition 8.4. The *Kontsevich integral* $Z(K)$ of a strict Morse knot K is given by the following formula:

$$Z(K) = \sum_{m=0}^{\infty} \frac{1}{(2\pi i)^m} \int\limits_{\substack{t_{\min} < t_m < \cdots < t_1 < t_{\max} \\ t_j \text{ are noncritical}}} \sum_{P=\{(z_j, z_j')\}} (-1)^{\downarrow P} D_P \bigwedge_{j=1}^{m} \frac{dz_j - dz_j'}{z_j - z_j'}.$$

The ingredients of this formula have the following meaning.

The real numbers $t_{\min}$ and $t_{\max}$ are the minimum and the maximum of the function t on K.

The integration domain is the set of all points of the m-dimensional simplex $t_{\min} < t_m < \cdots < t_1 < t_{\max}$ none of whose coordinates t_i is a critical value of t. The m-simplex is divided by the critical values into several *connected components*. For example, for the following embedding of the unknot and $m = 2$, the corresponding integration domain has six connected components and looks like

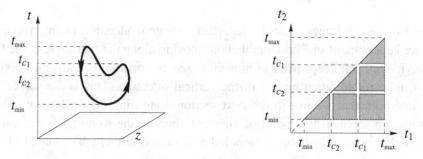

The number of summands in the integrand is constant in each connected component of the integration domain, but can be different for different components. In each plane $\{t = t_j\} \subset \mathbb{R}^3$ choose an unordered pair of distinct points (z_j, t_j) and (z_j', t_j) on K, so that $z_j(t_j)$ and $z_j'(t_j)$ are continuous functions. We denote by $P = \{(z_j, z_j')\}$ the set of such pairs for $j = 1, \ldots, m$ and call it a *pairing*.

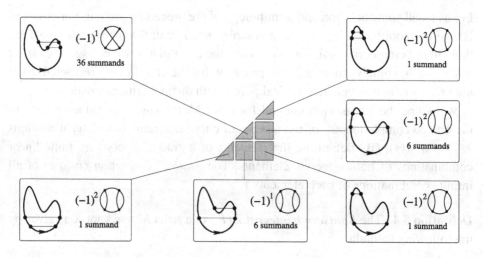

The integrand is the sum over all choices of the pairing P. In the example above for the component $\{t_{c_1} < t_1 < t_{max}, t_{min} < t_2 < t_{c_2}\}$, in the bottom right corner, we have only one possible pair of points on the levels $\{t = t_1\}$ and $\{t = t_2\}$. Therefore, the sum over P for this component consists of only one summand. In contrast, in the component next to it, $\{t_{c_2} < t_1 < t_{c_1}, t_{min} < t_2 < t_{c_2}\}$, we still have only one possibility for the chord (z_2, z_2') on the level $\{t = t_2\}$, but the plane $\{t = t_1\}$ intersects our knot K in four points. So we have $\binom{4}{2} = 6$ possible pairs (z_1, z_1') and the total number of summands here is six (see the picture above).

For a pairing P the symbol '$\downarrow_P$' denotes the number of points (z_j, t_j) or (z_j', t_j) in P where the coordinate t decreases as one goes along K.

Fix a pairing P. Consider the knot K as an oriented circle and connect the points (z_j, t_j) and (z_j', t_j) by a chord. We obtain a chord diagram with m chords. (Thus, intuitively, one can think of a pairing as a way of inscribing a chord diagram into a knot in such a way that all chords are horizontal and are placed on different levels.) The corresponding element of the algebra $\mathscr{A}'$ is denoted by D_P. In the second picture on page 220, for each connected component in our example, we show one of the possible pairings, the corresponding chord diagram with the sign $(-1)^{\downarrow_P}$ and the number of summands of the integrand (some of which are equal to zero in $\mathscr{A}'$ due to the one-term relation).

Over each connected component, z_j and z_j' are smooth functions in t_j. By

$$\bigwedge_{j=1}^{m} \frac{dz_j - dz_j'}{z_j - z_j'}$$

we mean the pullback of this form to the integration domain of the variables $t_1, \ldots, t_m$. The integration domain is considered with the orientation of the space $\mathbb{R}^m$ defined by the natural order of the coordinates $t_1, \ldots, t_m$.

By convention, the term in the Kontsevich integral corresponding to $m = 0$ is the (only) chord diagram of order 0 taken with coefficient one. It is the unit of the algebra $\widehat{\mathscr{A}'}$.

8.2.2 Basic properties

We shall see later in this chapter that the Kontsevich integral has the following basic properties:

- $Z(K)$ converges for any strict Morse knot K.
- It is invariant under the deformations of the knot in the class of (not necessarily strict) Morse knots.

- It behaves in a predictable way under the deformations that add a pair of new critical points to a Morse knot.

Let us explain the last item in more detail. While the Kontsevich integral is indeed an invariant of Morse knots, it is not preserved by deformations that change the number of critical points of t. However, the following formula shows how the integral changes when a new pair of critical points is added to the knot:

$$Z\left(\begin{array}{c}\end{array}\right) = Z(H) \cdot Z\left(\begin{array}{c}\end{array}\right). \tag{8.1}$$

Here the first and the third terms represent two embeddings of an arbitrary knot that coincide outside the fragment shown in the dashed rectangle:

$$H := \begin{array}{c}\end{array}$$

is the *hump* (an unknot with two maxima), and the product is the product in the completed algebra $\widehat{\mathscr{A}}$ of chord diagrams. The equality (8.1) allows one to define a genuine knot invariant by the formula

$$I(K) = \frac{Z(K)}{Z(H)^{c/2}},$$

where c denotes the number of critical points of K and the ratio means the division in the algebra $\widehat{\mathscr{A}}$ according to the rule $(1+a)^{-1} = 1 - a + a^2 - a^3 + \dots$ The knot invariant $I(K)$ is sometimes referred to as the *final* Kontsevich integral as opposed to the *preliminary* Kontsevich integral $Z(K)$.

The central importance of the final Kontsevich integral in the theory of finite type invariants is that it is a universal Vassiliev invariant in the following sense.

Consider an unframed weight system w of degree m (that is, a function on the set of chord diagrams with m chords satisfying one- and four-term relations). Applying w to the m-homogeneous part of the series $I(K)$, we get a numerical knot invariant $w(I(K))$. This is a Vassiliev invariant of order m and such invariants span the space of all finite type invariants. This argument will be used to prove the Fundamental Theorem on Vassiliev invariants; see Section 8.8.

The Kontsevich integral has many interesting properties that we shall describe in this and in the subsequent chapters. Among these are its behaviour with respect to the connected sum of knots (Sections 8.4 and 8.7.1) to the coproduct in the Hopf algebra of chord diagrams (Section 8.9.2), cablings (Chapter 9), mutation (Section 11.1.2). We shall see that it can be computed combinatorially (Section 10.3) and has rational coefficients (Section 10.4.3).

8.3 Example of calculation

Here we shall calculate the coefficient of the chord diagram 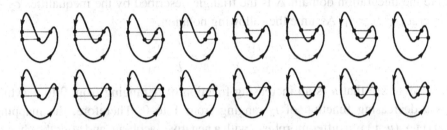 in $Z(H)$, where H is the hump (plane curve with four critical points, as in the previous section) directly from the definition of the Kontsevich integral. The following computation is valid for an arbitrary shape of the curve, provided that the length of the segments a_1a_2 and a_3a_4 (see picture below) decreases with t_1, while that of the segment a_2a_3 increases.

First of all, note that out of the total number of 51 pairings shown in the picture on page 220, the following 16 contribute to the coefficient of ⊗:

We are, therefore, interested only in the band between the critical values c_1 and c_2. Denote by a_1, a_2, a_3, a_4 (resp. b_1, b_2, b_3, b_4) the four points of intersection of the knot with the level $\{t = t_1\}$ (respectively, $\{t = t_2\}$):

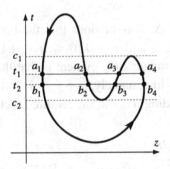

The sixteen pairings shown in the picture above correspond to the differential forms

$$(-1)^{j+k+l+m} d \ln a_{jk} \wedge d \ln b_{lm},$$

where $a_{jk} = a_k - a_j$, $b_{lm} = b_m - b_l$, and the pairs (jk) and (lm) can take four different values each: $(jk) \in \{(13), (23), (14), (24)\} =: A$, $(lm) \in \{(12), (13), (24), (34)\} =: B$. The sign $(-1)^{j+k+l+m}$ is equal to $(-1)^{\downarrow p}$, because in our case the upward oriented strings have even numbers, while the downward oriented strings have odd numbers.

The coefficient of 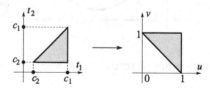 is, therefore, equal to

$$\frac{1}{(2\pi i)^2} \int_\Delta \sum_{(jk)\in A} \sum_{(lm)\in B} (-1)^{j+k+l+m} d\ln a_{jk} \wedge d\ln b_{lm}$$

$$= -\frac{1}{4\pi^2} \int_\Delta \sum_{(jk)\in A} (-1)^{j+k+1} d\ln a_{jk} \wedge \sum_{(lm)\in B} (-1)^{l+m-1} d\ln b_{lm}$$

$$= -\frac{1}{4\pi^2} \int_\Delta d\ln \frac{a_{14}a_{23}}{a_{13}a_{24}} \wedge d\ln \frac{b_{12}b_{34}}{b_{13}b_{24}},$$

where the integration domain Δ is the triangle described by the inequalities $c_2 < t_1 < c_1, c_2 < t_2 < t_1$. Assume the following notation:

$$u = \frac{a_{14}a_{23}}{a_{13}a_{24}}, \qquad v = \frac{b_{12}b_{34}}{b_{13}b_{24}}.$$

It is easy to see that u is an increasing function of t_1 ranging from 0 to 1, while v is a decreasing function of t_2 ranging from 1 to 0. Therefore, the mapping $(t_1, t_2) \mapsto (u, v)$ is a diffeomorphism with a negative Jacobian, and after the change of variables, the integral we are computing becomes

$$\frac{1}{4\pi^2} \int_{\Delta'} d\ln u \wedge d\ln v$$

where Δ' is the image of Δ. It is obvious that the boundary of Δ' contains the segments $u = 1, 0 \leqslant v \leqslant 1$ and $v = 1, 0 \leqslant u \leqslant 1$ that correspond to $t_1 = c_1$ and $t_2 = c_2$. What is not immediately evident is that the third side of the triangle Δ also goes into a straight line, namely, $u + v = 1$. Indeed, if $t_1 = t_2$, then all b's are equal to the corresponding a's and the required fact follows from the identity $a_{12}a_{34} + a_{14}a_{23} = a_{13}a_{24}$.

Therefore,

$$\frac{1}{4\pi^2} \int_{\Delta'} d\ln u \wedge d\ln v = \frac{1}{4\pi^2} \int_0^1 \left(\int_{1-u}^1 d\ln v \right) \frac{du}{u}$$

$$= -\frac{1}{4\pi^2} \int_0^1 \ln(1-u) \frac{du}{u}.$$

Taking the Taylor expansion of the logarithm we get

$$\frac{1}{4\pi^2} \sum_{k=1}^{\infty} \int_0^1 \frac{u^k}{k} \frac{du}{u} = \frac{1}{4\pi^2} \sum_{k=1}^{\infty} \frac{1}{k^2} = \frac{1}{4\pi^2} \zeta(2) = \frac{1}{24}.$$

Two things are quite remarkable in this answer: (1) that it is expressed via a value of the zeta function, and (2) that the answer is rational. In fact, for any knot K, the coefficient of any chord diagram in $Z(K)$ is rational and can be computed in terms of the values of multivariate ζ-functions:

$$\zeta(a_1, \ldots, a_n) = \sum_{0 < k_1 < k_2 < \cdots < k_n} k_1^{-a_1} \ldots k_n^{-a_n}.$$

We shall speak about that in more detail in Section 10.2.

For a complete formula for $Z(H)$, see Section 11.4.

8.4 The Kontsevich integral for tangles

8.4.1 Definition and multiplicativity

The definition of the preliminary Kontsevich integral for knots (see Section 8.2) makes sense for an arbitrary strict Morse tangle T. One only needs to replace the completed algebra $\widehat{\mathscr{A}}'$ of chord diagrams by the graded completion of the vector space of tangle chord diagrams on the skeleton of T, and take $t_{\min}$ and $t_{\max}$ to correspond to the bottom and the top of T, respectively. In Section 8.5 we shall show that the coefficients of the chord diagrams in the Kontsevich integral of any (strict Morse) tangle actually converge.

In particular, one can speak of the Kontsevich integral of links or braids.

Exercise. For a two-component link, what is the coefficient in the Kontsevich integral of the chord diagram of degree 1 whose chord has ends on both components?

Hint: See Section 8.1.2.

Exercise. Compute the integrals

$$Z\left(\overset{\text{-----}}{\bigtimes}\right) \quad \text{and} \quad Z\left(\overset{\text{-----}}{\bigtimes}\right).$$

Answer:

$$\bigtimes \cdot \exp\!\left(\frac{\text{---}}{2}\right) \quad \text{and} \quad \bigtimes \cdot \exp\!\left(-\frac{\text{---}}{2}\right), \quad \text{respectively,}$$

where $\exp a$ is the series $1 + a + \frac{a^2}{2!} + \frac{a^3}{3!} + \ldots$.

Strictly speaking, before describing the properties of the Kontsevich integral, we need to show that it is always well-defined. This will be done in the following section. Meanwhile, we shall assume that this is indeed the case for all the tangles in question.

Proposition 8.5. *The Kontsevich integral for tangles is multiplicative:*

$$Z(T_1 \cdot T_2) = Z(T_1) \cdot Z(T_2)$$

whenever the product $T_1 \cdot T_2$ is defined.

Proof Let t_{min} and t_{max} correspond to the bottom and the top of $T_1 \cdot T_2$, respectively, and let t_{mid} be the level of the top of T_2 (or the bottom of T_1, which is the same). In the expression for the Kontsevich integral of the tangle $T_1 \cdot T_2$ let us remove from the domain of integration all points with at least one coordinate t equal to t_{mid}. This set is of codimension one, so the value of integral remains unchanged. On the other hand, the connected components of the new domain of integration are precisely all products of the connected components for T_1 and T_2, and the integrand for $T_1 \cdot T_2$ is the exterior product of the integrands for T_1 and T_2. The Fubini theorem on multiple integrals implies that $Z(T_1 \cdot T_2) = Z(T_1) \cdot Z(T_2)$. $\square$

The behaviour of the Kontsevich integral under the tensor product of tangles is more complicated. In the expression for $Z(T_1 \otimes T_2)$ indeed there are terms that add up to the tensor product $Z(T_1) \otimes Z(T_2)$: they involve pairings without chords that connect T_1 with T_2. However, the terms with pairings that do have such chords are not necessarily zero and we have no effective way of describing them. Still, there is something we can say but we need a new definition for this.

8.4.2 Parametrized tensor products

By a (horizontal) ε-*rescaling* of $\mathbb{R}^3$ we mean the map sending (z,t) to $(\varepsilon z, t)$. For $\varepsilon > 0$ it induces an operation on tangles; we denote by εT the result of an ε-rescaling applied to T. Note that ε-rescaling of a tangle does not change its Kontsevich integral.

Let T_1 and T_2 be two tangles such that $T_1 \otimes T_2$ is defined. For $0 < \varepsilon \leqslant 1$ we define the ε-*parametrized tensor product* $T_1 \otimes_\varepsilon T_2$ as the result of placing εT_1 next to εT_2 on the left, with the distance of $1 - \varepsilon$ between the two tangles:

More precisely, let $\mathbf{0}_{1-\varepsilon}$ be the empty tangle of width $1 - \varepsilon$ and the same height and depth as εT_1 and εT_2. Then

$$T_1 \otimes_\varepsilon T_2 = \varepsilon T_1 \otimes \mathbf{0}_{1-\varepsilon} \otimes \varepsilon T_2.$$

When $\varepsilon = 1$ we get the usual tensor product. Note that when $\varepsilon < 1$, the parametrized tensor product is, in general, not associative.

Proposition 8.6. *The Kontsevich integral for tangles is asymptotically multiplicative with respect to the parametrized tensor product:*

$$\lim_{\varepsilon \to 0} Z(T_1 \otimes_\varepsilon T_2) = Z(T_1) \otimes Z(T_2)$$

whenever the product $T_1 \otimes T_2$ is defined. Moreover, the difference

$$Z(T_1 \otimes_\varepsilon T_2) - Z(T_1) \otimes Z(T_2)$$

as ε tends to 0 is of the same or smaller order of magnitude as ε.

Proof As we have already noted before, $Z(T_1 \otimes_\varepsilon T_2)$ consists of two parts: the terms that do not involve chords that connect εT_1 with εT_2, and the terms that do. The first part does not depend on ε and is equal to $Z(T_1) \otimes Z(T_2)$, and the second part tends to 0 as $\varepsilon \to 0$.

Indeed, each pairing $P = \{(z_j, z_j')\}$ for $T_1 \otimes T_2$ gives rise to a continuous family of pairings $P_\varepsilon = \{(z_j(\varepsilon), z_j'(\varepsilon))\}$ for $T_1 \otimes_\varepsilon T_2$. Consider one such family P_ε. For all k

$$dz_k(\varepsilon) - dz_k'(\varepsilon) = \varepsilon(dz_k - dz_k').$$

If the kth chord has both ends on εT_1 or on εT_2, we have

$$z_k(\varepsilon) - z_k'(\varepsilon) = \varepsilon(z_k - z_k')$$

for all ε. Therefore, the limit of the first part is equal to $Z(T_1) \otimes Z(T_2)$.

On the other hand, if P_ε has at least one chord connecting the two factors, we have $|z_k(\varepsilon) - z_k'(\varepsilon)| \to 1$ as $\varepsilon \to 0$. Thus the integral corresponding to the pairing P_ε tends to zero as ε gets smaller, and we see that the whole second part of the Kontsevich integral of $T_1 \otimes_\varepsilon T_2$ vanishes in the limit at least as fast as ε:

$$Z(T_1 \otimes_\varepsilon T_2) = Z(T_1) \otimes Z(T_2) + O(\varepsilon).$$

$\square$

8.5 Convergence of the integral

Proposition 8.7. *For any strict Morse tangle T, the Kontsevich integral $Z(T)$ converges.*

Proof The integrand of the Kontsevich integral may have singularities near the boundaries of the connected components. This happens near a critical point of a tangle when the pairing includes a "short" chord whose ends are on the branches of the tangle that come together at a critical point.

Let us assume that the tangle T has exactly one critical point. This is sufficient since any strict Morse tangle can be decomposed as a product of such tangles. The argument in the proof of the multiplicativity (page 226) shows that the Kontsevich integral of a product converges whenever the integral of the factors does.

Suppose, without loss of generality, that T has a critical point which is a maximum with the value t_c. Then we only need to consider pairings with no chords above t_c. Indeed, for any pairing its coefficient in the Kontsevich integral of T is a product of two integrals: one corresponding to the chords above t_c, and the other to the chords below t_c. The first integral obviously converges since the integrand has no singularities, so it is sufficient to consider the factor with chords below t_c.

Essentially, there are two cases.

(1) An isolated chord (z_1, z_1') tends to zero:

In this case, the corresponding chord diagram D_P is equal to zero in $\mathcal{A}'$ by the one-term relation.

(2) A chord (z_j, z_j') tends to zero near a critical point but is separated from that point by one or more other chords:

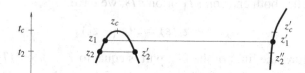

Consider, for example, the case shown on the figure, where the "short" chord (z_2, z_2') is separated from the critical point by another, "long" chord (z_1, z_1'). We have:

$$\left| \int_{t_2}^{t_c} \frac{dz_1 - dz_1'}{z_1 - z_1'} \right| \leqslant C \left| \int_{t_2}^{t_c} d(z_1 - z_1') \right|$$

$$= C \left| (z_c - z_2) - (z_c' - z_2'') \right| \leqslant C' |z_2 - z_2'|$$

for some positive constants C and C'. This integral is of the same order as $z_2 - z_2'$ and this compensates the denominator corresponding to the second chord.

More generally, one shows by induction that if a "short" chord (z_j, z'_j) is separated from the maximum by $j - 1$ chords, the first (that is, the nearest to the maximum) of which is "long," the integral

$$\int_{t_j < t_{j-1} < \cdots < t_1 < t_c} \bigwedge_{i=1}^{j-1} \frac{dz_i - dz'_i}{z_i - z'_i}$$

is of the same order as $z_j - z'_j$. This implies the convergence of the Kontsevich integral. $\square$

8.6 Invariance of the integral

Theorem 8.8. *The Kontsevich integral is invariant under the deformations in the class of (not necessarily strict) Morse knots.*

The proof of this theorem spans the whole of this section.

Any deformation of a knot within the class of Morse knots can be approximated by a sequence of deformations of three types: orientation-preserving reparametrizations, *horizontal deformations* and *movements of critical points*.

The invariance of the Kontsevich integral under orientation-preserving reparametrizations is immediate since the parameter plays no role in the definition of the integral apart from determining the orientation of the knot.

A horizontal deformation is an isotopy of a knot in $\mathbb{R}^3$ which preserves all horizontal planes $\{t = \text{const}\}$ and leaves all the critical points (together with some small neighbourhoods) fixed. The invariance under horizontal deformations is the most essential point of the theory. We prove it in the next subsection.

A movement of a critical point C is an isotopy which is identical everywhere outside a small neighbourhood of C and does not introduce new critical points on the knot. The invariance of the Kontsevich integral under the movements of critical points will be considered in Section 8.6.2.

As we mentioned before, the Kontsevich integral is not invariant under isotopies that change the number of critical points. Its behaviour under such deformations will be discussed in Section 8.7.

8.6.1 Invariance under horizontal deformations

Let us decompose the given knot into a product of tangles without critical points of the function t and very thin tangles containing the critical levels. A horizontal deformation keeps fixed the neighbourhoods of the critical points, so, due to multiplicativity, it is enough to prove that the Kontsevich integral for a tangle without critical points is invariant under horizontal deformations that preserve the boundary pointwise.

Proposition 8.9. *Let T_0 be a tangle without critical points and T_λ a horizontal deformation of T_0 to T_1 (preserving the top and the bottom of the tangle). Then $Z(T_0) = Z(T_1)$.*

Proof Denote by ω the integrand form in the mth term of the Kontsevich integral:

$$\omega = \sum_{P=\{(z_j,z_j')\}} (-1)^{\downarrow} D_P \bigwedge_{j=1}^{m} \frac{dz_j - dz_j'}{z_j - z_j'}.$$

Here the functions z_j, z_j' depend not only on $t_1, ..., t_m$, but also on λ, and *all differentials are understood as complete differentials with respect to all these variables.* This means that the form ω is not exactly the form which appears in the Kontsevich integral (it has some additional $d\lambda$'s), but this does not change the integrals over the simplices

$$\Delta_\lambda = \{t_{\min} < t_m < \cdots < t_1 < t_{\max}\} \times \{\lambda\},$$

because the value of λ on such a simplex is fixed.

We must prove that the integral of ω over Δ_0 is equal to its integral over Δ_1.

Consider the product polytope

$$\Delta = \Delta_0 \times [0, 1] =$$.

By Stokes' theorem, we have $\displaystyle\int_{\partial\Delta} \omega = \int_\Delta d\omega$.

The form ω is closed: $d\omega = 0$. The boundary of the integration domain is $\partial\Delta = \Delta_0 - \Delta_1 + \sum \{faces\}$. The theorem will follow from the fact that $\omega|_{\{face\}} = 0$. To show this, consider two types of faces.

The first type corresponds to $t_m = t_{\min}$ or $t_1 = t_{\max}$. In this situation, $dz_j = dz_j' = 0$ for $j = 1$ or m, since z_j and z_j' do not depend on λ.

The faces of the second type are those where we have $t_k = t_{k+1}$ for some k. In this case we have to choose the kth and $(k+1)$st chords on the same level $\{t = t_k\}$. In general, the endpoints of these chords may coincide and we do not get a chord diagram at all. Strictly speaking, D_P does not extend automatically to such a face so we have to be careful. Extend D_P to this face in the following manner: if some endpoints of kth and $(k + 1)$st chords belong to the same string (and therefore coincide) we place the kth chord a little higher than the $(k + 1)$st chord, so that its endpoint differs from the endpoint of the $(k+1)$st chord. This trick yields a well-defined prolongation of D_P to the face, and we use it here.

All summands of ω are divided into three parts:

(1) kth and $(k+1)$st chords connect the same two strings;

(2) kth and $(k+1)$st chords are chosen in such a way that their endpoints belong to four different strings;

(3) kth and $(k+1)$st chords are chosen in such a way that there exist exactly three different strings containing their endpoints.

Consider all these cases one by one.

(1) We have $z_k = z_{k+1}$ and $z'_k = z'_{k+1}$ or vice versa. So

$$d(z_k - z'_k) \wedge d(z_{k+1} - z'_{k+1}) = 0$$

and, therefore, the restriction of ω to the face is zero.

(2) All choices of chords in this part of ω appear in mutually canceling pairs. Fix four strings and number them by 1, 2, 3, 4. Suppose that for a certain choice of the pairing, the kth chord connects the first two strings and $(k+1)$st chord connects the last two strings. Then there exists another choice for which, on the contrary, the kth chord connects the last two strings and $(k+1)$st chord connects the first two strings. These two choices give two summands of ω differing by a sign:

$$\cdots d(z_k - z'_k) \wedge d(z_{k+1} - z'_{k+1}) \cdots + \cdots d(z_{k+1} - z'_{k+1}) \wedge d(z_k - z'_k) \cdots = 0.$$

(3) This is the most difficult case. The endpoints of kth and $(k+1)$st chords have exactly one string in common. Call the three relevant strings 1, 2, 3 and denote by ω_{ij} the 1-form $\dfrac{dz_i - dz_j}{z_i - z_j}$. Then ω is the product of a certain $(m-2)$-form and the sum of the following six 2-forms:

$$(-1)^{\downarrow}\,\omega_{12} \wedge \omega_{23} \;+\; (-1)^{\downarrow}\,\omega_{12} \wedge \omega_{13}$$

$$+(-1)^{\downarrow}\,\omega_{13} \wedge \omega_{12} \;+\; (-1)^{\downarrow}\,\omega_{13} \wedge \omega_{23}$$

$$+(-1)^{\downarrow}\,\omega_{23} \wedge \omega_{12} \;+\; (-1)^{\downarrow}\,\omega_{23} \wedge \omega_{13}.$$

Using the fact that $\omega_{ij} = \omega_{ji}$, we can rewrite this as follows:

$$\left((-1)^{\downarrow} \; \vcenter{\hbox{\includegraphics{}}} - (-1)^{\downarrow} \; \vcenter{\hbox{\includegraphics{}}}\right) \omega_{12} \wedge \omega_{23}$$

$$+\left((-1)^{\downarrow} \; \vcenter{\hbox{\includegraphics{}}} - (-1)^{\downarrow} \; \vcenter{\hbox{\includegraphics{}}}\right) \omega_{23} \wedge \omega_{31}$$

$$+\left((-1)^{\downarrow} \; \vcenter{\hbox{\includegraphics{}}} - (-1)^{\downarrow} \; \vcenter{\hbox{\includegraphics{}}}\right) \omega_{31} \wedge \omega_{12}.$$

The four-term relations in horizontal form (page 86) say that the expressions in parentheses are one and the same element of $\mathscr{A}'$, hence, the whole sum is equal to

$$\left((-1)^{\downarrow} \; \vcenter{\hbox{\includegraphics{}}} - (-1)^{\downarrow} \; \vcenter{\hbox{\includegraphics{}}}\right) (\omega_{12} \wedge \omega_{23} + \omega_{23} \wedge \omega_{31} + \omega_{31} \wedge \omega_{12}).$$

The 2-form that appears here is actually zero! This simple but remarkable fact, known as *Arnold's identity* (see Arnol'd 1969), can be put into the following form:

$$f + g + h = 0 \implies \frac{df}{f} \wedge \frac{dg}{g} + \frac{dg}{g} \wedge \frac{dh}{h} + \frac{dh}{h} \wedge \frac{df}{f} = 0$$

(in our case $f = z_1 - z_2$, $g = z_2 - z_3$, $h = z_3 - z_1$) and verified by a direct computation.

This finishes the proof. $\qquad\square$

Remark 8.10. The Kontsevich integral of a tangle may change, if the boundary points are moved. Examples may be found below in Exercises 8.8–8.11.

8.6.2 *Moving the critical points*

Let T_0 and T_1 be two tangles which are identical except a sharp "tail" of width ε, which may be twisted:

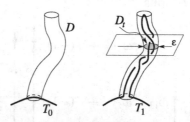

More exactly, we assume that (1) T_1 is different from T_0 only inside a region D which is the union of disks D_t of diameter ε lying in horizontal planes with fixed $t \in [t_1, t_2]$, (2) each tangle T_0 and T_1 has exactly one critical point in D, and (3)

each tangle T_0 and T_1 intersects every disk D_t at most in two points. We call the passage from T_0 to T_1 a *special movement* of the critical point. In order to prove the main result of this section, it is sufficient to show the invariance of the Kontsevich integral under such movements. Note that special movements of critical points may take a Morse knot out of the class of strict Morse knots.

Proposition 8.11. *The Kontsevich integral remains unchanged under a special movement of the critical point: $Z(T_0) = Z(T_1)$.*

Proof The difference between $Z(T_0)$ and $Z(T_1)$ can come only from the terms with a chord ending on the tail.

Consider the tangle T_1. If the highest of such chords connects the two sides of the tail, then the corresponding tangle chord diagram is zero by a one-term relation. So we can assume that the highest, say, the kth, chord is a "long" chord, which means that it connects the tail with another part of T_1. Suppose the endpoint of the chord belonging to the tail is (z'_k, t_k). Then there exists another choice for kth chord which is almost the same but ends at another point of the tail (z''_k, t_k) on the same horizontal level:

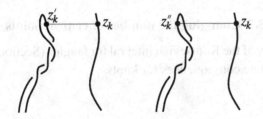

The corresponding two terms appear in $Z(T_1)$ with the opposite signs due to the sign $(-1)^{\downarrow}$.

Let us estimate the difference of the integrals corresponding to such kth chords:

$$\left| \int_{t_{k+1}}^{t_c} d(\ln(z'_k - z_k)) - \int_{t_{k+1}}^{t_c} d(\ln(z''_k - z_k)) \right| = \left| \ln\left(\frac{z''_{k+1} - z_{k+1}}{z'_{k+1} - z_{k+1}} \right) \right|$$

$$= \left| \ln\left(1 + \frac{z''_{k+1} - z'_{k+1}}{z'_{k+1} - z_{k+1}} \right) \right| \sim \left| z''_{k+1} - z'_{k+1} \right| \leqslant \varepsilon$$

(here t_c is the value of t at the uppermost point of the tail).

Now, if the next $(k+1)$st chord is also long, then, similarly, it can be paired with another long chord so that they give a contribution to the integral proportional to $\left| z''_{k+2} - z'_{k+2} \right| \leqslant \varepsilon$.

In the case the $(k+1)$st chord is short (that is, it connects two points z''_{k+1}, z'_{k+1} of the tail), we have the following estimate for the double integral corresponding to kth and $(k+1)$st chords:

$$\left| \int_{t_{k+2}}^{t_c} \left(\int_{t_{k+1}}^{t_c} d(\ln(z'_k - z_k)) - \int_{t_{k+1}}^{t_c} d(\ln(z''_k - z_k)) \right) \frac{dz''_{k+1} - dz'_{k+1}}{z''_{k+1} - z'_{k+1}} \right|$$

$$\leqslant \text{const} \cdot \left| \int_{t_{k+2}}^{t_c} |z''_{k+1} - z'_{k+1}| \frac{dz''_{k+1} - dz'_{k+1}}{|z''_{k+1} - z'_{k+1}|} \right|$$

$$= \text{const} \cdot \left| \int_{t_{k+2}}^{t_c} d(z''_{k+1} - z'_{k+1}) \right| \sim |z''_{k+2} - z'_{k+2}| \leqslant \varepsilon.$$

Continuing this argument, we see that the difference between $Z(T_0)$ and $Z(T_1)$ is $O(\varepsilon)$. Now, by horizontal deformations we can make ε tend to zero. This proves the theorem and completes the proof of the Kontsevich integral's invariance in the class of knots with nondegenerate critical points. $\qquad\square$

8.7 Changing the number of critical points

The multiplicativity of the Kontsevich integral for tangles (Sections 8.4.1 and 8.4.2) has several immediate consequences for knots.

8.7.1 From long knots to usual knots

A long (Morse) knot can be closed up so as to produce a usual (Morse) knot:

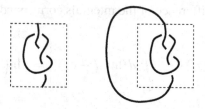

Recall that the algebras of chord diagrams for long knots and for usual knots are essentially the same; the isomorphism is given by closing up a linear chord diagram.

Proposition 8.12. *The Kontsevich integral of a long knot T coincides with that of its closure K_T.*

Proof Denote by id the tangle consisting of one vertical strand. Then K_T can be written as $T_{max} \cdot (T \otimes_\varepsilon \text{id}) \cdot T_{min}$ where T_{max} and T_{min} are elementary tangles consisting of a maximum and a minimum respectively, and $0 < \varepsilon \leqslant 1$.

Since the Kontsevich integral of K_T does not depend on ε, we can take $\varepsilon \to 0$. Therefore,

$$Z(K_T) = Z(T_{max}) \cdot (Z(T) \otimes Z(\text{id})) \cdot Z(T_{min}).$$

However, the Kontsevich integrals of T_{max}, T_{min} and id consist of one diagram with no chords, and the proposition follows. □

A corollary of this is Equation (8.1) (page 222) which describes the behaviour of the Kontsevich integral under the addition of a pair of critical points. Indeed, adding a pair of critical points to a long knot T is the same as multiplying it by

and (8.1) then follows from the multiplicativity of the Kontsevich integral for tangles.

8.7.2 *The universal Vassiliev invariant*

Equation (8.1) allows one to define the *universal Vassiliev invariant* by either

$$I(K) = \frac{Z(K)}{Z(H)^{c/2}}$$

or

$$I'(K) = \frac{Z(K)}{Z(H)^{c/2-1}},$$

where c denotes the number of critical points of K in an arbitrary Morse representation, and the quotient means division in the algebra $\widehat{\mathscr{A}}$: $(1 + a)^{-1} = 1 - a + a^2 - a^3 + \ldots$.

Any isotopy of a knot in $\mathbb{R}^3$ can be approximated by a sequence consisting of isotopies within the class of (not necessarily strict) Morse knots and insertions/deletions of "humps," that is, pairs of adjacent maxima and minima. Hence, the invariance of $Z(K)$ in the class of Morse knots and the formula (8.1) imply that both $I(K)$ and $I'(K)$ are invariant under an arbitrary deformation of K. (The meaning of the "universality" will be explained in Section 8.8.2.)

The version $I'(K)$ has the advantage of being multiplicative with respect to the connected sum of knots; in particular, it takes the value 1 on the unknot. However, the version $I(K)$ is also used as it has a direct relationship with the quantum invariants (see Ohtsuki 2002). In particular, we shall use the term "Kontsevich integral of the unknot"; this, of course, refers to I, and not I'.

8.8 The universal Vassiliev invariant

8.8.1 Proof of the Kontsevich theorem

First of all, we reformulate the Kontsevich theorem (or, more exactly, Kontsevich's part of the Vassiliev–Kontsevich theorem of Section 4.2.1) as follows.

Theorem 8.13. *Let w be an unframed weight system of order n. Then there exists a Vassiliev invariant of order n whose symbol is w.*

Proof The desired knot invariant is given by the formula

$$K \longmapsto w(I(K)).$$

Let D be a chord diagram of order n and let K_D be a singular knot with chord diagram D. The theorem follows from the fact that $I(K_D) = D +$ (terms of order $> n$). Since the denominator of $I(K)$ starts with the unit of the algebra $\mathscr{A}'$, it is sufficient to prove that

$$Z(K_D) = D + \text{(terms of order} > n). \tag{8.2}$$

In fact, we shall establish (8.2) for D an arbitrary *tangle* chord diagram and $K_D = T_D$ a singular tangle with the diagram D.

If $n = 0$, the diagram D has no chords and T_D is nonsingular. For a nonsingular tangle, the Kontsevich integral starts with a tangle chord diagram with no chords, and (8.2) clearly holds. Note that the Kontsevich integral of any singular tangle (with at least one double point) necessarily starts with terms of degree at least 1.

Consider now the case $n = 1$. If T_D is a singular 2-braid, there is only one possible term of degree 1, namely the chord diagram with the chord connecting the two strands. The coefficients of this diagram in $Z(T_+)$ and $Z(T_-)$, where $T_+ - T_-$ is a resolution of the double point of T_D, simply measure the number of full twists in T_+ and T_- respectively (see Section 8.1). The difference of these numbers is 1, so in this case (8.2) is also true.

Now, let T_D be an arbitrary singular tangle with exactly one double point, and V_ε be the ε-neighbourhood of the singularity. We can assume that the intersection of T_D with V_ε is a singular 2-braid, and that the double point of T_D is resolved as $T_D = T_+^\varepsilon - T_-^\varepsilon$ where T_+^ε and T_-^ε coincide with T outside V_ε.

Let us write the degree 1 part of $Z(T_{\pm}^{\varepsilon})$ as a sum $Z'_{\pm} + Z''_{\pm}$ where $Z'_{\pm}$ is the integral over all chords whose both ends are contained in V_{ε} and $Z''_{\pm}$ is the rest, that is, the integral over the chords with at least one end outside V_{ε}. As ε tends to 0, $Z''_{+} - Z''_{-}$ vanishes. On the other hand, for all ε we have that $Z''_{+} - Z''_{-}$ is equal to the diagram D with the coefficient 1. This settles the case $n = 1$.

Finally, if $n > 1$, using a suitable deformation, if necessary, we can always achieve that T_D is a product of n singular tangles with one double point each. Now (8.2) follows from the multiplicativity of the Kontsevich integral for tangles. $\qquad\square$

8.8.2 Universality of $I(K)$

In the proof of the Kontsevich theorem we have seen that for a singular knot K with n double points, $I(K)$ starts with terms of degree n. This means that if $I_n(K)$ denotes the nth graded component of the series $I(K)$, then the function $K \mapsto I_n(K)$ is a Vassiliev invariant of order n.

In some sense, all Vassiliev invariants are of this type:

Proposition 8.14. *Any Vassiliev invariant can be factored through I: for any $v \in \mathscr{V}$ there exists a linear function f on $\widehat{\mathscr{A}}$ such that $v = f \circ I$.*

Proof Let $v \in \mathscr{V}_n$. By the Kontsevich theorem we know that there is a function f_0 such that v and $f_0 \circ I_n$ have the same symbol. Therefore, the highest degree part of the difference $v - f_0 \circ I_n$ belongs to $\mathscr{V}_{n-1}$ and is thus representable as $f_1 \circ I_{n-1}$. Proceeding in this way, we obtain:

$$v = \sum_{i=0}^{n} f_i \circ I_{n-i} = \left(\sum_{i=0}^{n} f_i\right) \circ I.$$

$\qquad\square$

Remark 8.15. The construction of the foregoing proof shows that the universal Vassiliev invariant induces a splitting of the filtered space $\mathscr{V}$ into a direct sum with summands isomorphic to the factors $\mathscr{V}_n/\mathscr{V}_{n-1}$. Elements of these subspaces are referred to as *canonical Vassiliev invariants*. We shall speak about them in more detail later in Section 11.2.

As a corollary, we get the following statement:

Theorem 8.16. *The universal Vassiliev invariant I is exactly as strong as the set of all Vassiliev invariants: for any two knots K_1 and K_2 we have*

$$I(K_1) = I(K_2) \quad \Longleftrightarrow \quad \forall v \in \mathscr{V} \quad v(K_1) = v(K_2).$$

8.9 Symmetries and the group-like property of $Z(K)$

8.9.1 Reality

Choose a basis in the vector space $\mathcal{A}'$ consisting of chord diagrams. A priori, the coefficients of the Kontsevich integral of a knot K with respect to this basis are complex numbers.

Theorem 8.17. *All coefficients of the Kontsevich integral with respect to a basis of chord diagrams are real.*

Remark 8.18. Of course, this fact is a consequence of the Le–Murakami theorem stating that these coefficients are rational (Section 10.4.3). However, the rationality of the Kontsevich integral is a very nontrivial fact, while the proof that its coefficients are real is quite simple.

Proof Rotate the coordinate frame in $\mathbb{R}^3$ around the real axis x by $180°$; denote the new coordinates by $t^* = -t$, $z^* = \bar{z}$. If K is a Morse knot, it will still be a Morse knot, with the same number of maxima, with respect to the new coordinates, and its Kontsevich integral, both preliminary and final, will be the same in both coordinate systems. Let us denote the preliminary Kontsevich integral with respect to the starred coordinates by $Z^*(K)$.

For each pairing $P = \{(z_j, z'_j)\}$ with m chords that appears in the formula for $Z(K)$, there is a pairing $P^* = \{(z_j^*, z_j^{*'})\}$ that appears in the formula for $Z^*(K)$. It consists of the very same chords as P but taken in the starred coordinate system: $z_j^* = \bar{z}_{m-j+1}$ and $z_j^{*'} = \bar{z}'_{m-j+1}$. The corresponding chord diagrams are, obviously, equal: $D_P = D_{P^*}$. Moreover, $\downarrow^* = 2m - \downarrow$ and, hence, $(-1)^{\downarrow^*} = (-1)^{\downarrow}$. The simplex $\Delta = t_{\min} < t_1 < \cdots < t_m < t_{\max}$ for the variables t_i corresponds to the simplex $\Delta^* = -t_{\max} < t_m^* < \cdots < t_1^* < -t_{\min}$ for the variables t_i^*. The coefficient of D_{P^*} in $Z^*(K)$ is

$$c(D_{P^*}) = \frac{(-1)^{\downarrow}}{(2\pi i)^m} \int \bigwedge_{j=1}^{m} d\ln(z_j^* - z_j^{*'}),$$

where z_j^* and $z_j^{*'}$ are understood as functions in $t_1^*, \ldots, t_m^*$ and the integral is taken over a connected component in the simplex Δ^*. In the last integral we make the change of variables according to the formula $t_j^* = -t_{m-j+1}$. The Jacobian of this transformation is equal to $(-1)^{m(m+1)/2}$. Therefore,

$$c(D_{P^*}) = \frac{(-1)^{\downarrow}}{(2\pi i)^m} \int (-1)^{m(m+1)/2} \bigwedge_{j=1}^{m} d\ln(\bar{z}_{m-j+1} - \bar{z}'_{m-j+1})$$

(integral over the corresponding connected component in the simplex Δ). Now permute the differentials to arrange the subscripts in the increasing order. The sign of this permutation is $(-1)^{m(m-1)/2}$. Note that $(-1)^{m(m+1)/2} \cdot (-1)^{m(m-1)/2} = (-1)^m$. Hence,

$$
\begin{aligned}
c(D_{P^*}) &= \frac{(-1)^{\downarrow}}{(2\pi i)^m} (-1)^m \int \bigwedge_{j=1}^m d \ln(\overline{z}_j - \overline{z}'_j) \\
&= \frac{(-1)^{\downarrow}}{(2\pi i)^m} \int \bigwedge_{j=1}^m \overline{d \ln(z_j - z'_j)} = \overline{c(D_P)}.
\end{aligned}
$$

Since any chord diagram D_P can be expressed as a combination of the basis diagrams with real coefficients, this proves the theorem. $\qquad\square$

8.9.2 The group-like property

Theorem 8.19. *For any Morse tangle T with skeleton X, the Kontsevich integral $Z(T)$ is a group-like element in the graded completion of the coalgebra $\mathscr{C}(X)$:*

$$
\delta(Z(T)) = Z(T) \otimes Z(T).
$$

In particular, if K is a knot, $Z(K)$ is a group-like element in $\widehat{\mathscr{A}}'$.

Proof The proof consists of a direct comparison of the coefficients on both sides. Write the Kontsevich integral as

$$
Z(K) = \sum_P c_P D_P,
$$

where the sum is over all possible pairings P for K. Then, for given pairings P_1 and P_2, we have

$$
\sum_{P = P_1 \sqcup P_2} c_P = c_{P_1} c_{P_2}.
$$

Indeed, the chords in P_1 and P_2 vary over the simplices $\Delta_1 = \{t_{min} < t_{|P_1|} < \ldots < t_1 < t_{max}\}$ and $\Delta_2 = \{t_{min} < \tilde{t}_{|P_2|} < \ldots < \tilde{t}_1 < t_{max}\}$ respectively. The product $\Delta_1 \times \Delta_2$ is subdivided into simplices $\Delta_{P,J}$, each of dimension $|P_1| + |P_2|$, by hyperplanes $t_i = \tilde{t}_k$; each of these simplices can be thought of as a way of fusing P_1 and P_2 into one pairing P, with $P \backslash J = P_1$ and $J = P_2$. Then each term on the left-hand side is the integral

$$
\frac{(-1)^{\downarrow}}{(2\pi i)^{|P_1|+|P_2|}} \int_{\Delta_{P,J}} \bigwedge_{j=1}^{|P_1|+|P_2|} d \ln(z_j - z'_j),
$$

while $c_{P_1} c_{P_2}$ is the integral of the same expression over the product $\Delta_1 \times \Delta_2 = \cup \Delta_{P,J}$.

It follows now that

$$\delta(Z(K)) = \sum_{P, J \subseteq P} c_P D_{P \setminus J} \otimes D_J,$$

is equal to

$$Z(K) \otimes Z(K) = \sum_{P_1, P_2} c_{P_1} c_{P_2} D_{P_1} \otimes D_{P_2}.$$

$\square$

Group-like elements in a bialgebra form a group and this implies that the final Kontsevich integral is also group-like.

8.9.3 Change of orientation

Theorem 8.20. *The Kontsevich integral commutes with the operation τ of orientation reversal:*

$$Z(\tau(K)) = \tau(Z(K)).$$

Proof Changing the orientation of K has the following effect on the formula for the Kontsevich integral in Definition 8.4: each diagram D is replaced by $\tau(D)$ and the factor $(-1)^{\downarrow}$ is replaced by $(-1)^{\uparrow}$, where by $\uparrow$ we mean, of course, the number of points (z_j, t_j) or (z'_j, t_j) in a pairing P where the coordinate t grows along the parameter of K. Since the number of points in a pairing is always even, $(-1)^{\downarrow} = (-1)^{\uparrow}$, so that $\tau(D)$ appears in $Z(\tau(K))$ with the same coefficient as D in $Z(K)$. The theorem is proved. $\square$

Corollary 8.21. *The following two assertions are equivalent:*

- *Vassiliev invariants do not distinguish the orientation of knots,*
- *all chord diagrams are symmetric: $D = \tau(D)$ modulo one- and four-term relations.*

The calculations of Kneissler (1997) show that up to order 12 all chord diagrams are symmetric. For bigger orders the problem is still open.

8.9.4 Mirror images

Recall that σ is the operation sending a knot to its mirror image (see 1.3). Define the corresponding operation $\sigma : \mathscr{A}' \to \mathscr{A}'$ by sending a chord diagram D to $(-1)^{\deg D} D$. It extends to a map $\widehat{\mathscr{A}}' \to \widehat{\mathscr{A}}'$ which we also denote by σ.

Theorem 8.22. *The Kontsevich integral commutes with σ:*

$$Z(\sigma(K)) = \sigma(Z(K)).$$

Proof Let us realize the operation σ on knots by the reflection of $\mathbb{R}^3$ coming from the complex conjugation in $\mathbb{C}$: $(z, t) \mapsto (\bar{z}, t)$. Then the Kontsevich integral for $\sigma(K)$ can be written as

$$Z(\sigma(K)) = \sum_{m=0}^{\infty} \frac{1}{(2\pi i)^m} \int \sum_P (-1)^{\downarrow} D_P \bigwedge_{j=1}^{m} d \ln(\overline{z_j - z_j'})$$

$$= \sum_{m=0}^{\infty} (-1)^m \frac{1}{(2\pi i)^m} \overline{\int \sum_P (-1)^{\downarrow} D_P \bigwedge_{j=1}^{m} d \ln(z_j - z_j')}.$$

Comparing this with the formula for $Z(K)$ we see that the terms of $Z(\sigma(K))$ with an even number of chords coincide with those of $\overline{Z(K)}$ and terms of $Z(\sigma(K))$ with an odd number of chords differ from the corresponding terms of $\overline{Z(K)}$ by a sign. $\square$

Since the Kontsevich integral is equivalent to the totality of all finite type invariants, the above theorem implies that if v is a Vassiliev invariant of degree n, K is a singular knot with n double points and $\overline{K} = \sigma(K)$ its mirror image, then $v(K) = v(\overline{K})$ for even n and $v(K) = -v(\overline{K})$ for odd n.

Exercise. Prove this statement without using the Kontsevich integral.

Recall (page 7) that a knot K is called *plus-amphicheiral*, if it is equivalent to its mirror image as an oriented knot: $K = \sigma(K)$, and *minus-amphicheiral* if it is equivalent to the inverse of the mirror image: $K = \tau(\sigma(K))$. Write τ for the mirror reflection on chord diagrams (see Section 5.5.3), and recall that an element of $\mathscr{A}'$ is called *symmetric* (*antisymmetric*), if τ acts on it as identity (as multiplication by -1, respectively).

Corollary 8.23. *The Kontsevich integral $Z(K)$ of a plus-amphicheiral knot K consists only of even order terms. For a minus-amphicheiral knot K the Kontsevich integral $Z(K)$ has the following property: its even-degree part consists only of symmetric chord diagrams, while the odd-degree part consists only of anti-symmetric elements. The same is true for the universal Vassiliev invariant $I(K)$.*

Proof For a plus-amphicheiral knot, the theorem implies that $Z(K) = \sigma(Z(K))$, hence all the odd order terms in the series $Z(K)$ vanish. The quotient of two even series in the graded completion $\widehat{\mathscr{A}}'$ is obviously even, therefore the same property holds for $I(K) = Z(K)/Z(H)^{c/2}$.

For a minus-amphicheiral knot K, we have $Z(K) = \tau(\sigma(Z(K)))$, which implies the second assertion. $\square$

Note that it is an open question whether non-symmetric chord diagrams exist. If they do not, then, of course, both assertions of the theorem, for plus- and minus-amphicheiral knots, coincide.

8.10 Towards the combinatorial Kontsevich integral

Since the Kontsevich integral comprises all Vassiliev invariants, calculating it explicitly is a very important problem. Knots are, essentially, combinatorial objects so it is not surprising that the Kontsevich integral, which we have defined analytically, can be calculated combinatorially from the knot diagram. Different versions of such combinatorial definition were proposed in several papers (Bar-Natan 1997a; Cartier 1993; Le and Murakami 1995a, 1996a; Piunikhin 1995) and treated in several books (Kassel 1995; Ohtsuki 2002). Such a definition will be given in Chapter 10; here we shall explain the idea behind it.

The multiplicativity of the Kontsevich integral hints at the following method of computing it: present a knot as a product of several standard tangles whose Kontsevich integral is known and then multiply the corresponding values of the integral. This method works well for the quantum invariants, see Section 2.6.3; however, for the Kontsevich integral it turns out to be too naïve to be of direct use.

Indeed, in the case of quantum invariants we decompose the knot into elementary tangles, that is, crossings, max/min events and pieces of vertical strands using both the usual product and the tensor product of tangles. While the Kontsevich integral behaves well with respect to the usual product of tangles, there is no simple expression for the integral of the tensor product of two tangles, even if one of the factors is a trivial tangle. As a consequence, the Kontsevich integral is hard to calculate even for the generators of the braid group, not to mention other possible candidates for "standard" tangles.

Still, we know that the Kontsevich integral is *asymptotically* multiplicative with respect to the parametrized tensor product. This suggests the following procedure.

Write a knot K as a product of tangles $K = T_1 \cdot \ldots \cdot T_n$ where each T_i is simple, that is, a tensor product of several elementary tangles. Let us think of each T_i as of an ε-parametrized tensor product of elementary tangles with $\varepsilon = 1$. We want to vary this ε to make it very small. There are two issues here that should be taken care of.

First, the ε-parametrized tensor product is not associative for $\varepsilon \neq 1$, so we need a *parenthesizing* on the factors in T_i. We choose the parenthesizing arbitrarily on

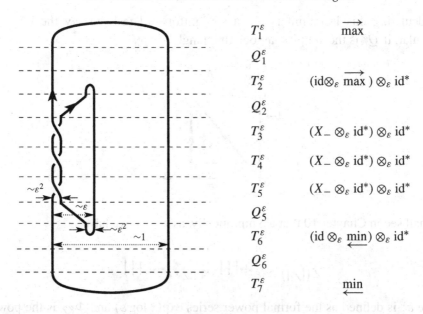

T_1^ε	$\overrightarrow{\max}$
Q_1^ε	
T_2^ε	$(\mathrm{id} \otimes_\varepsilon \overrightarrow{\max}) \otimes_\varepsilon \mathrm{id}^*$
Q_2^ε	
T_3^ε	$(X_- \otimes_\varepsilon \mathrm{id}^*) \otimes_\varepsilon \mathrm{id}^*$
T_4^ε	$(X_- \otimes_\varepsilon \mathrm{id}^*) \otimes_\varepsilon \mathrm{id}^*$
T_5^ε	$(X_- \otimes_\varepsilon \mathrm{id}^*) \otimes_\varepsilon \mathrm{id}^*$
Q_5^ε	
T_6^ε	$(\mathrm{id} \otimes_\varepsilon \overleftarrow{\min}) \otimes_\varepsilon \mathrm{id}^*$
Q_6^ε	
T_7^ε	$\overleftarrow{\min}$

Figure 8.1 A decomposition of the trefoil into associating tangles and ε-parametrized tensor products of elementary tangles, with the notations from Section 1.6.6. The associating tangles between T_3^ε, T_4^ε and T_5^ε are omitted since these tangles are composable for all ε.

each T_i and denote by T_i^ε the tangle obtained from T_i by replacing $\varepsilon = 1$ by an arbitrary positive $\varepsilon \leqslant 1$.

Second, even though the tangles T_i and T_{i+1} are composable, the tangles T_i^ε and T_{i+1}^ε may fail to be composable for $\varepsilon < 1$. Therefore, for each i we have to choose a family of *associating* tangles without crossings Q_i^ε which connect the bottom endpoints of T_i^ε with the corresponding top endpoints of T_{i+1}^ε.

Now we can define a family of knots K^ε as

$$K^\varepsilon = T_1^\varepsilon \cdot Q_1^\varepsilon \cdot T_2^\varepsilon \cdot \ldots \cdot Q_{n-1}^\varepsilon \cdot T_n^\varepsilon.$$

Figure 8.1 illustrates this construction on the example of a trefoil knot.

Since for each ε the knot K^ε is isotopic to K it is tempting to take $\varepsilon \to 0$, calculate the limits of the Kontsevich integrals of the factors and then take their product. The Kontsevich integral of any elementary tangle, and, hence, of the limit

$$\lim_{\varepsilon \to 0} Z(T_i^\varepsilon)$$

is easily evaluated, so it only remains to calculate the limit of $Z(Q_i^\varepsilon)$ as ε tends to zero.

Calculating this last limit is not a straightforward task, to say the least. In particular, if Q^ε is the simplest associating tangle

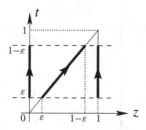

we shall see in Chapter 10 that asymptotically, as $\varepsilon \to 0$ we have

$$Z(\text{\Large ↑↗↑}) \simeq \varepsilon^{\frac{1}{2\pi i}|H|} \cdot \Phi_{KZ} \cdot \varepsilon^{-\frac{1}{2\pi i}|H|},$$

where ε^x is defined as the formal power series $\exp(x \log \varepsilon)$ and Φ_{KZ} is the power series known as the *Knizhnik–Zamolodchikov associator*. Similar formulae can be written for other associating tangles.

There are two difficulties here. One is that the integral $Z(Q^\varepsilon)$ does not converge as ε tends to 0. However, all the divergence is hidden in the terms $\varepsilon^{\frac{1}{2\pi i}|H|}$ and $\varepsilon^{-\frac{1}{2\pi i}|H|}$ and careful analysis shows that all such terms from all associating tangles cancel each other out in the limit, and can be omitted. The second problem is to calculate the associator. This a highly nontrivial task, and is the main subject of Chapter 10.

Exercises

8.1 For the link with two components K and L shown on the right, draw the configuration space of horizontal chords joining K and L as in the proof of Theorem 8.2 (see page 217). Compute the linking number of K and L using this theorem.

8.2 Is it true that $Z(\overline{H}) = Z(H)$, where H is the hump as shown on page 222 and $\overline{H}$ is the same hump reflected in a horizontal line?

8.3 M. Kontsevich in his pioneering paper (1993) and some of his followers (for example, Bar-Natan (1995a) and Chmutov and Duzhin (2001)) defined the Kontsevich integral slightly differently, numbering the chords upwards.

Namely, $Z_{Kont}(K) =$

$$= \sum_{m=0}^{\infty} \frac{1}{(2\pi i)^m} \int_{\substack{t_{min} < t_1 < \cdots < t_m < t_{max} \\ t_j \text{ are noncritical}}} \sum_{P=\{(z_j, z'_j)\}} (-1)^{\downarrow P} D_P \bigwedge_{j=1}^{m} \frac{dz_j - dz'_j}{z_j - z'_j}.$$

Prove that for any tangle T, $Z_{Kont}(T) = Z(T)$, as series of tangle chord diagrams.

Hint. Change of variables in multiple integrals.

8.4 Express the integral over the cube

$$Z_\square(K) := \sum_{m=0}^{\infty} \frac{1}{(2\pi i)^m} \int_{\substack{t_{min} < t_1, \dots, t_m < t_{max} \\ t_j \text{ are noncritical}}} \sum_{P=\{(z_j, z'_j)\}} (-1)^{\downarrow P} D_P \bigwedge_{j=1}^{m} \frac{dz_j - dz'_j}{z_j - z'_j}$$

in terms of $Z(K)$.

8.5 Compute the Kontsevich integral of the tangles $\not\!\!\nearrow$ and $\not\!\!\nwarrow$.

8.6 Prove that for the tangle $\uparrow\!\!\searrow$ shown on the

right, $Z(\uparrow\!\!\searrow) = \exp\left(\frac{\scriptstyle\Pi}{2\pi i} \cdot \ln \varepsilon\right).$

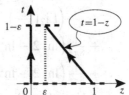

8.7 The *Euler dilogarithm* is defined by the power series

$$\text{Li}_2(z) = \sum_{k=1}^{\infty} \frac{z^k}{k^2}$$

for $|z| \leqslant 1$. Prove the following identities:

$$\text{Li}_2(0) = 0; \qquad \text{Li}_2(1) = \frac{\pi^2}{6}; \qquad \text{Li}'_2(z) = -\frac{\ln(1-z)}{z};$$
$$\frac{d}{dz}\left(\text{Li}_2(1-z) + \text{Li}_2(z) + \ln z \ln(1-z)\right) = 0;$$
$$\text{Li}_2(1-z) + \text{Li}_2(z) + \ln z \ln(1-z) = \frac{\pi^2}{6}.$$

About these and other remarkable properties of $\text{Li}_2(z)$ see Lewin (1981); Kirillov (1995); and Zagier (1988).

8.8 Consider the associating tangle $\uparrow\!\!\nearrow\!\!\uparrow$ shown on the right. Compute $Z(\uparrow\!\!\nearrow\!\!\uparrow)$ up to the second order.

Answer. $\uparrow\uparrow\uparrow \quad - \quad \frac{1}{2\pi i} \ln\left(\frac{1-\varepsilon}{\varepsilon}\right) \left(\Pi\!\uparrow - \uparrow\!\Pi\right)$

$- \frac{1}{8\pi^2} \ln^2\left(\frac{1-\varepsilon}{\varepsilon}\right) \left(\Pi\!\uparrow + \uparrow\!\Pi\right)$

$+ \frac{1}{4\pi^2}\left(\ln(1-\varepsilon)\ln\left(\frac{1-\varepsilon}{\varepsilon}\right) + \text{Li}_2(1-\varepsilon) - \text{Li}_2(\varepsilon)\right)\Pi\!\uparrow$

$- \frac{1}{4\pi^2}\left(\ln(\varepsilon)\ln\left(\frac{1-\varepsilon}{\varepsilon}\right) + \text{Li}_2(1-\varepsilon) - \text{Li}_2(\varepsilon)\right)\uparrow\!\Pi$

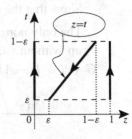

The calculation here uses the dilogarithm function defined in Exercise 8.7. Note that the Kontsevich integral diverges as $\varepsilon \to 0$.

8.9 Make the similar computation $Z(\uparrow\!\!\diagdown\!\!\uparrow)$ for the reflected tangle. Describe the difference with the answer to the previous problem.

8.10 Compute the Kontsevich integral $Z(\uparrow\cap)$ of the maximum tangle shown on the right.

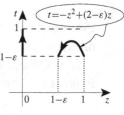

Answer. $\uparrow\cap + \frac{1}{2\pi i}\ln(1-\varepsilon)\,\text{\scriptsize$\bigsqcap$}\cap$

$+ \frac{1}{4\pi^2}\left(\text{Li}_2\left(\frac{\varepsilon}{2-\varepsilon}\right) - \text{Li}_2\left(\frac{-\varepsilon}{2-\varepsilon}\right)\right)\text{\scriptsize$\bigsqcap$}\cap$

$+ \frac{1}{8\pi^2}\left(\ln^2 2 - \ln^2\left(\frac{1-\varepsilon}{2-\varepsilon}\right) + 2\text{Li}_2\left(\frac{1}{2}\right) - 2\text{Li}_2\left(\frac{1-\varepsilon}{2-\varepsilon}\right)\right)\text{\scriptsize$\bigsqcap$}\cap$

$+ \frac{1}{8\pi^2}\left(\ln^2 2 - \ln^2(2-\varepsilon) + 2\text{Li}_2\left(\frac{1}{2}\right) - 2\text{Li}_2\left(\frac{1}{2-\varepsilon}\right)\right)\text{\scriptsize$\bigsqcap$}\cap$

8.11 Compute the Kontsevich integral $Z(\cup\,\uparrow)$ of the minimum tangle shown on the right.

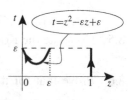

Answer. $\cup\,\uparrow - \frac{1}{2\pi i}\ln(1-\varepsilon)\text{\scriptsize$\bigcup$}\text{\scriptsize$\mid$}$

$+ \frac{1}{4\pi^2}\left(\text{Li}_2\left(\frac{\varepsilon}{2-\varepsilon}\right) - \text{Li}_2\left(\frac{-\varepsilon}{2-\varepsilon}\right)\right)\cup\text{\scriptsize$\mid$}$

$+ \frac{1}{8\pi^2}\left(\ln^2 2 - \ln^2\left(\frac{1-\varepsilon}{2-\varepsilon}\right) + 2\text{Li}_2\left(\frac{1}{2}\right) - 2\text{Li}_2\left(\frac{1-\varepsilon}{2-\varepsilon}\right)\right)\cup\text{\scriptsize$\mid$}$

$+ \frac{1}{8\pi^2}\left(\ln^2 2 - \ln^2(2-\varepsilon) + 2\text{Li}_2\left(\frac{1}{2}\right) - 2\text{Li}_2\left(\frac{1}{2-\varepsilon}\right)\right)\cup\text{\scriptsize$\mid$}$

Note that all nontrivial terms in the last two problems tend to zero as $\varepsilon \to 0$.

8.12 Express the Kontsevich integral of the hump as the product of tangle chord diagrams from Exercises 8.8, 8.10, 8.11:

$$Z\left(\bigsqcup\right) = Z(\uparrow\cap) \cdot Z(\uparrow\!\diagup\!\uparrow) \cdot Z(\cup\,\uparrow).$$

To do this, introduce shorthand notation for the coefficients:

$Z(\uparrow\cap) = \uparrow\cap + A\text{\scriptsize$\bigsqcap$}\cap + B\text{\scriptsize$\bigsqcap$}\cap + C\text{\scriptsize$\bigsqcap$}\cap + D\text{\scriptsize$\bigsqcap$}\cap$

$Z(\uparrow\!\diagup\!\uparrow) = \uparrow\!\diagup\!\uparrow + E\,(\text{\scriptsize$\sqcap$}\uparrow - \uparrow\text{\scriptsize$\sqcap$}) + F\,(\text{\scriptsize$\sqcap$}\uparrow + \uparrow\text{\scriptsize$\sqcap$}) + G\text{\scriptsize$\sqcap\!\sqcap$} + H\text{\scriptsize$\sqcap\!\sqcap$}$

$Z(\cup\,\uparrow) = \cup\,\uparrow + I\,\cup\text{\scriptsize$\mid$} + J\,\cup\text{\scriptsize$\mid$} + K\,\cup\text{\scriptsize$\mid$} + L\,\cup\text{\scriptsize$\mid$}.$

Show that the order 1 terms of the product vanish.

The only nonzero chord diagram of order 2 on the hump is the cross (diagram without isolated chords). The coefficient of this diagram is $B + D + G + J + L - AE + AI + EI$. Show that it is equal to

$$\frac{\text{Li}_2\left(\frac{\varepsilon}{2-\varepsilon}\right) - \text{Li}_2\left(\frac{-\varepsilon}{2-\varepsilon}\right) + \text{Li}_2\left(\frac{1}{2}\right) - \text{Li}_2\left(\frac{1}{2-\varepsilon}\right) - \text{Li}_2(\varepsilon)}{2\pi^2} + \frac{\ln^2 2 - \ln^2(2-\varepsilon)}{4\pi^2} + \frac{1}{24}.$$

Using the properties of the dilogarithm mentioned in Exercise 8.7, prove that the last expression equals $\frac{1}{24}$. This is also a consequence of the remarkable *Rogers five-term relation* (see, for example, Kirillov 1995):

$$\text{Li}_2 x + \text{Li}_2 y - \text{Li}_2 xy = \text{Li}_2 \frac{x(1-y)}{1-xy} + \text{Li}_2 \frac{y(1-x)}{1-xy} + \ln \frac{(1-x)}{1-xy} \ln \frac{(1-y)}{1-xy}$$

and the *Landen connection formula* (see, for example, Roos 2003):

$$\text{Li}_2 z + \text{Li}_2 \frac{-z}{1-z} = -\frac{1}{2} \ln^2 (1-z).$$

8.13 Let S_i be the operation of reversing the orientation of the ith component of a tangle T. Denote by the same symbol S_i the operation on tangle chord diagrams which (a) reverses the ith component of the skeleton of a diagram; (b) multiplies the diagram by -1 raised to the power equal to the number of chord endpoints lying on the ith component. Prove that

$$Z(S_i(T)) = S_i(Z(T)).$$

We shall use this operation in Chapter 10.

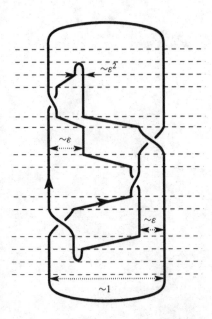

Figure 8.2 The figure eight knot in terms of elementary tangles and associating tangles.

8.14 Compute the Kontsevich integral $Z(AT_{b,w}^t)$ up to the order 2. Here ε is a small parameter, and w, t, b are natural numbers subject to inequalities $w < b$ and $w < t$.

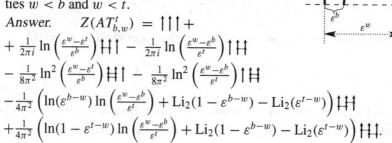

$$AT_{b,w}^t =$$

Answer. $\quad Z(AT_{b,w}^t) = \uparrow\uparrow\uparrow +$

$+ \frac{1}{2\pi i} \ln\left(\frac{\varepsilon^w - \varepsilon^t}{\varepsilon^b}\right) \includegraphics{H}\uparrow - \frac{1}{2\pi i} \ln\left(\frac{\varepsilon^w - \varepsilon^b}{\varepsilon^t}\right) \uparrow\includegraphics{H}$

$- \frac{1}{8\pi^2} \ln^2\left(\frac{\varepsilon^w - \varepsilon^t}{\varepsilon^b}\right) \includegraphics{H}\uparrow - \frac{1}{8\pi^2} \ln^2\left(\frac{\varepsilon^w - \varepsilon^b}{\varepsilon^t}\right) \uparrow\includegraphics{H}$

$- \frac{1}{4\pi^2} \left(\ln(\varepsilon^{b-w}) \ln\left(\frac{\varepsilon^w - \varepsilon^b}{\varepsilon^t}\right) + \mathrm{Li}_2(1 - \varepsilon^{b-w}) - \mathrm{Li}_2(\varepsilon^{t-w}) \right) \includegraphics{HH}$

$+ \frac{1}{4\pi^2} \left(\ln(1 - \varepsilon^{t-w}) \ln\left(\frac{\varepsilon^w - \varepsilon^b}{\varepsilon^t}\right) + \mathrm{Li}_2(1 - \varepsilon^{b-w}) - \mathrm{Li}_2(\varepsilon^{t-w}) \right) \includegraphics{HH}.$

8.15 The set of elementary tangles can be expanded by adding crossings with arbitrary orientations of strands. Express the figure eight knot 4_1 in terms of associating tangles and ε-parametrized tensor products of elementary (in this wider sense) tangles in the same manner as the trefoil 3_1 is described in Figure 8.1.

Answer: A possible answer is shown in Figure 8.2.

9

Framed knots and cabling operations

In this chapter we show how to associate to a framed knot K an infinite set of framed knots and links, called the (p, q)-*cables* of K. The operations of taking the (p, q)-cable respect the Vassiliev filtration, and give rise to operations on Vassiliev invariants and on chord diagrams. We shall give explicit formulae that describe how the Kontsevich integral of a framed knot changes under the cabling operations. As a corollary, this will give an expression for the Kontsevich integral of all torus knots.

9.1 Framed version of the Kontsevich integral

In order to describe a framed knot, one only needs to specify the corresponding unframed knot and the self-linking number. This suggests that there should be a simple formula to define the universal Vassiliev invariant for a framed knot via the Kontsevich integral of the corresponding unframed knot. This is, indeed, the case, as we shall see in Section 9.1.2. However, for our purposes it will be more convenient to use a definition of the framed Kontsevich integral given by V. Goryunov (1999) which is in the spirit of the original formula of Kontsevich described in Section 8.2.

Remark 9.1. For framed knots and links, the universal Vassiliev invariant was first defined by Le and Murakami (1996a), who gave a combinatorial construction of it using the Drinfeld associator (see Chapter 10). Goryunov used his framed Kontsevich integral in (Goryunov 1998) to study Arnold's J^+-theory of plane curves (or, equivalently, Legendrian knots in a solid torus).

9.1.1 The definition of the framed Kontsevich integral

Let K_ε be a *framing curve* of K, that is, a copy of K shifted by a small distance ε in the direction of the framing. We assume that both K and K_ε are in general position with respect to the height function t as in Section 8.2. Then we construct the (preliminary) integral $Z(K, K_\varepsilon)$ defined by the formula

$$Z(K, K_\varepsilon) = \sum_{m=0}^{\infty} \frac{1}{(4\pi i)^m} \int_{\substack{t_{\min} < t_m < \cdots < t_1 < t_{\max} \\ t_j \text{ are noncritical}}} \sum_{P=\{(z_j, z_j')\}} (-1)^{\downarrow} D_P \bigwedge_{j=1}^{m} \frac{dz_j - dz_j'}{z_j - z_j'},$$

whose only difference with the formula for the unframed Kontsevich integral is the numerical factor in front of the integral. However, the notation here has a different meaning. The class of the diagram D_P is taken in $\mathscr{A}$ rather than in $\mathscr{A}'$. We consider only those pairings $P = \{(z_j, z_j')\}$ where z_j lies on K while z_j' lies on K_ε. In order to obtain the chord diagram D_P, we project the chord endpoints that lie on K_ε back onto K along the framing. If an endpoint z_j' projects exactly to the point z_j on K, we place a "small" isolated chord in a neighbourhood of z_j. The following picture illustrates this definition:

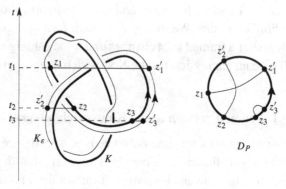

Now the framed Kontsevich integral can be defined as

$$Z^{fr}(K) = \lim_{\varepsilon \to 0} Z(K, K_\varepsilon).$$

V. Goryunov (1999) proved that the limit does exist and is invariant under the deformations of the framed knot K in the class of framed Morse knots; we refer to Goryunov (1999) for details.

Example 9.2. Let O^{+m} be the m-framed unknot:

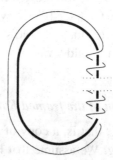

Then

$$Z^{fr}(O^{+m}) = \exp \frac{m\Theta}{2}.$$

Example 9.3. The integral formula for the linking number in Section 8.1.2 shows that the coefficient of the diagram Θ in $Z^{fr}(K)$ is equal to $w(K)/2$ where $w(K)$ is the self-linking number of K.

Define the final framed Kontsevich integral as

$$I^{fr}(K) = \frac{Z^{fr}(K)}{Z^{fr}(H)^{c/2}},$$

where H is the zero-framed hump unknot (see page 222). With its help one proves the framed version of Theorem 8.13:

Theorem 9.4. *Let w be a framed $\mathbb{C}$-valued weight system of order n. Then there exists a framed Vassiliev invariant of order $\leqslant n$ whose symbol is w.*

9.1.2 The relation with the unframed integral

Proposition 9.5. *The image of the framed Kontsevich integral $Z^{fr}(K)$ under the quotient map $\mathscr{A} \to \mathscr{A}'$ is the unframed Kontsevich integral $Z(K)$.*

Proof Each horizontal chord with endpoints on K can be lifted to a chord with one end on K and the other on K_ε in two possible ways. Therefore, each pairing P with m chords for the unframed Kontsevich integral comes from 2^m different pairings for the framed integral. As ε tends to zero, each of these pairings gives the same contribution to the integral as P and its coefficient is precisely $(2\pi i)^{-m}/2^m = (4\pi i)^{-m}$. $\qquad\square$

In fact, we can prove a much more precise statement. As we have seen in Section 4.4.4, the algebra of chord diagrams $\mathscr{A}'$ can be considered as a direct summand of $\mathscr{A}$. This allows us to compare the framed and the unframed Kontsevich integrals directly.

Theorem 9.6. *Let K be a framed knot with self-linking number $w(K)$. Then*

$$Z^{fr}(K) = Z(K) \cdot \exp \frac{w(K)\Theta}{2}$$

where $Z(K)$ is considered as an element of $\widehat{\mathscr{A}}$.

This statement can be taken as a definition of the framed Kontsevich integral.

Proof   Recall that $\mathscr{A}'$ is identified with a direct summand of $\mathscr{A}$ by means of the algebra homomorphism $p : \mathscr{A} \to \mathscr{A}$ whose kernel is the ideal generated by the diagram Θ, and which is defined on a diagram D as

$$p(D) = \sum_{J \subseteq [D]} (-\Theta)^{\deg D - |J|} \cdot D_J .$$

See Section 4.4.4. We shall prove that

$$p(Z^{fr}(K)) = Z^{fr}(K) \cdot \exp - \frac{w(K)\Theta}{2}, \tag{9.1}$$

which will imply the statement of the theorem.

Write $p(D)$ as a sum $\sum_k (-1)^k \Theta^k \cdot p_{(k)}(D)$ where the action of $p_{(k)}$ consists in omitting k chords from a diagram in all possible ways:

$$p_{(k)}(D) = \sum_{J \subseteq [D],\, \deg D - |J| = k} D_J.$$

We have $p_{(k)}(Z^{fr}(K)) = \sum c_P D_P$ where the sum is taken over all possible pairings P. The coefficient c_P is equal to the sum of all the coefficients in $Z^{fr}(K)$ that correspond to pairings P' obtained from P by adding k chords. These chords can be taken arbitrarily, so, writing m for the degree of P we have

$$c_P = \frac{1}{(4\pi i)^{m+k}} \int_{\substack{t_{\min} < t_m < \cdots < t_1 < t_{\max} \\ t_{\min} < \tau_k < \cdots < \tau_1 < t_{\max}}} \sum_{P = \{(z_j, z_j')\}} (-1)^{\downarrow} \bigwedge_{j=1}^{m} d\ln(z_j - z_j') \wedge \bigwedge_{i=1}^{k} d\ln(\zeta_i - \zeta_i'),$$

where all t_j and τ_i are noncritical and distinct, z_j and z_j' depend on t_j and ζ_i and ζ_i' depend on τ_i. This expression is readily seen to be a product of two factors: the coefficient at D_P in $Z^{fr}(K)$ and

$$\frac{1}{(4\pi i)^k} \int_{t_{\min} < \tau_k < \cdots < \tau_1 < t_{\max}} \sum_{P' = \{(\zeta_i, \zeta_i')\}} (-1)^{\downarrow} \bigwedge_{i=1}^{k} d\ln(\zeta_i - \zeta_i').$$

The latter expression is equal to

$$\frac{1}{k! \cdot (4\pi i)^k} \int_{t_{\min} < \tau_1, \ldots, \tau_k < t_{\max}} \sum_{P' = \{(\zeta_i, \zeta_i')\}} (-1)^{\downarrow} \bigwedge_{i=1}^{k} d\ln(\zeta_i - \zeta_i') = \frac{1}{k!} \left(\frac{w(K)}{2} \right)^k,$$

so that

$$p_{(k)}(Z^{fr}(K)) = \frac{1}{k!} \left(\frac{w(K)}{2} \right)^k \cdot Z^{fr}(K),$$

and (9.1) follows. $\square$

9.1.3 The case of framed tangles

The above methods produce the Kontsevich integral not just for framed knots, but for a more general class of framed tangles. Let T be a framed tangle each of whose components has the same number of maxima and minima. In other words, the boundary of each component of T is either empty or has points both on the top and on the bottom of T. The preliminary integral $Z^{fr}(T)$ can be constructed just as in the case of knots, and the final integral $I^{fr}(T)$ is defined as

$$I^{fr}(T) = Z^{fr}(H)^{-m_1}\#\ldots\#Z^{fr}(H)^{-m_k}\#Z^{fr}(T),$$

where m_i is the number of maxima on the ith component of T and $Z^{fr}(H)^{-m_i}$ acts on the ith component of $Z^{fr}(T)$ as defined in Section 5.10.3. Here k is the number of components of T.

Note that the final integral $I^{fr}(T)$ is multiplicative with respect to the tangle product, but not the connected sum of knots.

Exercise. Show that the Kontsevich integral of a single maximum, with an arbitrary framing, cannot be defined as above.

9.2 Cabling operations

9.2.1 Cabling operations on framed knots

Let p, q be two coprime integers with $p \neq 0$, and K be a framed knot given by an embedding $f : S^1 \to \mathbb{R}^3$ with the framing vector $v(\theta)$ for $\theta \in S^1$. Denote by $r_\alpha v(\theta)$ the rotation of the vector $v(\theta)$ by the angle α in the plane orthogonal to the knot. Then, for all sufficiently small values of ε, the map

$$\theta \to f(p\theta) + \varepsilon \cdot r_{q\theta} v(p\theta)$$

is actually a knot. This knot is called the (p, q)-*cable of* K; we denote it by $K^{(p,q)}$. Note that q is allowed to be zero: $K^{(1,0)}$ is K itself, and $K^{(-1,0)}$ is the inverse K^*.

Example 9.7. Here is the left trefoil with the blackboard framing and its $(3, 1)$-cable:

The (p, q)-cables can, in fact, be defined for *arbitrary* integers p, q with $p \neq 0$, as follows. Take a small tubular neighbourhood N of K. On its boundary there are two distinguished simple closed oriented curves: the *meridian*, which bounds a small disk perpendicular to the knot[1] and is oriented so as to have linking number one with K, and the *longitude*, which is obtained by shifting K to ∂N along the framing. The choice of a meridian and a longitude identifies ∂N with a torus (a, b) where a and b are real numbers mod 1 and the curves $a = 0$ and $b = 0$ define the meridian and the longitude respectively. The (p, q)-cable of K is the curve on ∂N given by the equation $qa = pb$ that represents p times the class of the longitude plus q times the class of the meridian in $H_1(\partial N)$. In general, $K^{(p,q)}$ is a knot if and only if p and q are relatively prime; otherwise, it is a link with more than one component. The number of components of the resulting links is precisely the greatest common divisor of p and q. Sometimes, the $(k, 0)$-cable of K is called *the kth disconnected cabling of K* and the $(k, 1)$-cable *the kth connected cabling of K*. We shall consider $K^{(p,q)}$ as a framed link with the framing normal to ∂N and pointing outwards:

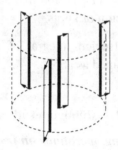

Example 9.8. The (p, q)-torus knot (link) can be defined as the (p, q)-cable of the zero-framed unknot.

9.2.2 Cables and Vassiliev invariants

Composing a link invariant with a cabling operation on knots, we obtain a new invariant of (framed) knots.

Proposition 9.9. *Let p, q be a pair of integers and r be their greatest common divisor. If v is a Vassiliev invariant of framed r-component links whose degree is at most n, the function $v^{(p,q)}$ sending a framed knot K to $v(K^{(p,q)})$ is an invariant of degree $\leqslant n$.*

Proof Indeed, the operation of taking the (p, q)-cable sends the singular knot filtration on $\mathbb{Z}\mathcal{K}^{fr}$, where $\mathcal{K}^{fr}$ is the set of the equivalence classes of framed knots, into the filtration by singular links on the free abelian group generated by the r-component framed links, since the difference

---

[1] This defines the meridian up to isotopy.

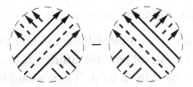

can be written as a sum of several double points. For instance,

$$\text{can be written as a sum of several double points.}$$

□

It is clear from the above argument what effect the cabling operation has on chord diagrams. Consider first the case of p and q coprime, when the (p,q)-cabling gives an operation on knot invariants. For a chord diagram D define $\psi^p(D)$ to be the sum of chord diagrams obtained by all possible ways of lifting the ends of the chords to the p-sheeted connected covering of the Wilson loop of D.

Example 9.10.

$$\psi^2\left(\bigcirc\right) = \bigcirc + \bigcirc + \bigcirc + \bigcirc + \bigcirc + \bigcirc$$
$$+ \bigcirc + \bigcirc + \bigcirc + \bigcirc + \bigcirc + \bigcirc$$
$$+ \bigcirc + \bigcirc + \bigcirc + \bigcirc$$
$$= 12\bigcirc + 4\bigotimes.$$

$$\psi^2\left(\bigotimes\right) = 8\bigcirc + 8\bigotimes.$$

It is a simple exercise to see that ψ^p respects the 4T relations; hence, it gives a linear map $\psi^p : \mathscr{A} \to \mathscr{A}$. We have the following

Proposition 9.11. $\operatorname{symb}(v^{p,q}) = \operatorname{symb}(v) \circ \psi^p$.

Note that the symbol of $v^{(p,q)}$ does not depend on q.

The case when p and q are not coprime and the (p,q)-cable is a link with at least two components is very similar. We shall treat this case in a slightly more general setting in Section 9.2.3.

9.2.3 Cabling operations in $\mathscr{C}$ and $\mathscr{B}$

The map ψ^p is defined on general closed diagrams in the very same way as on chord diagrams: it is the sum of all possible liftings of the legs of a diagram to the p-fold connected cover of its Wilson loop. It is not hard to see that ψ^p defined in this manner respects the STU relation. For instance,

$$\psi^2\left(\;\Upsilon\;\right) = \psi^2\left(\;)(\;\right) - \psi^2\left(\;\curlyvee\;\right)$$

$$= \; \text{⟨diagram⟩} + \text{⟨diagram⟩} + \text{⟨diagram⟩} + \text{⟨diagram⟩}$$

$$- \; \text{⟨diagram⟩} - \text{⟨diagram⟩} - \text{⟨diagram⟩} - \text{⟨diagram⟩}$$

$$= \; \text{⟨diagram⟩} - \text{⟨diagram⟩} + \text{⟨diagram⟩} - \text{⟨diagram⟩}$$

$$= \; \text{⟨diagram⟩} + \text{⟨diagram⟩} \, .$$

Therefore, ψ^p is a well-defined map of $\mathscr{C}$ to itself. Note that ψ^p is a coalgebra map; however, it does not respect the product in $\mathscr{C}$. This is hardly surprising since the cabling maps in general do not respect the connected sum of knots.

The algebra $\mathscr{B}$ is better suited for working with the cabling operations than $\mathscr{C}$: the map ψ^p applied to an open diagram with k legs simply multiplies this diagram by p^k. Indeed, the isomorphism $\chi : \mathscr{B} \cong \mathscr{C}$ takes an open diagram B with k legs into the average of the $k!$ closed diagrams obtained by all possible ways of attaching the legs of B to a Wilson loop. Lifting this average to the p-fold covering of the Wilson loop we get the same thing as $p^k \chi(B)$. We arrive at the following:

Proposition 9.12. *The operation $\psi^p : \mathscr{B} \to \mathscr{B}$ is an Hopf algebra map. In particular, the subspace $\mathscr{B}^k$ of diagrams with k legs is the eigenspace for ψ^p with the corresponding eigenvalue p^k.*

The fact that ψ^p respects the product on $\mathscr{B}$ follows from the second part of the proposition.

9.2.4 Cablings on tangle diagrams

So far we have only considered the effect of the (p, q)-cables on chord diagrams for coprime p and q. However, there is no difficulty in extending our results to the case of arbitrary p and q.

Given a framed tangle T with a closed component y, we can define its (p, q)-cable along y, denoted by $T_y^{(p,q)}$ in the same manner as for knots. If p, q are coprime the result will have the same skeleton as the original tangle, otherwise the component y will be replaced by several components whose number is the greatest common divisor of p and q.

If $p' = rp$ and $q' = rq$ with p and q coprime, the map $\psi_y^{r \cdot p}$ corresponding to the (p', q')-cable on the space of closed Jacobi diagrams with the skeleton $X \cup y$ can be described as follows. Consider the map

$$X \cup y_1 \cup \ldots \cup y_r \to X \cup y$$

where y_i are circles, which sends X to X by the identity map and maps each y_i to y as a p-fold covering. Then $\psi_y^{r \cdot p}$ of a closed diagram D is the sum of all the different ways of lifting the legs of D to $X \cup_i y_i$. For example,

$$\psi^{2 \cdot 1}\left(\;\right) = \;+\;+\;+\;+\;+$$

$$+\;+\;+\;+\;+\;+$$

$$+\;+\;+\;+$$

$$= 2\;+ 8\;+ 2\;+ 4$$

and

$$\psi^{2 \cdot 1}\left(\;\right) = 2\;+ 8$$

$$+ 2\;+ 4\;.$$

Here we have omitted the subscript indicating the component y, since the original diagram had only one component. In what follows, we shall write ψ_y^p instead of $\psi_y^{1 \cdot p}$.

As in Section 9.2.2, the (p, q)-cable along y composed with a Vassiliev invariant v of degree n is again a Vassiliev invariant of the same degree, whose symbol is obtained by composing $\psi_y^{r \cdot p}$ with the symbol of v. The map $\psi_y^{r \cdot p}$ satisfies the 4T relations and gives rise to a coalgebra map on the spaces of closed diagrams.

9.2.5 Disconnected cabling in $\mathcal{B}$

Just as with connected cabling, disconnected cabling looks very simple in the algebra $\mathcal{B}$. Composing $\psi_y^{r \cdot 1}$ with χ we immediately get the following:

Proposition 9.13. *The disconnected cabling operation* $\psi_y^{r\cdot 1}$ *sends an open diagram in* $\mathcal{B}(y)$ *with* k *legs to the sum of all* r^k *ways of replacing one label* y *by* r *labels* $y_1, \ldots, y_r$.

A similar statement holds, of course, for diagrams with more than one skeleton component.

9.3 Cabling operations and the Kontsevich integral

9.3.1 The Kontsevich integral of a (p, q)-cable

The Kontsevich integral is well-behaved with respect to taking (p, q)-cables for all values of p and q.

Theorem 9.14 (Le and Murakami 1997; Bar-Natan *et al.* 2003). *Let* T *be a framed tangle each of whose components is either closed or has boundary points both on the top and on the bottom of* T, *and let* y *be a closed component of* T. *If* p, q, r *are integers such that* r *is the greatest common divisor of* p *and* q, *we have*

$$I^{fr}(T_y^{(p,q)}) = \psi_y^{r\cdot p/r}\left(I^{fr}(T)\#_y \exp(\frac{q}{2p}\Theta)\right),$$

where $\#_y$ *denotes the action of* $\mathscr{C}$ *on the tangle chord diagrams by taking the connected sum along the component* y.

Remark 9.15. At the first sight the formula of Bar-Natan *et al.* (2003) for the Kontsevich integral of a $(p, 1)$-cable may seem to disagree with the above theorem. This is due to a different choice of framing on the $(p, 1)$-cable of a knot in their paper.

Proof For simplicity, we shall prove the theorem only for knots; the case of a general tangle is very similar. In the course of this proof it will be convenient to use the notion of the *parallel* of a tangle; this is an alternative way to define the cabling operations.

Let T be a tangle with the property that shifting T along the vector $(t, 0, 0)$ we obtain disjoint tangles for all sufficiently small non-negative t. Take $\delta > 0$ and define the pth parallel of T by taking T together with $p - 1$ copies of it shifted along the x-axis: the first copy shifted by δ, the second by 2δ and so on. We denote this tangle by $T^{(p)}$. If δ is sufficiently small, the pth parallel of T is well-defined up to isotopy. The tangle T and its parallels can be framed by taking the framing curve to be a small shift of T by some $\varepsilon > 0$ along the x-axis. Any framed knot can be embedded in $\mathbb{R}^3$ so that its $(p, 0)$-cable is the same thing as its pth parallel.

If T is a braid, we have

$$\lim_{\delta \to 0} Z(T^{(p)}) = \Delta^{(p)}(Z(T)),$$

where $\Delta^{(p)}$ is the composition of the operations $\Delta_x^{(p)}$ (see Section 5.10.4) for all the strands x of T. This can be seen by comparing the coefficients for each diagram on both sides as δ tends to zero. For a diagram of degree d its coefficient on the left-hand side differs from the corresponding coefficient on the right-hand side by $O(\delta^d)$.

The next logical step is now to consider the parallels of the maximum and the minimum tangles as shown in the figure for $p = 3$:

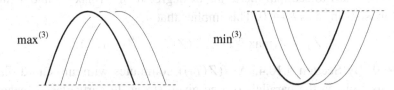

There is a difficulty here: the Kontsevich integrals of these expressions diverge as δ tends to zero. However, these divergencies can be made to cancel each other out, in the following sense.

The skeleta of $\min^{(p)}$ and $\max^{(p)}$ consist of interval components only and these intervals are naturally ordered. As a consequence, we can consider the expressions $Z(\min^{(p)})$ and $Z(\max^{(p)})$ as elements in the completion of the algebra $\mathscr{A}'(p)$, which is the quotient of $\mathscr{A}(p)$ by the diagrams with isolated chords; let ν_δ be their product in $\widehat{\mathscr{A}'}(p)$, in this order. Then $\lim_{\delta \to 0} \nu_\delta = \nu$ exists and can be calculated as follows.

Consider the hump unknot H and its kth parallel as in the figure:

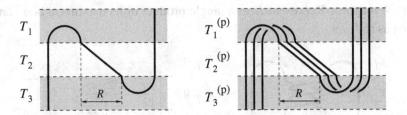

The unframed Kontsevich integral $Z(H)$ can be written as $Z_0 + Z_1$ where Z_1 is obtained from the pairings with at least one chord of length $\geqslant R$ in the shaded parts T_1 and T_3, and Z_0 comes from pairings no such chords. The series Z_0 can be obtained from $Z(T_2)$ by simply joining the upper ends of the first and the second strands and the lower ends of the second and the third strands in the skeleton of each chord diagram, since $Z(\max) = Z(\min) = 1$. On the other hand, keeping δ constant and increasing R we can make any coefficient in Z_1 arbitrarily small, since the chords of length $\geqslant R$ contribute terms of order $1/R$ and the terms of degree d in $Z(T_2)$ grow at most as $\ln^d R$. Hence, if $R \to \infty$ the Kontsevich integral

$Z(H)$ of the zero-framed hump unknot can be obtained from $Z(T_2)$ by glueing the components of the skeleton of each participating diagram into one Wilson loop.

A similar thing holds for the Kontsevich integral of the pth parallel of the hump. Write it as $Z_0 + Z_1$ where Z_1 contains pairings with at least one chord of length $\geqslant R$ in $T_1^{(p)}$ or $T_3^{(p)}$. We have that as δ tends to zero for the terms of degree d

$$Z(T_2^{(p)}) - \Delta^{(p)}(Z(T_2)) \sim \delta^d.$$

It is also not hard to see that the terms of degree d in $Z(\max^{(p)})$ and $Z(\min^{(p)})$ grow at most as $\ln^d \delta$ as $\delta \to 0$. This implies that

$$Z_0 \sim Z(\max^{(p)}) \cdot \Delta^{(p)}(Z(T_2)) \cdot Z(\min^{(p)})$$

as $\delta \to 0$. By Section 5.10.4, $\Delta^{(p)}(Z(T_2))$ commutes with any chord diagram that has its ends on the parallels of one given string. In particular, it means that $Z(\min^{(p)})$ and $Z(\max^{(p)})$ can be passed through all the strings of $\Delta^{(p)}(Z(T_2))$. By joining the appropriate endpoints of the skeleta of the diagrams in $\Delta^{(p)}(Z(T_2))$ we get the image of $\Delta^{(p)}(Z^{fr}(H))$ in the completion of $\mathscr{A}'(p)$; hence, Z_0 tends to $v \cdot \pi \Delta^{(p)}(Z^{fr}(H))$, where π is the projection from the completion of $\mathscr{A}(p)$ to that of $\mathscr{A}'(p)$. As before, Z_1 can be disregarded and we get that

$$\lim_{\delta \to 0} Z(H^{(p)}) = v \cdot \pi \Delta^{(p)}(Z^{fr}(H)).$$

On the other hand, $Z(H^{(p)})$ is easily seen to be equal to $Z(H)^{\otimes p}$. As a result,

$$v = Z(H)^{\otimes p} \cdot \Delta^{(p)}(Z^{fr}(H))^{-1} = Z(H)^{\otimes p} \cdot \pi \Delta^{(k)}(Z^{fr}(H)^{-1}).$$

Now we have the ingredients for calculating the pth parallel of an arbitrary knot K with m maxima. Represent K as a tangle product of its maxima, a braid and the minima as follows:

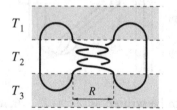
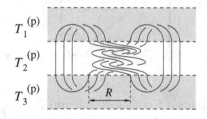

Reasoning as before, we see that the Kontsevich integral of its pth parallel as δ tends to zero is approximated by a product of three series: $Z(\max^{(p)})^{\otimes m}$, $\Delta^{(p)}(Z(T_2))$ and $Z(\min^{(p)})^{\otimes m}$. Each of the copy of $Z(\max^{(p)})$ or $Z(\min^{(p)})$ can be passed through any of the strings of $\Delta^{(p)}(Z(T_2))$ and we see that $Z(K^{(p)})$ can be obtained from $\Delta^{(p)}(Z(T_2))$ by inserting a copy of v^m into one p-tuple of strings and then glueing the components of the skeleton of each of the participating diagrams into p circles. In other words, we get that

$$Z(K^{(p)}) = Z(H)^m \# \dots \# Z(H)^m \# \pi \Delta^{(p)}(Z^{fr}(K) \# Z^{fr}(H)^{-m}),$$

where the ith copy of $Z(H)^m$ acts on the ith component of the skeleton. This gives an expression for the unframed integral $I(K^{(p,0)})$. The framed integral can possibly differ from $\Delta^{(p)}(I^{fr}(K))$ only by the framings on each component. However, erasing all the components apart from one we get the same results both for $I^{fr}(K^{(p,0)})$ and for $\Delta^{(p)}(I^{fr}(K))$ so that no additional correction of framing is necessary.

In the case $q \neq 0$ the link $K^{(p,q)}$ differs from $K^{(p,0)}$ by an insertion of a twisting:

 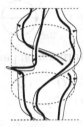

The effect of the insertion of the twisting is that the Kontsevich integral of the twisting braid should be inserted into one of the p-tuple of strings of $\Delta^{(p)}(Z(T_2))$ alongside ν^m. Now, the Kontsevich integral of the twisting braid is equal to

$$c \cdot \Delta^{(p)}(\exp(q/2p \cdot \Theta)) \cdot \tau \cdot c^{-1}$$

for some $c \in \widehat{\mathscr{A}}(p)$; here τ is the braid chord diagram with no chords whose role is to reconnect the strands. Here c and c^{-1} are the Kontsevich integrals for the upper and lower segments of the twisting braid, the fact that the central segment gives $\Delta^{(p)}(\exp(q/2p\Theta))$ is a straightforward calculation. Finally, c can be run around $K^{(p,q)}$ so as to cancel with c^{-1} and we get the theorem for arbitrary p, q. $\qquad\square$

9.3.2 Torus knots

The (p, q)-torus knot is the (p, q)-cable of the unknot, and, therefore, the formula for the cables of the Kontsevich integral as a particular case gives an expression for the Kontsevich integral of torus knots. An essential ingredient of this expression is the Kontsevich integral I^{fr} of the unknot, which will be treated later in Chapter 11.

To be precise, Theorem 9.14 gives the following expression for the Kontsevich integral of the (p, q)-torus knot:

$$I^{fr}(O^{(p,q)}) = \psi^p\left(I^{fr}(O) \# \exp\left(\frac{q}{2p}\Theta\right)\right),$$

where O is the zero-framed unknot. Marché (2004) gives a different formula for the Kontsevich integral of torus knots.

Example 9.16. By definition, $I^{fr}(O)$ is the inverse of the hump unknot, carried to $\mathscr{A}$ from $\mathscr{A}'$ by deframing. It follows from Example 9.10 and Section 8.3 that

$$I^{fr}(O) = 1 + \frac{1}{24}\,\bigcirc\hspace{-1.5em}\supset\hspace{0.4em} - \frac{1}{24}\,\otimes\hspace{0.4em} + \ldots$$

and, therefore,

$$I^{fr}(O^{(2,3)}) = 1 + 3\,\ominus\hspace{0.4em} + \frac{85}{24}\,\bigcirc\hspace{-1.5em}\supset\hspace{0.4em} + \frac{23}{24}\,\otimes\hspace{0.4em} + \ldots.$$

Compare this with the formula on page 295.

Exercise. Calculate $I^{fr}(O^{(3,2)})$ up to degree 2 using the formula of this section and compare it to $I^{fr}(O^{(2,3)})$.

9.4 Cablings of the Lie algebra weight systems

In Section 6.1 we have seen how a semi-simple Lie algebra $\mathfrak{g}$ gives rise to the universal Lie algebra weight system $\varphi_{\mathfrak{g}} : \mathscr{A} \to U(\mathfrak{g})$, and how a representation V of $\mathfrak{g}$ determines a numeric weight system $\varphi_{\mathfrak{g}}^V : \mathscr{A} \to \mathbb{C}$. The interaction of $\varphi_{\mathfrak{g}}$ with the operation ψ^p is rather straightforward.

Define $\mu^p : U(\mathfrak{g})^{\otimes p} \to U(\mathfrak{g})$ and $\delta^p : U(\mathfrak{g}) \to U(\mathfrak{g})^{\otimes p}$ by

$$\mu^p(x_1 \otimes x_2 \otimes \ldots \otimes x_p) = x_1 x_2 \ldots x_p$$

for $x_i \in U(\mathfrak{g})$ and

$$\delta^p(g) = g \otimes 1 \otimes \ldots \otimes 1 + 1 \otimes g \otimes \ldots \otimes 1 + \cdots + 1 \otimes 1 \otimes \ldots \otimes g,$$

where $g \in \mathfrak{g}$.

Proposition 9.17. *For $D \in \mathscr{A}$ we have*

$$(\varphi_{\mathfrak{g}} \circ \psi^p)(D) = (\mu^p \circ \delta^p)(\varphi_{\mathfrak{g}}(D)).$$

Proof   The construction of the universal Lie algebra weight system (Section 6.1.1) consists in assigning the basis vectors $e_{i_a} \in \mathfrak{g}$ to the endpoints of each chord a, then taking their product along the Wilson loop and summing up over each index i_a. For the weight system $\varphi_{\mathfrak{g}} \circ \psi^p$, to each endpoint of a chord we assign not only a basis vector, but also the sheet of the covering to which that particular point is lifted. (Since the construction of Lie algebra weight systems uses based diagrams, the sheets of the covering can actually be enumerated.) To form an element of the universal enveloping algebra, we must read the letters e_{i_a} along the circle n times. On the first pass we read only those letters which are related to the first sheet of the covering, omitting all the others. Then read the circle for the second time and now collect only the letters from the second sheet, etc. up to the pth reading. The products of e_{i_a}'s thus formed are summed up over all the i_a and over all the ways of lifting the endpoints to the covering.

On the other hand, the operation $\mu^p \circ \delta^p : U(\mathfrak{g}) \to U(\mathfrak{g})$ can be described as follows. If A is an ordered set of elements of $\mathfrak{g}$, let us write $\prod_A \in U(\mathfrak{g})$ for the product of all the elements of A, according to the order on A. Let $x = \prod_A$ for some A. To obtain $\mu^p \circ \delta^p(x)$ we take all possible decompositions of A into an ordered set of n disjoint subsets A_i, with $1 \leqslant i \leqslant n$, and take the sum of $\prod_{A_1} \prod_{A_2} \cdots \prod_{A_p}$ over all these decompositions.

When applied to $\varphi_\mathfrak{g}$, the sets A_k are the sets of the e_{i_a} corresponding to the endpoints that are lifted to the kth sheet of the p-fold covering. This establishes a bijection between the summands on the two sides of the formula. $\qquad \square$

Exercises

9.1 Define the connected sum of two framed knots as their usual connected sum with the framing whose self-linking number is the sum of the self-linking numbers of the summands. Show that the framed Kontsevich integral is multiplicative with respect to the connected sum.

9.2 Prove that the framed Kontsevich integral $Z^{fr}(K)$ is a group-like element of the Hopf algebra $\mathscr{A}$.

9.3 Let K be a framed knot. Consider the Kontsevich integral $I^{fr}(K)$ as an element of $\widehat{\mathscr{B}}$, and show that if at least one of the diagrams participating in it contains a strut (an interval component) then K has nonzero framing.

   Hint. Use the group-like property of Z^{fr}.

9.4 Check that the maps ψ^p, and, more generally, $\psi_y^{r \cdot p}$ are compatible with the four-term relations.

9.5 Compute $\psi^3\left(\bigcirc\!\!\!\bigcirc\right)$ and $\psi^3\left(\bigotimes\right)$.

9.6 Compute the eigenvalues and eigenvectors of $\psi^3|_{\mathscr{A}_2}$.

9.7 Compute $\psi^2 2\left(\bigoplus\right)$, $\psi^2\left(\bigoplus\right)$, $\psi^2\left(\bigoplus\right)$, $\psi^2\left(\bigoplus\right)$, and $\psi^2\left(\bigotimes\right)$.

9.8 Compute the eigenvalues and eigenvectors of $\psi^2|_{\mathscr{A}_3}$.

9.9 Compute $\psi^2(\Theta^m)$, where Θ^m is a chord diagram with m isolated chords, such as the one shown on the right.

$$\Theta^m = \bigcirc$$

m chords

9.10 Prove that ψ^p commutes with the comultiplication of chord diagrams. In other words, show that in the notation of Section 4.4.3, page 96, the identity

$$\delta(\psi^P(D)) = \sum_{J \subseteq [D]} \psi^P(D_J) \otimes \psi^P(D_{\bar{J}})$$

holds for any chord diagram D.

9.11   (Bar-Natan 1995a). Prove that $\psi^P \circ \psi^q = \psi^{pq}$.

9.12   Prove Proposition 9.11 from Section 9.2.2:

$$\text{symb}((v)^{(p,q)})(D) = \text{symb}(v)(\psi^P(D)).$$

9.13   Let T be a tangle with k numbered components, all of them intervals without critical points of the height function, and assume that the ith component
connects the ith point on the upper boundary with the ith point on the lower
boundary. (Pure braids are examples of such tangles.) Let $\Delta_\varepsilon^{n_1,\dots,n_k}$ be the
operation of replacing, for each i, the ith component of T by n_i parallel
copies of itself with the distance ε between each copy, as on Figure 9.1.
Denote by $\Delta^{n_1,\dots,n_k}$ the following operation on the corresponding tangle

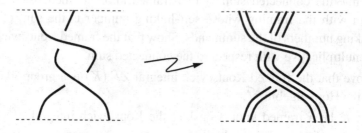

Figure 9.1 The effect of $\Delta_\varepsilon^{2,3}$

chord diagrams: for each i the ith strand is replaced by n_i copies of itself
and a chord diagram is sent to the sum of all of its liftings to the resulting
skeleton. Prove that

$$\lim_{\varepsilon \to 0} Z(\Delta_\varepsilon^{n_1,\dots,n_k}(T)) = \Delta^{n_1,\dots,n_k} Z(T).$$

9.14   Let T_ε be the following family of tangles depending on a parameter ε:

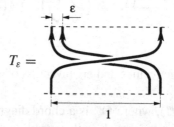

$$T_\varepsilon =$$

Show that

$$\lim_{\varepsilon \to 0} Z(T_\varepsilon) = \times \cdot \Delta^{1,2} \exp\left(\frac{\rightleftharpoons}{2}\right).$$

10

The Drinfeld associator

In this chapter we give the details of the combinatorial construction for the Kontsevich integral. The main ingredient of this construction is the power series known as the *Drinfeld associator* Φ_{KZ}. Here the subscript "KZ" indicates that the associator comes from the solutions to the Knizhnik–Zamolodchikov equation. The Drinfeld associator enters the theory as a (normalized) Kontsevich integral for a special tangle without crossings, which is the simplest *associating tangle*.

The associator Φ_{KZ} is an infinite series in two non-commuting variables whose coefficients are combinations of multiple zeta values. In the construction of the Kontsevich integral only some properties of Φ_{KZ} are used; adopting them as axioms, we arrive at the general notion of an associator that appeared in Drinfeld's papers (1989, 1990) in his study of quasi-Hopf algebras. These axioms actually describe a large collection of associators belonging to the completed algebra of chord diagrams on three strands. Some of these associators have rational coefficients, and this implies the rationality of the Kontsevich integral.

10.1 The KZ equation and iterated integrals

In this section, we give the original Drinfeld's definition of the associator in terms of the solutions of the simplest Knizhnik–Zamolodchikov equation.

The Knizhnik–Zamolodchikov (KZ) equation appears in the Wess–Zumino–Witten model of conformal field theory (Knizhnik and Zamolodchikov 1984). The theory of KZ type equations has been developed in the contexts of mathematical physics, representation theory and topology (Etingof *et al.* 1998; Varchenko 2003; Kassel 1995; Kohno 2002; and Ohtsuki 2002). Our exposition follows the topological approach and is close to that of the last three books.

10.1.1 General setup

Let X be a smooth manifold and $\widehat{\mathcal{A}}$ a completed graded algebra over the complex numbers. Choose a set $\omega_1, \ldots, \omega_p$ of $\mathbb{C}$-valued closed differential 1-forms on X and a set $c_1, \ldots, c_p$ of homogeneous elements of $\widehat{\mathcal{A}}$ of degree 1. Consider the closed 1-form

$$\omega = \sum_{j=1}^{p} \omega_j c_j$$

with values in $\widehat{\mathcal{A}}$. The Knizhnik-Zamolodchikov equation is a particular case of the following very general equation:

$$dI = \omega \cdot I, \tag{10.1}$$

where $I : X \to \mathcal{A}$ is the unknown function.

Exercise. One may be tempted to solve the above equation as follows: $d \log(I) = \omega$, therefore $I = \exp \int \omega$. Explain why this is wrong.

The form ω must satisfy certain conditions so that Equation (10.1) may have nonzero solutions. Indeed, taking the differential of both sides of (10.1), we get that $0 = d(\omega I)$. Applying the Leibniz rule, using the fact that $d\omega = 0$ and substituting $dI = \omega I$, we see that a necessary condition for integrability can be written as

$$\omega \wedge \omega = 0. \tag{10.2}$$

It turns out that this condition is not only necessary, but also sufficient for *local* integrability: if it holds near a point $x_0 \in X$, then (10.1) has the unique solution I_0 in a small neighbourhood of x_0, satisfying the initial condition $I_0(x_0) = a_0$ for any $a_0 \in \widehat{\mathcal{A}}$. This fact is standard in differential geometry, where it is called the *integrability of flat connections* (see, for instance, Kobayashi and Nomizu 1996). A direct *ad hoc* proof can be found in Ohtsuki (2002), proposition 5.2.

10.1.2 Monodromy

Assume that the integrability condition 10.2 is satisfied for all points of X. Given a (local) solution I of Equation (10.1) and $a \in \widehat{\mathcal{A}}$, the product Ia is also a (local) solution. Therefore, germs of local solutions at a point x_0 form an $\widehat{\mathcal{A}}$-module. This module is free of rank one; it is generated by the germ of a local solution taking value $1 \in \widehat{\mathcal{A}}$ at x_0.

The reason to consider germs rather than global solutions is that the global solutions of (10.1) are generally multivalued, unless X is simply connected. Indeed, one can extend a local solution at x_0 along any given path which starts at x_0 by patching together local solutions at the points on the path. (One can think of this extension as something like an analytic continuation of a holomorphic function.) Extending in this way a local solution I_0 at the point x_0 along a closed loop $\gamma : [0, 1] \rightarrow X$ we arrive to another local solution I_1, also defined in a neighbourhood of x_0.

Let $I_1(x_0) = a_\gamma$. Suppose that $a_0 = I_0(x_0)$ is an invertible element of $\widehat{\mathcal{A}}$. The fact that the local solutions form a free one-dimensional $\widehat{\mathcal{A}}$-module implies that the two solutions I_0 and I_1 are proportional to each other: $I_1 = I_0 a_0^{-1} a_\gamma$. The coefficient $a_0^{-1} a_\gamma$ does not depend on a particular choice of the invertible element $a_0 \in \widehat{\mathcal{A}}$ and the loop γ within a fixed homotopy class. Therefore, we get a homomorphism $\pi_1(X) \rightarrow \widehat{\mathcal{A}}^*$ from the fundamental group of X into the multiplicative group of the units of $\widehat{\mathcal{A}}$, called the *monodromy representation*.

10.1.3 Iterated integrals

Both the continuation of the solutions and the monodromy representation can be expressed in terms of the 1-form ω. Choose a path $\gamma : [0, 1] \rightarrow X$, not necessarily closed, and consider the composition $I \circ \gamma$. This is a function $[0, 1] \rightarrow \widehat{\mathcal{A}}$ which we denote by the same letter I; it satisfies the ordinary differential equation

$$\frac{d}{dt} I(t) = \omega(\dot{\gamma}(t)) \cdot I(t), \qquad I(0) = 1. \qquad (10.3)$$

The function I takes values in the completed graded algebra $\widehat{\mathcal{A}}$, and it can be expanded in an infinite series according to the grading:

$$I(t) = I_0(t) + I_1(t) + I_2(t) + \cdots ,$$

where each term $I_m(t)$ is the homogeneous degree m part of $I(t)$.

The form ω is homogeneous of degree 1 (recall that $\omega = \sum c_j \omega_j$, where ω_j are $\mathbb{C}$-valued 1-forms and c_j's are elements of $\mathcal{A}_1$). Therefore, Equation (10.3) is equivalent to an infinite system of ordinary differential equations

$$\begin{aligned} I_0'(t) &= 0, & I_0(0) &= 1, \\ I_1'(t) &= \omega(t) I_0(t), & I_1(0) &= 0, \\ I_2'(t) &= \omega(t) I_1(t), & I_2(0) &= 0, \\ &\cdots\cdots & &\cdots\cdots \end{aligned}$$

where $\omega(t) = \gamma^* \omega$ is the pull-back of the 1-form to the interval $[0, 1]$.

These equations can be solved iteratively, one by one. The first one gives $I_0 = $ const, and the initial condition implies $I_0(t) = 1$. Then, $I_1(t) = \int_0^t \omega(t_1)$. Here t_1 is an auxiliary variable that ranges from 0 to t. Coming to the next equation, we now get:

$$I_2(t) = \int_0^t \omega(t_2) \cdot I_1(t_2) = \int_0^t \omega(t_2) \left(\int_0^{t_2} \omega(t_1) \right) = \int\limits_{0 < t_1 < t_2 < t} \omega(t_2) \wedge \omega(t_1).$$

Proceeding in the same way, for an arbitrary m we obtain

$$I_m(t) = \int\limits_{0 < t_1 < t_2 < \cdots < t_m < t} \omega(t_m) \wedge \omega(t_{m-1}) \wedge \cdots \wedge \omega(t_1).$$

In what follows, it will be more convenient to use this formula with variables renumbered:

$$I(t) = 1 + \sum_{m=1}^\infty \int\limits_{0 < t_m < t_{m-1} < \cdots < t_1 < t} \omega(t_1) \wedge \omega(t_2) \wedge \cdots \wedge \omega(t_m). \qquad (10.4)$$

The value $I(1)$ represents the monodromy of the solution over the loop γ. Each iterated integral $I_m(1)$ is a homotopy invariant (of "order m") of γ. Note the resemblance of these expressions to the Kontsevich integral; we shall come back to that again later.

Remark 10.1. One may think of the closed 1-form ω as of an $\widehat{A}$-valued connection on X. Then the condition $\omega \wedge \omega = 0$ means that this connection is flat. The monodromy $I(t)$ represents the parallel transport. In this setting the presentation of the parallel transport as a series of iterated integrals was described by K.-T. Chen (1977b).

10.1.4 The Knizhnik–Zamolodchikov equation

Let $\mathcal{H} = \bigcup_{j=1}^p H_j$ be a collection of affine hyperplanes in $\mathbb{C}^n$. Each hyperplane H_j is defined by a (not necessarily homogeneous) linear equation $L_j = 0$. A *Knizhnik–Zamolodchikov*, or simply *KZ*, equation is an equation of the form (10.1) with

$$X = \mathbb{C}^n - \mathcal{H}$$

and

$$\omega_j = d \log L_j$$

for all j.

Many of the KZ equations are related to Lie algebras and their representations. This class of equations has attracted the most attention in the literature; see, for example, Knizhnik and Zamolodchikov (1994); Ohtsuki (2002); and Kohno (2002). We are specifically interested in the following situation.

Suppose that $X = \mathbb{C}^n \setminus \mathcal{H}$ where $\mathcal{H}$ is the union of the diagonal hyperplanes $\{z_j = z_k\}$, $1 \leqslant j < k \leqslant n$, and the algebra $\widehat{A} = \widehat{\mathscr{A}}^h(n)$ is the completed algebra of horizontal chord diagrams, see Section 5.11. Recall that $\widehat{\mathscr{A}}^h(n)$ is spanned by the diagrams on n vertical strands (which we assume to be oriented upwards) all of whose chords are horizontal. Multiplicatively, $\widehat{\mathscr{A}}^h(n)$ is generated by the degree-one elements u_{jk} for all $1 \leqslant j < k \leqslant n$, (which are simply the horizontal chords joining the jth and the kth strands) subject to the infinitesimal pure braid relations

$$[u_{jk}, u_{jl} + u_{kl}] = 0, \quad \text{if } j, k, l \text{ are different,}$$
$$[u_{jk}, u_{lm}] = 0, \quad \text{if } j, k, l, m \text{ are different,}$$

where, by definition, $u_{jk} = u_{kj}$.

Consider the $\mathscr{A}^h(n)$-valued 1-form $\omega = \dfrac{1}{2\pi i} \displaystyle\sum_{1 \leqslant j < k \leqslant n} u_{jk} \dfrac{dz_j - dz_k}{z_j - z_k}$ and the corresponding KZ equation

$$dI = \frac{1}{2\pi i}\left(\sum_{1 \leqslant j < k \leqslant n} u_{jk} \frac{dz_j - dz_k}{z_j - z_k} \right) \cdot I. \tag{10.5}$$

This case of the Knizhnik–Zamolodchikov equation is referred to as the *formal KZ equation*.

The integrability condition (10.2) for the formal KZ equation is the following identity on the 1-form ω on X with values in the algebra $\mathscr{A}^h(n)$:

$$\omega \wedge \omega = \sum_{\substack{1 \leqslant j < k \leqslant n \\ 1 \leqslant l < m \leqslant n}} u_{jk} u_{lm} \frac{dz_j - dz_k}{z_j - z_k} \wedge \frac{dz_l - dz_m}{z_l - z_m} = 0.$$

This identity, in a slightly different notation, was actually proved in Section 8.6.1 when we checked the horizontal invariance of the Kontsevich integral.

The space $X = \mathbb{C}^n \setminus \mathcal{H}$ is the configuration space of n distinct (and distinguishable) points in $\mathbb{C}$. A loop γ in this space may be identified with a pure braid (that is, a braid that does not permute the endpoints of the strands), and the iterated integral formula (10.4) gives

$$I(1) = \sum_{m=0}^{\infty} \frac{1}{(2\pi i)^m} \int_{0 < t_m < \cdots < t_1 < 1} \sum_{P = \{(z_j, z_j')\}} D_P \bigwedge_{j=1}^{m} \frac{dz_j - dz_j'}{z_j - z_j'},$$

where P (a *pairing*) is a choice of m pairs of points on the braid, with jth pair lying on the level $t = t_j$, and D_P is the product of m tangle chord diagrams of type $u_{jj'}$ corresponding to the pairing P. We can see that the monodromy of the KZ equation over γ coincides with the Kontsevich integral of the corresponding braid (see Section 8.4).

10.1.5 The cases $n = 2, 3$

For small values of n, Equation (10.5) is easier to handle. In the case $n = 2$, the algebra $\widehat{\mathscr{A}}^h(2)$ is free commutative on one generator and everything is very simple, as the following exercise shows.

Exercise. Solve explicitly Equation (10.5) and find the monodromy representation in the case $n = 2$.

For $n = 3$ the formal KZ equation has the form

$$dI = \frac{1}{2\pi i}\left(u_{12}d\log(z_2 - z_1) + u_{13}d\log(z_3 - z_1) + u_{23}d\log(z_3 - z_2)\right) \cdot I ,$$

which is a partial differential equation in three variables. It turns out that it can be reduced to an ordinary differential equation.

Indeed, make the substitution

$$I = (z_3 - z_1)^{\frac{u}{2\pi i}} \cdot G ,$$

where $u := u_{12} + u_{13} + u_{23}$ and we understand the factor in front of G as a multivalued analytic function with values in the algebra $\widehat{\mathscr{A}}^h(3)$:

$$(z_3 - z_1)^{\frac{u}{2\pi i}} = \exp\left(\frac{\log(z_3 - z_1)}{2\pi i}u\right)$$

$$= 1 + \frac{\log(z_3 - z_1)}{2\pi i}u + \frac{1}{2!}\frac{\log^2(z_3 - z_1)}{(2\pi i)^2}u^2 + \frac{1}{3!}\frac{\log^3(z_3 - z_1)}{(2\pi i)^3}u^3 + \dots .$$

By Proposition 5.50, the algebra $\widehat{\mathscr{A}}^h(3)$ is a direct product of the free algebra on u_{12} and u_{23}, and the free commutative algebra generated by u. In particular, u commutes with all elements of $\widehat{\mathscr{A}}^h(3)$. Taking this into account, we see that the differential equation for G can be simplified so as to become

$$dG = \frac{1}{2\pi i}\left(u_{12}d\log\left(\frac{z_2 - z_1}{z_3 - z_1}\right) + u_{23}d\log\left(1 - \frac{z_2 - z_1}{z_3 - z_1}\right)\right)G.$$

Denoting $\frac{z_2 - z_1}{z_3 - z_1}$ simply by z, we see that the function G depends only on z and satisfies the following ordinary differential equation (the *reduced KZ equation*):

$$\frac{dG}{dz} = \left(\frac{A}{z} + \frac{B}{z-1}\right)G \tag{10.6}$$

where $A := \frac{u_{12}}{2\pi i}$, $B := \frac{u_{23}}{2\pi i}$. As defined, G takes values in the algebra $\widehat{\mathscr{A}}^h(3)$ with three generators A, B, u. However, the space of local solutions of this equation is a free module over $\widehat{\mathscr{A}}^h(3)$ of rank 1, so the knowledge of just one solution is enough. Since the coefficients of Equation (10.6) do not involve u, the equation does have a solution with values in the ring of formal power series $\mathbb{C}\langle\langle A, B\rangle\rangle$ in two non-commuting variables A and B.

10.1.6 The reduced KZ equation

The reduced KZ Equation (10.6) is a particular case of the general KZ equation defined by the data $n = 1$, $X = \mathbb{C} \setminus \{0, 1\}$, $L_1 = z$, $L_2 = z - 1$, $\mathscr{A} = \mathbb{C}\langle\langle A, B\rangle\rangle$, $c_1 = A$, $c_2 = B$.

Although (10.6) is a first-order ordinary differential equation, it is hardly easier to solve than the general KZ equation. In the following exercises we invite the reader to try out two natural approaches to the reduced KZ equation.

Exercise. Try to find the general solution of Equation (10.6) by representing it as a series

$$G = G_0 + G_1 A + G_2 B + G_{11} A^2 + G_{12} AB + G_{21} BA + \ldots,$$

where the G's with subscripts are complex-valued functions of z.

Exercise. Try to find the general solution of Equation (10.6) in the form of a Taylor series $G = \sum_k G_k (z - \frac{1}{2})^k$, where the G_k's are elements of the algebra $\mathbb{C}\langle\langle A, B\rangle\rangle$. (Note that it is not possible to expand the solutions at $z = 0$ or $z = 1$, because they have essential singularities at these points.)

These exercises show that direct approaches do not give much insight into the nature of the solutions of (10.6). Luckily, we know that at least one solution exists (see Section 10.1.1) and that any solution can be obtained from one basic solution via multiplication by an element of the algebra $\mathbb{C}\langle\langle A, B\rangle\rangle$. The Drinfeld associator appears as a coefficient between two remarkable solutions.

Definition 10.2. The (Knizhnik–Zamolodchikov) *Drinfeld associator* Φ_{KZ} is the ratio $\Phi_{KZ} = G_1^{-1}(z) \cdot G_0(z)$ of two special solutions $G_0(z)$ and $G_1(z)$ of this equation described in the following lemma.

Lemma 10.3 (Drinfeld 1989, 1990). *There exist unique solutions $G_0(z)$ and $G_1(z)$ of Equation (10.6), analytic in the domain $\{z \in \mathbb{C} \mid |z| < 1, |z - 1| < 1\}$ and with the following asymptotic behaviour:*

$$G_0(z) \sim z^A \text{ as } z \to 0 \qquad \text{and} \qquad G_1(z) \sim (1 - z)^B \text{ as } z \to 1,$$

which means that

$$G_0(z) = f(z) \cdot z^A \quad and \quad G_1(z) = g(1-z) \cdot (1-z)^B \,,$$

where $f(z)$ and $g(z)$ are analytic functions in a neighbourhood of $0 \in \mathbb{C}$ with values in $\mathbb{C}\langle\langle A, B \rangle\rangle$ such that $f(0) = g(0) = 1$, and the (multivalued) exponential functions are understood as formal power series, that is, $z^A = \exp(A \log z) = \sum_{k \geqslant 0}(A \log z)^k / k!$

Remark 10.4. It is sometimes said that the element Φ_{KZ} represents the monodromy of the KZ equation over the horizontal interval from 0 to 1. This phrase has the following meaning. In general, the monodromy along a path γ connecting two points p and q is the value at q of the solution, analytical over γ and taking value 1 at p. If f_p and f_q are two solutions analytical over γ with initial values $f_p(p) = f_q(q) = 1$, then the monodromy is the element $f_q^{-1} f_p$. The reduced KZ equation has no analytic solutions at the points $p = 0$ and $q = 1$, and the usual definition of the monodromy cannot be applied directly in this case. What we do is we choose some natural solutions with reasonable asymptotics at these points and define the monodromy as their ratio in the appropriate order.

Proof Plugging the expression $G_0(z) = f(z) \cdot z^A$ into Equation (10.6) we get

$$f'(z) \cdot z^A + f \cdot \frac{A}{z} \cdot z^A = \left(\frac{A}{z} + \frac{B}{z-1}\right) \cdot f \cdot z^A \,,$$

hence $f(z)$ satisfies the differential equation

$$f' - \frac{1}{z}[A, f] = \frac{-B}{1-z} \cdot f.$$

Let us look for a formal power series solution $f = 1 + \sum_{k=1}^{\infty} f_k z^k$ with coefficients $f_k \in \mathbb{C}\langle\langle A, B \rangle\rangle$. We have the following recurrence equation for the coefficient of z^{k-1}:

$$k f_k - [A, f_k] = (k - \mathrm{ad}_A)(f_k) = -B(1 + f_1 + f_2 + \cdots + f_{k-1}) \,,$$

where ad_A denotes the operator $x \mapsto [A, x]$. The operator $k - \mathrm{ad}_A$ is invertible:

$$(k - \mathrm{ad}_A)^{-1} = \sum_{s=0}^{\infty} \frac{\mathrm{ad}_A^s}{k^{s+1}}$$

(the sum is well-defined because the operator ad_A increases the grading), so the recurrence can be solved:

$$f_k = \sum_{s=0}^{\infty} \frac{\mathrm{ad}_A^s}{k^{s+1}} \left(-B(1 + f_1 + f_2 + \cdots + f_{k-1}) \right).$$

Therefore, the desired solution does exist among formal power series. Since the point 0 is a regular singular point of Equation (10.6), it follows from the general theory (Walter 1998) that this power series converges for $|z| < 1$. We thus get an analytic solution $f(z)$.

To prove the existence of the second solution, $G_1(z)$, it is best to make the change of variable $z \mapsto 1 - z$ which transforms Equation (10.6) into a similar equation with A and B swapped. □

Remark 10.5. If the variables A and B were commutative, then the function explicitly given as the product $z^A(1 - z)^B$ would be a solution of Equation (10.6) satisfying both asymptotic conditions of the above lemma at once, so that the analogs of G_0 and G_1 would coincide. Therefore, the image of Φ_{KZ} under the abelianization map $\mathbb{C}\langle\langle A, B \rangle\rangle \to \mathbb{C}[[A, B]]$ is equal to 1.

The next lemma gives another expression for the associator in terms of the solutions of Equation (10.6).

Lemma 10.6 (Le and Murakami 1996a). *Suppose that $\varepsilon \in \mathbb{R}$, $0 < \varepsilon < 1$. Let $G_\varepsilon(z)$ be the unique solution of Equation (10.6) satisfying the initial condition $G_\varepsilon(\varepsilon) = 1$. Then*

$$\Phi_{KZ} = \lim_{\varepsilon \to 0} \varepsilon^{-B} \cdot G_\varepsilon(1 - \varepsilon) \cdot \varepsilon^A.$$

Proof We rely on, and use the notation of, Lemma 10.3. The solution G_ε is proportional to the distinguished solution G_0:

$$G_\varepsilon(z) = G_0(z)G_0(\varepsilon)^{-1} = G_0(z) \cdot \varepsilon^{-A} f(\varepsilon)^{-1} = G_1(z) \cdot \Phi_{KZ} \cdot \varepsilon^{-A} f(\varepsilon)^{-1}$$

(the function f, as well as g mentioned below, was defined in Lemma 10.3). In particular,

$$G_\varepsilon(1 - \varepsilon) = G_1(1 - \varepsilon) \cdot \Phi_{KZ} \cdot \varepsilon^{-A} f(\varepsilon)^{-1} = g(\varepsilon)\varepsilon^B \cdot \Phi_{KZ} \cdot \varepsilon^{-A} f(\varepsilon)^{-1}.$$

We must compute the limit

$$\lim_{\varepsilon \to 0} \varepsilon^{-B} g(\varepsilon)\varepsilon^B \cdot \Phi_{KZ} \cdot \varepsilon^{-A} f(\varepsilon)^{-1}\varepsilon^A,$$

which obviously equals Φ_{KZ} because $f(0) = g(0) = 1$ and $f(z)$ and $g(z)$ are analytic functions in a neighbourhood of zero. The lemma is proved. $\qquad\square$

10.1.7 The Drinfeld associator and the Kontsevich integral

Consider the reduced KZ equation (10.6) on the real interval $[0, 1]$ and apply the techniques of iterated integrals from Section 10.1.3. Let the path γ be the identity inclusion $[\varepsilon, 1] \to \mathbb{C}$. Then the solution G_ε can be written as

$$G_\varepsilon(t) = 1 + \sum_{m=1}^{\infty} \int_{\varepsilon < t_m < \cdots < t_2 < t_1 < t} \omega(t_1) \wedge \omega(t_2) \wedge \cdots \wedge \omega(t_m).$$

The lower limit in the integrals is ε because the parameter on the path γ starts from this value.

We are interested in the value of this solution at $t = 1 - \varepsilon$:

$$G_\varepsilon(1 - \varepsilon) = 1 + \sum_{m=1}^{\infty} \int_{\varepsilon < t_m < \cdots < t_2 < t_1 < 1 - \varepsilon} \omega(t_1) \wedge \omega(t_2) \wedge \ldots \omega(t_m).$$

We claim that this series literally coincides with the Kontsevich integral of the following tangle:

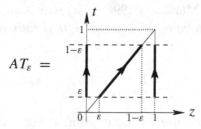

under the identification $A = \frac{1}{2\pi i} \sqcap\!\!\!\sqcap\, |$, $B = \frac{1}{2\pi i} |\,\sqcap\!\!\!\sqcap$. Indeed, on every level t_j the differential form $\omega(t_j)$ consists of two summands. The first summand $A\frac{dt_j}{t_j}$ corresponds to the choice of a pair $P = (0, t_j)$ on the first and the second strings and is related to the chord diagram $A = \sqcap\!\!\!\sqcap\, |$. The second summand $B\frac{d(1-t_j)}{1-t_j}$ corresponds to the choice of a pair $P = (t_j, 1)$ on the second and third strings and is related to the chord diagram $B = |\,\sqcap\!\!\!\sqcap$. The pairing of the first and the third strings does not contribute to the Kontsevich integral, because these strings are parallel and the corresponding differential vanishes. We have thus proved the following proposition.

Proposition 10.7. *The value of the solution G_ε at $1 - \varepsilon$ is equal to the Kontsevich integral:* $G_\varepsilon(1 - \varepsilon) = Z(AT_\varepsilon)$. *Consequently, by Lemma 10.6, the*

KZ associator coincides with the regularization of the Kontsevich integral of the tangle AT_ε:

$$\Phi_{KZ} = \lim_{\varepsilon \to 0} \varepsilon^{-B} \cdot Z(AT_\varepsilon) \cdot \varepsilon^A,$$

where $A = \frac{1}{2\pi i} \text{H} |$ and $B = \frac{1}{2\pi i} | \text{H}$.

10.2 Calculation of the KZ Drinfeld associator

In this section, following Le and Murakami (1995b and 1996b), we deduce an explicit formula for the Drinfeld associator $\Phi = \Phi_{KZ}$. It turns out that all the coefficients in the expansion of Φ_{KZ} as a power series in A and B are values of multiple zeta functions (see Section 10.2.4) divided by powers of $2\pi i$.

10.2.1 Convergent and divergent monomials

Put $\omega_0(z) = \frac{dz}{z}$ and $\omega_1(z) = \frac{d(1-z)}{1-z}$. Then the 1-form ω studied in Section 10.1.7 is the linear combination $\omega(z) = A\omega_0(z) + B\omega_1(z)$, where $A = \frac{\text{H}|}{2\pi i}$ and $B = \frac{|\text{H}}{2\pi i}$. By definition, the terms of the Kontsevich integral $Z(AT_\varepsilon)$ represent the monomials corresponding to all choices of one of the two summands of $\omega(t_j)$ for every level t_j. The coefficients of these monomials are integrals over the simplex $\varepsilon < t_m < \cdots < t_2 < t_1 < 1 - \varepsilon$ of all possible products of the forms ω_0 and ω_1. The coefficient of the monomial $B^{i_1} A^{j_1} \ldots B^{i_l} A^{j_l}$ ($i_1 \geqslant 0, j_1 > 0, \ldots, i_l > 0, j_l \geqslant 0$) is

$$\int_{\varepsilon < t_m < \cdots < t_2 < t_1 < 1-\varepsilon} \underbrace{\omega_1(t_1) \wedge \cdots \wedge \omega_1(t_{i_1})}_{i_1} \wedge \underbrace{\omega_0(t_{i_1+1}) \wedge \cdots \wedge \omega_0(t_{i_1+j_1})}_{j_1} \wedge \cdots$$

$$\wedge \underbrace{\omega_0(t_{i_1+\cdots+i_l+1}) \wedge \cdots \wedge \omega_0(t_{i_1+\cdots+j_l})}_{j_l},$$

where $m = i_1 + j_1 + \cdots + i_l + j_l$. For example, the coefficient of AB^2A equals

$$\int_{\varepsilon < t_4 < t_3 < t_2 < t_1 < 1-\varepsilon} \omega_0(t_1) \wedge \omega_1(t_2) \wedge \omega_1(t_3) \wedge \omega_0(t_4).$$

We are going to divide the sum of all monomials into two parts, "convergent" Z^{conv} and "divergent" Z^{div}, depending on the behaviour of the coefficients as $\varepsilon \to 0$. We shall have $Z(AT_\varepsilon) = Z^{conv} + Z^{div}$ and

$$\Phi = \lim_{\varepsilon \to 0} \varepsilon^{-B} \cdot Z^{conv} \cdot \varepsilon^{-A} + \lim_{\varepsilon \to 0} \varepsilon^{-B} \cdot Z^{div} \cdot \varepsilon^{-A}. \tag{10.7}$$

Then we shall prove that the second limit equals zero and find an explicit expression for the first one in terms of multiple zeta values. We shall see that, although the sum Z^{conv} does not contain any divergent monomials, the first limit in (10.7) does.

We pass to exact definitions.

Definition 10.8. A non-unit monomial in letters A and B with positive powers is said to be *convergent* if it starts with an A and ends with a B. Otherwise the monomial is said to be *divergent*. We regard the unit monomial 1 as convergent.

Example 10.9. The integral

$$\int_{a<t_p<\cdots<t_2<t_1<b} \omega_1(t_1) \wedge \cdots \wedge \omega_1(t_p) = \frac{1}{p!} \log^p \left(\frac{1-b}{1-a} \right)$$

diverges as $b \to 1$. It is the coefficient of the monomial B^p in $G_\varepsilon(1 - \varepsilon)$ when $a = \varepsilon$, $b = 1 - \varepsilon$, and this is the reason to call monomials that start with a B divergent.

Similarly, the integral

$$\int_{a<t_q<\cdots<t_2<t_1<b} \omega_0(t_1) \wedge \cdots \wedge \omega_0(t_q) = \frac{1}{q!} \log^q \left(\frac{b}{a} \right)$$

diverges as $a \to 0$. It is the coefficient of the monomial A^q in $G_\varepsilon(1 - \varepsilon)$ when $a = \varepsilon$, $b = 1 - \varepsilon$, and this is the reason to call monomials that end with an A divergent.

Now consider the general case: integral of a product that contains both ω_0 and ω_1. For $\delta_j = 0$ or 1 and $0 < a < b < 1$, introduce the notation

$$I^{a,b}_{\delta_1\ldots\delta_m} = \int_{a<t_m<\cdots<t_2<t_1<b} \omega_{\delta_1}(t_1) \wedge \cdots \wedge \omega_{\delta_m}(t_m).$$

Lemma 10.10.

(i) If $\delta_1 = 0$, then the integral $I^{a,b}_{\delta_1\ldots\delta_m}$ converges to a non-zero constant as $b \to 1$, and it grows as a power of $\log(1 - b)$ if $\delta_1 = 1$.

(ii) If $\delta_m = 1$, then the integral $I^{a,b}_{\delta_1\ldots\delta_m}$ converges to a non-zero constant as $a \to 0$, and it grows as a power of $\log a$ if $\delta_m = 0$.

Proof Induction on the number of chords m. If $m = 1$ then the integral can be calculated explicitly like in the previous example, and the lemma follows from the result. Now suppose that the lemma is proved for $m - 1$ chords. By the Fubini theorem, the integral can be represented as

$$I_{1\delta_2...\delta_m}^{a,b} = \int\limits_{a<t<b} I_{\delta_2...\delta_m}^{a,t} \cdot \frac{dt}{t-1}, \qquad I_{0\delta_2...\delta_m}^{a,b} = \int\limits_{a<t<b} I_{\delta_2...\delta_m}^{a,t} \cdot \frac{dt}{t},$$

for the cases $\delta_1 = 1$ and $\delta_1 = 0$ respectively. By the induction assumption, $0 < c < \left|I_{\delta_2...\delta_m}^{a,t}\right| < \left|\log^k(1-t)\right|$ for some constants c and k. The comparison test implies that the integral $I_{0\delta_2...\delta_m}^{a,b}$ converges as $b \to 1$ because $I_{\delta_2...\delta_m}^{a,t}$ grows slower than any power of $(1-t)$. Moreover, $\left|I_{0\delta_2...\delta_m}^{a,b}\right| > c \int_a^1 \frac{dt}{t} = -c\log(a) > 0$ because $0 < a < b < 1$.

In the case $\delta_1 = 1$ we have

$$c\log(1-b) = c\int_0^b \frac{dt}{t-1} < \left|I_{1\delta_2...\delta_m}^{a,b}\right| < \left|\int_0^b \log^k(1-t)d(\log(1-t))\right|$$

$$= \left|\frac{\log^{k+1}(1-b)}{k+1}\right|,$$

which proves assertion (i). The proof of assertion (ii) is similar. $\qquad\square$

10.2.2 The road map

Here is the plan of our subsequent actions.

Let $\widehat{A}^{conv}$ ($\widehat{A}^{div}$) be the subspace of $\widehat{A} = \mathbb{C}\langle\langle A, B\rangle\rangle$ spanned by all convergent (respectively, divergent) monomials. We are going to define a certain linear map $f : \widehat{A} \to \widehat{A}$ which kills divergent monomials and preserves the associator Φ. Applying f to both parts of Equation (10.7) we shall have

$$\Phi = f(\Phi) = f\left(\lim_{\varepsilon\to 0} \varepsilon^{-B} \cdot Z^{conv} \cdot \varepsilon^A\right) = f\left(\lim_{\varepsilon\to 0} Z^{conv}\right). \tag{10.8}$$

The last equality here follows from the fact that only the unit terms of ε^{-B} and ε^A are convergent and therefore survive under the action of f.

The convergent improper integral

$$\lim_{\varepsilon\to 0} Z^{conv} = 1 + \sum_{m=2}^{\infty} \sum_{\delta_2,...,\delta_{m-1}=0,1} I_{0\delta_2...\delta_{m-1}1}^{0,1} \cdot AC_{\delta_2}\dots C_{\delta_{m-1}}B \tag{10.9}$$

can be computed explicitly (here $C_j = A$ if $\delta_j = 0$ and $C_j = B$ if $\delta_j = 1$). Combining Equations (10.8) and (10.9) we get

$$\Phi = 1 + \sum_{m=2}^{\infty} \sum_{\delta_2,...,\delta_{m-1}=0,1} I_{1\delta_2...\delta_{m-1}0}^{0,1} \cdot f(AC_{\delta_2}\dots C_{\delta_{m-1}}B). \tag{10.10}$$

The knowledge of how f acts on the monomials from $\widehat{A}$ leads to the desired formula for the associator.

10.2.3 The linear map $f : \widehat{\mathcal{A}} \to \widehat{\mathcal{A}}$

Consider the algebra $\widehat{\mathcal{A}}[\alpha, \beta]$ of polynomials in two commuting variables α and β with coefficients in $\widehat{\mathcal{A}}$. Every monomial in $\widehat{\mathcal{A}}[\alpha, \beta]$ can be written uniquely as $\beta^p M \alpha^q$, where M is a monomial in $\widehat{\mathcal{A}}$. Define a $\mathbb{C}$-linear map $j : \widehat{\mathcal{A}}[\alpha, \beta] \to \widehat{\mathcal{A}}$ by $j(\beta^p M \alpha^q) = B^p M A^q$. Now for any element $\Gamma(A, B) \in \widehat{\mathcal{A}}$ let

$$f(\Gamma(A, B)) = j(\Gamma(A - \alpha, B - \beta)).$$

Lemma 10.11. *If M is a divergent monomial in $\widehat{\mathcal{A}}$, then $f(M) = 0$.*

Proof Consider the case where M starts with B, say $M = BC_2 \ldots C_m$, where each C_j is either A or B. Then

$$f(M) = j((B - \beta)M_2) = j(BM_2) - j(\beta M_2),$$

where $M_2 = (C_2 - \gamma_2) \ldots (C_m - \gamma_m)$ with $\gamma_j = \alpha$ or $\gamma_j = \beta$ depending on C_j. But $j(BM_2)$ equals $j(\beta M_2)$ by the definition of j above. The case where M ends with an A can be done similarly.   □

One may notice that for any monomial $M \in \widehat{\mathcal{A}}$ we have

$$f(M) = M + (\text{sum of divergent monomials}).$$

Therefore, by the lemma, f is an idempotent map, $f^2 = f$, that is, f is a projection along $\widehat{\mathcal{A}}^{div}$ (but not onto $\widehat{\mathcal{A}}^{conv}$).

Proposition 10.12. $f(\Phi) = \Phi$.

Proof We use the definition of the associator Φ as the KZ Drinfeld associator from Section 10.1.6 (Proposition 10.7).

It is the ratio $\Phi(A, B) = G_1^{-1} \cdot G_0$ of two solutions of the differential equation (10.6) from Section 10.1.6:

$$G' = \left(\frac{A}{z} + \frac{B}{z - 1} \right) \cdot G$$

with the asymptotics

$$G_0(z) \sim z^A \text{ as } z \to 0 \quad \text{and} \quad G_1(z) \sim (1 - z)^B \text{ as } z \to 1.$$

Consider the differential equation

$$H' = \left(\frac{A - \alpha}{z} + \frac{B - \beta}{z - 1} \right) \cdot H.$$

A direct substitution shows that the functions

$$H_0(z) = z^{-\alpha}(1 - z)^{-\beta} \cdot G_0(z) \quad \text{and} \quad H_1(z) = z^{-\alpha}(1 - z)^{-\beta} \cdot G_1(z)$$

are its solutions with the asymptotics

$$H_0(z) \sim z^{A-\alpha} \text{ as } z \to 0 \qquad \text{and} \qquad H_1(z) \sim (1-z)^{B-\beta} \text{ as } z \to 1.$$

Hence we have

$$\Phi(A - \alpha, B - \beta) = H_1^{-1} \cdot H_0 = G_1^{-1} \cdot G_0 = \Phi(A, B).$$

Therefore,

$$f(\Phi(A, B)) = j(\Phi(A - \alpha, B - \beta)) = j(\Phi(A, B)) = \Phi(A, B)$$

because j acts as the identity map on the subspace $\widehat{A} \subset \widehat{A}[\alpha, \beta]$. The proposition is proved. $\qquad\square$

10.2.4 The calculation

In order to compute Φ according to (10.10) we must find the integrals $I_{0\delta_2\ldots\delta_{m-1}1}^{0,1}$ and the action of f on the monomials. Let us compute $f(AC_{\delta_2}\ldots C_{\delta_{m-1}}B)$ first.

Represent the monomial $M = AC_{\delta_2}\ldots C_{\delta_{m-1}}B$ in the form

$$M = A^{p_1} B^{q_1} \ldots A^{p_l} B^{q_l}$$

for some positive integers $p_1, q_1, \ldots, p_l, q_l$. Then

$$f(M) = j((A - \alpha)^{p_1}(B - \beta)^{q_1} \ldots (A - \alpha)^{p_l}(B - \beta)^{q_l}).$$

We are going to expand the product, collect all β's on the left and all α's on the right, and then replace β by B and α by A. To this end, let us introduce the following multi-index notations:

$$\mathbf{r} = (r_1, \ldots, r_l); \quad \mathbf{i} = (i_1, \ldots, i_l); \quad \mathbf{s} = (s_1, \ldots, s_l); \quad \mathbf{j} = (j_1, \ldots, j_l);$$

$$\mathbf{p} = \mathbf{r} + \mathbf{i} = (r_1 + i_1, \ldots, r_l + i_l); \quad \mathbf{q} = \mathbf{s} + \mathbf{j} = (s_1 + j_1, \ldots, s_l + j_l);$$

$$|\mathbf{r}| = r_1 + \cdots + r_l; \qquad |\mathbf{s}| = s_1 + \cdots + s_l;$$

$$\binom{\mathbf{p}}{\mathbf{r}} = \binom{p_1}{r_1}\binom{p_2}{r_2}\ldots\binom{p_l}{r_l}; \qquad \binom{\mathbf{q}}{\mathbf{s}} = \binom{q_1}{s_1}\binom{q_2}{s_2}\ldots\binom{q_l}{s_l};$$

$$(A, B)^{(\mathbf{i},\mathbf{j})} = A^{i_1} \cdot B^{j_1} \cdot \ldots \cdot A^{i_l} \cdot B^{j_l}.$$

We have

$$(A - \alpha)^{p_1}(B - \beta)^{q_1} \ldots (A - \alpha)^{p_l}(B - \beta)^{q_l} =$$

$$\sum_{\substack{0 \leqslant \mathbf{r} \leqslant \mathbf{p} \\ 0 \leqslant \mathbf{s} \leqslant \mathbf{q}}} (-1)^{|\mathbf{r}| + |\mathbf{s}|} \binom{\mathbf{p}}{\mathbf{r}}\binom{\mathbf{q}}{\mathbf{s}} \cdot \beta^{|\mathbf{s}|}(A, B)^{(\mathbf{i},\mathbf{j})}\alpha^{|\mathbf{r}|},$$

where the inequalities $0 \leqslant \mathbf{r} \leqslant \mathbf{p}$ and $0 \leqslant \mathbf{s} \leqslant \mathbf{q}$ mean $0 \leqslant r_1 \leqslant p_1, \ldots,$ $0 \leqslant r_l \leqslant p_l$, and $0 \leqslant s_1 \leqslant q_1, \ldots, 0 \leqslant s_l \leqslant q_l$. Therefore

$$f(M) = \sum_{\substack{0 \leqslant \mathbf{r} \leqslant \mathbf{p} \\ 0 \leqslant \mathbf{s} \leqslant \mathbf{q}}} (-1)^{|\mathbf{r}|+|\mathbf{s}|} \binom{\mathbf{p}}{\mathbf{r}} \binom{\mathbf{q}}{\mathbf{s}} \cdot B^{|\mathbf{s}|} (A, B)^{\langle \mathbf{i}, \mathbf{j} \rangle} A^{|\mathbf{r}|}. \tag{10.11}$$

In order to complete the formula for the associator, we need to compute the coefficient $I^{0,1}_{1\delta_2 \ldots \delta_m-10}$ of $f(M)$. It turns out that, up to a sign, they are equal to some values of the multivariate ζ-function

$$\zeta(a_1, \ldots, a_n) = \sum_{0 < k_1 < k_2 < \cdots < k_n} k_1^{-a_1} \ldots k_n^{-a_n}$$

where $a_1, \ldots, a_n$ are positive integers (Le and Murakami 1995a). Namely, the coefficients in question are equal, up to a sign, to the values of ζ at integer points $(a_1, \ldots, a_n) \in \mathbb{Z}^n$, which are called *multiple zeta values*, or MZV for short. Multiple zeta values for $n = 2$ were first studied by L. Euler. His paper (Euler 1775) contains several dozen interesting relations between MZVs and values of the univariate (Riemann's) zeta function. This subject was almost forgotten for more than 200 years until M. Hoffman and D. Zagier revived a general interest in MZVs by their papers (Hoffman 1992; Zagier 1994).

Exercise. The sum in the definition of the multivariate ζ-function converges if and only if $a_n \geqslant 2$.

Remark 10.13. Two different conventions about the order of arguments in ζ are in use: we follow that of D. Zagier (1994), also used by P. Deligne, A. Goncharov and Le and Murakami (1995a, 1995b, 1996a, 1996b). The opposite school that goes back to L. Euler and includes J. Borwein, M. Hoffman, M. Petitot, writes $\zeta(2, 1)$ for what we would write as $\zeta(1, 2)$. (They use $k_1 > k_2 > \ldots > k_n > 0$ as the set of summation in the formula for $\zeta(a_1, \ldots, a_k)$.)

Proposition 10.14. *For* $\mathbf{p} > 0$ *and* $\mathbf{q} > 0$ *let*

$$\eta(\mathbf{p}, \mathbf{q}) := \zeta(\underbrace{1, \ldots, 1}_{q_l - 1}, \; p_l + 1, \; \underbrace{1, \ldots, 1}_{q_{l-1} - 1}, \; p_{l-1} + 1, \; \ldots \underbrace{1, \ldots, 1}_{q_1 - 1}, \; p_1 + 1).$$

$$\tag{10.12}$$

Then

$$I^{0,1}_{\underbrace{0 \ldots 0}_{p_l} \underbrace{1 \ldots 1}_{q_l} \ldots \ldots \underbrace{0 \ldots 0}_{p_l} \underbrace{1 \ldots 1}_{q_l}} = (-1)^{|\mathbf{q}|} \eta(\mathbf{p}, \mathbf{q}). \tag{10.13}$$

The calculations needed to prove the proposition are best organised in terms of the (univariate) *polylogarithm*[1] function defined by the series

$$\mathrm{Li}_{a_1,\ldots,a_n}(z) = \sum_{0 < k_1 < k_2 < \cdots < k_n} \frac{z^{k_n}}{k_1^{a_1} \ldots k_n^{a_n}}, \tag{10.14}$$

which obviously converges for $|z| < 1$.

Lemma 10.15. *For* $|z| < 1$

$$\mathrm{Li}_{a_1,\ldots,a_n}(z) = \begin{cases} \displaystyle\int_0^z \mathrm{Li}_{a_1,\ldots,a_n-1}(t)\frac{dt}{t}, & \text{if } a_n > 1 \\[2ex] \displaystyle-\int_0^z \mathrm{Li}_{a_1,\ldots,a_n-1}(t)\frac{d(1-t)}{1-t}, & \text{if } a_n = 1. \end{cases}$$

Proof of the Lemma The lemma follows from the identities below, whose proofs we leave to the reader as an exercise.

$$\frac{d}{dz}\mathrm{Li}_{a_1,\ldots,a_n}(z) = \begin{cases} \frac{1}{z} \cdot \mathrm{Li}_{a_1,\ldots,a_n-1}(z), & \text{if } a_n > 1 \\[2ex] \frac{1}{1-z} \cdot \mathrm{Li}_{a_1,\ldots,a_{n-1}}(z), & \text{if } a_n = 1; \end{cases}$$

$$\frac{d}{dz}\mathrm{Li}_1(z) = \frac{1}{1-z}. \qquad\qquad\qquad \square$$

Proof of the Proposition From the previous lemma we have

$$\mathrm{Li}_{\underbrace{1,1,\ldots,1}_{q_1-1},\, p_l+1,\, \underbrace{1,1,\ldots,1}_{q_{l-1}-1},\, p_{l-1}+1,\, \ldots,\, \underbrace{1,1,\ldots,1}_{q_1-1},\, p_1+1}(z) =$$

$$= (-1)^{q_1+\cdots+q_l} \int_{0 < t_m < \cdots < t_2 < t_1 < z} \underbrace{\omega_0(t_1) \wedge \cdots \wedge \omega_0(t_{p_1})}_{p_1} \wedge$$

$$\wedge \underbrace{\omega_1(t_{p_1+1}) \wedge \cdots \wedge \omega_1(t_{p_1+q_1})}_{q_1} \wedge \cdots \wedge \underbrace{\omega_1(t_{p_1+\cdots+p_l+1}) \wedge \cdots \wedge \omega_1(t_{p_1+\cdots+q_l})}_{q_l} =$$

$$= (-1)^{|\mathbf{q}|} I^{0,z}_{\underbrace{0\ldots0}_{p_1}\underbrace{1\ldots1}_{q_1}\underbrace{\ldots\ldots}_{}\underbrace{0\ldots0}_{p_l}\underbrace{1\ldots1}_{q_l}}.$$

---

[1] It is a generalization of Euler's dilogarithm $\mathrm{Li}_2(z)$ we used in Exercise 8.7, and a specialization of the multivariate polylogarithm

$$\mathrm{Li}_{a_1,\ldots,a_n}(z_1,\ldots,z_n) = \sum_{0 < k_1 < k_2 < \cdots < k_n} \frac{z_1^{k_1} \ldots z_n^{k_n}}{k_1^{a_1} \ldots k_n^{a_n}}$$

introduced by A. Goncharov (1998).

Note that the multiple polylogarithm series (10.14) converges for $z = 1$ in the case $a_n > 1$. This implies that if $p_1 \geqslant 1$ (which holds for a convergent monomial), then we have

$$\eta(\mathbf{p}, \mathbf{q}) = \zeta(\underbrace{1, \ldots, 1}_{q_l-1}, p_l+1, \underbrace{1, \ldots, 1}_{q_{l-1}-1}, p_{l-1}+1, \ldots \underbrace{1, \ldots, 1}_{q_1-1}, p_1+1)$$

$$= \mathrm{Li}_{\underbrace{1,1,\ldots,1}_{q_l-1}, p_l+1, \underbrace{1,1,\ldots,1}_{q_{l-1}-1}, p_{l-1}+1, \ldots, \underbrace{1,1,\ldots,1}_{q_1-1}, p_1+1}(1)$$

$$= (-1)^{|\mathbf{q}|} I^{0,1}_{\underbrace{0\ldots0}_{p_1}\underbrace{1\ldots1}_{q_1}\ldots\ldots\underbrace{0\ldots0}_{p_l}\underbrace{1\ldots1}_{q_l}}.$$

The proposition is proved.   □

10.2.5 Explicit formula for the associator

Combining equations (10.10), (10.11) and (10.13), we get the following formula for the associator:

$$\Phi_{KZ} = 1 + \sum_{\substack{m=2}}^{\infty} \sum_{\substack{0<\mathbf{p},0<\mathbf{q} \\ |\mathbf{p}|+|\mathbf{q}|=m}} \eta(\mathbf{p}, \mathbf{q}) \cdot \sum_{\substack{0\leqslant\mathbf{r}\leqslant\mathbf{p} \\ 0\leqslant\mathbf{s}\leqslant\mathbf{q}}} (-1)^{|\mathbf{r}|+|\mathbf{j}|} \binom{\mathbf{p}}{\mathbf{r}} \binom{\mathbf{q}}{\mathbf{s}} \cdot B^{|\mathbf{s}|}(A, B)^{(\mathbf{i},\mathbf{j})} A^{|\mathbf{r}|}$$

where $\mathbf{i}$ and $\mathbf{j}$ are multi-indices of the same length, $\mathbf{p} = \mathbf{r}+\mathbf{i}$, $\mathbf{q} = \mathbf{s}+\mathbf{j}$, and $\eta(\mathbf{p}, \mathbf{q})$ is the multiple zeta value given by (10.12).

This formula was obtained by Le and Murakami (1996b) .

Example 10.16. Degree 2 terms of the associator. There is only one possibility to represent $m = 2$ as the sum of two positive integers: $2 = 1+1$. So we have only one possibility for $\mathbf{p}$ and $\mathbf{q}$: $\mathbf{p} = (1)$, $\mathbf{q} = (1)$. In this case $\eta(\mathbf{p}, \mathbf{q}) = \zeta(2) = \pi^2/6$ according to (10.12). The multi-indices $\mathbf{r}$ and $\mathbf{s}$ have length 1 and thus consist of a single number $\mathbf{r} = (r_1)$ and $\mathbf{s} = (s_1)$. There are two possibilities for each of them: $r_1 = 0$ or $r_1 = 1$, and $s_1 = 0$ or $s_1 = 1$. In all these cases the binomial coefficients $\binom{\mathbf{p}}{\mathbf{r}}$ and $\binom{\mathbf{q}}{\mathbf{s}}$ are equal to 1. We arrange all the possibilities in the following table.

r_1	s_1	i_1	j_1	$(-1)^{|\mathbf{r}|+|\mathbf{j}|} \cdot B^{|\mathbf{s}|}(A, B)^{(\mathbf{i},\mathbf{j})} A^{|\mathbf{r}|}$
0	0	1	1	$-AB$
0	1	1	0	BA
1	0	0	1	BA
1	1	0	0	$-BA$

Hence, for the degree 2 terms of the associator we get the formula:

$$-\zeta(2)[A, B] = -\frac{\zeta(2)}{(2\pi i)^2}[a, b] = \frac{1}{24}[a, b],$$

where $a = (2\pi i)A = \hspace{1mm}\text{⊢⊣}$, and $b = (2\pi i)B = \text{⊤⊥}$.

Example 10.17. Degree 3 terms of the associator. There are two ways to represent $m = 3$ as the sum of two positive integers: $3 = 2 + 1$ and $3 = 1 + 2$. In each case either $\mathbf{p} = (1)$ or $\mathbf{q} = (1)$. Hence $l = 1$ and both multi-indices consist of just one number $\mathbf{p} = (p_1)$, $\mathbf{q} = (q_1)$. Therefore, all other multi-indices $\mathbf{r}$, $\mathbf{s}$, $\mathbf{i}$, $\mathbf{j}$ also consist of one number.

Here is the corresponding table.

p_1	q_1	$\eta(\mathbf{p},\mathbf{q})$	r_1	s_1	i_1	j_1	$(-1)^{|\mathbf{r}|+|\mathbf{j}|}\binom{\mathbf{p}}{\mathbf{r}}\binom{\mathbf{q}}{\mathbf{s}}\cdot B^{|\mathbf{s}|}(A,B)^{(\mathbf{i},\mathbf{j})}A^{|\mathbf{r}|}$
			0	0	2	1	$-AAB$
			0	1	2	0	BAA
2	1	$\zeta(3)$	1	0	1	1	$2ABA$
			1	1	1	0	$-2BAA$
			2	0	0	1	$-BAA$
			2	1	0	0	BAA
			0	0	1	2	ABB
			1	0	0	2	$-BBA$
1	2	$\zeta(1,2)$	0	1	1	1	$-2BAB$
			1	1	0	1	$2BBA$
			0	2	1	0	BBA
			1	2	0	0	$-BBA$

Using the Euler identity $\zeta(1, 2) = \zeta(3)$ (see Section 10.2.6) we can sum up the degree 3 part of Φ into the formula

$$\zeta(3)\left(-AAB + 2ABA - BAA + ABB - 2BAB + BBA\right)$$

$$= \zeta(3)\left(-\left[A, [A, B]\right] - \left[B, [A, B]\right]\right) = -\frac{\zeta(3)}{(2\pi i)^3}[a + b, [a, b]].$$

Example 10.18. Degree 4 terms of the associator. There are three ways to represent $m = 4$ as the sum of two positive integers: $4 = 3+1$, $4 = 1+3$ and $4 = 2+2$. So we have the following four possibilities for $\mathbf{p}$ and $\mathbf{q}$:

$$\frac{\mathbf{p} \quad (1) \quad (3) \quad (2) \quad (1,1)}{\mathbf{q} \quad (3) \quad (1) \quad (2) \quad (1,1)}$$

The table for the multi-indices $\mathbf{r}, \mathbf{s}, \mathbf{p}, \mathbf{q}$ and the corresponding term

$$T = (-1)^{|\mathbf{r}|+|\mathbf{j}|} \binom{\mathbf{p}}{\mathbf{r}} \binom{\mathbf{q}}{\mathbf{s}} \cdot B^{|\mathbf{s}|}(A, B)^{(\mathbf{i}, \mathbf{j})} A^{|\mathbf{r}|}$$

is shown in Figure 10.1.

Combining the terms into the commutators we get the degree 4 part of the associator Φ:

$$\zeta(1,1,2) \Big[B, \Big[B, [B, A] \Big] \Big] + \zeta(4) \Big[A, \Big[A, [B, A] \Big] \Big]$$

$$+ \zeta(1,3) \Big[B, \Big[A, [B, A] \Big] \Big] + (2\zeta(1,3) + \zeta(2,2))[B, A]^2.$$

Recalling that $A = \frac{1}{2\pi i} a$ and $B = \frac{1}{2\pi i} b$, where a and b are the basic chord diagrams with one chord, and using the identities from Section 10.2.6:

$$\zeta(1,1,2) = \zeta(4) = \pi^4/90, \qquad \zeta(1,3) = \pi^4/360, \qquad \zeta(2,2) = \pi^4/120,$$

we can write out the associator Φ up to degree 4:

$$\Phi_{KZ} = 1 + \frac{1}{24}[a, b] - \frac{\zeta(3)}{(2\pi i)^3}[a + b, [a, b]] - \frac{1}{1440}[a, [a, [a, b]]]$$

$$- \frac{1}{5760}[a, [b, [a, b]]] - \frac{1}{1440}[b, [b, [a, b]]] + \frac{1}{1152}[a, b]^2$$

$$+ \text{(terms of order} > 4).$$

10.2.6 Multiple zeta values

There are many relations among MZV's and powers of π. Some of them, like $\zeta(2) = \frac{\pi^2}{6}$ or $\zeta(1, 2) = \zeta(3)$, were already known to Euler. The last one can be obtained in the following way. According to (10.12) and (10.13) we have

$$\zeta(1, 2) = \eta((1), (2)) = I_{011}^{0,1} = \int\limits_{0 < t_3 < t_2 < t_1 < 1} \omega_0(t_1) \wedge \omega_1(t_2) \wedge \omega_1(t_3)$$

$$= \int\limits_{0 < t_3 < t_2 < t_1 < 1} \frac{dt_1}{t_1} \wedge \frac{d(1 - t_2)}{1 - t_2} \wedge \frac{d(1 - t_3)}{1 - t_3}.$$

p	q	$\eta(p,q)$	r	s	i	j	T
(1)	(3)	$\zeta(1,1,2)$	(0)	(0)	(1)	(3)	$-ABBB$
			(0)	(1)	(1)	(2)	$+3BABB$
			(0)	(2)	(1)	(1)	$-3BBAB$
			(0)	(3)	(1)	(0)	$+BBBA$
			(1)	(0)	(0)	(3)	$+BBBA$
			(1)	(1)	(0)	(2)	$-3BBBA$
			(1)	(2)	(0)	(1)	$+3BBBA$
			(1)	(3)	(0)	(0)	$-BBBA$
(3)	(1)	$\zeta(4)$	(0)	(0)	(3)	(1)	$-AAAB$
			(0)	(1)	(3)	(0)	$+BAAA$
			(1)	(0)	(2)	(1)	$+3AABA$
			(1)	(1)	(2)	(0)	$-3BAAA$
			(2)	(0)	(1)	(1)	$-3ABAA$
			(2)	(1)	(1)	(0)	$+3BAAA$
			(3)	(0)	(0)	(1)	$+BAAA$
			(3)	(1)	(0)	(0)	$-BAAA$
(2)	(2)	$\zeta(1,3)$	(0)	(0)	(2)	(2)	$+AABB$
			(0)	(1)	(2)	(1)	$-2BAAB$
			(0)	(2)	(2)	(0)	$+BBAA$
			(1)	(0)	(1)	(2)	$-2ABBA$
			(1)	(1)	(1)	(1)	$+4BABA$
			(1)	(2)	(1)	(0)	$-2BBAA$
			(2)	(0)	(0)	(2)	$+BBAA$
			(2)	(1)	(0)	(1)	$-2BBAA$
			(2)	(2)	(0)	(0)	$+BBAA$
(1,1)	(1,1)	$\zeta(2,2)$	(0,0)	(0,0)	(1,1)	(1,1)	$+ABAB$
			(0,0)	(0,1)	(1,1)	(1,0)	$-BABA$
			(0,0)	(1,0)	(1,1)	(0,1)	$-BAAB$
			(0,0)	(1,1)	(1,1)	(0,0)	$+BBAA$
			(0,1)	(0,0)	(1,0)	(1,1)	$-ABBA$
			(0,1)	(0,1)	(1,0)	(1,0)	$+BABA$
			(0,1)	(1,0)	(1,0)	(0,1)	$+BABA$
			(0,1)	(1,1)	(1,0)	(0,0)	$-BBAA$
			(1,0)	(0,0)	(0,1)	(1,1)	$-BABA$
			(1,0)	(0,1)	(0,1)	(1,0)	$+BBAA$
			(1,0)	(1,0)	(0,1)	(0,1)	$+BABA$
			(1,0)	(1,1)	(0,1)	(0,0)	$-BBAA$
			(1,1)	(0,0)	(0,0)	(1,1)	$+BBAA$
			(1,1)	(0,1)	(0,0)	(1,0)	$-BBAA$
			(1,1)	(1,0)	(0,0)	(0,1)	$-BBAA$
			(1,1)	(1,1)	(0,0)	(0,0)	$+BBAA$

Figure 10.1 Degree 4 terms of the associator

The change of variables $(t_1, t_2, t_3) \mapsto (1 - t_3, 1 - t_2, 1 - t_1)$ transforms the last integral to

$$\int\limits_{0<t_3<t_2<t_1<1} \frac{d(1 - t_3)}{1 - t_3} \wedge \frac{dt_2}{t_2} \wedge \frac{dt_1}{t_1}$$

$$= - \int\limits_{0<t_3<t_2<t_1<1} \omega_0(t_1) \wedge \omega_0(t_2) \wedge \omega_1(t_3) = -I_{001}^{0,1} = \eta((2), (1)) = \zeta(3).$$

In the general case a similar change of variables

$$(t_1, t_2, \ldots, t_m) \mapsto (1 - t_m, \ldots, 1 - t_2, 1 - t_1)$$

gives the identity

$$I^{0,1}_{\underbrace{0...0}_{p_1} \underbrace{1...1}_{q_1} \cdots \underbrace{0...0}_{p_l} \underbrace{1...1}_{q_l}} = (-1)^m I^{0,1}_{\underbrace{0...0}_{q_l} \underbrace{1...1}_{p_l} \cdots \underbrace{0...0}_{q_1} \underbrace{1...1}_{p_1}}.$$

By (10.13), we have

$$I^{0,1}_{\underbrace{0...0}_{p_1} \underbrace{1...1}_{q_1} \cdots \underbrace{0...0}_{p_l} \underbrace{1...1}_{q_l}} = (-1)^{|q|} \eta(\mathbf{p}, \mathbf{q}),$$

$$I^{0,1}_{\underbrace{0...0}_{q_l} \underbrace{1...1}_{p_l} \cdots \underbrace{0...0}_{q_1} \underbrace{1...1}_{p_1}} = (-1)^{|p|} \eta(\overline{\mathbf{q}}, \overline{\mathbf{p}}),$$

where the bar denotes the inversion of a sequence: $\overline{\mathbf{p}} = (p_l, p_{l-1}, \ldots, p_1)$, $\overline{\mathbf{q}} = (q_l, q_{l-1}, \ldots, q_1)$.

Since $|p| + |q| = m$, we deduce that

$$\eta(\mathbf{p}, \mathbf{q}) = \eta(\overline{\mathbf{q}}, \overline{\mathbf{p}}).$$

This relation is called the *duality relation* between MZV's. After the conversion from η to ζ according to Equation (10.12), the duality relations become picturesque and unexpected.

Exercise. Relate the duality to the rotation of a chord diagram by 180° as in Figure 10.2.

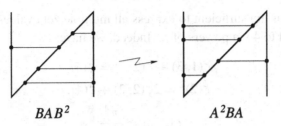

$$BAB^2 \qquad\qquad A^2BA$$

Figure 10.2

As an example, we give a table of all nontrivial duality relations of weight $m \leqslant 5$:

p	q	q̄	p̄	relation
(1)	(2)	(2)	(1)	$\zeta(1, 2) = \zeta(3)$
(1)	(3)	(3)	(1)	$\zeta(1, 1, 2) = \zeta(4)$
(1)	(4)	(4)	(1)	$\zeta(1, 1, 1, 2) = \zeta(5)$
(2)	(3)	(3)	(2)	$\zeta(1, 1, 3) = \zeta(1, 4)$
(1, 1)	(1, 2)	(2, 1)	(1, 1)	$\zeta(1, 2, 2) = \zeta(2, 3)$
(1, 1)	(2, 1)	(1, 2)	(1, 1)	$\zeta(2, 1, 2) = \zeta(3, 2)$

The reader may want to check this table by way of exercise.

There are other relations between the multiple zeta values that do not follow from the duality law. Let us quote just a few.

1. Euler's relations:

$$\zeta(1, n - 1) + \zeta(2, n - 2) + \cdots + \zeta(n - 2, 2) = \zeta(n), \qquad (10.15)$$

$$\zeta(m) \cdot \zeta(n) = \zeta(m, n) + \zeta(n, m) + \zeta(m + n). \qquad (10.16)$$

2. Relations obtained by Le and Murakami (1995a) computing the Kontsevich integral of the unknot by the combinatorial procedure explained below in Section 10.3 (the first one was earlier proved by M. Hoffman (1992)):

$$\zeta(\underbrace{2, 2, \ldots, 2}_{m}) = \frac{\pi^{2m}}{(2m + 1)!} \qquad (10.17)$$

$$\left(\frac{1}{2^{2n-2}} - 1\right)\zeta(2n) - \zeta(1, 2n - 1) + \zeta(1, 1, 2n - 1) - \ldots$$

$$+ \zeta(\underbrace{1, 1, \ldots, 1}_{2n-2}, 2) = 0. \qquad (10.18)$$

These relations are sufficient to express all multiple zeta values with the sum of arguments equal to 4 via powers of π. Indeed, we have:

$$\zeta(1, 3) + \zeta(2, 2) = \zeta(4),$$
$$\zeta(2)^2 = 2\zeta(2, 2) + \zeta(4),$$
$$\zeta(2, 2) = \frac{\pi^4}{120},$$
$$-\frac{3}{4}\zeta(4) - \zeta(1, 3) + \zeta(1, 1, 2) = 0.$$

Solving these equations one by one and using the identity $\zeta(2) = \pi^2/6$, we find the values of all MZVs of weight 4: $\zeta(2, 2) = \pi^4/120$, $\zeta(1, 3) = \pi^4/360$, $\zeta(1, 1, 2) = \zeta(4) = \pi^4/90$.

There exists an extensive literature about the relations between MZV's, for instance (Borwein *et al.* 1998; Cartier 1999; Hoffman 1992; Hoffman and Ohno 2003; and Okuda and Ueno 2004), and the interested reader is invited to consult these.

An attempt to overview the whole variety of relations between MZV's was undertaken by D. Zagier (1994). Call the *weight* of a multiple zeta value $\zeta(n_1, \ldots, n_k)$ the sum of all its arguments $w = n_1 + \cdots + n_k$. Let $\mathcal{Z}_w$ be the vector subspace of the reals $\mathbb{R}$ over the rationals $\mathbb{Q}$ spanned by all MZV's of a fixed weight w. For completeness we put $\mathcal{Z}_0 = \mathbb{Q}$ and $\mathcal{Z}_1 = 0$. No inhomogeneous relations between the MZV's of different weight are known, so that conjecturally the sum of all $\mathcal{Z}_i$'s is direct. In any case, we can consider the *formal* direct sum of all $\mathcal{Z}_w$

$$\mathcal{Z}_\bullet := \bigoplus_{w \geqslant 0} \mathcal{Z}_w.$$

Proposition 10.19. *The vector space $\mathcal{Z}_\bullet$ forms a graded algebra over $\mathbb{Q}$, i.e. $\mathcal{Z}_u \cdot \mathcal{Z}_v \subseteq \mathcal{Z}_{u+v}$.*

Euler's product formula (10.16) illustrates this statement. A proof can be found in Goncharov (1998). D. Zagier made a conjecture about the Poincaré series of this algebra.

Conjecture 10.20 (Zagier 1994).

$$\sum_{w=0}^{\infty} \dim_{\mathbb{Q}}(\mathcal{Z}_w) \cdot t^w = \frac{1}{1 - t^2 - t^3},$$

which is equivalent to say that $\dim \mathcal{Z}_0 = \dim \mathcal{Z}_2 = 1$, $\dim \mathcal{Z}_1 = 0$ *and* $\dim \mathcal{Z}_w = \dim \mathcal{Z}_{w-2} + \dim \mathcal{Z}_{w-3}$ *for all* $w \geqslant 3$.

This conjecture turns out to be related to the dimensions of various subspaces in the primitive space of the chord diagram algebra $\mathscr{A}$ (see Broadhurst 1997; Kreimer 2000) and also to Drinfeld's conjecture about the structure of the Lie algebra of the Grothendieck–Teichmüller group (Etingof and Schiffmann 1998).

It is known (Goncharov 2002; Terasoma 2002) that Zagier's sequence gives an upper bound on the dimension of $\mathcal{Z}_w$; in fact, up to weight 12 any zeta-number can be written as a rational polynomial in

$$\zeta(2), \zeta(3), \zeta(5), \zeta(7), \zeta(2, 6), \zeta(9), \zeta(2, 8), \zeta(11), \zeta(1, 2, 8),$$
$$\zeta(2, 10), \zeta(1, 1, 2, 8).$$

More information about the generators of the algebra $\mathcal{Z}$ is available on the web pages of M. Petitot and J. Vermaseren (see References).

10.2.7 Logarithm of the KZ associator

The associator Φ_{KZ} is group-like (see Exercise 10.3 at the end of the chapter). Therefore, its logarithm can be expressed as a Lie series in variables A and B. Let L be the completion of a free Lie algebra generated by A and B.

An explicit expression for $\log \Phi_{KZ}$ up to degree 6 was first written out in Minh *et al.* (2000), and up to degree 12, in Duzhin (2009). The last formula truncated up to degree 7 in the variables A and B is shown on page 290. We use the shorthand notations

$$\zeta_n = \zeta(n) , \quad C_{kl} = \operatorname{ad}_B^{k-1} \operatorname{ad}_A^{l-1} [A, B].$$

Remark 10.21. We have expanded the associator with respect to the *Lyndon basis* of the free Lie algebra (see Reutenauer 1993). There is a remarkable one-to-one correspondence between the Lyndon words and the irreducible polynomials over the field of two elements $\mathbb{F}_2$, so that the associator may be thought of as a mapping from the set of irreducible polynomials over $\mathbb{F}_2$ into the algebra of multiple zeta values.

Now let $L'' := [[L, L], [L, L]]$ be the second commutant of the algebra L. We can consider L as a subspace of $\mathbb{C}\langle\langle A, B \rangle\rangle$. V. Drinfeld (1990) proved the following formula:

$$\log \Phi_{KZ} = \sum_{k,l \geqslant 1} c_{kl} \, C_{kl} \qquad (\operatorname{mod} L''),$$

where the coefficients c_{kl} are defined by the generating function

$$1 + \sum_{k,l \geqslant 1} c_{kl} u^k v^l = \exp\left(\sum_{n=2}^{\infty} \frac{\zeta(n)}{n} (u^n + v^n - (u+v)^n)\right)$$

$$\log(\Phi_{KZ}) = -\zeta_2 C_{11} - \zeta_3 (C_{12} + C_{21})$$

$$- \frac{2}{5}\zeta_2^2 (C_{13} + C_{31}) - \frac{1}{10}\zeta_2^2 C_{22}$$

$$- \zeta_5 (C_{14} + C_{41}) + (\zeta_2\zeta_3 - 2\zeta_5) (C_{23} + C_{32})$$

$$+ \frac{\zeta_2\zeta_3 - \zeta_5}{2} [C_{11}, C_{12}] + \frac{\zeta_2\zeta_3 - 3\zeta_5}{2} [C_{11}, C_{21}]$$

$$- \frac{8}{35}\zeta_2^3 (C_{15} + C_{51}) + \left(\frac{1}{2}\zeta_3^2 - \frac{6}{35}\zeta_2^3\right) (C_{24} + C_{42}) + \left(\zeta_3^2 - \frac{23}{70}\zeta_2^3\right) C_{33}$$

$$+ \left(-\frac{19}{105}\zeta_2^3 + \zeta_3^2\right) [C_{11}, C_{13}] + \left(-\frac{69}{140}\zeta_2^3 + \frac{3}{2}\zeta_3^2\right) [C_{11}, C_{22}]$$

$$+ \left(-\frac{17}{105}\zeta_2^3\right) [C_{11}, C_{31}] + \left(\frac{2}{105}\zeta_2^3 - \frac{1}{2}\zeta_3^2\right) [C_{12}, C_{21}]$$

$$- \zeta_7 (C_{16} + C_{61}) + \left(\frac{2}{5}\zeta_3\zeta_2^2 + \zeta_2\zeta_5 - 3\zeta_7\right) (C_{25} + C_{52})$$

$$+ \left(\frac{1}{2}\zeta_3\zeta_2^2 + 2\zeta_2\zeta_5 - 5\zeta_7\right) (C_{34} + C_{43})$$

$$+ \left(\frac{6}{5}\zeta_3\zeta_2^2 + \frac{1}{2}\zeta_2\zeta_5 - 4\zeta_7\right) [C_{11}, C_{14}]$$

$$+ \left(\frac{11}{5}\zeta_3\zeta_2^2 + \frac{7}{2}\zeta_2\zeta_5 - 13\zeta_7\right) [C_{11}, C_{23}]$$

$$+ \left(\frac{3}{10}\zeta_3\zeta_2^2 + \frac{13}{2}\zeta_2\zeta_5 - 12\zeta_7\right) [C_{11}, C_{32}] + \left(\frac{5}{2}\zeta_2\zeta_5 - 5\zeta_7\right) [C_{11}, C_{41}]$$

$$+ \left(\zeta_3\zeta_2^2 - 3\zeta_7\right) [C_{12}, C_{13}] + \left(\frac{23}{20}\zeta_3\zeta_2^2 - \frac{61}{16}\zeta_7\right) [C_{12}, C_{22}]$$

$$+ \left(-\frac{3}{10}\zeta_3\zeta_2^2 - \frac{1}{2}\zeta_2\zeta_5 + \frac{19}{16}\zeta_7\right) [C_{12}, C_{31}]$$

$$+ \left(\frac{4}{5}\zeta_3\zeta_2^2 + \frac{5}{2}\zeta_2\zeta_5 - \frac{99}{16}\zeta_7\right) [C_{21}, C_{13}]$$

$$+ \left(\frac{7}{20}\zeta_3\zeta_2^2 + 6\zeta_2\zeta_5 - \frac{179}{16}\zeta_7\right) [C_{21}, C_{22}]$$

$$+ \left(-\frac{1}{5}\zeta_3\zeta_2^2 + 2\zeta_2\zeta_5 - 3\zeta_7\right) [C_{21}, C_{31}]$$

$$+ \left(\frac{67}{60}\zeta_3\zeta_2^2 + \frac{1}{4}\zeta_2\zeta_5 - \frac{65}{16}\zeta_7\right) [C_{11}, [C_{11}, C_{12}]]$$

$$+ \left(-\frac{1}{12}\zeta_3\zeta_2^2 + \frac{3}{4}\zeta_2\zeta_5 - \frac{17}{16}\zeta_7\right) [C_{11}, [C_{11}, C_{21}]] + \ldots$$

Figure 10.3 Logarithm of the Drinfeld associator up to degree 7

expressed in terms of the univariate zeta function $\zeta(n) := \sum_{k=1}^{\infty} k^{-n}$. In particular, $c_{kl} = c_{lk}$ and $c_{k1} = c_{1k} = -\zeta(k+2)$.

10.3 Combinatorial construction of the Kontsevich integral

In this section we fulfil the promise of Section 8.10 and describe in detail a combinatorial construction for the Kontsevich integral of knots and links. The associator Φ_{KZ} is an essential part of this construction. In Section 10.2 we gave formulae for Φ_{KZ}; using these expressions one can perform explicit calculations, at least in low degrees.

10.3.1 Non-associative monomials

A *non-associative monomial* in one variable is simply a choice of an order (that is, a choice of parentheses) of multiplying n factors; the number n is referred to as the *degree* of a non-associative monomial. The only such monomial in x of degree 1 is x itself. In degree 2 there is also only one monomial, namely xx, in degree 3 there are two monomials $(xx)x$ and $x(xx)$, in degree 4 we have $((xx)x)x$, $(x(xx))x$, $(xx)(xx)$, $x((xx)x)$ and $x(x(xx))$, etc. Define the product $u \cdot v$ of two non-associative monomials u and v as their concatenation with each factor of length more than one surrounded by an extra pair of parentheses, for instance $x \cdot x = xx$, $xx \cdot xx = (xx)(xx)$.

For each pair u, v of non-associative monomials of the same degree n one can define the element $\Phi(u, v) \in \widehat{\mathscr{A}}^h(n)$ as follows. If $n < 3$ we set $\Phi(u, v) = \mathbf{1}_n$, the unit in $\widehat{\mathscr{A}}^h(n)$. Assume $n \geqslant 3$. Then $\Phi(u, v)$ is determined by the following properties:

1. If $u = w_1 \cdot (w_2 \cdot w_3)$ and $v = (w_1 \cdot w_2) \cdot w_3$ where w_1, w_2, w_3 are monomials of degrees n_1, n_2 and n_3 respectively, then

$$\Phi(u, v) = \Delta^{n_1,n_2,n_3} \Phi_{KZ},$$

where Δ^{n_1,n_2,n_3} is the cabling-type operation defined in Exercise 9.13 to Chapter 9.

2. If w is a monomial of degree m,

$$\Phi(w \cdot u, w \cdot v) = \mathbf{1}_m \otimes \Phi(u, v)$$

and

$$\Phi(u \cdot w, v \cdot w) = \Phi(u, v) \otimes \mathbf{1}_m;$$

3. If u, v, w are monomials of the same degree, then

$$\Phi(u, v) = \Phi(u, w)\Phi(w, v).$$

These properties are sufficient to determine $\Phi(u, v)$ since each non-associative monomial in one variable can be obtained from any other such monomial of the same degree by moving the parentheses in triple products. It is not immediate that $\Phi(u, v)$ is well-defined, however. Indeed, according to (3), we can define $\Phi(u, v)$ by choosing a sequence of moves that shift one pair of parentheses at a time, and have the effect of changing u into v. A potential problem is that there may be more than one such sequence; however, let us postpone this matter for the moment and work under the assumption that $\Phi(u, v)$ may be multivalued (which it is not, see page 299).

Recall from Section 1.6 the notion of an elementary tangle: basically, these are maxima, minima, crossings and vertical segments.

Take a tensor product of several elementary tangles and choose the brackets in it, enclosing each elementary tangle other than a vertical segment in its own pair of parentheses. This choice of parentheses is encoded by a non-associative monomial w, where each vertical segment is represented by an x and each crossing or a critical point – by the product xx. Further, we have two more non-associative monomials, $\overline{w}$ and $\underline{w}$: $\overline{w}$ is formed by the top boundary points of the tangle, and $\underline{w}$ is formed by the bottom endpoints. For example, consider the following tensor product, parenthesized as indicated, of three elementary tangles:

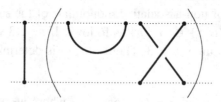

The parentheses in the product are coded by $w = x((xx)(xx))$. The top part of the boundary gives $\overline{w} = w$, and the bottom part produces $\underline{w} = x(xx)$.

Note that here it is important that the factors in the product are not arbitrary, but elementary tangles, since each elementary tangle has at most two upper and at most two lower boundary points.

10.3.2 The construction

First, recall that in Exercise 8.13 we defined the operations S_k which describe how the Kontsevich integral changes when one of the components of a tangle is reversed. Assume that the components of the diagram skeleton are numbered.

Then S_k changes the direction of the kth component and multiplies the diagram by -1 if the number of chord endpoints lying on the kth component is odd.

Represent a given knot K as a product of tangles

$$K = T_1 T_2 \ldots T_n$$

so that each T_i is a tensor product of elementary tangles:

$$T_i = T_{i,1} \otimes \cdots \otimes T_{i,k_i}.$$

Write Z_i for the tensor product of the Kontsevich integrals of the elementary tangles $T_{i,j}$:

$$Z_i = Z(T_{i,1}) \otimes \cdots \otimes Z(T_{i,k_i}).$$

Note that the only elementary tangles for which the Kontsevich integral is non-trivial are the crossings X_- and X_+, and for them

$$Z(X_+) = \diagup\!\!\!\!\diagdown \cdot \exp\left(\frac{H}{2}\right), \qquad Z(X_-) = \diagup\!\!\!\!\diagdown \cdot \exp\left(-\frac{H}{2}\right).$$

For all other elementary tangles the Kontsevich integral consists of a diagram with no chords:

$$Z(\overrightarrow{\max}) = \cap, \qquad Z(\mathrm{id}) = \uparrow,$$

and so on. We remind that Z_i in general does not coincide with $Z(T_i)$.

For each simple tangle T_i choose the parentheses in the tensor product, and represent this choice by a non-associative monomial w_i. Then the *combinatorial Kontsevich integral* $Z_{comb}(K)$ is defined as

$$Z_{comb}(K) = Z_1 \cdot \Phi(\underline{w}_1, \overline{w}_2)_{\downarrow^1} \cdot Z_2 \cdot \ldots \cdot Z_{n-1} \cdot \Phi(\underline{w}_{n-1}, \overline{w}_n)_{\downarrow^{n-1}} \cdot Z_n,$$

where $\Phi(\underline{w}_i, \overline{w}_{i+1})_{\downarrow^i}$ is the result of applying to $\Phi(\underline{w}_i, \overline{w}_{i+1})$ all the operations S_k such that at the kth point on the bottom of T_i (or on the top of T_{i+1}) the corresponding strand is oriented downwards.

The combinatorial Kontsevich integral at first glance may seem to be a complicated expression. However, it is built of only two types of elements: the exponential of $H/2$ and the Drinfeld associator Φ_{KZ} which produces all the $\Phi(\underline{w}_i, \overline{w}_{i+1})$.

Remark 10.22. The definition of an elementary tangle in Section 1.6 is somewhat restrictive. In particular, of all types of crossings only X_+ and X_- are considered to be elementary tangles. Note that rotating X_+ and X_- by $\pm\pi/2$ and by π we get tangles whose Kontsevich integral is an exponential of the same kind as for X_+ and X_-. It will be clear from our argument that we can count these tangles as elementary for the definition of the combinatorial Kontsevich integral.

10.3.3 Example of computation

Let us see how the combinatorial Kontsevich integral can be computed, up to order 2, on the example of the left trefoil 3_1. Explicit formulae for the associator were proved in Section 10.2. In particular, we shall see that

$$\Phi_{KZ} = 1 + \frac{1}{24}(\,\mathsf{H\!H} - \mathsf{H\!H}\,) + \cdots.$$

Decompose the left trefoil into elementary tangles as shown below and choose the parentheses in the tensor product as shown in the second column:

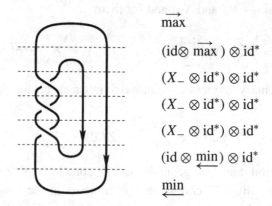

$$\overrightarrow{\mathrm{max}}$$
$$(\mathrm{id}\otimes \overrightarrow{\mathrm{max}})\otimes \mathrm{id}^*$$
$$(X_- \otimes \mathrm{id}^*)\otimes \mathrm{id}^*$$
$$(X_- \otimes \mathrm{id}^*)\otimes \mathrm{id}^*$$
$$(X_- \otimes \mathrm{id}^*)\otimes \mathrm{id}^*$$
$$(\mathrm{id} \otimes \underleftarrow{\mathrm{min}})\otimes \mathrm{id}^*$$
$$\underleftarrow{\mathrm{min}}$$

The combinatorial Kontsevich integral may then be represented as

$$Z_{comb}(3_1) \quad = \quad$$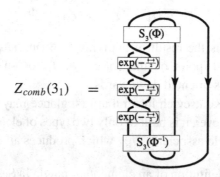

where $S_3(\mathsf{H}\!|\,) = \mathsf{H}\!|$ and $S_3(|\,\mathsf{H}) = -|\,\mathsf{H}$. The crossings in the above picture are, of course, irrelevant since it shows chord diagrams and not knot diagrams.

We have that

$$S_3(\Phi_{KZ}^{\pm 1}) = 1 \pm \frac{1}{24}(\,\mathsf{H}\!\mathsf{H} - \mathsf{H}\!\mathsf{H}\,) + \cdots$$

and

$$\exp\left(\pm\frac{\mathsf{H}}{2}\right) = 1 \pm \frac{\mathsf{H}}{2} + \frac{\mathsf{H}^2}{8} + \cdots.$$

Plugging these expressions into the diagram above we see that, up to degree 2, the combinatorial Kontsevich integral of the left trefoil is

$$Z_{comb}(3_1) = 1 + \frac{25}{24}\bigotimes + \cdots.$$

Representing the hump H as

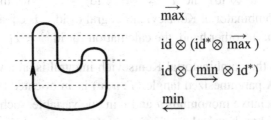

$$\overrightarrow{\max}$$
$$\mathrm{id} \otimes (\mathrm{id}^* \otimes \overrightarrow{\max})$$
$$\mathrm{id} \otimes (\underrightarrow{\min} \otimes \mathrm{id}^*)$$
$$\underleftarrow{\min}$$

we have

$$Z_{comb}(H) = 1 + \frac{1}{24}\bigotimes + \cdots$$

and we can speak of the *final* combinatorial Kontsevich integral of the trefoil (for instance, in the multiplicative normalization as on page 235):

$$I'_{comb}(3_1) = Z_{comb}(3_1)/Z_{comb}(H)$$

$$= \left(1 + \frac{25}{24}\bigotimes + \cdots\right)\left(1 + \frac{1}{24}\bigotimes + \cdots\right)^{-1} = 1 + \bigotimes + \cdots.$$

10.3.4 Equivalence of the combinatorial and analytic definitions

The main result about the combinatorial Kontsevich integral is the following theorem:

Theorem 10.23 (Le and Murakami 1995b). *The combinatorial Kontsevich integral of a knot or a link is equal to the usual Kontsevich integral:*

$$Z_{comb}(K) = Z(K).$$

The idea of the proof was sketched in Section 8.10. The most important part of the argument consists in expressing the Kontsevich integral of an associating tangle via Φ_{KZ}. As often happens in our setting, this argument will give results about objects more general than links, and we shall prove it in this greater generality.

10.3.5 The combinatorial integral for parenthesized tangles

The combinatorial construction for the Kontsevich integral (see Section 10.3.2) can also be performed for arbitrary oriented tangles, in the very same manner as for knots or links. However, the result of this construction can be manifestly non-invariant.

Example 10.24. Take the trivial tangle $\mathrm{id}^{\otimes 3}$ on 3 strands and write it as $\mathrm{id}^{\otimes 3} = T_1 T_2$ where $T_1 = \mathrm{id} \otimes (\mathrm{id} \otimes \mathrm{id})$ and $T_2 = (\mathrm{id} \otimes \mathrm{id}) \otimes \mathrm{id}$. With this choice of the parentheses, the combinatorial Kontsevich integral of $\mathrm{id}^{\otimes 3}$ is equal to the Drinfeld associator Φ_{KZ}. On the other hand, the calculation for $\mathrm{id}^{\otimes 3} = T_1$ simply gives $\mathbf{1}_3$.

It turns out that the combinatorial Kontsevich integral is an invariant of *parenthesized tangles*. A parenthesized tangle (T, u, v) is an oriented tangle T together with two non-associative monomials u and v in one variable, such that the degrees of u and v are equal to the number of points in the upper, and, respectively, lower, parts of the boundary of T. One can think of these monomials as sets of parentheses on the boundary of T.

The combinatorial Kontsevich integral Z_{comb} of a parenthesized tangle (T, u, v) is defined in the same way as the Kontsevich integral of knots or links, by decomposing T into a product of simple tangles $T_1 \ldots T_n$ and choosing parentheses on the T_i. The only difference is that now we require that the bracketing chosen on T_1 give rise to the monomial u on the top part of T_1 and that the parentheses of T_n produce v on the bottom of T_n. As usual (for instance, in Section 9.1.3), we can define

$$I_{comb}(T) = Z(H)^{-m_1} \# \ldots \# Z(H)^{-m_k} \# Z_{comb}(T),$$

where m_i is the number of maxima on the ith component of T.

It turns out that the combinatorial Kontsevich integral $I_{comb}(T, u, v)$ depends only on the isotopy class of T and on the monomials u, v. This can be proved by relating $Z_{comb}(T, u, v)$ to the Kontsevich integral of a certain family of tangles. In order to write down the exact formula expressing this relation, we need to define its ingredients first.

Remarks 10.25. For tangles whose upper and lower boundary consist of at most two points, there is only one way to choose the parentheses on the top and on the bottom, and therefore, in this case I_{comb} is an invariant of usual (not parenthesized) tangles.

Note also that I_{comb} of parenthesized tangles is preserved by *all* isotopies, while the analytic Kontsevich integral is only constant under fixed-end isotopies.

10.3.6 Deformations associated with monomials and regularizing factors

Let t be a set of n distinct points in an interval $[a, b]$. To each non-associative monomial w of degree n we can associate a deformation t_ε^w, with $0 < \varepsilon \leqslant 1$, as follows.

If t_1 and t_2 are two configurations of distinct points, in the intervals $[a_1, b_1]$ and $[a_2, b_2]$ respectively, we can speak of their ε-parametrized tensor product: it is obtained by rescaling both t_1 and t_2 by ε and placing the resulting intervals at the distance $1 - \varepsilon$ from each other:

This is completely analogous to the ε-parametrized tensor product of tangles in Section 8.4.2. Just as for tangles, the ε-parametrized tensor product of configurations of distinct points is not associative, and defined only up to a translation.

Now let us consider our configuration of points t. Divide the interval $[a, b]$ into n smaller intervals so that there is exactly one point of t in each of them, and take their ε-parametrized tensor product in the order prescribed by the monomial w. Call the result t_ε^w.

Exercise. Show that t_ε^w only depends on t, w and ε. In particular, it does not depend on the choice of the decomposition of t into n intervals.

Now, let (T, u, v) be a parenthesized tangle. Denote by s and t the sets of top and bottom boundary points, respectively. A continuous deformation of the boundary of a tangle can always be extended to an horizontal deformation of the whole tangle. We shall denote by $T_{\varepsilon,u,v}$ the family of (non-parenthesized) tangles obtained by deforming s by means of s_ε^u and, at the same time, deforming t by means of t_ε^v.

The second ingredient we shall need is a certain function from non-associative monomials of degree n in one variable to $\widehat{\mathscr{A}}^h(n)$.

First, we define for each integer $i \geqslant 0$ and each non-associative monomial w in x the element $c_i(w) \in \widehat{\mathscr{A}}^h(n)$, where n is the degree of w, by setting

- $c_i(x) = 0$ for all i;
- $c_0(w_1 w_2) = \Delta^{n_1, n_2}\left(\frac{\mathord{\text{⊟}}}{2\pi i}\right)$ if $w_1, w_2 \neq 1$, where n_1 and n_2 are the degrees of w_1 and w_2 respectively;
- $c_i(w_1 w_2) = c_{i-1}(w_1) \otimes \mathbf{1}_{n_2} + \mathbf{1}_{n_1} \otimes c_{i-1}(w_2)$ if $w_1, w_2 \neq 1$ with $\deg w_1 = n_1$, $\deg w_2 = n_2$ and $i > 0$.

It is easy to see that for each w all the $c_i(w)$ commute with each other (this follows directly from Lemma 5.51) and that only a finite number of the c_i is non-zero. Now, we set

$$\rho^\varepsilon(w) = \prod_{k=1}^{\infty} \varepsilon^{k c_k(w)}.$$

This product is, of course, finite since almost all terms in it are equal to the unit in $\mathscr{A}^h(n)$. The element $\rho^\varepsilon(w)$ is called the *regularizing factor* of w.

10.3.7 The comparison of the analytic and combinatorial integrals

Theorem 10.26. *For a parenthesized tangle* (T, u, v)

$$\lim_{\varepsilon \to 0} \rho^\varepsilon(u)_{\downarrow^t}^{-1} \cdot Z(T_{\varepsilon, u, v}) \cdot \rho^\varepsilon(v)_{\downarrow^b} = Z_{comb}(T, u, v),$$

where $\rho^\varepsilon(u)_{\downarrow^t}$ *is the result of applying to* $\rho^\varepsilon(u)$ *all the operations* S_k *such that at the kth point on the top of* T *the corresponding strand is oriented downwards, and* $\downarrow^b$ *denotes the same operation at the bottom of the tangle* T.

In particular, it follows that for a knot or a link both definitions of the Kontsevich integral coincide, since the boundary of a link is empty.

Example 10.27. Let (T, u, v) be the parenthesized tangle with T being the trivial braid on 4 strands, all oriented upwards, $u = (x(xx))x$ and $v = ((xx)x)x$. We have

$$c_1(u) = \frac{1}{2\pi i}\left(\mathord{\text{⊓⎮⎮}} + \mathord{\text{⊓⊓⎮}} \right), \qquad c_1(v) = \frac{1}{2\pi i}\left(\mathord{\text{⎮⊓⎮}} + \mathord{\text{⊓⊓⎮}} \right),$$

$$c_2(u) = \frac{1}{2\pi i}\mathord{\text{⎮⊓⎮}}, \qquad c_2(v) = \frac{1}{2\pi i}\mathord{\text{⊓⎮⎮}}.$$

The combinatorial Kontsevich integral of (T, u, v) equals $\Phi_{KZ} \otimes \mathrm{id}$, and we have

$$\lim_{\varepsilon \to 0} \varepsilon^{-\frac{1}{2\pi i}\left(\mathord{\text{⊓⎮⎮}} + \mathord{\text{⊓⊓⎮}} \right)} \varepsilon^{-\frac{1}{\pi i}\mathord{\text{⎮⊓⎮}}} \cdot Z(T_{\varepsilon, u, v}) \cdot \varepsilon^{\frac{1}{\pi i}\mathord{\text{⊓⎮⎮}}} \varepsilon^{\frac{1}{2\pi i}\left(\mathord{\text{⎮⊓⎮}} + \mathord{\text{⊓⊓⎮}} \right)}$$

$$= \Phi_{KZ} \otimes \mathrm{id}.$$

This is a particular case of Exercise 10.11 on page 309.

Proof of the Theorem Let (T_1, u_1, v_1) and (T_2, u_2, v_2) be two parenthesized tangles with $v_1 = u_2$ and such that the orientations of the strands on the bottom of T_1 agree with those on top of T_2. Then we can define their *product* to be the parenthesized tangle $(T_1 T_2, u_1, v_2)$. Every parenthesized tangle is a product of tangles (T, u, v) of three types:

1. *associating* tangles with T trivial (all strands vertical, though with arbitrary orientations) and $u \neq v$;
2. tangles where T is a tensor product, in some order, of one crossing and several vertical strands;
3. tangles where T is a tensor product, in some order, of one critical point and several vertical strands.

In the latter two cases we require that u and v come from the same choice of brackets on the elementary factors of T. For the tangles of type (2) this implies that $u = v$; in the case of type (3) tangles one monomial is obtained from the other by deleting one factor of the form (xx).

Both sides of the equality in the statement of the theorem are multiplicative with respect to this product, so it is sufficient to consider the three cases separately.

Let us introduce, for this proof only, the following notation. If x and y are two elements of $\widehat{\mathscr{A}}^h(n)$ that depend on a parameter ε, by saying that $x \sim y$ as $\varepsilon \to 0$ we shall mean that in some fixed basis of $\widehat{\mathscr{A}}^h(n)$ (and, hence, in any basis of this algebra) the coefficient of each diagram in $x - y$ is of the same or smaller order of magnitude than $\varepsilon \ln^N \varepsilon$ for some non-negative integer N that may depend on the diagram. Note that for any non-negative N the limit of $\varepsilon \ln^N \varepsilon$ as $\varepsilon \to 0$ is equal to 0.

First let us consider the associating tangles. Without loss of generality we can assume that all the strands of the tangle are oriented upwards. We need to show that if $I = \mathrm{id}^{\otimes n}$ is a trivial tangle,

$$\rho^\varepsilon(u)^{-1} \cdot Z(I_{\varepsilon, u, v}) \cdot \rho^\varepsilon(v) \sim \Phi(u, v) \tag{10.19}$$

as $\varepsilon \to 0$.

Remark 10.28. An important corollary of the above formula is that $\Phi(u, v)$ is well-defined, since the left-hand side is.

Let w be a non-associative word and t, a configuration of distinct points in an interval. We denote by εt a configuration of the same cardinality and in the same interval as t but whose distances between points are equal to the corresponding distances in t, multiplied by ε. (There are many such configurations, of course, but this is of no importance in what follows.)

Write $N_\varepsilon(w)$ for a tangle with no crossings which has $\varepsilon t_\varepsilon^w$ and t_ε^w as its top and bottom configurations of boundary points respectively, and each of whose strands connects one point on the top to one on the bottom:

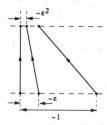

As ε tends to 0, the Kontsevich integral of $N_\varepsilon(w)$ diverges. We have the following asymptotic formula:

$$Z(N_\varepsilon(w)) \sim \prod_{k=0}^{\infty} \varepsilon^{c_k(w)}. \tag{10.20}$$

If t is a two-point configuration, this formula is exact and amounts to a straightforward computation (see Exercise 8.6). In general, if $w = w_1 w_2$ and $n_i = \deg w_i$, we can write $N_\varepsilon(w)$ as a product in the following way:

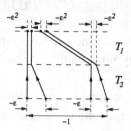

As ε tends to 0, we have

$$Z(T_1) \sim \Delta^{n_1,n_2}\varepsilon^{H/2\pi i} = \varepsilon^{c_0(w_1 w_2)},$$

see Exercise 9.13, and

$$Z(T_2) \sim Z(N_\varepsilon(w_1)) \otimes Z(N_\varepsilon(w_2)).$$

Using induction and the definition of the c_i we arrive at the formula (10.20).

Now, notice that it is sufficient to prove (10.19) in the case when $u = w_1(w_2 w_3)$ and $v = (w_1 w_2)w_3$. Let us draw $\mathbf{1}_{\varepsilon,u,v}$ as a product $T_1 \cdot T_2 \cdot T_3$ as in the picture:

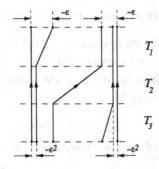

As $\varepsilon \to 0$ we have:

- $Z(T_1) \sim Z(N_\varepsilon(w_1))^{-1} \otimes \mathbf{1}_{n_2+n_3}$;
- $Z(T_2) \sim (\mathbf{1}_{n_1} \otimes c_0(w_2 w_3)) \cdot \Delta^{n_1, n_2, n_3} \Phi_{KZ} \cdot (c_0(w_1 w_2) \otimes \mathbf{1}_{n_3})^{-1}$;
- $Z(T_3) \sim \mathbf{1}_{n_1+n_2} \otimes Z(N_\varepsilon(w_3))$,

where $n_i = \deg w_i$. Notice that these asymptotic expressions for $Z(T_1)$, $Z(T_2)$ and $Z(T_3)$ all commute with each other. Now (10.19) follows from (10.20) and the definition of $\rho^\varepsilon(w)$.

Let us now consider the case when T is a tensor product of one crossing and several vertical strands. In this case $T_{\varepsilon,u,u}$ is an iterated ε-parametrized tensor product, so the proposition of Section 8.4.2 gives $Z(T_{\varepsilon,u,u}) \sim Z_{comb}(T, u, u)$. From the definition of $c_k(u)$ we see that

$$\rho^\varepsilon(u)_{\downarrow^t} \cdot Z_{comb}(T, u, u) = Z_{comb}(T, u, u) \cdot \rho^\varepsilon(u)_{\downarrow^b},$$

and we are done.

Finally, let T be a tensor product of one critical point (say, minimum) and several vertical strands. For example, assume that $T = \mathrm{id}^{\otimes n} \otimes \underset{\longrightarrow}{\min}$, and that $u = v \cdot (xx)$. Now $T_{\varepsilon,u,v} = T'_\varepsilon \cdot N_\varepsilon(v)$ where T'_ε is the iterated ε-parametrized tensor product corresponding to the monomial u, so that, in particular, $T'_1 = T$, and $N_\varepsilon(v)$ is as on page 300. As $\varepsilon \to 0$ we have

$$T'_\varepsilon \sim Z_{comb}(T).$$

As for the regularizing factors,

$$\rho^\varepsilon(u) = \rho^\varepsilon(v(xx)) = \left(\rho^\varepsilon(v) \otimes \mathrm{id}^{\otimes 2} \right) \cdot \left(\prod_k \varepsilon^{c_k(v)} \otimes \varepsilon^{\frac{\smile}{2\pi i}} \right)$$

and we see that they cancel out together with $Z(N_\varepsilon(u))$.

The general case for a tangle of type (3) is entirely similar. $\qquad\square$

10.4 General associators

We have seen that the Kontsevich integral of a knot is assembled from Knizhnik–Zamolodchikov associators and exponentials of a one-chord diagram on two strands. Given that the coefficients of Φ_{KZ} are multiple ζ-values, the following theorem may come as a surprise:

Theorem 10.29. *For any knot (or link) K the coefficients of the Kontsevich integral $Z(K)$, in an arbitrary basis of $\mathscr{A}'$ consisting of chord diagrams, are rational.*

The proof of this important theorem is rather involved, and we shall not give it here. Nevertheless, in this section we sketch very briefly some ideas central to the argument of the proof.

10.4.1 Axioms for associators

One may ask what properties of Φ_{KZ} imply that the combinatorial construction indeed produces the Kontsevich integral for links. Here we shall give a list of such properties.

Consider the algebra $\mathscr{A}(n)$ of tangle chord diagrams on n vertical strands. (Recall that, unlike in $\mathscr{A}^h(n)$, the chords of diagrams in $\mathscr{A}(n)$ need not be horizontal.) There are various homomorphisms between the algebras $\mathscr{A}(n)$, some of which we have already seen. Let us introduce some notation.

Definition 10.30. The operation $\varepsilon_i : \mathscr{A}(n) \to \mathscr{A}(n-1)$ sends a tangle chord diagram D to 0 if at least one chord of D has an endpoint on the ith strand; otherwise $\varepsilon_i(D)$ is obtained from D by removing the ith strand.

Examples 10.31. $\varepsilon_i(|\,|\,|) = |\,|$ for any i, $\varepsilon_1(\mathsf{H}\,|) = \varepsilon_2(\mathsf{H}\,|) = 0$, $\varepsilon_3(\mathsf{H}\,|) = \mathsf{H}$.

The following notation is simply shorthand for $\Delta^{1,...,1,2,1,...,1}$:

Definition 10.32. The operation $\Delta_i : A(n) \to \mathscr{A}(n+1)$ consists in doubling the ith strand of a tangle chord diagram D and taking the sum over all possible lifts of the chord endpoints of D from the ith strand to one of the two new strands.

The symmetric group on 3 letters acts on $\mathscr{A}(3)$ by permuting the strands. The action of σ can be thought of as conjugation

$$D \to \sigma D \sigma^{-1}$$

by a strand-permuting diagram with no chords whose ith point on the bottom is connected with the $\sigma(i)$th point on top. For $D \in \mathscr{A}(3)$ and $\{i, j, k\} = \{1, 2, 3\}$ we shall write D^{ijk} for D conjugated by the permutation that sends $(1\,2\,3)$ to $(i\,j\,k)$.

All the above operations can be extended to the graded completion $\widehat{\mathscr{A}}(n)$ of the algebra $\mathscr{A}(n)$ with respect to the number of chords.

Finally, we write R for $\exp\left(\frac{H}{2}\right) \in \widehat{\mathscr{A}}(2)$ and R^{ij} for $\exp(u_{ij}/2) \in \widehat{\mathscr{A}}(3)$ where u_{ij} has only one chord that connects the strands i and j.

Definition 10.33. An *associator* Φ is an element of the algebra $\widehat{\mathscr{A}}(3)$ satisfying the following axioms:

- (*strong invertibility*)

$$\varepsilon_1(\Phi) = \varepsilon_2(\Phi) = \varepsilon_3(\Phi) = 1$$

(this property, in particular, implies that the series Φ starts with 1 and thus represents an invertible element of the algebra $\widehat{\mathscr{A}}(3)$);

- (*skew symmetry*)

$$\Phi^{-1} = \Phi^{321};$$

- (*pentagon relation*)

$$(\mathrm{id} \otimes \Phi) \cdot (\Delta_2 \Phi) \cdot (\Phi \otimes \mathrm{id}) = (\Delta_3 \Phi) \cdot (\Delta_1 \Phi);$$

- (*hexagon relation*)

$$\Phi^{231} \cdot (\Delta_2 R) \cdot \Phi = R^{13} \cdot \Phi^{213} \cdot R^{12}.$$

A version of the last two relations appears in abstract category theory where they form part of the definition of a monoidal category (MacLane 1988, sec.XI.1).

Theorem 10.34. *The Knizhnik–Zamolodchikov Drinfeld associator Φ_{KZ} satisfies the axioms above.*

Proof The main observation is that the pentagon and the hexagon relations hold for Φ_{KZ} as it can be expressed via the Kontsevich integral. The details of the proof are as follows.

Property 1 immediately follows from the explicit formula for the associator Φ_{KZ} given in Section 10.2.5, which shows that the series starts with 1 and every term appearing with non-zero coefficient has endpoints of chords on each of the three strands.

Property 2. Notice that Φ^{321} is obtained from Φ simply by flipping Φ about a vertical axis. Now, consider the following tangle:

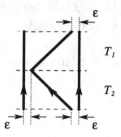

It is isotopic to a tangle whose all strands are vertical, so its Kontsevich integral is equal to 1. As we know from Section 10.1.7, the Kontsevich integral of the two halves of this tangle can be expressed as

$$Z(T_1) = \lim_{\varepsilon \to 0} \varepsilon^{\frac{1}{2\pi i}} \!\uparrow\!\!H\!\cdot \Phi_{\mathrm{KZ}} \cdot \varepsilon^{-\frac{1}{2\pi i}}\!H\!\!\uparrow,$$

and

$$Z(T_2) = \lim_{\varepsilon \to 0} \varepsilon^{\frac{1}{2\pi i}} H\!\!\uparrow \cdot \Phi_{\mathrm{KZ}}^{321} \cdot \varepsilon^{-\frac{1}{2\pi i}}\!\uparrow\!\!H,$$

since T_2 is obtained from T_1 by flipping it about a vertical axis. We see that the regularizing factors cancel out and

$$\Phi_{\mathrm{KZ}} \cdot \Phi_{\mathrm{KZ}}^{321} = 1.$$

Property 3. The pentagon relation for Φ_{KZ} can be represented by the following diagram:

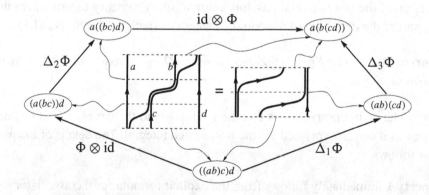

Both sides of this relation are, actually, two expressions for the combinatorial Kontsevich integral of the trivial tangle parenthesized by $x(x(xx))$ at the top and

$((xx)x)x$ at the bottom. On the left-hand side it is written as a product of three trivial tangles with the monomials $x((xx)x)$ and $(x(xx))x$ in the middle. On the right-hand side it is a product of two trivial tangles, the monomial in the middle being $(xx)(xx)$.

Property 4, the hexagon relation, is illustrated by the following diagram:

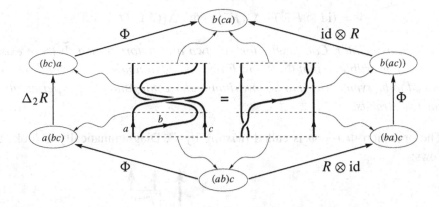

On both sides we have the combinatorial Kontsevich integral of the tangle

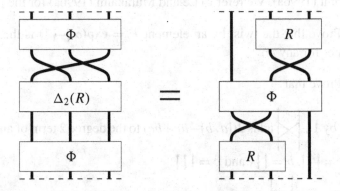

parenthesized at the top as $x(xx)$ and at the bottom as $(xx)x$. On the right-hand side this integral is calculated by decomposing T into a product of two crossings. On the left-hand side we use Theorem 10.26 and the expression for the Kontsevich integral of T_ε from Exercise 9.14. We have

which gives the hexagon relation. ☐

10.4.2 The set of all associators

Interestingly, the axioms do not define the associator uniquely. The following theorem describes the totality of all associators.

Theorem 10.35 (Drinfeld 1989; Le and Murakami 1996a). *Let Φ be an associator and $F \in \widehat{\mathscr{A}}(2)$ an invertible element. Then*

$$\widetilde{\Phi} = (\text{id} \otimes F^{-1}) \cdot \Delta_2(F^{-1}) \cdot \Phi \cdot \Delta_1(F) \cdot (F \otimes \text{id})$$

is also an associator. Conversely, for any two associators Φ and $\widetilde{\Phi}$ there exists $F \in \widehat{\mathscr{A}}(2)$ invertible so that $\Phi, \widetilde{\Phi}$ and F are related as above. Moreover, F can be assumed to be symmetric, that is, invariant under conjugation by the permutation of the two strands.

The operation $\Phi \mapsto \widetilde{\Phi}$ is called *twisting* by F. Diagrammatically, it looks as follows:

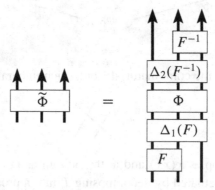

Twisting and the above theorem were discovered by V. Drinfeld (1989) in the context of quasi-triangular quasi-Hopf algebras, and adapted for chord diagrams in Le and Murakami (1996a). We refer to Le and Murakami (1996a) for the proof.

Exercise. Prove that the twist by an element $F = \exp(\alpha \, \text{H}^m)$ is the identity on any associator for any m.

Exercise. Prove that

1. twisting by $1 + \bowtie$ adds $2([a, b] - ac + bc)$ to the degree 2 term of an associator, where $a = \text{H}\,\text{I}, b = \text{I}\,\text{H}$ and $c = \text{H}\,\text{I}$.

2. twisting by $1 + \text{)(}$ does not change the degree 3 term of an associator.

Example 10.36. Let Φ_{BN} be the rational associator described in the next section. It is remarkable that both Φ_{BN} and Φ_{KZ} are horizontal, that is, they belong to the subalgebra $\mathscr{A}^h(3)$ of horizontal diagrams, but can be converted into one another only by a non-horizontal twist. For example, twisting Φ_{BN} by the element

$$F = 1 + \alpha \;\; \boxed{}$$

with an appropriate constant α ensures the coincidence with Φ_{KZ} up to degree 4.

On the other hand, the set of all *horizontal* associators can also be described in terms of the action of the so-called *Grothendieck–Techmüller group(s)*; see Drinfeld (1990) and Bar-Natan (1998).

V. Kurlin (2004) described all *group-like* associators modulo the second commutant.

10.4.3 Rationality of the Kontsevich integral

Let us replace Φ_{KZ} in the combinatorial construction of the Kontsevich integral for parenthesized tangles by an arbitrary associator Φ; denote the result by $Z_\Phi(K)$.

Theorem 10.37 (Le and Murakami 1996a). *For any two associators Φ and $\widetilde{\Phi}$ the corresponding combinatorial integrals coincide for any link K: $Z_\Phi(K) = Z_{\widetilde{\Phi}}(K)$.*

A more precise statement is that for any parenthesized tangle (T, u, v) the integrals $Z_\Phi(T)$ and $Z_{\widetilde{\Phi}}(T)$ are conjugate in the sense that $Z_{\widetilde{\Phi}}(T) = \mathcal{F}_u \cdot Z_\Phi(T) \cdot \mathcal{F}_v^{-1}$, where the elements $\mathcal{F}_u$ and $\mathcal{F}_v$ depend only on u and v respectively. This can be proved in the same spirit as Theorem 10.26 by decomposing a parenthesized tangle into building blocks for which the statement is easy to verify. Then, since a link has an empty boundary, the corresponding combinatorial integrals are equal.

The fact that the Kontsevich integral does not depend on the associator used to compute it implies its rationality. Indeed, V. Drinfeld (1990) (see also Bar-Natan 1998) showed that there exists an associator $\Phi_{\mathbb{Q}}$ with rational coefficients. Therefore, $Z(K) = Z_{comb}(K) = Z_{\Phi_{\mathbb{Q}}}(K)$. The last combinatorial integral has rational coefficients.

We should stress here that the existence of a rational associator $\Phi_{\mathbb{Q}}$ is a highly nontrivial fact, and that computing it is a difficult task. In Bar-Natan (1997a) D. Bar-Natan, following Drinfeld (1990), gave a construction of $\Phi_{\mathbb{Q}}$ by induction on the degree. He implemented the inductive procedure in `Mathematica` (Bar-Natan 1998) and computed the logarithm of the associator up to degree 7.

With the notation $a = $ ⊣↑, $b = $ ↑⊢ his answer, which we denote by Φ_{BN}, is as follows:

$$\log \Phi_{BN} = \frac{1}{48}[ab] - \frac{1}{1440}[a[a[ab]]] - \frac{1}{11520}[a[b[ab]]]$$

$$+ \frac{1}{60480}[a[a[a[a[ab]]]]] + \frac{1}{1451520}[a[a[a[b[ab]]]]]$$

$$+ \frac{13}{1161216}[a[a[b[b[ab]]]]] + \frac{17}{1451520}[a[b[a[a[ab]]]]]$$

$$+ \frac{1}{1451520}[a[b[a[b[ab]]]]]$$

$$-(\text{similar terms with } a \text{ and } b \text{ interchanged}) + \dots.$$

Remark 10.38. This expression is obtained from Φ_{KZ} expanded to degree 7 (see the formula in Figure 10.3) by substitutions $\zeta(3) \to 0$, $\zeta(5) \to 0$, $\zeta(7) \to 0$.

Exercises

10.1 Find the monodromy of the reduced KZ equation (Section 10.1.6) around the points 0, 1 and ∞.

10.2 Using the action of the symmetric group S_n on the configuration space $X = \mathbb{C}^n \setminus \mathcal{H}$ determine the algebra of values and the KZ equation on the quotient space X/S_n in such a way that the monodromy gives the Kontsevich integral of (not necessarily pure) braids. Compute the result for $n = 2$ and compare it with the second exercise on page 225.

10.3 Prove that the associator Φ_{KZ} is group-like.

  Hint. Use the fact the Kontsevich integral is group-like.

10.4 Find $Z_{comb}(3_1)$ up to degree 4 using the parenthesized presentation of the trefoil knot given in Figure 8.1 (page 243).

10.5 Compute the Kontsevich integral of the knot 4_1 up to degree 4, using the parenthesized presentation of the knot 4_1 from Exercise 8.15.

10.6 Prove that the condition $\varepsilon_2(\Phi) = 1$ and the pentagon relation imply the other two equalities for strong invertibility: $\varepsilon_1(\Phi) = 1$ and $\varepsilon_3(\Phi) = 1$.

10.7 Prove the second hexagon relation

$$(\Phi^{312})^{-1} \cdot (\Delta_1 R) \cdot \Phi^{-1} = R^{13} \cdot (\Phi^{132})^{-1} \cdot R^{23}$$

for an arbitrary associator Φ.

10.8 Any associator Φ in the algebra of horizontal diagrams $\mathcal{A}^h(3)$ can be written as a power series in non-commuting variables $a = $ ⊣↑, $b = $ ↑⊢, $c = $ ⊣⊢: $\Phi = \Phi(a, b, c)$.

(a) Check that the skew-symmetry axiom is equivalent to the identity $\Phi^{-1}(a, b, c) = \Phi(b, a, c)$. In particular, for an associator $\Phi(A, B)$ with values in $\mathbb{C}\langle\!\langle A, B \rangle\!\rangle$ (like Φ_{BN}, or Φ_{KZ}), we have $\Phi^{-1}(A, B) = \Phi(B, A)$.

(b) Prove that the hexagon relation from Definition 10.33 can be written in the form

$$\Phi(a, b, c) \exp\left(\frac{b+c}{2}\right)\Phi(c, a, b) = \exp\left(\frac{b}{2}\right)\Phi(c, b, a) \exp\left(\frac{c}{2}\right).$$

(c) (Kurlin 2004) Prove that for a horizontal associator the hexagon relation is equivalent to the relation

$$\Phi(a, b, c)e^{\frac{-a}{2}}\Phi(c, a, b)e^{\frac{-c}{2}}\Phi(b, c, a)e^{\frac{-b}{2}} = e^{\frac{-a-b-c}{2}}.$$

(d) Show that for a horizontal associator Φ,

$$\Phi \cdot \Delta_2(R) \cdot \Phi \cdot \Delta_2(R) \cdot \Phi \cdot \Delta_2(R) = \exp(a + b + c).$$

10.9 Express $Z(H)$ via Φ_{KZ}. Is it true that the resulting power series contains only even degree terms?

10.10 Prove that

$$\Phi = \lim_{\varepsilon \to 0} \varepsilon^{-\frac{w}{2\pi i} \cdot (\text{H}\text{I} + \text{H}\text{H})} \varepsilon^{-\frac{t}{2\pi i} \cdot \text{I}\text{H}} \cdot Z(AT_{b,w}^t) \cdot \varepsilon^{\frac{b}{2\pi i}\text{H}\text{I}} \cdot \varepsilon^{\frac{w}{2\pi i}(\text{H}\text{H} + \text{I}\text{H})},$$

where the tangle $AT_{b,w}^t$ is as in Exercise 8.14.

10.11 Prove that for the tangle $T_{m,b,w}^t$ in the picture on the right

$$\lim_{\varepsilon \to 0} \varepsilon^{-\frac{b-w}{2\pi i}(\text{H}\text{I}\text{I} + \text{H}\text{H}\text{I})} \cdot \varepsilon^{-\frac{t-w}{2\pi i}\text{I}\text{H}\text{I}} \cdot$$
$$\cdot Z(T_{m,b,w}^t) \cdot$$
$$\cdot \varepsilon^{\frac{m-w}{2\pi i}\text{H}\text{I}\text{I}} \cdot \varepsilon^{\frac{b-w}{2\pi i}(\text{H}\text{H}\text{I} + \text{I}\text{H}\text{I})} = \Phi \otimes \text{id}.$$

10.12 Prove that for the tangle $T_{b,w}^{t,m}$ in the picture on the right

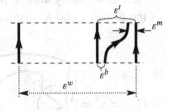

$$\lim_{\varepsilon \to 0} \varepsilon^{-\frac{t-w}{2\pi i}(\text{I}\text{H}\text{I} + \text{I}\text{H}\text{H})} \cdot \varepsilon^{-\frac{m-w}{2\pi i}\text{I}\text{I}\text{H}} \cdot$$
$$\cdot Z(T_{b,w}^{t,m}) \cdot$$
$$\cdot \varepsilon^{\frac{b-w}{2\pi i}\text{I}\text{H}\text{I}} \cdot \varepsilon^{\frac{t-w}{2\pi i}(\text{I}\text{H}\text{H} + \text{I}\text{I}\text{H})} = \text{id} \otimes \Phi.$$

11

The Kontsevich integral: advanced features

11.1 Mutation

The purpose of this section is to prove that the Kontsevich integral commutes with the operation of mutation (this fact was first noticed by T. Le). As an application, we construct a counterexample to the original intersection graph conjecture (Section 4.8.3) and describe, following Chmutov and Lando (2007), all Vassiliev invariants which do not distinguish mutants.

11.1.1 Mutation of knots

Suppose we have a knot K with a distinguished tangle T whose boundary consists of two points at the bottom and two points at the top. If the orientations of the strands of T agree both at the top and the bottom of T, we can cut out the tangle, rotate it by $180°$ around a vertical axis and insert it back. This operation M_T is called *mutation* and the knot $M_T(K)$ thus obtained is called a *mutant* of K.

Here is a widely known pair of mutant knots, 11^n_{34} and 11^n_{42}, which are mirrors of the *Conway* and *Kinoshita–Terasaka* knots respectively:

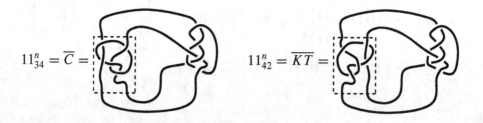

$$11^n_{34} = \overline{C} = \qquad\qquad 11^n_{42} = \overline{KT} =$$

Theorem 11.1 (Morton and Cromwell 1996). *There exists a Vassiliev invariant v of order 11 such that $v(C) \neq v(KT)$.*

Morton and Cromwell manufactured the invariant v using the Lie algebra $\mathfrak{gl}_N$ with a nonstandard representation (or, in other words, the HOMFLY polynomial of certain cablings of the knots).

J. Murakami (2000) showed that any invariant or order at most 10 does not distinguish mutants. So order 11 is the smallest where Vassiliev invariants detect mutants.

11.1.2 Mutation of the Kontsevich integral

Let us describe the behaviour of the Kontsevich integral with respect to knot mutation.

First, recall the definition of a *share* from Section 4.8.4: it is a part of the Wilson loop of a chord diagram, consisting of two arcs, such that each chord has either both or no endpoints on it. A mutation of a chord diagram is an operation of flipping the share with all the chords on it.

In the construction of the Kontsevich integral of a knot K the Wilson loop of the chord diagrams is parametrized by the same circle as K. For each chord diagram participating in $Z(K)$, the mutation of K with respect to a subtangle T gives rise to a flip of two arcs on the Wilson loop.

Theorem 11.2 (Le 1994). *Let $M_T(K)$ be the mutant of a knot K with respect to a subtangle T. Then $Z(K)$ consists only of diagrams for which the part of the Wilson loop that corresponds to T is a share. Moreover, if $M_T(Z(K))$ is obtained from $Z(K)$ by flipping the T-share of each diagram, we have*

$$Z(M_T(K)) = M_T(Z(K)).$$

Proof The proof is a straightforward application of the combinatorial construction of the Kontsevich integral. Write K as a product $K = A \cdot (T \otimes B) \cdot C$ where A, B, C are some tangles. Then the mutation operation consists in replacing T in this expression by its flip T' about a vertical axis.

First, observe that rotating a parenthesized tangle with two points at the top and two points at the bottom by $180°$ about a vertical axis results in the same operation on its combinatorial Kontsevich integral. Moreover, since there is only one choice of parentheses for a product of two factors, the non-associative monomials on the boundary of T are the same as those of T' (all are equal to (xx)). Choose the non-associative monomials for B to be u at the top and v at the bottom. Then

$$Z(K) = Z(A, 1, (xx)u) \cdot \Big(Z(T, (xx), (xx)) \otimes Z(B, u, v) \Big) \cdot Z(C, (xx)v, 1),$$

where we write simply Z for Z_{comb}, and

$$Z(M_T(K)) = Z(A, 1, (xx)u) \cdot \Big(Z(T', (xx), (xx)) \otimes Z(B, u, v) \Big) \cdot Z(C, (xx)v, 1).$$

Both expressions only involve diagrams for which the part of the Wilson loop that corresponds to T is a share; they differ exactly by the mutation of all the T-shares of the diagrams. □

11.1.3 Counterexample to the intersection graph conjecture

It is easy to see that the mutation of chord diagrams does not change the intersection graph. Thus, if the intersection graph conjecture (see Section 4.8.3) were true, the Kontsevich integrals of mutant knots would coincide, and all Vassiliev invariants would take the same value on mutant knots. But this contradicts Theorem 11.1 of Cromwell and Morton.

11.1.4 Vassiliev invariants that do not distinguish mutants

Now we can prove Theorem 4.25:

Theorem 11.3 (Chmutov and Lando 2007). *The symbol of a Vassiliev invariant that does not distinguish mutant knots depends on the intersection graph only.*

Proof The idea of the proof can be summarized in one sentence: a mutation of a chord diagram is always induced by a mutation of a singular knot.

Let D_1 and D_2 be chord diagrams of degree n with the same intersection graph. We must prove that if a Vassiliev knot invariant v, of order at most n, does not distinguish mutants, then the symbol of v takes the same value on D_1 and D_2.

According to Theorem 4.28, D_2 can be obtained from D_1 by a sequence of mutations. It is sufficient to consider the case when D_1 and D_2 differ by a single mutation in a share S.

Let K_1 be a singular knot with n double points whose chord diagram is D_1. The share S corresponds to two arcs on K_1; the double points on these two arcs correspond to the chords with endpoints on S. Now, shrinking and deforming the two arcs, if necessary, we can find a ball in $\mathbb{R}^3$ whose intersection with K_1 consists of these two arcs and a finite number of other arcs. These other arcs can be pushed out of the ball, though not necessarily by an isotopy, that is, passing through self-intersections. The result is a new singular knot K_1' with the same chord diagram D_1, whose double points corresponding to S are collected in a tangle T_S. Performing an appropriate rotation of T_S we obtain a singular knot K_2 with the chord diagram D_2. Since v does not distinguish mutants, its values on K_1 and K_2 are equal. The theorem is proved. □

To illustrate the proof, consider the chord diagram D_1 below. Pick a singular knot K_1 representing D_1.

$$D_1 = \qquad K_1 = $$

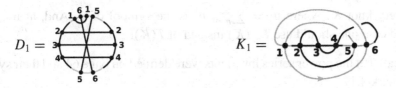

By deforming K_1 we achieve that the two arcs of the share form a tangle (placed on its side in the pictures below), and then push all other segments of the knot out of this subtangle:

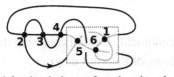 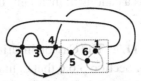

deforming the knot to form the subtangle   pushing out other segments

Combining the last theorem with Theorem 11.2 we get the following corollary.

Corollary 11.4. *Let w be a weight system on chord diagrams with n chords. Consider a Vassiliev invariant $v(K) := w \circ I(K)$. Then v does not distinguish mutants if and only if the weight system w depends only on the intersection graph.*

11.2 Canonical Vassiliev invariants

11.2.1 The definition of canonical Vassiliev invariants

The theorem on the universality of the Kontsevich integral in Section 8.8.1 and its framed version in Section 9.4 provide a means to recover a Vassiliev invariant of order $\leqslant n$ from its symbol, up to invariants of smaller order. It is natural to consider those remarkable Vassiliev invariants whose recovery gives precisely the original invariant.

Definition 11.5 (Bar-Natan and Garoufalidis 1996). A (framed) Vassiliev invariant v of order $\leqslant n$ is called *canonical* if for every (framed) knot K,

$$v(K) = \mathrm{symb}(v) \circ I(K).$$

In the case of framed invariants one should write $I^{fr}(K)$ instead of $I(K)$.

A power series invariant $f = \sum_{n=0}^{\infty} f_n h^n$, with f_n of order $\leqslant n$, is called *canonical* if

$$f(K) = \sum_{n=0}^{\infty} \bigl(w_n(I(K))\bigr)h^n$$

for every knot K, where $w = \sum_{n=0}^{\infty} w_n$ is the symbol of f. And, again, in the framed case one should use $I^{fr}(K)$ instead of $I(K)$.

Recall that the power series invariants were defined on page 64 and their symbols in Remark 4.17.

Canonical invariants define a grading in the filtered space of Vassiliev invariants which is consistent with the filtration.

Example 11.6. The trivial invariant of order 0 which is identically equal to 1 on all knots is a canonical invariant. Its weight system is equal to $\mathbf{I}_0$ in the notation of Section 4.5.

Example 11.7. The Casson invariant c_2 is canonical. This follows from the explicit formula in Section 3.6.6 that defines it in terms of the knot diagram.

Exercise 11.8. Prove that the invariant j_3 (see page 70) is canonical.

Surprisingly many of the classical knot invariants discussed in Chapters 2 and 3 turn out to be canonical.

The notion of a canonical invariant allows one to reduce various relations between Vassiliev knot invariants to some combinatorial relations between their symbols, which gives a powerful tool to study knot invariants. This approach will be used in Section 14.1 to prove the Melvin–Morton conjecture. Now we shall give examples of canonical invariants following Bar-Natan and Garoufalidis (1996).

11.2.2 Quantum invariants

Building on the work of Drinfeld (1989, 1990) and Kohno (1987), T. Le and J. Murakami (1995b; theorem 10), and C. Kassel (1995; theorem XX.8.3) (see also Ohtsuki 2002; theorem 6.14) proved that the quantum knot invariants $\theta^{fr}(K)$ and $\theta(K)$ introduced in Section 2.6 become canonical series after substitution $q = e^h$ and expansion into a power series in h.

The initial data for these invariants are a semi-simple Lie algebra $\mathfrak{g}$ and its finite-dimensional irreducible representation V_λ, where λ is its highest weight (see Humphreys 1980). To emphasize this data, we shall write $\theta_{\mathfrak{g}}^{V_\lambda}(K)$ for $\theta(K)$ and $\theta_{\mathfrak{g}}^{fr,V_\lambda}(K)$ for $\theta^{fr}(K)$.

The quadratic Casimir element c (see Section 6.1) acts on V_λ as multiplication by a constant, call it c_λ. The relation between the framed and unframed quantum invariants is

$$\theta_{\mathfrak{g}}^{fr,V_\lambda}(K) = q^{\frac{c_\lambda \cdot w(K)}{2}} \theta_{\mathfrak{g}}^{V_\lambda}(K),$$

where $w(K)$ is the writhe of K.

Set $q = e^h$. Write $\theta_{\mathfrak{g}}^{fr,V_\lambda}$ and $\theta_{\mathfrak{g}}^{V_\lambda}$ as power series in h:

$$\theta_{\mathfrak{g}}^{fr,V_\lambda} = \sum_{n=0}^{\infty} \theta_{\mathfrak{g},n}^{fr,\lambda} h^n \qquad \theta_{\mathfrak{g}}^{V_\lambda} = \sum_{n=0}^{\infty} \theta_{\mathfrak{g},n}^{\lambda} h^n.$$

According to Theorem 3.27, the coefficients $\theta_{\mathfrak{g},n}^{fr,\lambda}$ and $\theta_{\mathfrak{g},n}^{\lambda}$ are Vassiliev invariants of order n. The Le–Murakami–Kassel theorem states that they both are canonical series.

It is important that the symbol of $\theta_{\mathfrak{g}}^{fr,V_\lambda}$ is precisely the weight system $\varphi_{\mathfrak{g}}^{V_\lambda}$ described in Chapter 6. The symbol of $\theta_{\mathfrak{g}}^{V_\lambda}$ equals $\varphi_{\mathfrak{g}}^{\prime V_\lambda}$. In other words, it is obtained from $\varphi_{\mathfrak{g}}^{V_\lambda}$ by the deframing procedure of Section 4.5.4. Hence, knowing the Kontsevich integral allows us to restore the quantum invariants $\theta_{\mathfrak{g}}^{fr,V_\lambda}$ and $\theta_{\mathfrak{g}}^{V_\lambda}$ from these weight systems without the quantum procedure of Section 2.6.

11.2.3 Coloured Jones polynomial

The *coloured Jones polynomials* $J^k := \theta_{\mathfrak{sl}_2}^{V_\lambda}$ and $J^{fr,k} := \theta_{\mathfrak{sl}_2}^{fr,V_\lambda}$ are particular cases of quantum invariants for $\mathfrak{g} = \mathfrak{sl}_2$. For this Lie algebra, the highest weight is an integer $\lambda = k - 1$, where k is the dimension of the representation, so in our notation we may use k instead of λ. The quadratic Casimir number in this case is $c_\lambda = \frac{k^2-1}{2}$, and the relation between the framed and unframed coloured Jones polynomials is

$$J^{fr,k}(K) = q^{\frac{k^2-1}{4} \cdot w(K)} J^k(K).$$

The ordinary Jones polynomial of Section 2.4 corresponds to the case $k = 2$, that is, to the standard two-dimensional representation of the Lie algebra $\mathfrak{sl}_2$.

Set $q = e^h$. Write $J^{fr,k}$ and J^k as power series in h:

$$J^{fr,k} = \sum_{n=0}^{\infty} J_n^{fr,k} h^n \qquad J^k = \sum_{n=0}^{\infty} J_n^k h^n.$$

Both series are canonical with the symbols

$$\mathrm{symb}(J^{fr,k}) = \varphi_{\mathfrak{sl}_2}^{V_k}, \qquad \mathrm{symb}(J^k) = \varphi_{\mathfrak{sl}_2}^{\prime V_k}$$

defined in Sections 6.1.3 and 6.2.3.

11.2.4 Alexander–Conway polynomial

Consider the unframed quantum invariant $\theta_{\mathfrak{sl}_N}^{St}$ as a function of the parameter N. Let us think of N not as a discrete parameter but rather as a continuous variable, where for non-integer N the invariant $\theta^{St}_{\mathfrak{sl}_N}$ is defined by the skein and initial relations above. Its symbol $\varphi'^{St}_{\mathfrak{sl}_N} = \varphi'^{St}_{\mathfrak{gl}_N}$ (see Exercise 6.12) also makes sense for all real values of N, because for every chord diagram D, $\varphi'^{St}_{\mathfrak{gl}_N}(D)$ is a polynomial of N. Even more, since this polynomial is divisible by N, we may consider the limit

$$\lim_{N \to 0} \frac{\varphi'^{St}_{\mathfrak{sl}_N}}{N}.$$

Exercise. Prove that the weight system defined by this limit coincides with the symbol of the Conway polynomial, $\mathrm{symb}(C) = \sum_{n=0}^{\infty} \mathrm{symb}(c_n)$.
  Hint. Use Exercise 3.16.

Make the substitution $\theta^{St}_{\mathfrak{sl}_N}\big|_{q=e^h}$. The skein and initial relations for $\theta^{St}_{\mathfrak{sl}_N}$ allow us to show (see Exercise 11.5 to this chapter) that the limit

$$A := \lim_{N \to 0} \frac{\theta^{St}_{\mathfrak{sl}_N}\big|_{q=e^h}}{N}$$

does exist and satisfies the relations

$$A\left(\!\begin{array}{c}\includegraphics\end{array}\!\right) - A\left(\!\begin{array}{c}\includegraphics\end{array}\!\right) = (e^{h/2} - e^{-h/2})A\left(\!\begin{array}{c}\includegraphics\end{array}\!\right); \qquad (11.1)$$

$$A\left(\!\bigcirc\!\right) = \frac{h}{e^{h/2} - e^{-h/2}}. \qquad (11.2)$$

A comparison of these relations with the defining relation for the Conway polynomial in Section 2.3 shows that

$$A = \frac{h}{e^{h/2} - e^{-h/2}} C\big|_{t=e^{h/2}-e^{-h/2}}.$$

Despite the fact that the Conway polynomial C itself is not a canonical series, it becomes canonical after the substitution $t = e^{h/2} - e^{-h/2}$ and multiplication by $\frac{h}{e^{h/2}-e^{-h/2}}$. The weight system of this canonical series is the same as for the Conway polynomial. Or, in other words,

$$\frac{h}{e^{h/2} - e^{-h/2}} C\big|_{t=e^{h/2}-e^{-h/2}}(K) = \sum_{n=0}^{\infty} (\mathrm{symb}(c_n) \circ I(K)) h^n.$$

Remark 11.9. We cannot do the same for framed invariants because none of the limits

$$\lim_{N \to 0} \frac{\theta_{\mathfrak{sl}_N}^{fr,St}\big|_{q=e^h}}{N}, \qquad \lim_{N \to 0} \frac{\varphi_{\mathfrak{sl}_N}^{St}}{N}$$

exists.

11.3 Wheeling

We mentioned in Section 5.8 that the relation between the algebras $\mathscr{B}$ and $\mathscr{C}$ is similar to the relation between the invariants in the symmetric algebra of a Lie algebra and the centre of its universal enveloping algebra. One may then expect that there exists an algebra isomorphism between $\mathscr{B}$ and $\mathscr{C}$ similar to the Duflo isomorphism for Lie algebras (see Section 11.3.3).

This isomorphism indeed exists. It is called *wheeling* and we describe it in this section. It will be used in the next section to deduce an explicit formula for the Kontsevich integral of the unknot.

11.3.1 Diagrammatic differential operators and the wheeling map

For an open diagram C with n legs, let us define the *diagrammatic differential operator*

$$\partial_C : \mathscr{B} \to \mathscr{B}.$$

Take an open diagram D. If D has at most n legs, set $\partial_C D = 0$. If D has more than n legs, we define $\partial_C(D) \in \mathscr{B}$ as the sum of all those ways of glueing all the legs of C to some legs of D that produce diagrams having at least one leg on each connected component. For example, if w_2 stands for the diagram , we have

$$\partial_{w_2}(\,\text{}\,) = 8\,\text{} + 4\,\text{} = 10\,\text{}.$$

Also,

$$\partial_{w_2}(\,\text{}\,) = 8\,\text{},$$

since the other four ways of glueing w_2 into produce diagrams one of whose components has no legs.

Extending the definition by linearity, we can replace the diagram C in the definition of ∂_C by any linear combination of diagrams. Moreover, C can be taken to be a formal power series in diagrams, *with respect to the grading by the number of legs*. Indeed, for any given diagram D almost all terms in such formal power series would have at least as many legs as D.

Recall that the *wheel* w_n in the algebra $\mathscr{B}$ is the diagram

$$w_n = \qquad$$

n spokes

The wheels w_n with n odd are equal to zero; this follows directly from Lemma 5.28 in Section 5.6.2.

Definition 11.10. The *wheels element* Ω is the formal power series

$$\Omega = \exp \sum_{n=1}^{\infty} b_{2n} w_{2n}$$

where b_{2n} are the *modified Bernoulli numbers*, and the products are understood to be in the algebra $\mathscr{B}$.

The modified Bernoulli numbers b_{2n} are the coefficients at x^{2n} in the Taylor expansion of the function

$$\frac{1}{2} \ln \frac{\sinh x/2}{x/2}.$$

We have $b_2 = 1/48$, $b_4 = -1/5760$ and $b_6 = 1/362880$. In general,

$$b_{2n} = \frac{B_{2n}}{4n \cdot (2n)!},$$

where B_{2n} are the usual Bernoulli numbers.

Definition 11.11. The *wheeling map* is the map

$$\partial_{\Omega} = \exp \sum_{n=1}^{\infty} b_{2n} \partial_{w_{2n}}.$$

The wheeling map is a degree-preserving linear map $\mathscr{B} \to \mathscr{B}$. It is, clearly, a vector space isomorphism since $\partial_{\Omega^{-1}}$ is an inverse for it.

Theorem 11.12 (Wheeling Theorem). *The map* $\chi \circ \partial_{\Omega} : \mathscr{B} \to \mathscr{C}$ *is an algebra isomorphism.*

There are several approaches to the proof of the above theorem. It has been noted by Kontsevich (2003) that the Duflo–Kirillov isomorphism holds for a Lie algebra in any rigid tensor category; Hinich and Vaintrob showed (2002) that the wheeling map can be interpreted as a particular case of such a situation. Here, we shall follow the proof of Bar-Natan *et al.* (2003).

Example 11.13. At the beginning of Section 5.8 we saw that χ is not compatible with the multiplication. Let us check the multiplicativity of $\chi \circ \partial_\Omega$ on the same example:

$$\chi \circ \partial_\Omega(\ \rightleftharpoons\ ) \;=\; \chi \circ (1 + b_2 \partial_{w_2})(\ \rightleftharpoons\ )$$

$$=\; \chi\left(\ \rightleftharpoons\ +\ \tfrac{1}{48}\cdot 8\cdot\ \multimap\!\!\bigcirc\!\!\multimap\ \right) \;=\; \chi\left(\ \rightleftharpoons\ +\ \tfrac{1}{6}\cdot\ \multimap\!\!\bigcirc\!\!\multimap\ \right)$$

$$=\; \tfrac{1}{3}\ \bigotimes\ +\ \tfrac{2}{3}\ \bigcirc\!\!\!\mid\ +\ \tfrac{1}{6}\ \ominus\!\!\!\bigcirc\ \;=\;\ \bigcirc\!\!\!\mid\ ,$$

which is the square of the element $\chi \circ \partial_\Omega(\ \bullet\!\!-\!\!\bullet\ ) = \ominus$ in the algebra $\mathscr{C}$.

11.3.2 The algebra $\mathscr{B}^\circ$

For what follows it will be convenient to enlarge the algebras $\mathscr{B}$ and $\mathscr{C}$ by allowing diagrams with components that have no legs.

A diagram in the enlarged algebra $\mathscr{B}^\circ$ is a union of a unitrivalent graph with a finite number of circles with no vertices on them; a cyclic order of half-edges at every trivalent vertex is given. The algebra $\mathscr{B}^\circ$ is spanned by all such diagrams modulo IHX and antisymmetry relations. The multiplication in $\mathscr{B}^\circ$ is the disjoint union. The algebra $\mathscr{B}$ is the subalgebra of $\mathscr{B}^\circ$ spanned by graphs which have at least one univalent vertex in each connected component. Killing all diagrams which have components with no legs, we get a homomorphism $\mathscr{B}^\circ \to \mathscr{B}$, which restricts to the identity map on $\mathscr{B} \subset \mathscr{B}^\circ$.

The algebra of 3-graphs Γ from Chapter 7 is also a subspace of $\mathscr{B}^\circ$. In fact, the algebra $\mathscr{B}^\circ$ is the tensor product of $\mathscr{B}$ and the symmetric algebra $S(\Gamma)$ of the vector space Γ and the polynomial algebra in one variable (which counts the circles with no vertices on them).

The reason to consider $\mathscr{B}^\circ$ instead of $\mathscr{B}$ can be roughly explained as follows. One of our main tools for the study of $\mathscr{B}$ is the universal weight system

$$\rho_{\mathfrak{g}} : \mathscr{B} \to S(\mathfrak{g})$$

with the values in the symmetric algebra of a Lie algebra $\mathfrak{g}$. In fact, much of our intuition about $\mathscr{B}$ comes from Lie algebras, since $\rho_{\mathfrak{g}}$ respects some basic constructions. For instance, glueing together the legs of two diagrams corresponds to a contraction of the corresponding tensors. However, there is a very simple operation in $S(\mathfrak{g})$ that cannot be lifted to $\mathscr{B}$ via $\rho_{\mathfrak{g}}$. Namely, there is a pairing $S^n(\mathfrak{g}) \otimes S^n(\mathfrak{g}) \to \mathbb{C}$ which extends the invariant form $\mathfrak{g} \otimes \mathfrak{g} \to \mathbb{C}$. Roughly, if the elements of $S(\mathfrak{g})$ are thought of as symmetric tensors (see Section A.1.5), this pairing consists in taking the sum of all possible contractions of two tensors of the same rank. This operation is essential if one works with differential operators on $S(\mathfrak{g})$, and it cannot be lifted to $\mathscr{B}$ since glueing two diagrams with the same number of legs produces a diagram with no univalent vertices. The introduction of $\mathscr{B}^\circ$ remedies this problem.

Indeed, the map ρ_g naturally extends to $\mathscr{B}^\circ$. On a connected diagram with no legs it coincides with the $\mathbb{C}$-valued weight system $\varphi_\mathfrak{g}$ for 3-graphs described in Section 7.5; in particular, it takes value $\dim \mathfrak{g}$ on a circle with no vertices. Finally, ρ_g is multiplicative with respect to the disjoint union of diagrams.

There is a bilinear symmetric pairing on $\mathscr{B}^\circ$ whose image lies in the subspace spanned by legless diagrams.

Definition 11.14. For two diagrams $C, D \in \mathscr{B}^\circ$ with the same number of legs, we define $\langle C, D \rangle$ to be the sum of all ways of glueing all legs of C to those of D. If the numbers of legs of C and D do not coincide, we set $\langle C, D \rangle = 0$.

Now, if C and D are two diagrams with the same number of legs, $\rho_\mathfrak{g}(\langle C, D \rangle)$ is the sum of all possible contractions of $\rho_\mathfrak{g}(C)$ and $\rho_\mathfrak{g}(D)$ considered as symmetric tensors.

Definition 11.15. Let C be an open diagram. The *diagrammatic differential operator*

$$\partial^\circ_C : \mathscr{B}^\circ \to \mathscr{B}^\circ$$

sends $D \in \mathscr{B}^\circ$ to the sum of all ways of glueing the legs of C to those of D, if D has at least as many legs as C; if D has less legs than C, then $\partial^\circ_C(D) = 0$.

For example,

$$\partial^\circ_{w_2}(\text{⊟}) = 8\,\text{⊸O⊸} + 4\,\text{⊖}.$$

This definition of diagrammatic operators is consistent with the definition of diagrammatic operators in $\mathscr{B}$. Namely, if C is a diagram in $\mathscr{B}$ and $p : \mathscr{B}^\circ \to \mathscr{B}$ is the projection, we have

$$\partial_C \circ p = p \circ \partial^\circ_C.$$

Note that while ∂_C and ∂°_C are compatible with the projection p, they are not compatible with the inclusion $\mathscr{B} \to \mathscr{B}^\circ$.

Similar to the algebra $\mathscr{B}^\circ$ one defines the algebra $\mathscr{C}^\circ$ by considering not necessarily connected trivalent graphs in the definition of $\mathscr{C}$. The vector space isomorphism $\chi : \mathscr{B} \to \mathscr{C}$ extends to an isomorphism $\mathscr{C}^\circ \simeq \mathscr{B}^\circ$ whose definition literally coincides with that of χ (and which we also denote by χ). In particular, for a legless diagram in $\mathscr{B}^\circ$ the map χ consists in simply adding the Wilson loop.

Our method of proving the Wheeling theorem will be to prove it for the algebras $\mathscr{B}^\circ$ and $\mathscr{C}^\circ$, with the diagrammatic operator $\partial^\circ_\Omega : \mathscr{B}^\circ \to \mathscr{B}^\circ$. Then the version for $\mathscr{B}$ and $\mathscr{C}$ will follow immediately by applying the projection map. First, however, let us explain the connection of the Wheeling theorem with the Duflo isomorphism for Lie algebras.

11.3.3 The Duflo isomorphism

The wheeling map is a diagrammatic analogue of the *Duflo–Kirillov map* for metrized Lie algebras.

Recall that for a Lie algebra $\mathfrak{g}$ the Poincaré–Birkhoff–Witt isomorphism

$$S(\mathfrak{g}) \simeq U(\mathfrak{g})$$

sends a commutative monomial in n variables to the average of all possible *noncommutative* monomials in the same variables, see page 462. It is not an algebra isomorphism, of course, since $S(\mathfrak{g})$ is commutative and $U(\mathfrak{g})$ is not (unless $\mathfrak{g}$ is abelian); however, it is an isomorphism of $\mathfrak{g}$-modules. In particular, we have an isomorphism of vector spaces

$$S(\mathfrak{g})^\mathfrak{g} \simeq U(\mathfrak{g})^\mathfrak{g} = Z(U(\mathfrak{g}))$$

between the subalgebra of invariants in the symmetric algebra and the centre of the universal enveloping algebra. This map does not respect the product either, but it turns out that $S(\mathfrak{g})^\mathfrak{g}$ and $Z(U(\mathfrak{g}))$ are actually isomorphic as commutative algebras. The isomorphism between them, known as the *Duflo isomorphism*, is given by the *Duflo–Kirillov map*, which is described in the Appendix, see page 462.

Lemma 11.16. *The wheeling map* $\partial^\circ_\Omega : \mathscr{B}^\circ \to \mathscr{B}^\circ$ *is taken by the universal Lie algebra weight system* $\rho_\mathfrak{g}$ *to the Duflo–Kirillov map:*

$$\sqrt{j} \circ \rho_\mathfrak{g} = \rho_\mathfrak{g} \circ \partial^\circ_\Omega.$$

Proof Observe that a diagrammatic operator $\partial^\circ_C : \mathscr{B}^\circ \to \mathscr{B}^\circ$ is taken by $\rho_\mathfrak{g}$ to the corresponding differential operator $\partial^\circ_{\rho_\mathfrak{g}(C)} : S(\mathfrak{g}) \to S(\mathfrak{g})$ in the sense that

$$\rho_\mathfrak{g} \circ \partial^\circ_C = \partial_{\rho_\mathfrak{g}(C)} \circ \rho_\mathfrak{g}.$$

This simply reflects the fact that glueing the legs of two diagrams corresponds to a contraction of the corresponding tensors.

In Example 6.30 we have calculated the value of $\rho_{\mathfrak{g}}$ on the wheel w_k:

$$\rho_{\mathfrak{g}}(w_k) = \sum_{i_1,\dots,i_k} \mathrm{Tr}\,(\mathrm{ad}\,e_{i_1}\dots\mathrm{ad}\,e_{i_k})\cdot e_{i_1}\dots e_{i_k},$$

where $\{e_i\}$ is a basis for $\mathfrak{g}$. In order to interpret this expression as an element of $S^k(\mathfrak{g}^*)$, we must contract it with k copies of $x \in \mathfrak{g}$. The resulting homogeneous polynomial of degree k on $\mathfrak{g}$ sends $x \in \mathfrak{g}$ to $\mathrm{Tr}\,(\mathrm{ad}\,x)^k$.

Since $\rho_{\mathfrak{g}}$ is multiplicative, it carries the wheeling map ∂_Ω° to

$$\exp\left(\sum_n b_{2n}\,\mathrm{Tr}\,(\mathrm{ad}\,x)^{2n}\right) = \exp\mathrm{Tr}\left(\sum_n b_{2n}(\mathrm{ad}\,x)^{2n}\right)$$

$$= \det\exp\left(\frac{1}{2}\ln\frac{\sinh\frac{1}{2}\mathrm{ad}\,x}{\frac{1}{2}\mathrm{ad}\,x}\right) = \sqrt{j},$$

that is, to the Duflo–Kirillov map.   □

The Duflo isomorphism is a rather mysterious fact. Remarkably, more than one of its proofs involve diagrammatic techniques: apart from being a consequence of the Wheeling theorem, it follows from Kontsevich's work (2003) on deformation quantization. In fact, in Kontsevich (2003) the Duflo isomorphism is generalized to a sequence of isomorphisms between $H^i(\mathfrak{g}, S(\mathfrak{g}))$ and $H^i(\mathfrak{g}, U(\mathfrak{g}))$ with the usual Duflo isomorphism being the case $i = 0$.

11.3.4 Pairings on diagram spaces and cabling operations

Everything we said about the algebras $\mathscr{B}^\circ$ and $\mathscr{C}^\circ$ can be generalized for the case of tangles with several components. In particular, there is a bilinear pairing

$$\mathscr{C}^\circ(x \mid y) \otimes \mathscr{B}^\circ(y) \to \mathscr{C}^\circ(x).$$

For diagrams $C \in \mathscr{C}^\circ(x \mid y)$ and $D \in \mathscr{B}^\circ(y)$ define the diagram $\langle C, D\rangle_y \in \mathscr{C}^\circ(x)$ as the sum of all ways of glueing all the y-legs of C to the y-legs of D. If the numbers of y-legs of C and D are not equal, we set $\langle C, D\rangle_y$ to be zero. This is a version of the inner product for diagram spaces defined in Section 5.10.2; Lemma 5.47 shows that it is well-defined. In what follows, we shall indicate by a subscript the component along which the inner product is taken.

The inner product can be used to express the diagrammatic differential operators in $\mathscr{B}^\circ$ via the disconnected cabling operations. These are defined for $\mathscr{B}^\circ$ in the same way as for $\mathscr{B}$; for instance, $\psi^{2\cdot 1}(D)$ is the sum of all diagrams obtained from D by replacing the label (say, y) on its univalent vertices by one of the two labels y_1

or y_2. If $D \in \mathscr{B}(y)$, the labels y_1, y_2 are obtained by doubling y and the diagram C is considered as an element of $\mathscr{B}(y_1)$, we have

$$\partial_C^\circ(D) = \langle C, \psi^{2 \cdot 1}(D) \rangle_{y_1}.$$

The proof consists in simply comparing the diagrams on both sides.

The cabling operation $\psi^{2 \cdot 1}$ can be thought of as being dual to the disjoint union with respect to the inner product:

$$\langle C, D_1 \cup D_2 \rangle = \langle \psi^{2 \cdot 1}(C), D_1 \otimes D_2 \rangle_{y_1, y_2}, \tag{11.3}$$

where on the right-hand side y_1 and y_2 are the two labels for the legs of $\psi^{2 \cdot 1}(C)$, and $D_1 \otimes D_2$ belongs to $\mathscr{B}^\circ(y_1) \otimes \mathscr{B}^\circ(y_2) \subset \mathscr{B}^\circ(y_1, y_2)$. The proof of this last formula is also by inspection of both sides.

11.3.5 The Hopf link and the map Φ_0

In what follows we shall often write # for the connected sum and $\cup$ — for the disjoint union product, in order to avoid confusion.

Consider the framed Hopf link Φ with one interval component labelled x, one closed component labelled y, zero framing, and orientations as indicated:

The framed Kontsevich integral $I^{fr}(\Phi)$ lives in $\mathscr{C}(x, y)$ or, via the isomorphism

$$\chi_y^{-1} : \mathscr{C}(x, y) \to \mathscr{C}(x \mid y),$$

in $\mathscr{C}(x \mid y)$.

Let us write $\mathscr{Z}(\Phi)$ for the image of $I^{fr}(\Phi)$ in $\mathscr{C}(x \mid y)$. For any diagram $D \in \mathscr{B}^\circ(y)$, the pairing $\langle \mathscr{Z}(\Phi), D \rangle_y$ is well-defined and lives in $\mathscr{C}^\circ(x)$. Identifying $\mathscr{B}^\circ(y)$ with $\mathscr{B}^\circ$ and $\mathscr{C}^\circ(x)$ with $\mathscr{C}^\circ$, we obtain a map

$$\Phi : \mathscr{B}^\circ \to \mathscr{C}^\circ$$

defined by

$$D \to \langle \mathscr{Z}(\Phi), D \rangle_y.$$

Lemma 11.17. *The map $\Phi : \mathscr{B}^\circ \to \mathscr{C}^\circ$ is a homomorphism of algebras.*

Proof Taking the disconnected cabling of the Hopf link ϕ along the component y, we obtain a link ϕ with one interval component labelled x and two closed parallel components labelled y_1 and y_2:

In the same spirit as Φ, we define the map

$$\Phi_2 : \mathscr{B}^\circ \otimes \mathscr{B}^\circ \to \mathscr{C}^\circ$$

using the link ϕ instead of ϕ. Namely, given two diagrams, $D_1 \in \mathscr{B}^\circ(y_1)$ and $D_2 \in \mathscr{B}^\circ(y_2)$ we have

$$D_1 \otimes D_2 \in \mathscr{B}^\circ(y_1, y_2).$$

Write $\mathscr{Z}(\phi)$ for the image of the Kontsevich integral $I^{fr}(\phi)$ under the map

$$\chi_{y_1, y_2}^{-1} : \mathscr{C}(x, y_1, y_2) \to \mathscr{C}(x \mid y_1, y_2).$$

Identify $\mathscr{B}^\circ(y_1, y_2)$ with $\mathscr{B}^\circ \otimes \mathscr{B}^\circ$ and $\mathscr{C}^\circ(x)$ with $\mathscr{C}^\circ$, and define Φ_2 as

$$D_1 \otimes D_2 \xrightarrow{\Phi_2} \langle \mathscr{Z}(\phi), D_1 \otimes D_2 \rangle_{y_1, y_2}.$$

The map Φ_2 glues the legs of the diagram D_1 to the y_1-legs of $\mathscr{Z}(\phi)$, and the legs of D_2-to the y_2-legs of $\mathscr{Z}(\phi)$.

There are two ways of expressing $\Phi_2(D_1 \otimes D_2)$ in terms of $\Phi(D_i)$. First, we can use the fact that ϕ is a product (as tangles) of two copies of the Hopf link ϕ. Since the legs of D_1 and D_2 are glued independently to the legs corresponding to y_1 and y_2, it follows that

$$\Phi_2(D_1 \otimes D_2) = \Phi(D_1)\#\Phi(D_2).$$

On the other hand, we can apply the formula in Equation (11.3) that relates the disjoint union multiplication with disconnected cabling. We have

$$\Phi_2(D_1 \otimes D_2) = \langle \psi_y^{2\cdot 1}(\mathscr{Z}(\Phi)), D_1 \otimes D_2 \rangle_{y_1, y_2}$$
$$= \langle \mathscr{Z}(\Phi), D_1 \cup D_2 \rangle_y$$
$$= \Phi(D_1 \cup D_2),$$

and, therefore,

$$\Phi(D_1)\#\Phi(D_2) = \Phi(D_1 \cup D_2).$$

$\square$

Given a diagram $D \in \mathscr{C}(x \mid y)$, the map $\mathscr{B}^\circ \to \mathscr{C}^\circ$ given by sending $C \in \mathscr{B}^\circ(y)$ to $\langle D, C \rangle_y \in \mathscr{C}^\circ(x)$ shifts the degree of C by the amount equal to the degree of D minus the number of y-legs of D. If D appears in $\mathscr{Z}(\Phi)$ with a non-zero coefficient, this difference is non-negative. Indeed, the diagrams participating in $\mathscr{Z}(\Phi)$ contain no struts (interval components) both of whose ends are labelled with y, since the y component of Φ comes with zero framing (see Exercise 9.3). Also, if two y-legs are attached to the same internal vertex, the diagram is zero, because of the antisymmetry relation, and therefore, the number of inner vertices of D is at least as big as the number of y-legs.

It follows that the Kontsevich integral $\mathscr{Z}(\Phi)$ can be written as $\mathscr{Z}_0(\Phi) + \mathscr{Z}_1(\Phi) + \ldots$, where $\mathscr{Z}_i(\Phi)$ is the part consisting of diagrams whose degree exceeds the number of y legs by i. We shall be interested in the term $\mathscr{Z}_0(\Phi)$ of this sum.

Each diagram that appears in this term is a union of a *comb* with some wheels:

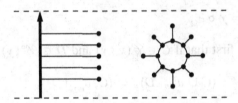

Indeed, each vertex of such diagram is either a y-leg, or is adjacent to exactly one y-leg.

Denote a comb with n teeth by u^n. Strictly speaking, u^n is not really a product of n copies of u since $\mathscr{C}(x \mid y)$ is not an algebra. However, we can introduce a Hopf algebra structure in the space of all diagrams in $\mathscr{C}(x \mid y)$ that consist of combs and wheels. The product of two diagrams is the disjoint union of all components followed by the concatenation of the combs; in particular $u^k u^m = u^{k+m}$. The coproduct is the same as in $\mathscr{C}(x \mid y)$. This Hopf algebra is nothing else but the free commutative Hopf algebra on a countable number of generators.

The Kontsevich integral is group-like, and this implies that

$$\delta(\mathscr{Z}_0(\Phi)) = \mathscr{Z}_0(\Phi) \otimes \mathscr{Z}_0(\Phi).$$

Group-like elements in the completion of the free commutative Hopf algebra are the exponentials of linear combinations of generators and, therefore

$$\mathcal{Z}_0(\Phi) = \exp(cu \cup \sum_n a_{2n} w_{2n}),$$

where c and a_{2n} are some constants.

In fact, the constant c is precisely the linking number of the components x and y, and, hence is equal to 1. We can write

$$\mathcal{Z}_0(\Phi) = \sum_n \frac{u^n}{n!} \cup \Omega',$$

where Ω' the part of $\mathcal{Z}_0(\Phi)$ containing wheels:

$$\Omega' = \exp \sum_n a_{2n} w_{2n}.$$

Define the map $\Phi_0 : \mathcal{B}^\circ \to \mathcal{C}^\circ$ by taking the pairing of a diagram in $\mathcal{B}^\circ(y)$ with $\mathcal{Z}_0(\Phi)$:

$$D \to \langle \mathcal{Z}_0(\Phi), D \rangle_y.$$

The map Φ_0 can be thought of as the part of Φ that shifts the degree by zero. Since Φ is multiplicative, Φ_0 also is. In fact, we shall see later that $\Phi_0 = \Phi$.

Lemma 11.18. $\Phi_0 = \chi \circ \partial_{\Omega'}^\circ$.

Proof Let us notice first that if $C \in \mathcal{C}(x \mid y)$ and $D \in \mathcal{B}^\circ(y)$, we have

$$\langle C \cup w_{2n}, D \rangle_y = \langle C, \partial_{w_{2n}}^\circ (D) \rangle_y.$$

Also, for any $D \in \mathcal{B}^\circ$ the expression

$$\left\langle \sum_n \frac{u^n}{n!}, D \right\rangle_y$$

is precisely the average of all possible ways of attaching the legs of D to the line x.

Therefore, for $D \in \mathcal{B}^\circ(y)$

$$\Phi_0(D) = \left\langle \sum_n \frac{u^n}{n!} \cup \Omega', D \right\rangle_y = \left\langle \sum_n \frac{u^n}{n!}, \partial_{\Omega'}^\circ D \right\rangle_y = \chi \circ \partial_{\Omega'}^\circ D.$$

$\square$

11.3.6 The coefficients of the wheels in Φ_0

If $D \in \mathcal{B}$ is a diagram, we shall denote by D_y the result of decorating all the legs of D with the label y; the same notation will be used for linear combinations of diagrams.

First, let us observe that Ω' is group-like with respect to the coproduct $\psi^{2 \cdot 1}$:

Lemma 11.19.

$$\psi_y^{2 \cdot 1} \Omega'_y = \Omega'_{y_1} \otimes \Omega'_{y_2},$$

where y_1 and y_2 are obtained by doubling y.

Proof We again use the fact that the sum of the Hopf link Φ with itself coincides with its two-fold disconnected cabling along the closed component y. Since the Kontsevich integral is multiplicative, we see that

$$\psi_y^{2 \cdot 1} \mathcal{Z}_0(\Phi)_{x, y} = \mathcal{Z}_0(\Phi)_{x, y_1} \mathcal{Z}_0(\Phi)_{x, y_2},$$

where the subscripts indicate the labels of the components and product on the right-hand side lives in the graded completion of $\mathcal{C}(x \mid y_1, y_2)$. Now, if we factor out on both sides the diagrams that have at least one vertex on the x component, we obtain the statement of the lemma. □

Lemma 11.20. *For any $D \in \mathcal{B}^\circ$*

$$\partial_D^\circ(\Omega') = \langle D, \Omega' \rangle \, \Omega'.$$

Proof It is clear from the definitions and from the preceding lemma that

$$\partial_D^\circ(\Omega') = \langle D_{y_1}, \psi_y^{2 \cdot 1} \Omega' \rangle = \langle D_{y_1}, \Omega'_{y_1} \Omega'_{y_2} \rangle = \langle D, \Omega' \rangle \, \Omega'.$$

□

Lemma 11.21. *The following holds in $\mathcal{B}^\circ$:*

$$\langle \, \Omega', (\text{———}\,)^n \, \rangle = \left(\frac{1}{24} \bigcirc \right)^n.$$

Proof According to Exercise 11.16, the Kontsevich integral of the Hopf link Φ up to degree two is equal to

$$\mathcal{Z}_0(\Phi) = \Big| + \dfrac{}{} + \frac{1}{2} \, \dfrac{}{} + \frac{1}{48} \, \dfrac{}{} .$$

It follows that the coefficient a_2 in Ω' is equal to $1/48$ and that

$$\langle \Omega', \text{———} \rangle = \frac{1}{24} \bigcirc .$$

This establishes the lemma for $n = 1$. Now, use induction:

$$\langle \Omega', (\text{---})^n \rangle = \langle \partial^\circ_{\text{---}} \Omega', (\text{---})^{n-1} \rangle$$

$$= \frac{1}{24}\ \ominus\ \cdot\ \langle \Omega', (\text{---})^{n-1} \rangle$$

$$= \left(\frac{1}{24}\ \ominus\right)^n.$$

The first equality follows from the obvious identity valid for arbitrary $A, B, C \in \mathcal{B}^\circ$:

$$\langle C, A \cup B \rangle = \langle \partial^\circ_B(C), A \rangle.$$

The second equality follows from the preceding lemma, and the third is the induction step.   □

In order to establish that $\Omega' = \Omega$ we have to show that the coefficients a_{2n} in the expression $\Omega' = \exp \sum_n a_{2n} w_{2n}$ are equal to the modified Bernoulli numbers b_{2n}. In other words, we have to prove that

$$\sum_n a_{2n} x^{2n} = \frac{1}{2} \ln \frac{\sinh x/2}{x/2}, \quad \text{or} \quad \exp\left(2 \sum_n a_{2n} x^{2n}\right) = \frac{\sinh x/2}{x/2}. \quad (11.4)$$

To do this we compute the value of the $\mathfrak{sl}_2$-weight system $\varphi_{\mathfrak{sl}_2}$ from Section 7.5 on the 3-graph $\langle \Omega', (\text{---})^n \rangle \in \Gamma$ in two different ways.

Using the last lemma and Theorem 6.17 we have

$$\varphi_{\mathfrak{sl}_2}\left(\langle \Omega', (\text{---})^n \rangle\right) = \varphi_{\mathfrak{sl}_2}\left(\left(\frac{1}{24}\ \ominus\right)^n\right) = \frac{1}{2^n}.$$

On the other hand, according to Exercise 6.23, the value of the $\mathfrak{sl}_2$ weight system on the wheel w_{2n} is equal to 2^{n+1} times its value on $(\text{---})^n$. Therefore,

$$\varphi_{\mathfrak{sl}_2}\left(\langle \Omega', (\text{---})^n \rangle\right) = \varphi_{\mathfrak{sl}_2}\left(\langle \exp \sum_m a_{2m} 2^{m+1} (\text{---})^m, (\text{---})^n \rangle\right).$$

Denote by f_n the coefficient at z^n of the power series expansion of the function $\exp(2 \sum_n a_{2n} z^n) = \sum_n f_n z^n$. We get

$$\varphi_{\mathfrak{sl}_2}\Big(\langle\,\Omega',\,(\!\bullet\!-\!\!\bullet)^n\,\rangle\Big) = 2^n f_n\,\varphi_{\mathfrak{sl}_2}\Big(\langle(\!\bullet\!-\!\!\bullet)^n,\,(\!\bullet\!-\!\!\bullet)^n\rangle\Big).$$

Now, using Exercise 11.19 and the fact that for the circle without vertices

$$\varphi_{\mathfrak{sl}_2}(\bigcirc) = 3,$$

(see page 206) we obtain

$$
\begin{aligned}
\varphi_{\mathfrak{sl}_2}\Big(\langle(\!\bullet\!-\!\!\bullet)^n,\,(\!\bullet\!-\!\!\bullet)^n\rangle\Big) &= (2n+1)(2n)\,\varphi_{\mathfrak{sl}_2}\Big(\langle(\!\bullet\!-\!\!\bullet)^{n-1},\,(\!\bullet\!-\!\!\bullet)^{n=1}\rangle\Big) = \dots \\
&= (2n+1)!\,.
\end{aligned}
$$

Comparing these two calculations we find that

$$f_n = \frac{1}{4^n(2n+1)!},$$

which is the coefficient of z^n of the power series expansion of

$$\frac{\sinh\sqrt{z}/2}{\sqrt{z}/2}.$$

Hence,

$$\exp\Big(2\sum_n a_{2n}z^n\Big) = \frac{\sinh\sqrt{z}/2}{\sqrt{z}/2}.$$

Substituting $z = x^2$ we get the equality (11.4) which establishes that $\Omega' = \Omega$ and completes the proof of the Wheeling theorem.

11.3.7 *Wheeling for tangle diagrams*

A version of the Wheeling theorem exists for more general spaces of tangle diagrams. For our purposes it is sufficient to consider the spaces of diagrams for links with two closed components x and y.

For $D \in \mathcal{B}$ define the operator

$$(\partial_D)_x : \mathcal{B}(x, y) \to \mathcal{B}(x, y)$$

as the sum of all possible ways of glueing all the legs of D to some of the x-legs of a diagram in $\mathcal{B}(x, y)$ that do not produce components without legs.

Exercise. Show that $(\partial_D)_x$ respects the link relations, and, therefore, is well-defined.

Define the wheeling map Φ_x as $\chi_x \circ (\partial_\Omega)_x$. The Wheeling theorem can now be generalized as follows:

Theorem 11.22. *The map*

$$\Phi_x : \mathcal{B}(x, y) \to \mathcal{C}(x \mid y)$$

identifies the $\mathcal{B}(x)$-module $\mathcal{B}(x, y)$ with the $\mathcal{C}(x)$-module $\mathcal{C}(x \mid y)$.

The proof is, essentially, the same as the proof of the Wheeling theorem, and we leave it to the reader.

11.4 The unknot and the Hopf link

The arguments similar to those used in the proof of the Wheeling theorem allow us to write down an explicit expression for the framed Kontsevich integral of the zero-framed unknot O. Let us denote by $\mathscr{L}(O)$ the image $\chi^{-1}I^{fr}(O)$ of the Kontsevich integral of O in the graded completion of $\mathcal{B}$. (Note that we use the notation $\mathscr{L}(\phi)$ in a similar, but not exactly the same context.)

Theorem 11.23.

$$\mathscr{L}(O) = \Omega = \exp \sum_{n=1}^{\infty} b_{2n} w_{2n}. \tag{11.5}$$

A very similar formula holds for the Kontsevich integral of the Hopf link ϕ:

Theorem 11.24.

$$\mathscr{L}(\phi) = \sum_{n} \frac{u^n}{n!} \cup \Omega. \tag{11.6}$$

This formula implies that the maps Φ and Φ_0 of the previous section, in fact, coincide.

We start the proof with a lemma.

Lemma 11.25. *If $C_1, \ldots, C_n$ are non-trivial elements of the algebra $\mathcal{C}$, then $\chi^{-1}(C_1 \# \ldots \# C_n)$ is a combination of diagrams in $\mathcal{B}$ with at least n legs.*

Proof We shall use the same notation as before. If $D \in \mathcal{B}$ is a diagram, we denote by D_y the result of decorating all the legs of D with the label y. The components of the skeleton obtained from y by applying $\psi_y^{2 \cdot 1}$ will be called y_1 and y_2.

Let $D_i = (\chi \circ \partial_\Omega)^{-1}(C_i) \in \mathcal{B}$ be the inverse of C_i under the wheeling map. By the Wheeling Theorem we have that

$$\chi^{-1}(C_1 \# \ldots \# C_n) = \partial_\Omega(D_1 \cup \ldots \cup D_n) = \langle \Omega_{y_1}, \psi_y^{2 \cdot 1}(D_1 \cup \ldots \cup D_n)\rangle.$$

Decompose $\psi_y^{2 \cdot 1}(D_i)$ as a sum $(D_i)_{y_1} + D_i'$ where D_i' contains only diagrams with at least one leg labelled by y_2.

Recall that in the completion of the algebra $\mathscr{B}^\circ$ we have $\partial_D^\circ(\Omega) = \langle D, \Omega \rangle \, \Omega$. By projecting this equality to $\mathscr{B}$ we see that $\partial_D(\Omega)$ vanishes unless D is empty. Hence,

$$\langle \Omega_{\mathbf{y}_1}, (D_1)_{\mathbf{y}_1} \cup \psi_{\mathbf{y}}^{2\cdot 1}(D_2 \cup \ldots \cup D_n) \rangle = \langle (\partial_{D_1}\Omega)_{\mathbf{y}_1}, \psi_{\mathbf{y}}^{2\cdot 1}(D_2 \cup \ldots \cup D_n) \rangle = 0.$$

As a result we have

$$\begin{aligned}
\partial_\Omega(D_1 \cup \ldots \cup D_n) &= \langle \Omega_{\mathbf{y}_1}, (D_1)_{\mathbf{y}_1} \cup \psi_{\mathbf{y}}^{2\cdot 1}(D_2 \cup \ldots \cup D_n) \rangle \\
&\quad + \langle \Omega_{\mathbf{y}_1}, D_1' \cup \psi_{\mathbf{y}}^{2\cdot 1}(D_2 \cup \ldots \cup D_n) \rangle \\
&= \langle \Omega_{\mathbf{y}_1}, D_1' \cup \psi_{\mathbf{y}}^{2\cdot 1}(D_2 \cup \ldots \cup D_n) \rangle \\
&= \langle \Omega_{\mathbf{y}_1}, D_1' \cup \ldots \cup D_n' \rangle.
\end{aligned}$$

Each of the D_i' has at least one leg labelled $\mathbf{y}_2$, and these legs are preserved by taking the pairing with respect to the label $\mathbf{y}_1$. $\qquad\square$

11.4.1 The Kontsevich integral of the unknot

The calculation of the Kontsevich integral for the unknot is based on the following geometric fact: the nth connected cabling of the unknot is again an unknot.

The cabling formula on page 258 in this case reads

$$\psi^n\left(I^{fr}(O) \# \exp\left(\tfrac{1}{2n}\,\mathbf{\theta}\right)\right) = I^{fr}(O^{(p,1)}) = I^{fr}(O) \# \exp\left(\tfrac{n}{2}\,\mathbf{\theta}\right). \qquad (11.7)$$

In each degree, the right-hand side of this formula depends on n polynomially. The term of degree 0 in n is precisely the Kontsevich integral of the unknot $I^{fr}(O)$.

As a consequence, the left-hand side also contains only non-negative powers of n. We shall be specifically interested in the terms that are of degree 0 in n.

The operator ψ^n has a particularly simple form in the algebra $\mathscr{B}$ (see Section 9.2.3): it multiplies a diagram with k legs by n^k. Let us expand the argument of ψ^n into a power series and convert it to $\mathscr{B}$ term by term.

It follows from Lemma 11.25 that if a diagram D is contained in

$$\chi^{-1}\left(I^{fr}(O) \# \left(\tfrac{1}{2n}\,\mathbf{\theta}\right)^k\right),$$

then it has $k' \geqslant k$ legs. Moreover, by the same lemma it can have precisely k legs only if it is contained in

$$\chi^{-1}\left(\tfrac{1}{2n}\,\mathbf{\theta}\right)^k.$$

Applying ψ^n, we multiply D by $n^{k'}$, hence the coefficient of D on the left-hand side of (11.7) depends on n as $n^{k'-k}$. We see that if the coefficient of D is of degree 0 as a function of n, then the number of legs of D must be equal to the degree of D.

Thus we have proved that $\mathscr{Z}(O)$ contains only diagrams whose number of legs is equal to their degree. We have seen in Section 11.3.5 that the part of the Kontsevich integral of the Hopf link that consists of such diagrams has the form $\sum \frac{u^n}{n!} \cup \Omega$. Deleting from this expression the diagrams with legs attached to the interval component, we obtain Ω. On the other hand, this is the Kontsevich integral of the unknot $\mathscr{Z}(O)$.

11.4.2 The Kontsevich integral of the Hopf link

The Kontsevich integral of the Hopf link both of whose components are closed with zero framing is computed in Bar-Natan *et al.* (2003). Such Hopf link can be obtained from the zero-framed unknot in three steps: first, change the framing of the unknot from 0 to $+1$, then take the disconnected twofold cabling, and, finally, change the framings of the resulting components from $+1$ to 0. We know how the Kontsevich integral behaves under all these operations and this gives us the following theorem (see Section 11.3.7 for notation):

Theorem 11.26. *Let $\mathbf{\infty}$ be the Hopf link both of whose components are closed with zero framing and oriented counterclockwise. Then*

$$I^{fr}(\mathbf{\infty}) = (\Phi_x \circ \Phi_y)(\exp |_y^x),$$

where $|_y^x \in \mathcal{B}(x, y)$ is an interval with one x leg and one y-leg.

We shall obtain the expression (11.6) in Theorem 11.24 from the above statement.

Proof Let O^{+1} be the unknot with $+1$ framing. Its Kontsevich integral is related to that of the zero-framed unknot as in Theorem 9.6:

$$I^{fr}(O^{+1}) = I^{fr}(O) \# \exp \left(\tfrac{1}{2} \mathbf{\flat} \right).$$

Applying the inverse of the wheeling map we get

$$\partial_\Omega^{-1} \mathscr{Z}(O^{+1}) = \partial_\Omega^{-1} \mathscr{Z}(O) \cup \exp \left(\frac{1}{2} \partial_\Omega^{-1} x^{-1} \left(\mathbf{\flat} \right) \right)$$

$$= \Omega \cup \exp \left(\frac{1}{2} \partial_\Omega^{-1} (\text{\textbullet---\textbullet}) \right).$$

Recall that in the proof of Lemma 11.25 we have seen that $\partial_D(\Omega) = 0$ unless D is empty. In particular, $\partial_\Omega^{-1}(\Omega) = \Omega$. We see that

$$\mathscr{Z}(O^{+1}) = \partial_\Omega\big(\Omega \cup \exp(\tfrac{1}{2}\,\text{—})\big), \tag{11.8}$$

since $\partial_\Omega^{-1}(\text{——}) = \text{——}$.

Our next goal is the following formula:

$$\partial_\Omega^{-2}(\mathscr{Z}(O^{+1})) = \exp(\tfrac{1}{2}\,\text{—}) \,. \tag{11.9}$$

Applying ∂_Ω to both sides of this equation and using (11.8), we obtain an equivalent form of (11.9):

$$\partial_\Omega\big(\exp(\tfrac{1}{2}\,\text{—})\big) = \Omega \cup \exp(\tfrac{1}{2}\,\text{—}) \,.$$

To prove it, we observe that

$$\partial_\Omega\big(\exp(\tfrac{1}{2}\,\text{—})\big) = \langle \Omega_{y_1}, \psi_y^{2\cdot1}\exp(\tfrac{1}{2}\,\text{—})\rangle_{y_1}$$

$$= \langle \Omega_{y_1}, \exp(\tfrac{1}{2}\,|_{y_1}^{y_1})\exp(|_{y_1}^{y_2})\exp(\tfrac{1}{2}\,|_{y_2}^{y_2})\rangle_{y_1}.$$

The pairing $\mathscr{B}(y_1, y_2) \otimes \mathscr{B}(y_1) \to \mathscr{B}(y_2)$ satisfies

$$\langle C, A \cup B\rangle_{y_1} = \langle \partial_B(C), A\rangle_{y_1}.$$

for all $A, B \in \mathscr{B}(y_1)$, $C \in \mathscr{B}(y_1, y_2)$. Therefore, the last expression can be re-written as

$$\langle \partial_{\exp(\frac{1}{2}|_{y_1}^{y_1})}\Omega_{y_1}, \exp(|_{y_1}^{y_2})\rangle_{y_1} \cup \exp(\tfrac{1}{2}\,|_{y_2}^{y_2}) \,.$$

Taking into the account the fact that $\partial_D(\Omega) = 0$ unless D is empty, we see that this is the same thing as

$$\langle \Omega_{y_1}, \exp(|_{y_1}^{y_2})\rangle_{y_1} \cup \exp(\tfrac{1}{2}\,|_{y_2}^{y_2}) = \Omega \cup \exp(\tfrac{1}{2}\,\text{—}),$$

and this proves (11.9).

To proceed, we need the following simple observation:

Lemma 11.27.

$$\psi_y^{2\cdot1}\partial_C(D) = (\partial_C)_{y_1}(\psi_y^{2\cdot1}(D)) = (\partial_C)_{y_2}(\psi_y^{2\cdot1}(D)).$$

Now, let $\mathbf{\infty}^{+1}$ be the Hopf link both of whose components are closed with $+1$ framing. Since $\partial_\Omega^{-1} = \partial_{\Omega^{-1}}$, the above lemma and the cabling formula on page 258 imply that

$$\psi^{2\cdot1}\partial_\Omega^{-2}(\mathscr{Z}(O^{+1})) = (\partial_\Omega)_{y_1}^{-1}(\partial_\Omega)_{y_2}^{-1}\big(\chi_{y_1,y_2}^{-1}I^{fr}(\mathbf{\infty}^{+1})\big).$$

On the other hand, this, by (11.9), is equal to

$$\psi^{2\cdot 1} \exp\left(\tfrac{1}{2}\,\text{\textbullet\!---\!\textbullet}\right) = \exp\left(\left|\begin{smallmatrix} y_2 \\ y_1 \end{smallmatrix}\right.\right) \cdot \exp\left(\tfrac{1}{2}\left|\begin{smallmatrix} y_1 \\ y_1 \end{smallmatrix}\right.\right) \cdot \exp\left(\tfrac{1}{2}\left|\begin{smallmatrix} y_2 \\ y_2 \end{smallmatrix}\right.\right).$$

Applying $\Phi_{y_1} \circ \Phi_{y_2}$ to the first expression, we get exactly $I^{fr}(\text{\textcircled{O}}^{+1})$. On the second expression, this evaluates to

$$\Phi_{y_1}\left(\Phi_{y_2}\left(\exp\left(\left|\begin{smallmatrix} y_2 \\ y_1 \end{smallmatrix}\right.\right)\right)\right) \# \exp_\#\left(\tfrac{1}{2}\,\text{\large$\mathsf{D}$}_{y_1}\right) \# \exp_\#\left(\tfrac{1}{2}\,\text{\large$\mathsf{D}$}_{y_2}\right).$$

Changing the framing, we see that

$$I^{fr}(\text{\textcircled{O}}) = (\Phi_{y_1} \circ \Phi_{y_2})\left(\exp\left|\begin{smallmatrix} y_1 \\ y_2 \end{smallmatrix}\right.\right).$$

The statement of the theorem follows by a simple change of notation. $\qquad\square$

Now we are in a position to prove formula (11.6) in Theorem 11.24. First, let us observe that for any diagram $D \in \mathcal{B}$ we have

$$(\partial_D)_x \exp\left|\begin{smallmatrix} y \\ x \end{smallmatrix}\right. = D_y \cup \exp\left|\begin{smallmatrix} y \\ x \end{smallmatrix}\right..$$

Now, we have

$$I^{fr}(\phi)\#\chi_x(\Omega_x) = I^{fr}(\text{\textcircled{O}})$$
$$= \Phi_x\left(\Phi_y\left(\exp\left|\begin{smallmatrix} y \\ x \end{smallmatrix}\right.\right)\right)$$
$$= \Phi_x\left(\exp\left|\begin{smallmatrix} y \\ x \end{smallmatrix}\right. \cup \Omega_x\right) \qquad \text{by the observation above.}$$

Since $\partial_\Omega = \Omega$, it follows that $\chi_x(\Omega_x = \Phi_x(\Omega_x))$ and

$$I^{fr}(\phi) = \Phi_x(\Omega_x^{-1})\#\Phi_x\left(\exp\left|\begin{smallmatrix} y \\ x \end{smallmatrix}\right. \cup \Omega_x\right)$$
$$= \Phi_x\left(\exp\left|\begin{smallmatrix} y \\ x \end{smallmatrix}\right. \cup \Omega_x \cup \Omega_x^{-1}\right) \qquad \text{by the Wheeling Theorem}$$
$$= \Phi_x\left(\exp\left|\begin{smallmatrix} y \\ x \end{smallmatrix}\right.\right)$$
$$= \chi_x\left(\Omega \cup \exp\left|\begin{smallmatrix} y \\ x \end{smallmatrix}\right.\right).$$

11.5 Rozansky's rationality conjecture

This section concerns a generalization of the formula for the Kontsevich integral of the unknot to arbitrary knots. The generalization is, however, not complete – the Rozansky–Kricker theorem does not give an explicit formula, it only suggests that $I^{fr}(K)$ can be written in a certain form.

It turns out that the terms of the Kontsevich integral $I^{fr}(K)$ with values in $\mathcal{B}$ can be rearranged into lines corresponding to the number of loops in open diagrams from $\mathcal{B}$. Namely, for any term of $I^{fr}(K)$, shaving off all legs of the corresponding diagram $G \in \mathcal{B}$, we get a trivalent graph γ. Infinitely many terms of $I^{fr}(K)$ give rise to the same γ. It turns out that these terms behave in a regular fashion, so that it is possible to recover all of them from γ and some finite information.

To make this statement precise, we introduce *marked open diagrams* which are represented by a trivalent graph whose edges are marked by power series (it does matter on which side of the edge the mark is located, and we shall indicate the side in question by a small leg near the mark). We use such marked open diagrams to represent infinite series of open diagrams which differ by the number of legs. More specifically, an edge marked by a power series $f(x) = c_0 + c_1 x + c_2 x^2 + c_3 x^3 + \dots$ stands for the following series of open diagrams:

$$\left.\begin{matrix}\\\\\end{matrix}\right\}_{f(x)} := c_0 \left.\begin{matrix}\\\\\end{matrix}\right) + c_1 \left.\begin{matrix}\\\\\end{matrix}\right\} + c_2 \left.\begin{matrix}\\\\\end{matrix}\right\} + c_3 \left.\begin{matrix}\\\\\end{matrix}\right\} + \dots$$

In this notation the formula for the Kontsevich integral of the unknot (page 330) can be written as

$$\ln I^{fr}(O) = \bigcirc^{\frac{1}{2}\ln \frac{e^{x/2}-e^{-x/2}}{x}} .$$

Now we can state the

Rozansky's rationality conjecture (Rozansky 2003).

$$\ln I^{fr}(K) = \boxed{\frac{1}{2}\ln \frac{e^{x/2}-e^{-x/2}}{x} - \frac{1}{2}\ln A_K(e^x)} + \sum_{i}^{\text{finite}} \boxed{\begin{matrix} p_{i,1}(e^x)/A_K(e^x) \\ p_{i,2}(e^x)/A_K(e^x) \\ p_{i,3}(e^x)/A_K(e^x) \end{matrix}}$$

$$+ \text{ (terms with } \geqslant 3 \text{ loops) },$$

where $A_K(t)$ is the Alexander polynomial of K normalized so that $A_K(t) = A_K(t^{-1})$ and $A_K(1) = 1$, $p_{i,j}(t)$ are polynomials, and the higher loop terms mean the sum over marked trivalent graphs (with finitely many copies of each graph) whose edges are marked by a polynomial in e^x divided by $A_K(t)$.

The word "rationality" refers to the fact that the labels on all 3-graphs of degree $\geqslant 1$ are rational functions of e^x. The conjecture was proved by A. Kricker (2000). Due to AS and IHX relations the specified presentation of the

Kontsevich integral is not unique. Hence the polynomials $p_{i,j}(t)$ themselves cannot be knot invariants. However, there are certain combinations of these polynomials that are genuine knot invariants. For example, consider the polynomial

$$\Theta'_K(t_1, t_2, t_3) = \sum_i p_{i,1}(t_1) p_{i,2}(t_2) p_{i,3}(t_3) .$$

Its symmetrization,

$$\Theta_K(t_1, t_2, t_3) = \sum_{\substack{\varepsilon=\pm 1 \\ \{i,j,k\}=\{1,2,3\}}} \Theta'_K(t_i^\varepsilon, t_j^\varepsilon, t_k^\varepsilon) \in \mathbb{Q}[t_1^{\pm 1}, t_2^{\pm 1}, t_3^{\pm 1}]/(t_1 t_2 t_3 = 1) ,$$

over the order 12 group of symmetries of the theta graph, is a knot invariant. It is called the *2-loop polynomial* of K. Its values on knots with few crossings are tabulated in Rozansky (2003). T. Ohtsuki (2004a) found a cabling formula for the 2-loop polynomial and its values on torus knots $T(p, q)$.

Exercises

11.1   * Find two chord diagrams with 11 chords which have the same intersection graph but are not equal modulo four- and one-term relations. According to Section 11.1.1, eleven is the least number of chords for such diagrams. Their existence is known, but no explicit examples were found yet.

11.2   * In the algebra $\mathscr{A}'$ consider the subspace $\mathscr{A}^M$ generated by those chord diagrams whose class in $\mathscr{A}'$ is determined by their intersection graph only. It is natural to regard the quotient space $\mathscr{A}'/\mathscr{A}^M$ as the space of chord diagram *distinguishing mutants*. Find the dimension of $\mathscr{A}'_n/\mathscr{A}^M_n$. It is known that it is zero for $n \leqslant 10$ and greater than zero for $n = 11$. Is it true that $\dim(\mathscr{A}'_{11}/\mathscr{A}^M_{11}) = 1$?

11.3   Find a basis in the space of canonical invariants of degree 4.

Answer: $j_4, c_4 + c_2/6, c_2^2$.

11.4   Show that the self-linking number defined in Section 2.2.3 is a canonical framed Vassiliev invariant of order 1.

11.5   Show the existence of the limit from Section 11.2.4

$$A = \lim_{N \to 0} \frac{\theta_{\mathfrak{sl}_N, V}\big|_{q=e^h}}{N} .$$

Hint. Choose a complexity function on link diagrams in such a way that two of the diagrams participating in the skein relation for $\theta_{\mathfrak{sl}_N, V}$ are strictly simpler then the third one. Then use induction on complexity.

11.6 Let $f(h) = \sum\limits_{n=0}^{\infty} f_n h^n$ and $g(h) = \sum\limits_{n=0}^{\infty} g_n h^n$ be two power series Vassiliev invariants.

(a) Prove that their product $f(h) \cdot g(h)$ as formal power series in h is a Vassiliev power series invariant, and

$$\text{symb}(f \cdot g) = \text{symb}(f) \cdot \text{symb}(g).$$

(b) Suppose that f and g are related to each other via substitution and multiplication:

$$f(h) = \beta(h) \cdot g\big(\alpha(h)\big),$$

where $\alpha(h)$ and $\beta(h)$ are formal power series in h, and

$$\alpha(h) = ah + \text{(terms of degree} \geqslant 2), \qquad \beta(h) = 1 + \text{(terms of degree} \geqslant 1).$$

Prove that $\text{symb}(f_n) = a^n \text{symb}(g_n)$.

11.7 Prove that a canonical Vassiliev invariant is primitive if its symbol is primitive.

11.8 Prove that the product of any two canonical Vassiliev power series is a canonical Vassiliev power series.

11.9 If v is a canonical Vassiliev invariant of odd order and K an amphicheiral knot, then $v(K) = 0$.

11.10 Let $\kappa \in \mathcal{W}_n$ be a weight system of degree n. Construct another weight system $(\kappa \circ \psi^p)' \in \mathcal{W}_n$, where ψ^p is the pth connected cabling operator, and $(\cdot)'$ is the deframing operator from Section 4.5.4. We get a function $f_\kappa: p \mapsto (\kappa \circ \psi^p)'$ with values in $\mathcal{W}_n$. Prove that

(a) $f_\kappa(p)$ is a polynomial in p of degree $\leqslant n$ if n is even, and of degree $\leqslant n - 1$ if n is odd.

(b) The nth degree term of the polynomial $f_\kappa(p)$ is equal to $-\dfrac{\kappa(\overline{w}_n)}{2} \text{symb}(c_n) p^n$, where $\overline{w}_n$ is the wheel with n spokes, and c_n is the nth coefficient of the Conway polynomial.

$$\overline{w}_n = \quad \text{}$$

n spokes

11.11 Find the framed Kontsevich integrals $Z^{fr}(H)$ and $I^{fr}(H)$ for the hump unknot with zero framing up to order 4.

Answer.

$$Z^{fr}\left(\text{⊌}\right) \;=\; 1 - \tfrac{1}{24}\,\text{⬭} + \tfrac{1}{24}\,\text{⊗} + \tfrac{7}{5760}\,\text{⬭} - \tfrac{17}{5760}\,\text{⬭}$$

$$+\,\tfrac{7}{2880}\,\text{⬭} - \tfrac{1}{720}\,\text{⬭} + \tfrac{1}{1920}\,\text{⬭} + \tfrac{1}{5760}\,\text{⊞} \;.$$

$$I^{fr}\left(\text{⊌}\right) \;=\; 1\big/ Z^{fr}\left(\text{⊌}\right)\;.$$

11.12 Using Exercise 5.4 show that up to degree 4

$$Z^{fr}\left(\text{⊌}\right) \;=\; 1 - \tfrac{1}{48}\,\text{⊖} + \tfrac{1}{4608}\,\text{⊟} + \tfrac{1}{46080}\,\text{⦁⦁⦁} + \tfrac{1}{5760}\,\text{⊗}\,,$$

$$I^{fr}\left(\text{⊌}\right) \;=\; 1 + \tfrac{1}{48}\,\text{⊖} + \tfrac{1}{4608}\,\text{⊟} - \tfrac{1}{46080}\,\text{⦁⦁⦁} - \tfrac{1}{5760}\,\text{⊗}\;.$$

11.13 Using the previous problem and Exercise 5.24 prove that up to degree 4

$$\mathscr{L}(O) = \chi^{-1} I^{fr}(O) = 1 + \tfrac{1}{48}\,\text{•⊖•} + \tfrac{1}{4608}\,\text{•⊟•} - \tfrac{1}{5760}\,\text{⋈}\;.$$

This result confirms Theorem 11.23 up to degree 4.

11.14 Compute the framed Kontsevich integral $Z^{fr}(\phi)$ up to degree 4 for the Hopf link ϕ with one vertical interval component x and one closed component y represented by the tangle on the picture. Write the result as an element of $\mathscr{C}(x, y)$.

$$\phi =$$

Answer.

$$Z^{fr}(\phi) = \;\Big|\,\text{O} + \text{⊢O} + \tfrac{1}{2}\,\text{⊢O} + \tfrac{1}{6}\,\text{⊞O} - \tfrac{1}{24}\,\text{⊢OO} + \tfrac{1}{24}\,\text{⊞O} - \tfrac{1}{48}\,\text{⊢OO}\,\Big)\;.$$

11.15 Prove that the final framed Kontsevich integral $I^{fr}(\phi)$ up to degree 4 has the form

$$I^{fr}(\phi) \;=\; \Big|\,\text{O} + \text{⊢O} + \tfrac{1}{2}\,\text{⊞O} + \tfrac{1}{48}\,\big|\text{⊝} + \tfrac{1}{6}\,\text{⊞O} - \tfrac{1}{24}\,\text{⊢OO} + \tfrac{1}{48}\,\text{⊢⊝}$$

$$+\,\tfrac{1}{24}\,\text{⊞O} - \tfrac{1}{48}\,\text{⊢OO} + \tfrac{1}{4608}\,\big|\text{⊝} - \tfrac{1}{46080}\,\big|\text{⊝} - \tfrac{1}{5760}\,\big|\text{⊛} + \tfrac{1}{96}\,\text{⊢⊝}\;.$$

11.16 Using the previous problem and Exercise 5.24 prove that up to degree 4

$$\mathscr{Z}(\Phi) = \chi_y^{-1} I^{fr}(\Phi) = \left| + \vdash + \frac{1}{2} \Big[+ \frac{1}{6} \Big[+ \frac{1}{24} \Big[+ \frac{1}{48} \Big| \xi + \frac{1}{48} \Big[\xi \right.$$

$$+ \frac{1}{96} \Big[\xi - \frac{1}{5760} \Big| \Big[+ \frac{1}{4608} \Big| \xi + \frac{1}{384} \Big[\xi \, .$$

Indicate the parts of this expression forming $\mathscr{Z}_0(\Phi)$, $\mathscr{Z}_1(\Phi)$, $\mathscr{Z}_2(\Phi)$ up to degree 4. This result confirms formula (11.6) in Theorem 11.24 up to degree 4.

11.17 Prove that $\chi \circ \partial_\Omega : \mathscr{B} \to \mathscr{C}$ is a bialgebra isomorphism.

11.18 Compute $\chi \circ \partial_\Omega(\boxminus)$, $\quad \chi \circ \partial_\Omega(\mathrel{\underset{\bullet\bullet}{\bullet}})$, $\quad \chi \circ \partial_\Omega(w_6)$.

11.19 Show that the pairing $\langle (\!\!\bullet\!\!-\!\!\!\bullet)^n, (\!\!\bullet\!\!-\!\!\!\bullet)^n \rangle$ satisfies the recurrence relation

$$\langle (\!\!\bullet\!\!-\!\!\!\bullet)^n, (\!\!\bullet\!\!-\!\!\!\bullet)^n \rangle = 2n \cdot (O + 2n - 2) \cdot \langle (\!\!\bullet\!\!-\!\!\!\bullet)^{n-1}, (\!\!\bullet\!\!-\!\!\!\bullet)^{n-1} \rangle \, ,$$

where O is a 3-graph in $\Gamma_0 \subset \Gamma \subset \mathscr{B}^\circ$ of degree 0 represented by a circle without vertices and multiplication is understood in algebra $\mathscr{B}^\circ$ (disjoint union).

11.20 Prove that, after being carried over from $\mathscr{B}$ to $\mathscr{A}$, the right-hand side of Equation (11.5) belongs in fact to the subalgebra $\mathscr{A}' \subset \mathscr{A}$. Find an explicit expression of the series in some basis of $\mathscr{A}'$ up to degree 4.

Answer. The first terms of the infinite series giving the Kontsevich integral of the unknot, are:

$$I(O) = 1 - \frac{1}{24} \bigotimes - \frac{1}{5760} \boxplus + \frac{1}{1152} \bigotimes + \frac{1}{2880} \bigotimes + \cdots$$

Check that this agrees with the answer to Exercise 11.11.

12

Braids and string links

Essentially, the theory of Vassiliev invariants of braids is a particular case of the Vassiliev theory for tangles, and the main constructions are very similar to the case of knots. There is, however, one big difference: many of the questions that are still open for knots are rather easy to answer in the case of braids. This, in part, can be explained by the fact that braids form groups, and it turns out that the whole Vassiliev theory for braids can be described in group-theoretic terms. In this chapter we shall see that the Vassiliev filtration on the pure braid groups coincides with the filtrations coming from the nilpotency theory of groups. In fact, for any given group the nilpotency theory could be thought of as a theory of finite type invariants.

The group-theoretic techniques of this chapter can be used to study knots and links. One such application is the theorem of Goussarov which says that n-equivalence classes of string links on m strands form a group. Another application of the same methods is a proof that Vassiliev invariants of pure braids extend to invariants of string links of the same order. In order to make these connections, we shall describe a certain braid closure that produces string links out of pure braids.

The theory of the finite type invariants for braids was first developed by T. Kohno (1985, 1987) several years before Vassiliev knot invariants were introduced. The connection between the theory of commutators in braid groups and the Vassiliev knot invariants was first made by T. Stanford (1996b).

12.1 Basics of the theory of nilpotent groups

We shall start by reviewing some basic notions related to nilpotency in groups. These will not only serve us a technical tool: we shall see that the theory of Vassiliev knot invariants has the logic and structure similar to those of the theory of nilpotent groups. The classical reference for nilpotent groups is found in the lecture notes of P. Hall (1957). For modern aspects of the theory, see Mikhailov and

Passi (2009). Iterated integrals are described in the papers of K. T. Chen (1977a, 1977b).

12.1.1 The dimension series

Let G be an arbitrary group. The *group algebra* $\mathbb{Z}G$ consists of finite linear combinations $\sum a_i g_i$ where g_i are elements of G and a_i are integers. The product in $\mathbb{Z}G$ is the linear extension of the product in G. The group itself can be considered as a subset of the group algebra if we identify g with $1g$. The identity in G is the unit in $\mathbb{Z}G$ and we shall denote it simply by 1.

Let $JG \subset \mathbb{Z}G$ be the *augmentation ideal*, that is, the kernel of the homomorphism $\mathbb{Z}G \to \mathbb{Z}$ that sends each $g \in G$ to 1. Elements of JG are the linear combinations $\sum a_i g_i$ with $\sum a_i = 0$. The powers $J^n G$ of the augmentation ideal form a descending filtration on $\mathbb{Z}G$.

Let $\mathscr{D}_k(G)$ be the subset of G consisting of all $g \in G$ such that

$$g - 1 \in J^k G.$$

Obviously, the neutral element of G always belongs to $\mathscr{D}_k(G)$. Also, for all $g, h \in \mathscr{D}_k(G)$ we have

$$gh - 1 = (g - 1)(h - 1) + (g - 1) + (h - 1)$$

and, hence, $\mathscr{D}_k(G)$ is closed under the product. Finally,

$$g^{-1} - 1 = -(g - 1)g^{-1}$$

which shows that $\mathscr{D}_k(G)$ is a subgroup of G; it is called the kth *dimension subgroup* of G. Clearly, $\mathscr{D}_1(G) = G$ and for each k the subgroup $\mathscr{D}_{k+1}(G)$ is contained in $\mathscr{D}_k(G)$.

Exercise. Show that $\mathscr{D}_k(G)$ is invariant under all automorphisms of G. In particular, it is a normal subgroup of G.

The descending series of subgroups

$$G = \mathscr{D}_1(G) \supseteq \mathscr{D}_2(G) \supseteq \mathscr{D}_3(G) \supseteq \ldots$$

is called the *dimension series of G*.

Consider the *group commutator* which can be defined as[1]

$$[g, h] := g^{-1}h^{-1}gh.$$

---

[1] There are other, equally good, options, such as $[g, h] = ghg^{-1}h^{-1}$.

If $g \in \mathscr{D}_p(G)$ and $h \in \mathscr{D}_q(G)$, we have

$$g^{-1}h^{-1}gh - 1 = g^{-1}h^{-1}\big((g-1)(h-1) - (h-1)(g-1)\big),$$

and, hence, $[g, h] \in \mathscr{D}_{p+q}(G)$. It follows that the group commutator descends to a bilinear bracket on

$$\mathscr{D}(G) := \bigoplus_k \mathscr{D}_k(G)/\mathscr{D}_{k+1}(G).$$

Exercise. Show that this bracket on $\mathscr{D}(G)$ is antisymmetric and satisfies the Jacobi identity. In other words, show that the commutator endows $\mathscr{D}(G)$ with the structure of a *Lie ring*.

This exercise implies that $\mathscr{D}(G) \otimes \mathbb{Q}$ is a Lie algebra over the rationals. The universal enveloping algebra of $\mathscr{D}(G) \otimes \mathbb{Q}$ admits a very simple description. Denote by $\mathscr{A}_k(G)$ the quotient $J^k G/J^{k+1}G$. Then the direct sum

$$\mathscr{A}(G) := \bigoplus_k \mathscr{A}_k(G)$$

is a graded ring, with the product induced by that of $\mathbb{Z}G$. Quillen (1968) shows that $\mathscr{A}(G) \otimes \mathbb{Q}$ is the universal enveloping algebra of $\mathscr{D}(G) \otimes \mathbb{Q}$.

The dimension series can be generalized by replacing the integer coefficients in the definition of the group algebra by coefficients in an arbitrary ring. The augmentation ideal is defined in the same fashion as consisting of linear combinations whose coefficients add up to zero, and the arguments given in this section for integer coefficients remain unchanged. We denote the kth dimension subgroup of G over a ring $\mathscr{R}$ by $\mathscr{D}_k(G, \mathscr{R})$. Since there is a canonical homomorphism of the integers to any ring, $\mathscr{D}_k(G)$ is contained in $\mathscr{D}_k(G, \mathscr{R})$ for any ring $\mathscr{R}$.

Exercise. Show that $\mathscr{D}_k(G, \mathbb{Q})/\mathscr{D}_{k+1}(G, \mathbb{Q})$ is a torsion-free abelian group for any k.

12.1.2 Commutators and the lower central series

For many groups, the dimension subgroups can be described entirely in terms of group commutators. For H, K normal subgroups of G, denote by $[H, K]$ the subgroup of G generated by all the commutators of the form $[h, k]$ with $h \in H$ and $k \in K$. The *lower central series* subgroups $\gamma_k G$ of a group G are defined inductively by setting $\gamma_1 G = G$ and

$$\gamma_k G = [\gamma_{k-1}G, G].$$

A group G is called *nilpotent* if $\gamma_n G = 1$ for some n. The maximal n such that $\gamma_n G \neq 1$ is called the *nilpotency class* of G. If the intersection of all $\gamma_n G$ is trivial, the group G is called *residually nilpotent*.

Exercise. Show that $\gamma_k G$ is invariant under all automorphisms of G. In particular, it is a normal subgroup of G.

Exercise. Show that for any p and q the subgroup $[\gamma_p G, \gamma_q G]$ is contained in $\gamma_{p+q} G$.

We have already seen that the commutator of any two elements of G belongs to $\mathscr{D}_2(G)$. Using induction, it is not hard to show that $\gamma_k G$ is always contained in $\mathscr{D}_k(G)$. If $\gamma_k G$ is actually the same thing as the kth dimension subgroup of G over the integers, it is said that G has the *dimension subgroup property*. Many groups have the dimension subgroup property. In fact, it was conjectured that *all* groups have this property until E. Rips found a counterexample (1972). His counterexample was later simplified; we refer to Mikhailov and Passi (2009) for the current state of knowledge in this field. In general, if $x \in \mathscr{D}_k(G)$, there exists q such that $x^q \in \gamma_k G$, and the group $\mathscr{D}_k(G, \mathbb{Q})$ consists of all x with this property, see Jennings' theorem (Theorem 12.4).

The subtlety of the difference between the lower central series and the dimension subgroups is underlined by the fact that for all groups $\gamma_k G = \mathscr{D}_k(G)$ when $k < 4$. In order to give the reader some feeling of the subject, let us treat one simple case here:

Proposition 12.1. *For any group G we have $\gamma_2 G = \mathscr{D}_2(G)$.*

Proof First let us assume that G is abelian, that is, $\gamma_2 G = 1$ (or, in additive notation, $\gamma_2 G = 0$). In this case there is a homomorphism of abelian groups $s : \mathbb{Z}G \to G$ defined by replacing a formal linear combination by a linear combination in G. The homomorphism s sends $g - 1 \in \mathbb{Z}G$ to $g \in G$. On the other hand, $s(J^2 G) = 0$. Indeed, it is easy to check that $J^2 G$ is additively spanned by products of the type $(x - 1)(y - 1)$ with $x, y \in G$; we have

$$s\big((x - 1)(y - 1)\big) = s(xy - x - y + 1) = xy - x - y = 0$$

since G is abelian. It follows that $g \in \mathscr{D}_2(G)$ implies that $g = 0$.

Now, let G be an arbitrary group. It can be seen from the definitions that group homomorphisms respect both the dimension series and the lower central series. Moreover, it is clear that a surjective homomorphism of groups induces surjections on the corresponding terms of the lower central series. This means that if $\mathscr{D}_2(G)$ is strictly greater than $\gamma_2 G$, the same is true for $G/\gamma_2 G$. On the other hand, $G/\gamma_2 G$ is abelian. $\qquad\square$

Recall that $H_1(G)$ is the *abelianization* of G, that is, its maximal abelian quotient $G/\gamma_2 G$, and that $H_1(G, \mathbb{Q}) = H_1(G) \otimes \mathbb{Q}$.

Exercise. Show that $H_1(G, \mathbb{Q})$ is canonically isomorphic to $\mathscr{A}_1(G) \otimes \mathbb{Q}$.

12.1.3 Filtrations induced by series of subgroups

Let $\{G_i\}$ be a descending series of subgroups

$$G = G_1 \supseteq G_2 \supseteq \ldots$$

of a group G with the property that $[G_p, G_q] \subseteq G_{p+q}$. For $x \in G$ denote by $\mu(x)$ the maximal k such that $x \in G_k$. Let $\mathbb{Q}G$ be the group algebra of G with rational coefficients and $E_n G$ its ideal spanned by the products of the form $(x_1 - 1) \cdot \ldots \cdot (x_s - 1)$ with $\sum_{i=1}^{s} \mu(x_i) \geqslant n$. We have the filtration of $\mathbb{Q}G$:

$$\mathbb{Q}G \supset JG = E_1 G \supseteq E_2 G \supseteq \ldots.$$

This filtration is referred to as the *canonical filtration induced by the series* $\{G_n\}$.

Theorem 12.2. *Let*

$$G = G_1 \supseteq G_2 \supseteq \ldots \supseteq G_N = \{1\}$$

be a finite series of subgroups of a group G with the property that $[G_p, G_q] \subseteq G_{p+q}$, and such that G_i/G_{i+1} is torsion-free for all $1 \leqslant i < N$. Then for all $i \geqslant 1$

$$G_i = G \cap (1 + E_i G),$$

where $\{E_i G\}$ is the canonical filtration of $\mathbb{Q}G$ induced by $\{G_i\}$.

As stated above, this theorem can be found in Passi (1979) and Passman (1977). The most important case of it has been proved by Jennings (1955), see also Hall (1957). It clarifies the relationship between the dimension series and the lower central series.

For a subset H of a group G, let $\sqrt{H}$ be the set of all $x \in G$ such that $x^p \in H$ for some $p > 0$. If H is a normal subgroup, and G/H is nilpotent, then $\sqrt{H}$ is again a normal subgroup of G. The set $\sqrt{\{1\}}$ is precisely the set of all periodic (torsion) elements of G; it is a subgroup if G is nilpotent.

Theorem 12.3. *Let*

$$G = G_1 \supseteq G_2 \supseteq \ldots \supseteq G_N = \{1\}$$

be a finite series of subgroups of a group G with the property that $[G_p, G_q] \subseteq G_{p+q}$. Then $[\sqrt{G_p}, \sqrt{G_q}] \subseteq \sqrt{G_{p+q}}$ and the canonical filtration of $\mathbb{Q}G$ induced by $\{\sqrt{G_i}\}$ coincides with the filtration induced by $\{G_i\}$.

For the proof, see the proofs of lemmas 1.3 and 1.4 in chapter IV of Passi (1979).

Now, consider a nilpotent group G. We have mentioned that $\gamma_n G$ is always contained in $\mathscr{D}_n(G)$, and, hence, $E_n G$ in this case coincides with $J^n G$. It follows from the above two theorems that

$$\mathscr{D}_n(G) = \sqrt{\gamma_n G}$$

for all n. The assumption that G is nilpotent can be removed by considering the group $G/\gamma_n G$ instead of G, and we get the following characterization of the dimension series over $\mathbb{Q}$:

Theorem 12.4 (Jennings 1955). *For an arbitrary group G, an element x of G belongs to $\mathscr{D}_n(G, \mathbb{Q})$ if and only if $x^r \in \gamma_n G$ for some $r > 0$.*

12.1.4 Semi-direct products

The augmentation ideals, the dimension series and the lower central series behave in a predictable way under direct products of groups. When $G = G_1 \times G_2$ we have

$$\mathbb{Z}G = \mathbb{Z}G_1 \otimes \mathbb{Z}G_2.$$

Moreover,

$$J^k G = \sum_{i+j=k} J^i G_1 \otimes J^j G_2,$$

and this implies

$$\mathscr{D}_k(G) = \mathscr{D}_k(G_1) \times \mathscr{D}_k(G_2).$$

It is also easy to see that

$$\gamma_k G = \gamma_k G_1 \times \gamma_k G_2.$$

When G is a *semi-direct*, rather than direct, product of G_1 and G_2 these isomorphisms break down in general. However, they do extend to one particular case of semi-direct products, namely, the *almost direct product* defined as follows.

Having a semi-direct product $A \ltimes B$ is the same as having an action of B on A by automorphisms. An action of B on A gives rise to an action of B on the abelianization of A; we say that a semi-direct product $A \ltimes B$ is almost direct if this latter action is trivial.

Proposition 12.5. *For an almost direct product $G = G_1 \ltimes G_2$*

$$\gamma_k G = \gamma_k G_1 \ltimes \gamma_k G_2$$

for all k. Moreover,

$$J^k(G) = \sum_{i+j=k} J^i(G_1) \otimes J^j(G_2),$$

inside $\mathbb{Z}G$ and, hence,

$$\mathscr{D}_k G = \mathscr{D}_k G_1 \ltimes \mathscr{D}_k G_2$$

and

$$\mathscr{A}(G) = \mathscr{A}(G_1) \otimes \mathscr{A}(G_2)$$

as a graded $\mathbb{Z}$-module.

The proof is not difficult and we leave it as an exercise. The case of the lower central series is proved in Falk and Randell (1985); for the dimension subgroups see Papadima (2002) (or Mostovoy and Willerton 2002 for the case when G is a pure braid group).

12.1.5 The free group

Let $x_1, \ldots, x_m$ be a set of free generators of the free group F_m and set $X_i = x_i - 1 \in \mathbb{Z}F_m$. Then, for any $k > 0$ each element $w \in F_m$ can be uniquely expressed inside $\mathbb{Z}F_m$ as

$$w = 1 + \sum_{1 \leqslant i \leqslant m} a_i X_i + \ldots + \sum_{1 \leqslant i_1, \ldots, i_k \leqslant m} a_{i_1, \ldots, i_k} X_{i_1} \ldots X_{i_k} + r(w),$$

where $a_{i_1, \ldots, i_j}$ are integers and $r(w) \in J^{k+1} F_m$. This formula can be considered as a Taylor formula for the free group. In fact, the coefficients $a_{i_1, \ldots, i_j}$ can be interpreted as some kind of derivatives, see Fox (1953).

To show that such formula exists, it is enough to have it for the generators of F_m and their inverses:

$$x_i = 1 + X_i$$

and

$$x_i^{-1} = 1 - X_i + X_i^2 - \ldots + (-1)^k X_i^k + (-1)^{k+1} X_i^{k+1} x_i^{-1}.$$

The uniqueness of the coefficients $a_{i_1, \ldots, i_j}$ will be clear from the construction below.

Having defined the Taylor formula, we can go further and define something like the Taylor series.

Let $\mathbb{Z}\langle\langle X_1, \ldots, X_m \rangle\rangle$ be the algebra of formal power series in m non-commuting variables X_i. Consider the homomorphism of F_m into the group of units of this algebra

$$\mathscr{M} : F_m \to \mathbb{Z}\langle\!\langle X_1, \ldots, X_m \rangle\!\rangle,$$

which sends the ith generator x_i of F_m to $1 + X_i$. In particular,

$$\mathscr{M}(x_i^{-1}) = 1 - X_i + X_i^2 - X_i^3 + \ldots.$$

This homomorphism is called the *Magnus expansion*. It is injective: the Magnus expansion of a reduced word $x_{\alpha_1}^{\varepsilon_1} \ldots x_{\alpha_k}^{\varepsilon_k}$ contains the monomial $X_{\alpha_1} \ldots X_{\alpha_k}$ with the coefficient $\varepsilon_1 \ldots \varepsilon_k$, and, hence, the kernel of $\mathscr{M}$ is trivial.

The Magnus expansion is very useful since it gives a simple test for an element of the free group to belong to a given dimension subgroup.

Lemma 12.6. *For $w \in F_m$ the power series $\mathscr{M}(w) - 1$ starts with terms of degree k if and only if $w \in \mathscr{D}_k(F_m)$ and $w \notin \mathscr{D}_{k+1}(F_m)$.*

Proof Extend the Magnus expansion by linearity to the group algebra $\mathbb{Z}F_m$. The augmentation ideal is sent by $\mathscr{M}$ to the set of power series with no constant term and, hence, the Magnus expansion of anything in $J^{k+1}F_m$ starts with terms of degree at least $k + 1$. It follows that the first k terms of the Magnus expansion coincide with the first k terms of the Taylor formula. Notice that this implies the uniqueness of the coefficients in the Taylor formula. Now, the term of lowest non-zero degree on right-hand side of the Taylor formula has degree k only if $w - 1 \in J^k F_m$. $\qquad\square$

One can easily see that the non-commutative monomials of degree k in the X_i give a basis for $J^k F_m / J^{k+1} F_m$. The Magnus expansion gives a map

$$\mathscr{M}_k : J^k F_m \to \mathscr{A}_k(F_m) = J^k F_m / J^{k+1} F_m$$

which sends $x \in J^k F_m$ to the degree k term of $\mathscr{M}(x)$. The following is straightforward:

Lemma 12.7. *The map $\mathscr{M}_k$ is the quotient map $J^k F_m \to \mathscr{A}_k(F_m)$.*

The dimension subgroups for the free group coincide with the corresponding terms of the lower central series. In other words, the free group F_m has the dimension subgroup property. A proof can be found, for example, in section 5.7 of Magnus, Karrass and Solitar (1976). Note also that the free groups are residually nilpotent since the kernel of the Magnus expansion is trivial.

12.1.6 Chen's iterated integrals

The Magnus expansion for the free group does not generalize readily to arbitrary groups. However, there is a general geometric construction which works in the same way for all finitely generated groups and detects the terms of the dimension series for any group, just as the Magnus expansion detects them for the free group. This construction is given by *Chen's iterated integrals* (Chen 1977a, 1977b). We shall only describe it very briefly here since we shall not need it in the sequel. An accessible introduction to Chen's integrals can be found in Harris (2004).

Let us assume that the group G is the fundamental group of a smooth manifold M. Let $X_1, \ldots, X_m$ be a basis for $H_1(M, \mathbb{R})$ and $w_1, \ldots, w_m$ be a set of real closed 1-forms on M representing the basis of $H^1(M, \mathbb{R})$ dual to the basis $\{X_i\}$.

Consider an expression

$$\alpha = \sum_i \alpha_i X_i + \sum_{i,j} \alpha_{ij} X_i X_j + \sum_{i,j,k} \alpha_{ijk} X_i X_j X_k + \ldots$$

where all the coefficients α_* are 1-forms on M. We shall say that α is a $\mathbb{R}\langle\langle X_1, \ldots, X_m \rangle\rangle$-*valued 1-form* on M. We refer to $\sum_i \alpha_i X_i$ as the *linear part* of α. Denote by $\mathfrak{x}$ the ideal in $\mathbb{R}\langle\langle X_1, \ldots, X_m \rangle\rangle$ consisting of the power series with no constant term.

K. T. Chen (1977a) proves the following fact:

Theorem 12.8. *There exists a* $\mathbb{R}\langle\langle X_1, \ldots, X_m \rangle\rangle$-*valued 1-form w on M whose linear part is* $\sum_i w_i X_i$ *and an ideal $\mathfrak{j}$ of* $\mathbb{R}\langle\langle X_1, \ldots, X_m \rangle\rangle$ *such that there is a ring homomorphism*

$$Z : \mathbb{R}\pi_1 M \to \mathbb{R}\langle\langle X_1, \ldots, X_m \rangle\rangle/\mathfrak{j}$$

given by

$$Z(g) = \sum_{\substack{0 \leqslant k \\ 1 \leqslant i_1, \ldots, i_k \leqslant m}} \int_{0 < t_k < \cdots < t_1 < 1} w(t_1) \wedge \ldots \wedge w(t_k) \,,$$

where $w(t)$ is the pull-back to the interval $[0, 1]$ of the 1-form w under a map $\gamma : [0, 1] \to M$ representing g, with the property that the kernel of the composite map is

$$Z : \mathbb{R}\pi_1 M \to \mathbb{R}\langle\langle X_1, \ldots, X_m \rangle\rangle/\mathfrak{j} \to \mathbb{R}\langle\langle X_1, \ldots, X_m \rangle\rangle/(\mathfrak{j} + \mathfrak{x}^n)$$

is precisely $J^n(\pi_1 M) \otimes \mathbb{R}$.

We shall call the map Z the *Chen expansion*.

In certain important situations the algebra $\mathbb{R}\langle\langle X_1, \ldots, X_m \rangle\rangle/\mathfrak{j}$ can be replaced by the algebra $\mathscr{A}(\pi_1 M) \otimes \mathbb{R}$. Suppose that the algebra ΛM of the differential forms on M has a differential graded subalgebra A with the following properties:

- the inclusion $A \to \Lambda M$ induces isomorphisms in cohomology in all dimensions;
- each element in $H^*(M, \mathbb{R})$ can be represented by a closed form in A so that there is a direct sum decomposition

$$A = H^*(M, \mathbb{R}) \oplus A'$$

where A' is an ideal.

Chen shows (1977b, lemma 3.4.2) that in this situation the ideal $\mathfrak{j}$ is actually homogeneous. As a consequence, the algebra $\mathbb{R}\langle\langle X_1, \ldots, X_m \rangle\rangle/\mathfrak{j}$ is graded. Since the Chen expansion sends $J^n(\pi_1 M) \otimes \mathbb{R}$ to the terms of degree n and higher, it induces an injective map

$$\mathscr{A}(\pi_1 M) \otimes \mathbb{R} \to \mathbb{R}\langle\langle X_1, \ldots, X_m \rangle\rangle/\mathfrak{j}$$

and the image of the Chen expansion is contained in the graded completion of the image of this map. This means that the 1-form

$$w = \sum_i w_i X_i + \sum_{i,j} w_{ij} X_i X_j + \ldots$$

is, actually, $\mathscr{A}(\pi_1 M) \otimes \mathbb{R}$-valued and we can think of X_i as the generators of $\mathscr{A}_1(\pi_1 M)$.

Examples of manifolds with a subalgebra A satisfying the above conditions include all compact Kähler manifolds. Another example which is of importance for us is the configuration space of k distinct ordered particles $z_1, \ldots, z_k$ in $\mathbb{C}$: its fundamental group is the pure braid group on n strands. If we allow complex, rather than real coefficients in the Chen expansion, we obtain a particularly simple form w which only contains linear terms:

$$w = \frac{1}{2\pi i} \sum d \log (z_i - z_j) \cdot X_{ij},$$

where X_{ij} can be thought of as a chord diagram with one horizontal chord connecting the ith and the jth strands. Comparing the definitions, we see that the Chen expansion of a pure braid coincides exactly with its Kontsevich integral.

12.2 Vassiliev invariants for free groups

The main subject of this chapter are the Vassiliev braid invariants, and, more specifically, the invariants of *pure braids*, that is, the braids whose associated permutation is trivial. Pure braids are a particular case of tangles and thus we have a general

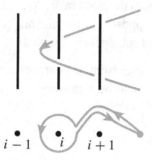

Figure 12.1 The generator x_i of F_m as a braid and as a path in a plane with m punctures

recipe for constructing their Vassiliev invariants. The only special feature of braids is the requirement that the tangent vector to a strand is nowhere horizontal. This leads to the fact that the chord diagrams for braids have only horizontal chords on a skeleton consisting of vertical lines; the relations they satisfy are the horizontal 4T-relations.

We shall start by treating what may seem to be a very particular case: braids on $m + 1$ strands whose all strands, apart from the last (the rightmost) one, are straight. Such a braid can be thought of as the graph of a path of a particle in a plane with m punctures. (The punctures correspond to the vertical strands.) The set of equivalence classes of such braids can be identified with the fundamental group of the punctured plane, that is, with the free group F_m on m generators x_i, where $1 \leqslant i \leqslant m$.

A *singular path* in the m-punctured plane is represented by a braid with a finite number of transversal double points, whose first m strands are vertical. Resolving the double points of a singular path with the help of the Vassiliev skein relation we obtain an element of the group algebra $\mathbb{Z}F_m$. Singular paths with k double points span the kth term of a descending filtration on $\mathbb{Z}F_m$ which is analogous to the singular knot filtration on $\mathbb{Z}\mathcal{K}$, defined in Section 3.2.1. A Vassiliev invariant of order k for the free group F_m is, of course, just a linear map from $\mathbb{Z}F_m$ to some abelian group that vanishes on singular paths with more than k double points.

Tangle chord diagrams which correspond to singular paths have a very specific form: these are horizontal chord diagrams (see Section 5.11) on $m + 1$ strands whose all chords have one endpoint on the last strand. Such diagrams form an algebra, which we denote temporarily by $\mathscr{A}^*(F_m)$, freely generated by m diagrams of degree 1. We shall see in Section 12.3.2 that this algebra is a subalgebra of $\mathscr{A}(m + 1)$, or, equivalently, that the horizontal 4T relations do not imply any relations in $\mathscr{A}^*(F_m)$.

The radical difference between the singular knots and singular paths (and, for that matter, arbitrary singular braids) lies in the following:

Lemma 12.9. *A singular path in the m-punctured plane with k double points is a product of k singular paths with one double point each.*

This is clear from the picture:

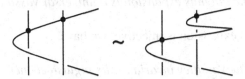

The above lemma allows us to describe the singular path filtration in purely algebraic terms. Namely, singular paths span the augmentation ideal $J F_m$ in $\mathbb{Z} F_m$ and singular paths with k double points span the kth power of this ideal.

Indeed, each singular path is an alternating sum of non-singular paths, and, hence, it defines an element of the augmentation ideal of F_m. On the other hand, the augmentation ideal of F_m is spanned by differences of the form $g - 1$ where g is some path. By successive crossing changes on its braid diagram, the path g can be made trivial. Let $g_1, \ldots, g_s$ be the sequence of paths obtained in the process of changing the crossings from g to 1. Then

$$g - 1 = (g - g_1) + (g_1 - g_2) + \ldots + (g_s - 1),$$

where the difference enclosed by each pair of brackets is a singular path with one double point.

We see that the Vassiliev invariants are those that vanish on some power of the augmentation ideal of F_m. The dimension subgroups of F_m are the counterpart of the Goussarov filtration: $\mathscr{D}_k F_m$ consists of elements that cannot be distinguished from the unit by Vassiliev invariants of order less than k. We shall refer to these as to being $k - 1$-*trivial*.

The algebra $\mathscr{A}^*(F_m)$ of chord diagrams for paths is the same thing as the algebra

$$\mathscr{A}(F_m) = \bigoplus_k J^k F_m / J^{k+1} F_m.$$

Indeed, the set of chord diagrams of degree k is the space of paths with k double points modulo those with $k + 1$ double points. The generator of $\mathscr{A}(F_m)$ which is the class of the element $x_i - 1$, where x_i is the ith generator of F_m, is represented by a chord joining the ith and the $m + 1$st strands:

$$x_i - 1 \quad = \quad \left|\begin{matrix} \cdots \end{matrix}\right|\begin{matrix} \bullet\!-\!\bullet \\ \cdots \end{matrix}\left|\begin{matrix} \; \end{matrix}\right|$$

$$1 \qquad i \quad m + 1$$

In fact, the Magnus expansion identifies the algebra $\mathbb{Z}\langle\langle X_1, \ldots, X_m \rangle\rangle$ with the completion $\widehat{\mathscr{A}}(F_m)$ of the algebra of the chord diagrams $\mathscr{A}(F_m)$. The following statement is a reformulation of Lemma 12.7:

Theorem 12.10. *The Magnus expansion is a universal Vassiliev invariant.*

Since the Magnus expansion is injective, we have

Corollary 12.11. *The Vassiliev invariants distinguish elements of the free group.*

Remark 12.12. If a word $w \in F_m$ contains only positive powers of the generators x_i, the Magnus expansion of w has a transparent combinatorial meaning: $\mathcal{M}(w)$ is simply the sum of all subwords of w, with the letters capitalized. This is also the logic behind the construction of the universal invariant for virtual knots discussed in Chapter 13: it associates to a diagram the sum of all its subdiagrams.

Remark 12.13. The Magnus expansion is not the only universal Vassiliev invariant (see Exercise 12.2). Another important universal invariant is, of course, the Kontsevich integral. In this case, the Kontsevich integral is nothing but the Chen expansion of F_m where the manifold M is taken to be the plane $\mathbb{C}$ with m punctures $z_1, \ldots, z_m$ and

$$w = \frac{1}{2\pi i} \cdot \frac{dz}{z - z_j} \cdot X_j.$$

Note that the Kontsevich integral depends on the positions of the punctures z_j (Exercise 12.3).

In contrast to the Kontsevich integral, the Magnus expansion has integer coefficients. We shall see that it also gives rise to a universal Vassiliev invariant of pure braids with integer coefficients; however, unlike the Kontsevich integral, this invariant fails to be multiplicative.

12.3 Vassiliev invariants of pure braids

The interpretation of the Vassiliev invariants for the free group F_m in terms of the powers of the augmentation ideal in $\mathbb{Z}F_m$ remains valid if the free groups are replaced by the pure braid groups. One new difficulty is that instead of the free algebra $\mathscr{A}(F_m)$ we have to study the algebra $\mathscr{A}(P_m) = \mathscr{A}^h(m)$ of horizontal chord diagrams (see Section 5.11). The multiplicative structure of $\mathscr{A}^h(m)$ is rather complex, but an explicit additive basis for this algebra can be easily described. This is due to the very particular structure of the pure braid groups.

12.3.1 Pure braids and free groups

Pure braid groups are, in some sense, very close to being direct products of free groups.

Erasing one (say, the rightmost) strand of a pure braid on m strands produces a pure braid on $m - 1$ strands. This procedure respects braid multiplication, so, in fact, it gives a homomorphism $P_m \to P_{m-1}$. Note that this homomorphism has a section $P_{m-1} \to P_m$ defined by adding a vertical non-interacting strand on the right.

The kernel of erasing the rightmost strand consists of braids on m strands whose first $m - 1$ strands are vertical. Such braids are graphs of paths in a plane with $m - 1$ punctures, and they form a group isomorphic to the free group on $m - 1$ letters F_{m-1}.

All the above can be re-stated as follows: there is a split extension

$$1 \to F_{m-1} \to P_m \leftrightarrows P_{m-1} \to 1.$$

It follows that P_m is a semi-direct product $F_{m-1} \ltimes P_{m-1}$, and, proceeding inductively, we see that

$$P_m \cong F_{m-1} \ltimes \ldots F_2 \ltimes F_1.$$

Here F_{k-1} can be identified with the free subgroup of P_m formed by pure braids which can be made to be totally straight apart from the kth strand which is allowed to braid around the strands to the left. As a consequence, every braid in P_n can be written uniquely as a product $\beta_{m-1}\beta_{m-2} \ldots \beta_1$, where $\beta_k \in F_k$. This decomposition is called the *combing* of a pure braid.

One can show that the above semi-direct products are not direct (see Exercise 12.4 at the end of the chapter). However, they are almost direct (see the definition in Section 12.1.4).

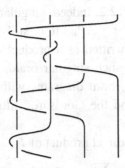

Figure 12.2 An example of a combed braid

Lemma 12.14. *The semi-direct product $P_m = F_{m-1} \ltimes P_{m-1}$ is almost direct.*

Proof The abelianization F_{m-1}^{ab} of F_{m-1} is a direct sum of $m-1$ copies of $\mathbb{Z}$. Given a path $x \in F_{m-1}$, its image in F_{m-1}^{ab} is given by the $m-1$ linking numbers with each puncture. The action of a braid $b \in P_{m-1}$ on a generator $x_i \in F_{m-1}$ consists in "pushing" the x_i through the braid:

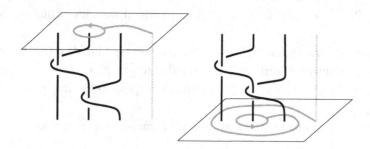

It is clear the linking numbers of the path $b^{-1}x_i b$ with the punctures in the plane are the same as those of x_i, therefore the action of P_{m-1} on F_{m-1}^{ab} is trivial. $\square$

Remark 12.15. Strictly speaking, in Section 2.2 we have only defined the linking number for two curves in space, while in the above proof we use the linking number of a point and a loop in a plane. This number can be defined as the intersection (or incidence) number of the point with an immersed disk whose boundary is the loop.

Generally, the linking number is defined for two disjoint cycles (for instance, oriented submanifolds) in $\mathbb{R}^n$ when the sum of the dimensions of the cycles is one less than n, see, for instance, Dold (1995). This linking number is crucial for the definition of the Alexander duality which we shall use in Chapter 15.

12.3.2 Vassiliev invariants and the Magnus expansion

The Vassiliev filtration on the group algebra $\mathbb{Z}P_m$ can be described in the same algebraic terms as in Section 12.2. Indeed, singular braids can be identified with the augmentation ideal $JP_m \subset \mathbb{Z}P_m$. It is still true that each singular braid with k double points can be written as a product of k singular braids with one double point each; therefore, such singular braids span the kth power of JP_m. The (linear combinations of) chord diagrams with k chords are identified with $J^k P_m / J^{k+1} P_m = \mathscr{A}_k(P_m)$ and the Goussarov filtration on P_m is given by the dimension subgroups $\mathscr{D}_k(P_m)$.

Now, since P_m is an almost direct product of F_{m-1} and P_{m-1} we have that

$$J^k(P_m) = \sum_{i+j=k} J^i(F_{m-1}) \otimes J^j(P_{m-1}),$$

$$\mathscr{A}_k(P_m) = \bigoplus_{i+j=k} \mathscr{A}_i(F_{m-1}) \otimes \mathscr{A}_j(P_{m-1}),$$

and

$$\mathscr{D}_k(P_m) = \mathscr{D}_k(F_{m-1}) \ltimes \mathscr{D}_k(P_{m-1}),$$

see Section 12.1.4.

These algebraic facts can be re-stated in the language of Vassiliev invariants as follows.

First, each singular braid with k double points is a linear combination of *combed singular braids* with the same number of double points. A combed singular braid with k double points is a product $b_{m-1}b_{m-2}\dots b_1$ where b_i is a singular path in $\mathbb{Z}F_i$ with k_i double points, and $k_{m-1} + \dots + k_1 = k$.

Second, *combed diagrams* form a basis in the space of all horizontal chord diagrams. A combed diagram D is a product $D_{m-1}D_{m-2}\dots D_1$ where D_i is a diagram whose all chords have their rightmost end on the ith strand.

Third, a pure braid is n-trivial if and only if, when combed, it becomes a product of n-trivial elements of free groups. In particular, the only braid that is n-trivial for all n is the trivial braid.

Let $\beta \in P_m$ be a combed braid: $\beta = \beta_{m-1}\beta_{m-2}\dots\beta_1$, where $\beta_k \in F_k$. The Magnus expansions of the elements β_i can be "glued together." Let $i_k : \mathscr{A}(F_k) \hookrightarrow \mathscr{A}^h(m)$ be the map that adds $m - k - 1$ vertical strands, with no chords on them, to the right:

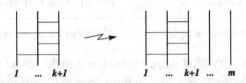

The maps i_k extend to the completions of the algebras $\mathscr{A}(F_k)$ and $\mathscr{A}^h(m)$. Define the Magnus expansion

$$\mathscr{M} : P_m \to \widehat{\mathscr{A}^h}(m)$$

as the map sending β to $i_{m-1}\mathscr{M}(\beta_{m-1})\dots i_1\mathscr{M}(\beta_1)$. For example:

$$\mathscr{M}\left(\begin{array}{c}\end{array}\right) = \left(1 + \left|\;\right|\right)\left(1 - \left|\right| + \left|\right| - \left|\right| + \dots\right)$$

$$= 1 + \left|\right| - \left|\right| - \left|\right| + \left|\right| + \left|\right| - \left|\right| + \dots.$$

Theorem 12.16. *The Magnus expansion is a universal Vassiliev invariant of pure braids.*

As in the case of free groups, the Magnus expansion is injective, and, therefore, Vassiliev invariants distinguish pure braids. Note that combing is not multiplicative so the Magnus expansion is not multiplicative either.

12.3.3 A dictionary

The theory of finite type invariants for the pure braids suggests the following dictionary between the nilpotency theory for groups and the theory of Vassiliev invariants:

Nilpotency theory for groups	Vassiliev theory
a group G	a class of tangles with a fixed skeleton X
$\mathscr{A}(G) = \oplus J^k G / J^{k-1} G$	diagram space $\mathscr{C}(X)$
functions $\mathbb{Z}G \to \mathscr{R}$ that vanish on $J^{n+1}G$	$\mathscr{R}$-valued Vassiliev invariants of order n
Chen expansion	Kontsevich integral
dimension series $\mathscr{D}_n G$	filtration by n-trivial tangles
lower central series $\gamma_n G$	filtration by γ_n-trivial tangles

The notion of γ_n-triviality (that is, γ_n-equivalence to the trivial tangle) that appears in the last line will be discussed later in this chapter, for string links rather than for general tangles. Note that we do not have a general definition for the trivial tangle with a given skeleton X, so in the last two lines we should restrict our attention to knots or (string) links.

The above dictionary must be used with certain care, as illustrated in the following paragraph.

12.3.4 Invariants for the full braid group

The finite type invariants for braids, considered as tangles, are defined separately for each permutation. The set of braids on m strands corresponding to the same permutation is in one-to-one (non-canonical) correspondence with the pure braid group P_m: given a braid b the subset $bP_m \subset B_m$ consists of all the braids with the same permutation as b. This correspondence also identifies the Vassiliev invariants for P_m with those of bP_m. In particular, the Vassiliev invariants separate braids.

On the other hand, the dimension series for the full braid group contains very little information. Indeed, it is known from Gorin and Lin (1969) that for $m \geqslant 5$ the lower central series of B_m stabilizes at $k = 2$:

$$\gamma_k B_m = \gamma_2 B_m$$

for $k \geqslant 2$.

Exercise. Show that for all m the quotient $B_m / \gamma_2 B_m$ is an infinite cyclic group.

12.4 String links as closures of pure braids

The Vassiliev invariants for pure braids can be used to prove some facts about the invariants of knots, and, more generally, string links.

12.4.1 The short-circuit closure

String links can be obtained from pure braids by a procedure called *short-circuit closure*. Essentially, it is a modification of the *plat closure* construction described in Birman (1976).

In the simplest case when string links have one component, the short-circuit closure produces a long knot out of a pure braid on an odd number of strands by joining the endpoints of the strands in turn at the bottom and at the top:

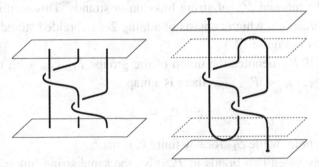

In order to get a string link with m components we have to start with a pure braid on $(2k + 1)m$ strands and proceed as follows.

Draw a braid in such a way that its top and bottom consist of the integer points of the rectangle $[0, 2k] \times [1, m]$ in the plane. A string link on m strands can be obtained from such a braid by joining the points $(2j - 1, i)$ and $(2j, i)$ (with $0 < j \leqslant k$) in the top plane and $(2j, i)$ and $(2j + 1, i)$ (with $0 \leqslant j < k$) in the bottom plane by

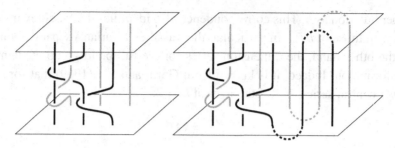

Figure 12.3 The stabilization map

little arcs, and extending the strands at the points $(0, i)$ in the top plane and $(2k, i)$ in the bottom plane. Here is an example with $m = 2$ and $k = 1$:

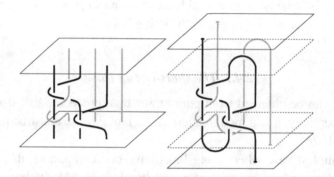

The short-circuit closure can be thought of as a map $\mathcal{S}_k$ from the pure braid group $P_{(2k+1)m}$ to the monoid $\mathcal{L}_m$ of string links on m strands. This map is compatible with the *stabilization*, which consists of adding $2m$ unbraided strands to the braid on the right, as in Figure 12.3.

Therefore, if P_∞ denotes the union of the groups $P_{(2k+1)m}$ with respect to the inclusions $P_{(2k+1)m} \to P_{(2k+3)m}$, there is a map

$$S : P_\infty \to \mathcal{L}_m.$$

The map S is onto, while $\mathcal{S}_k$, for any finite k, is not.[2]

One can say when two braids in P_∞ give the same string link after the short-circuit closure:

Theorem 12.17. *There exist two subgroups H^T and H^B of P_∞ such that the map S_n is constant on the double cosets of the form $H^T x H^B$. The preimage of every string link is a coset of this form.*

----

[2] To show this, one has to use the *bridge number* (see Exercise 2.1) of knots.

In other words, $\mathscr{L}_m = H^T \backslash P_\infty / H^B$.

The above theorem generalizes a similar statement for knots (the case $m = 1$), which was proved for the first time by J. Birman (1976) in the setting of the plat closure. Below we sketch a proof which closely follows the argument given for knots in Mostovoy and Stanford (2003).

First, notice that the short-circuit closure of a braid in $P_{(2k+1)m}$ is not just a string link, but a *Morse* string link: the height in the 3-space is a function on the link with a finite number of isolated critical points, none of which is on the boundary. We shall say that two Morse string links are *Morse equivalent* if one of them can be deformed into the other through Morse string links.

Lemma 12.18. *Assume that the short-circuit closures of $b_1, b_2 \in P_{(2k+1)m}$ are isotopic. There exist $k' \geqslant k$ such that the short-circuit closures of the images of b_1 and b_2 in $P_{(2k'+1)m}$ under the (iterated) stabilization map are Morse equivalent.*

The proof of this lemma is not difficult; it is identical to the proof of lemma 4 in Mostovoy and Stanford (2003) and we omit it.

Let us now describe the groups H^T and H^B. The group H^T is generated by elements of two kinds. For each pair of strands joined on top by the short-circuit map, take (a) the full twist of this pair of strands, (b) the braid obtained by taking this pair of strands around some strand, as in Figure 12.4:

The group H^B is defined similarly, but instead of pairs of strands joined on top, we consider those joined at the bottom. Clearly, multiplying a braid x on the left by an element of H^T and on the right by an element of H^B does not change the string link $S(x)$.

Now, given a Morse string link with the same numbers of maxima of the height function on each component (say, k), we can reconstruct a braid whose short-circuit closure it is, as follows.

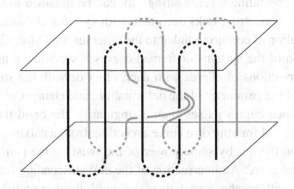

Figure 12.4 A generator of H^T

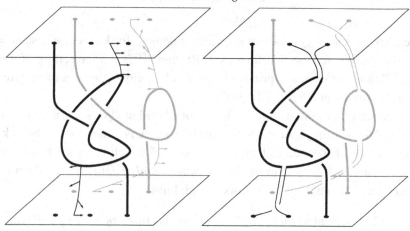

Figure 12.5 Obtaining a braid from a string link

Suppose that the string link is situated between the top and the bottom planes of the braid. Without loss of generality we can also assume that the top point of ith strand is the point $(0, i)$ in the top plane and the bottom point of the same strand is $(2k, i)$ in the bottom plane. For the jth maximum on the ith strand, choose an ascending curve that joins it with the point $(2j - 1/2, i)$ in the top plane, and for the jth minimum choose a descending curve joining it to the point $(2k - 3/2, i)$ in the bottom plane. We choose the curves in such a way that they are all disjoint from each other and only have common points with the string link at the corresponding maxima and minima. On each of these curves choose a framing that is tangent to the knot at one end and is equal to $(1, 0, 0)$ at the other end. Then, doubling each of these curves in the direction of its framing, we obtain a braid as in Figure 12.5.

Each braid representing a given string link can be obtained in this way. Given two Morse equivalent string links decorated with systems of framed curves, there exists a deformation of one string link into the other through Morse links. It extends to a deformation of the systems of framed curves if we allow a finite number of transversal intersections of curves with each other or with the string link, all at distinct values of the parameter of the deformation, and changes of framing. When a system of framed curves passes such a singularity, the braid that it represents changes. A change of framing on a curve ascending from a maximum produces the multiplication on the left by some power of the twist on the pair of strands corresponding to the curve. An intersection of the curve ascending from a maximum with the link or with another curve gives the multiplication on the left by a braid in H^T obtained by taking the pair of strands corresponding to the curve around some other strands. Similarly, singularities involving a curve descending from a minimum produce multiplications on the right by elements of H^B.

Remark 12.19. The subgroups H^T and H^B can be described in the following terms. The short-circuit map S can be thought of as consisting of two independent steps: joining the top ends of the strands and joining the bottom ends. A braid belongs to H^T if and only if the tangle obtained from it after joining the top strands only is "trivial," that is, equivalent to the tangle obtained in this way from the trivial braid. The subgroup H^B is described in the same way.

12.4.2 Vassiliev knot invariants as pure braid invariants

A knot invariant v gives rise to a pure braid invariant $v \circ S$ which is just the pull-back of v with respect to the short-circuit map. It is clear that if v is of order n the same is true for $v \circ S$ since the short-circuit map sends braids with double points to singular knots with the same number of double points.

An example is provided by the Conway polynomial. Each of its coefficients gives rise to an invariant of pure braids; these invariants factor through the Magnus expansion since the latter is the universal Vassiliev invariant. As a result, we get a function on the chord diagram algebra $\mathscr{A}(2m + 1)$ which can be explicitly described, at least for $m = 1$.

Recall that the algebra $\mathscr{A}(3)$ has a basis consisting of diagrams of the form $w(u_{13}, u_{23}) \cdot u_{12}^m$ where the u_{ij} are the generators (horizontal chords connecting the strands i and j) and w is some non-commutative monomial in two variables. Let

$$\chi : \mathscr{A}(3) \to \mathbb{Z}[t]$$

be the map such that for all $x \in P_3$ the Conway polynomial of $S(x)$ coincides with $\chi(\mathscr{M}(x))$. The following description of χ is given in Duzhin (2010).

First, it can be shown that χ vanishes on all the basis elements of the form uu_{12} and $u_{23}u$ for any u, and on all uu_{23}^2u' for any u and u'. This leaves us with just two kinds of basis elements:

$$[c_1, \ldots, c_k] := u_{13}^{c_1} u_{23} \ldots u_{13}^{c_{k-1}} u_{23} u_{13}^{c_k}$$

and

$$[c_1, \ldots, c_k]' := u_{13}^{c_1} u_{23} \cdot \ldots \cdot u_{23} u_{13}^{c_{k-1}} u_{23} u_{13}^{c_k} u_{23}.$$

The values of χ on the elements of the second kind are expressed via those on the elements of the first kind:

$$\chi([c_1, \ldots, c_k]') = t^{-2} \cdot \chi([c_1, \ldots, c_k, 1]).$$

As for the elements of the first kind, we have

$$\chi([c_1, \ldots, c_k]) = (-1)^{k-1} \left(\prod_{i=1}^{k-1} p_1 p_{c_i-1}\right) \cdot p_{c_k},$$

where $p_s = \chi([s])$ is a sequence of polynomials in t that are defined recursively by $p_0 = 1$, $p_1 = t^2$ and $p_{s+2} = t^2(p_s + p_{s+1})$ for $s \geqslant 0$.

12.5 Goussarov groups of knots

There are several facts about the Vassiliev string link invariants that can be proved by studying the interaction between the short-circuit closure and the dimension/lower central series for the pure braid groups. (In view of the results in Section 12.1.4 these two series on P_m always coincide.) In this section we shall consider the case of knots which is slightly simpler than the general case of string links.

12.5.1 *Vassiliev invariants and γ_n-equivalence classes of knots*

Definition 12.20. Two knots K_1 and K_2 are γ_n-*equivalent* if there are $x_1, x_2 \in P_\infty$ such that $K_i = S(x_i)$ and $x_1 x_2^{-1} \in \gamma_n P_\infty$.

Exercise. Show that the connected sum of knots descends to their γ_n-equivalence classes.

Theorem 12.21 (Goussarov 1991; Habiro 2000). *For each n, the γ_n-equivalence classes of knots form an abelian group under the connected sum.*

The group of knots modulo γ_{n+1}-equivalence is called the *nth Goussarov group* and is denoted by $\mathscr{K}(n)$.

Theorem 12.22. *Two knots cannot be distinguished by Vassiliev invariants (with values in any abelian group) of degree at most n if and only if they define the same element in $\mathscr{K}(n)$.*

In other words, two knots are γ_{n+1}-equivalent if and only if they are n-equivalent (see Section 3.2.1).

These two results are the main content of what is sometimes referred to as the *Goussarov–Habiro theory*; we shall discuss them again from a slightly different point of view in Chapter 14. The rest of this section is dedicated to their proofs. The idea of the argument, which is due to T. Stanford (1999), is to interpret knot invariants as pure braid invariants.

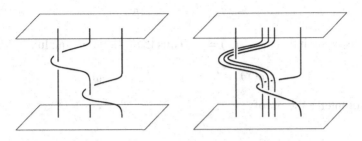

Figure 12.6 Tripling a strand of a pure braid

12.5.2 The shifting endomorphisms

For $k > 0$, define τ_k to be the endomorphism of P_∞ which replaces the kth strand by three parallel copies of itself as in Figure 12.6:

Denote by τ_0 the endomorphism of P_∞ which adds two non-interacting strands to the left of the braid (this is in contrast to the stabilization map, which adds two strands to the right and is defined only for P_{2k+1} with finite k).

Strand-tripling preserves both H^T and H^B. Also, since τ_k is an endomorphism, it respects the lower central series of P_∞.

Lemma 12.23 (Conant *et al.* 2010). *For any n and any $x \in \gamma_n P_{2N-1}$ there exist $t \in H^T \cap \gamma_n P_{2N+1}$ and $b \in H^B \cap \gamma_n P_{2N+1}$ such that $\tau_0(x) = txb$.*

Proof Let $t_{2k-1} = \tau_{2k-1}(x)(\tau_{2k}(x))^{-1}$, and let $b_{2k} = (\tau_{2k+1}(x))^{-1}\tau_{2k}(x)$. Notice that $t_{2k-1}, b_{2k} \in \gamma_n P_\infty$. Moreover, t_{2k-1} looks as in Figure 12.7 and, by Remark 12.19, lies in H^T. Similarly, $b_{2k} \in H^B$. We have

$$\tau_{2k-1}(x) = t_{2k-1}\tau_{2k}(x),$$

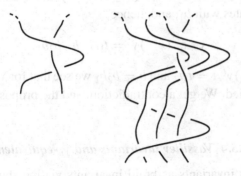

Figure 12.7 Braids x and t_{2k-1}

$$\tau_{2k}(x) = \tau_{2k+1}(x)b_{2k}.$$

There exists N such that $\tau_{2N+1}(x) = x$. Thus the following equality holds:

$$\tau_0(x) = t_1 \cdots t_{2N-1}xb_{2N}\cdots b_0,$$

and this completes the proof. $\qquad\square$

12.5.3 *Existence of inverses*

The existence of inverses modulo γ_n-equivalence is a consequence of the following, stronger, statement:

Proposition 12.24. *For any $x \in \gamma_k P_{2N-1}$ and any n there exists $y \in \gamma_k P_\infty$ such that:*

- *y is contained in the image of τ_0^N;*
- *$xy = thb$ with $h \in \gamma_n P_\infty$ and $t, b \in \gamma_k P_\infty$.*

The first condition implies that $S(xy) = S(x)\#S(y)$. It follows from the second condition that the class of $S(y)$ is the inverse for $S(x)$. The fact that t and b lie in $\gamma_k P_\infty$ is not needed here, but will be useful for Theorem 12.28.

Proof Fix n. For $k \geqslant n$ there is nothing to prove.

Assume there exist braids for which the statement of the proposition fails; among such braids choose x with the maximal possible value of k. By Lemma 12.23 we have $\tau_0^N(x^{-1}) = t_1x^{-1}b_1$ with $t_1 \in H^T \cap \gamma_k P_{4N-1}$ and $b_1 \in H^B \cap \gamma_k P_{4N-1}$. Then

$$x\tau_0^N(x^{-1}) = xt_1x^{-1}b_1 = t_1 \cdot t_1^{-1}xt_1x^{-1} \cdot b_1.$$

Since $t_1^{-1}xt_1x^{-1} \in \gamma_{k+1}P_{4N-1}$, there exists $y' \in \gamma_{k+1}P_\infty \cap \mathrm{Im}\tau_0^{2N}$ such that $t_1^{-1}xt_1x^{-1} \cdot y' = t_2hb_2$ where $h \in \gamma_n P_\infty, t_2 \in H^T \cap \gamma_{k+1}P_\infty$ and $b_2 \in H^B \cap \gamma_{k+1}P_\infty$. Note that y' commutes with b_1, and, hence,

$$x \cdot \tau_0^N(x^{-1})y' = t_1t_2 \cdot h \cdot b_2b_1.$$

Setting $y = \tau_0^N(x^{-1})y', t = t_1t_2$ and $b = b_2b_1$ we see that for x the statement of the proposition is satisfied. We get a contradiction, and the proposition is proved. $\qquad\square$

12.5.4 *Vassiliev invariants and γ_n-equivalence*

If we consider knot invariants as braid invariants via the short-circuit closure, it becomes clear that the value of a knot invariant of order n or less only depends on the γ_{n+1}-equivalence class of the knot. Indeed, multiplying a pure braid by an element of $\gamma_{n+1} P_m$ amounts to adding an element of $J^{n+1} P_m$, and this does not affect the invariants of degree n or less.

Lemma 12.25. *The map*

$$\mathcal{K} \to \mathcal{K}(n)$$

that sends a knot to its γ_{n+1}-equivalence class is an invariant of degree n.

This lemma establishes Theorem 12.22 since it tautologically implies that Vassiliev invariants of degree at most n distinguish γ_{n+1}-equivalence classes of knots.

Proof Extend the map $\mathcal{K} \to \mathcal{K}(n)$ by linearity to a homomorphism of abelian groups $\mathbb{Z}\mathcal{K} \to \mathcal{K}(n)$. The kernel of this map is spanned by two types of elements:

- elements of the form $x - y$ where x and y are γ_{n+1}-equivalent;
- elements of the form $x_1 \# x_2 - x_1 - x_2$.

Note that the trivial knot is in the kernel since, up to sign, it is an element of the second type. The subspace of elements of the second type in $\mathbb{Z}\mathcal{K}$ coincides with $\mathcal{K}_1 \# \mathcal{K}_1$, where $\mathcal{K}_1$ is the ideal of singular knots.

We need to show that the composite map

$$\mathbb{Z}P_\infty \xrightarrow{\mathcal{S}} \mathbb{Z}\mathcal{K} \to \mathcal{K}(n)$$

sends $J^{n+1} P_\infty$ to zero.

Define a *relator of order d and length s* as an element of $\mathbb{Z}\mathcal{K}$ of the form

(∗) $$\mathcal{S}((x_1 - 1)(x_2 - 1) \ldots (x_s - 1)y)$$

with $y \in P_\infty$, $x_i \in \gamma_{d_i} P_\infty$ and $\sum d_i = d$. The greatest d such that a relator is of order d will be called the *exact order* of a relator. A *composite relator* is an element of $\mathcal{K}_1 \# \mathcal{K}_1 \subset \mathbb{Z}\mathcal{K}$.

As we noted, the kernel of the map $\mathbb{Z}\mathcal{K} \to \mathcal{K}(n)$ contains all the relators of length 1 and order $n + 1$ and all the composite relators. On the other hand, an element of $\mathcal{S}(J^{n+1} P_\infty)$ is a linear combination of relators of length $n + 1$ and, hence, of order $n + 1$. Thus we need to show that any relator of order $n + 1$ is a linear combination of relators of order $n + 1$ and length 1 and composite relators.

Suppose that there exist relators of order $n + 1$ which cannot be represented as linear combinations of the above form. Among such relators, choose the relator R of minimal length and, given the length, of maximal exact order.

Assume that R is of the form (∗) as above, with $y, x_i \in P_{2N-1}$. Choose $t \in H^T$ and $b \in H^B$ such that the braid $t x_1 b$ coincides with the braid obtained from x_1 by shifting it by $2N$ strands to the right, that is, with $\tau_0^N(x_1)$. By Lemma 12.23

the braids t and b can be taken to belong to the same term of the lower central series of P_∞ as the braid x. The relator

$$R' = \mathcal{S}((tx_1b - 1)(x_2 - 1)\ldots(x_s - 1)y)$$

is a connected sum of two relators and, hence, is a combination of composite relators. On the other hand,

$$
\begin{aligned}
R' - R &= \mathcal{S}((tx_1b - x_1)(x_2 - 1)\ldots(x_s - 1)y) \\
&= \mathcal{S}(x_1(b - 1)(x_2 - 1)\ldots(x_s - 1)y).
\end{aligned}
$$

Notice now that $(b - 1)$ can be exchanged with $(x_i - 1)$ and y modulo relators of shorter length or higher order. Indeed,

$$(b - 1)y = y(b - 1) + ([b, y] - 1)yb$$

and

$$(b - 1)(x_i - 1) = (x_i - 1)(b - 1) + ([b, x_i] - 1)(x_ib - 1) + ([b, x_i] - 1).$$

Thus, modulo relators of shorter length or higher order

$$\mathcal{S}(x_1(b - 1)(x_2 - 1)\ldots(x_m - 1)y) = \mathcal{S}(x_1(x_2 - 1)\ldots(x_m - 1)y(b - 1)) = 0,$$

and this means that R is a linear combination of composite relators and relators of length 1 and order n.   □

12.6 Goussarov groups of string links

12.6.1 Vassiliev invariants and γ_n-equivalence classes of string links

Much of what was said about the Goussarov groups of knots can be extended to string links without change. Just as in the case of knots, two string links L_1 and L_2 are said to be γ_n-*equivalent* if there are $x_1, x_2 \in P_\infty$ such that $L_i = \mathcal{S}(x_i)$ and $x_1x_2^{-1} \in \gamma_n P_\infty$. A string link is γ_n-*trivial* if it is γ_n-equivalent to the trivial link. The product of string links descends to their γ_n-equivalence classes.

Theorem 12.26 (Goussarov 1991; Habiro 2000). *For each m and n, the γ_n-equivalence classes of string links on m strands form a group under the string link product.*

These groups are also referred to as *Goussarov groups*. We shall denote the group of string links on m strands modulo γ_{n+1}-equivalence by $\mathscr{L}_m(n)$, or simply by $\mathscr{L}(n)$, dropping the reference to the number of strands. Let $\mathscr{L}(n)_k$ be the subgroup of $\mathscr{L}(n)$ consisting of the classes of k-trivial links. Note that $\mathscr{L}(n)_k = 1$ for $k > n$.

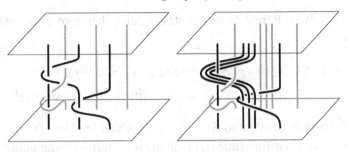

Figure 12.8 Tripling a row of strands

For string links with more than one component, the Goussarov groups need not be abelian. The most we can say is the following:

Theorem 12.27 (Goussarov 1991; Habiro 2000). *For all* p, q *we have*

$$[\mathscr{L}(n)_p, \mathscr{L}(n)_q] \subset \mathscr{L}(n)_{p+q}.$$

In particular, $\mathscr{L}(n)$ *is nilpotent of nilpotency class at most n.*

As for the relation between γ_{n+1}-equivalence and n-equivalence for string links, it is not known whether these two notions coincide as it is the case for knots. We shall prove a weaker statement:

Theorem 12.28 (Massuyeau 2007). *Two string links cannot be distinguished by* $\mathbb{Q}$-*valued Vassiliev invariants of degree n and smaller if and only if the elements they define in* $\mathscr{L}(n)$ *differ by an element of finite order.*

We refer the reader to Massuyeau (2007) for further results.

The proof of the fact that string links form groups modulo γ_n-equivalence repeats the proof of the same statement for knots word for word. The only modification necessary is in the definition of the shifting endomorphisms: rather than tripling the kth strand, τ_k triples the kth *row of strands*. In other words, τ_k replaces each strand with ends at the points $(k-1, i)$ in the top and bottom planes, with $1 \leqslant i \leqslant n$, by three parallel copies of itself as in Figure 12.8. Similarly, τ_0 adds $2m$ non-interacting strands, arranged in two rows, to the left of the braid.

12.6.2 The nilpotency of $\mathscr{L}(n)$

Let $x \in \gamma_p P_\infty$ and $x' \in \gamma_q P_\infty$. Choose the braids y and y' representing the inverses in $\mathscr{L}(n)$ of x and x', respectively, such that the conditions of Proposition 12.24 are satisfied, with n replaced by $n + 1$: $xy = t_1 h_1 b_1$ and $x'y' = t_2 h_2 b_2$ with $h_i \in \gamma_{n+1} P_\infty$, $t_1, b_1 \in \gamma_p P_\infty$ and $t_2, b_2 \in \gamma_q P_\infty$. Replacing the braids by their iterated shifts to the right, if necessary, we can achieve that the braids x, x',

y and y' all involve different blocks of strands, and, therefore, commute with each other. Then

$$\mathcal{S}(x) \cdot \mathcal{S}(x') \cdot \mathcal{S}(y) \cdot \mathcal{S}(y') = \mathcal{S}(xx'yy') = \mathcal{S}(xyx'y')$$
$$= \mathcal{S}(t_1 h_1 b_1 t_2 h_2 b_2) = \mathcal{S}(h_1 b_1 t_2 h_2).$$

The latter link is n-equivalent to $\mathcal{S}(t_2^{-1} b_1 t_2 b_1^{-1})$ which lives in $\mathcal{L}(n)_{p+q}$.

It follows that each n-fold (that is, involving $n+1$ terms) commutator in $\mathcal{L}(n)$ is trivial, which means that $\mathcal{L}(n)$ is nilpotent of nilpotency class at most n. Theorem 12.27 is proved.

12.6.3 Vassiliev invariants and γ_n-equivalence

As in the case of knots, the value of any order n Vassiliev invariant on a string link depends only on the γ_{n+1}-equivalence class of the link. The following proposition is the key to determining when two different γ_{n+1}-equivalence classes of string links cannot be distinguished by Vassiliev invariants of order n:

Proposition 12.29. *The filtration by the powers of the augmentation ideal $J P_\infty \subset \mathbb{Q} P_\infty$ is carried by short-circuit map to the canonical filtration $\{E_i \mathcal{L}(n)\}$ of the group algebra $\mathbb{Q}\mathcal{L}(n)$, induced by $\{\mathcal{L}(n)_i\}$.*

We remind that the canonical filtration was defined in Section 12.1.3.

Proof We use induction on the power k of $J P_\infty$. For $k = 1$ there is nothing to prove.

Any product of the form

$$(*) \qquad\qquad (x_1 - 1)(x_2 - 1) \dots (x_s - 1) y$$

with $y \in P_\infty$, $x_i \in \gamma_{d_i} P_\infty$ and $\sum d_i = d$ belongs to $J^d P_\infty$ since for any d_i we have $\gamma_{d_i} P_\infty - 1 \subset J^{d_i} P_\infty$. We shall refer to s as the *length* of such product, and to d as its *degree*. The maximal d such that a product of the form $(*)$ is of degree d, will be referred to as the *exact degree* of the product.

The short-circuit closure of a product of length 1 and degree k is in $E_k \mathcal{L}(n)$. Assume there exists a product of the form $(*)$ of degree k whose image R is not in $E_k \mathcal{L}(n)$; among such products choose one of minimal length, say r, and, given the length, of maximal exact degree.

There exists N such that

$$R' := \mathcal{S}((\tau_0^N(x_1) - 1)(x_2 - 1) \dots (x_r - 1) y) = \mathcal{S}((x_2 - 1) \dots (x_r - 1) y) \cdot \mathcal{S}(x_1 - 1).$$

The length of both factors on the right-hand side is smaller than k, so, by the induction assumption, $R' \in E_k \mathcal{L}(n)$. If $\tau_0^N(x_1) = t x_1 b$ we have

$$
\begin{aligned}
R' - R &= \mathcal{S}((t x_1 b - x_1)(x_2 - 1) \ldots (x_{m+1} - 1) y) \\
&= \mathcal{S}(x_1(b - 1)(x_2 - 1) \ldots (x_{m+1} - 1) y).
\end{aligned}
$$

Notice now that $(b - 1)$ can be exchanged with $(x_i - 1)$ and y modulo closures of products having shorter length or higher degree. Indeed,

$$
(b - 1)y = y(b - 1) + ([b, y] - 1)yb
$$

and

$$
(b - 1)(x_i - 1) = (x_i - 1)(b - 1) + ([b, x_i] - 1)(x_i b - 1) + ([b, x_i] - 1).
$$

Thus, modulo elements of $E_k \mathcal{L}(n)$

$$
\mathcal{S}(x_1(b - 1)(x_2 - 1) \ldots (x_{m+1} - 1) y) = \mathcal{S}(x_1(x_2 - 1) \ldots (x_{m+1} - 1) y(b - 1)) = 0.
$$

$\square$

By the above proposition the elements of $\mathcal{L}(n)$ that cannot be distinguished from the trivial link by the Vassiliev invariants of degree n form the subgroup

$$
\mathcal{L}(n) \cap (1 + E_{n+1} \mathcal{L}(n)).
$$

Since $\mathcal{L}(n)_{n+1} = 1$, by Theorems 12.2 and 12.3 this subgroup consists of all the elements of finite order in $\mathcal{L}(n)$. Finally, if the classes of two links L_1 and L_2 cannot be distinguished by invariants of order n, then $L_1 - L_2 \in E_{n+1} \mathcal{L}(n)$, and, hence, $L_1 L_2^{-1} - 1 \in E_{n+1} \mathcal{L}(n)$ and $L_1 L_2^{-1}$ is of finite order in $\mathcal{L}(n)$. Theorem 12.28 is proved.

12.6.4 Some comments

Remark 12.30. Rational-valued Vassiliev invariants separate pure braids, and the Goussarov group of γ_{n+1}-equivalence classes of pure braids on k strands is nothing but $P_k / \gamma_{n+1} P_k$, which is nilpotent of class n for $k > 2$. Since this group is a subgroup of $\mathcal{L}(n)$, we see that $\mathcal{L}(n)$ is nilpotent of class n for links on at least three strands. String links on one strand are knots, in this case $\mathcal{L}(n)$ is abelian. The nilpotency class of $\mathcal{L}(n)$ for links on two strands is unknown. Note that it follows from the results of Duzhin and Karev (2007) that $\mathcal{L}(n)$ for links on two strands is, in general, non-abelian.

Remark 12.31. The relation of the Goussarov groups of string links on more than one strand to integer-valued invariants seems to be a much more difficult problem.

While in Proposition 12.29 the field $\mathbb{Q}$ can be replaced by the integers with no changes in the proof, Theorem 12.2 fails over $\mathbb{Z}$.

Remark 12.32. Proposition 12.29 shows that the map

$$\mathscr{L}_m \to \mathscr{L}(n) \to \mathbb{Q}\mathscr{L}(n)/E_{n+1}\mathscr{L}(n)$$

is the *universal degree n Vassiliev invariant* in the following sense: each Vassiliev invariant of links in $\mathscr{L}_m$ of degree n can be extended uniquely to a linear function on $\mathbb{Q}\mathscr{L}(n)/E_{n+1}\mathscr{L}(n)$.

12.7 Braid invariants as string link invariants

12.7.1 Braid invariants extend to string links

A pure braid is a string link so every finite type string link invariant is also a braid invariant of the same order (at most). It turns out that the converse is true:

Theorem 12.33. *A finite type integer-valued pure braid invariant extends to a string link invariant of the same order.*

Corollary 12.34. *The natural map $\mathscr{A}^h(m) \to \mathscr{A}(m)$, where $\mathscr{A}^h(m)$ is the algebra of the horizontal chord diagrams and $\mathscr{A}(m)$ is the algebra of all string link chord diagrams, is injective.*

This was first proved by Bar-Natan (1996c). He considered quantum invariants of pure braids, which all extend to string link invariants, and showed that they span the space of all Vassiliev braid invariants.

Our approach will be somewhat different. We shall define a map

$$\mathscr{L}_m(n) \to P_m/\gamma_{n+1}P_m$$

from the Goussarov group of γ_{n+1}-equivalence classes of string links to the group of γ_{n+1}-equivalence classes of pure braids on m strands, together with a section $P_m/\gamma_{n+1}P_m \to \mathscr{L}_m(n)$. A Vassiliev invariant v of order n for pure braids is just a function on $P_m/\gamma_{n+1}P_m$, its pullback to $\mathscr{L}_m(n)$ gives the extension of v to string links.

Remark 12.35. Erasing one strand of a string link gives a homomorphism $\mathscr{L}_m \to \mathscr{L}_{m-1}$, which has a section. If $\mathscr{L}_m$ were a group, this would imply that string links can be combed, that is, that $\mathscr{L}_m$ splits as a semi-direct product of $\mathscr{L}_{m-1}$ with the kernel of the strand-erasing map. Of course, $\mathscr{L}_m$ is only a monoid, but it has many quotients that are groups, and these all split as iterated semi-direct products. For instance, string links form groups modulo concordance or link homotopy (Habegger and Lin 1990); here we are interested in the Goussarov groups.

Denote by $\mathscr{FL}_{m-1}(n)$ the kernel of the homomorphism $\mathscr{L}_m(n) \to \mathscr{L}_{m-1}(n)$ induced by erasing the last strand. We have semi-direct product decompositions

$$\mathscr{L}_m(n) \cong \mathscr{FL}_{m-1}(n) \ltimes \ldots \mathscr{FL}_2(n) \ltimes \mathscr{FL}_1(n).$$

We shall see that any element of $\mathscr{FL}_k(n)$ can be represented by a string link on $k+1$ strands whose first k strands are vertical. Moreover, taking the homotopy class of the last strand in the complement of the first k strands gives a well-defined map

$$\pi_k : \mathscr{FL}_k(n) \to F_k/\gamma_{n+1}F_k.$$

Modulo the $n + 1$st term of the lower central series, the pure braid group has a semi-direct product decomposition

$$P_\infty/\gamma_{n+1}P_\infty \cong F_{m-1}/\gamma_{n+1}F_{m-1} \ltimes \ldots \ltimes F_1/\gamma_{n+1}F_1.$$

The homomorphisms π_i with $i < m$ can now be assembled into one surjective map

$$\mathscr{L}_m(n) \to P_m/\gamma_{n+1}P_m.$$

Considering a braid as a string link gives a section of this map; this will establish the theorem stated above as soon as we justify our claims about the groups $\mathscr{FL}_k(n)$.

12.7.2 String links with one nontrivial component

The fundamental group of the complement of a string link certainly depends on the link. However, it turns out that all this dependence is hidden in the intersection of all the lower central series subgroups.

Let X be a string link on m strands and $\widetilde{X}$ be its complement. The inclusion of the top plane of X, punctured at the endpoints, into $\widetilde{X}$ gives a homomorphism i_t of F_m into $\pi_1\widetilde{X}$.

Lemma 12.36 (Habegger and Lin 1990). *For any n the homomorphism*

$$F_m/\gamma_n F_m \to \pi_1\widetilde{X}/\gamma_n\pi_1\widetilde{X}$$

induced by i_t is an isomorphism.

A corollary of this lemma is that for any n there is a well-defined map

$$\mathscr{L}_m(n) \to F_{m-1}/\gamma_{n+1}F_{m-1}$$

given by taking the homotopy class of the last strand of a string link in the complement of the first $m-1$ strands. We must prove that if two string links represent the same element of $\mathscr{FL}_{m-1}(n)$, their images under this map coincide.

In terms of braid closures, erasing the last strand of a string link corresponds to erasing all strands of P_∞ with ends at the points (i, m) for all $i \geqslant 0$. Erasing these strands, we obtain the group which we denote by P_∞^{m-1}; write Φ for the kernel of the erasing map. We have a semi-direct product decomposition

$$P_\infty = \Phi \ltimes P_\infty^{m-1},$$

and the product is almost direct. In particular, this means that

$$\gamma_k P_\infty = \gamma_k \Phi \ltimes \gamma_k P_\infty^{m-1}$$

for all k.

Lemma 12.37. *Let* $x \in \Phi$, *and* $h \in \gamma_{n+1}\Phi$. *The string links* $\mathcal{S}(x)$ *and* $\mathcal{S}(xh)$ *define the same element of* $F_{m-1}/\gamma_{n+1}F_{m-1}$.

Proof Each braid in Φ can be combed: Φ is an almost direct product of the free groups G_i which consist of braids all of whose strands, apart from the one with the endpoints at (i, m), are straight, and whose strands with endpoints at (j, m) with $j < i$ do not interact. Each element a of G_i gives a path in the complement of the first $m - 1$ strands of the string link, and, hence, an element $[a]$ of F_{m-1}. Notice that this correspondence is a homomorphism of G_i to F_{m-1}. (Strictly speaking, these copies of F_{m-1} for different i are only isomorphic, since these are fundamental groups of the same space with different basepoints. To identify these groups we need a choice of paths connecting the base points. Here we shall choose intervals of straight lines.)

Given $x \in \Phi$ we can write it as $x_1 x_2 \dots x_r$ with $x_i \in G_i$. Then the homotopy class of the last strand of $\mathcal{S}_n(x)$ produces the element

$$[x_1][x_2]^{-1} \dots [x_r]^{(-1)^{r-1}} \in F_{m-1}.$$

Let $x' = xh$ with $h \in \gamma_{n+1}\Phi$. Then the fact that Φ is an almost direct product of the G_i implies that if $x' = x_1' x_2' \dots x_r'$ with $x_i \in G_i$, then $x_i \equiv x_i' \mod \gamma_{n+1}G_i$. It follows that the elements of F_{m-1} defined by $\mathcal{S}(x)$ and $\mathcal{S}_n(x')$ differ by multiplication by an element of $\gamma_{n+1}F_{m-1}$. $\qquad\square$

Lemma 12.38. *Let* $x \in \Phi$, *and* $y \in \gamma_{n+1}P_\infty^{m-1}$. *The string links* $\mathcal{S}(x)$ *and* $\mathcal{S}(xy)$ *define the same element of* $F_{m-1}/\gamma_{n+1}F_{m-1}$.

Proof Denote by $\widetilde{X}$ the complement of $\mathcal{S}(y)$. We shall write a presentation for the fundamental group of $\widetilde{X}$. It will be clear from this presentation that the element of

$$F_{m-1}/\gamma_{n+1}F_{m-1} = \pi_1\widetilde{X}/\gamma_{n+1}\pi_1\widetilde{X}$$

given by the homotopy class of the last strand of $\mathcal{S}(xy)$ does not depend on y.

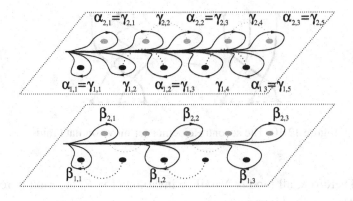

Figure 12.9

Let us assume that both x and y lie in the braid group $P_{m(2N+1)}$. Let H be the horizontal plane coinciding with the top plane of the braid y. The plane H cuts the space $\widetilde{X}$ into the upper part H_+ and the lower part H_-. The fundamental groups of H_+, H_- and $H_+ \cap H_-$ are free. Let us denote by $\{\alpha_{i,j}\}$, $\{\beta_{i,j}\}$ y $\{\gamma_{i,k}\}$ the corresponding free sets of generators (here $1 \leqslant i < m$, $1 \leqslant j \leqslant N+1$ and $1 \leqslant k \leqslant 2N+1$) as in Figure 12.9 by the Van Kampen theorem, $\pi_1 \widetilde{X}$ has a presentation

$$\langle \alpha_{i,j}, \beta_{i,j}, \gamma_{i,k} \mid \quad \theta_y^{-1}(\gamma_{i,2q-1}) = \beta_{i,q}, \quad \theta_y^{-1}(\gamma_{i,2q}) = \beta_{i,q}^{-1}$$
$$\gamma_{i,2q-1} = \alpha_{i,q}, \quad \gamma_{i,2q} = \alpha_{i,q+1}^{-1} \rangle,$$

where $1 \leqslant q \leqslant N+1$ and θ_y is the automorphism of $F_{(m-1)(2N+1)}$ given by the braid y. Since $y \in \gamma_{n+1} P_{(m-1)(2N+1)}$, it is easy to see that

$$\theta_y^{-1}(\gamma_{i,j}) \equiv \gamma_{i,j} \quad \mathrm{mod} \ \gamma_{m+1} \pi_1 \widetilde{X}.$$

Replacing $\theta_y^{-1}(\gamma_{i,j})$ by $\gamma_{i,j}$ in the presentation of $\pi_1 \widetilde{X}$ we obtain a presentation of the free group F_{m-1}. □

Now, a string link that gives rise to an element of $\mathscr{FL}_{m-1}(n)$ can be written as $S(xy)$ where $x \in \Phi$ and $y \in \gamma_{n+1} P_\infty^{m-1}$. Any link n-equivalent to it is of the form $S(txyb \cdot h)$ where $t \in H^T$, $b \in H^B$ and $h \in \gamma_{n+1} P_\infty$. We have

$$S(txyb \cdot h) = S(xy \cdot bhb^{-1}) = S(xh'yh''),$$

where $h' \in \gamma_{n+1} \Phi$ and $h'' \in \gamma_{n+1} P_\infty^{m-1}$. It follows from the two foregoing lemmas that $S(xh'yh'')$ and $S(xy)$ define the same element of $F_{m-1}/\gamma_{n+1} F_{m-1}$.

Exercises

12.1 Show that reducing the coefficients of the Magnus expansion of an element of F_n modulo m, we obtain the universal $\mathbb{Z}_m$-valued Vassiliev invariant for

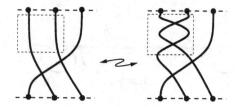

Figure 12.10 The second Reidemeister move on flat braids

F_n. Therefore, all mod m Vassiliev invariants for F_n are mod m reductions of integer-valued invariants.

12.2 Let $\mathcal{M}' : F_n \rightarrow \mathbb{Z}\langle\!\langle X_1, \ldots, X_n \rangle\!\rangle$ be any multiplicative map such that for all x_i we have $\mathcal{M}'(x_i) = 1 + \alpha_i X_i + \ldots$ with $\alpha_i \neq 0$. Show that $\mathcal{M}'$ is a universal Vassiliev invariant for F_n.

12.3 Show that the Kontsevich integral of an element of a free group F_m thought of as a path in a plane with m punctures depends on the positions of the punctures.

12.4 (a) Show that the semi-direct product in the decomposition $P_3 = F_2 \ltimes \mathbb{Z}$ given by combing is not direct.

(b) Find an isomorphism between P_3 and $F_2 \times \mathbb{Z}$.

12.5 A *twin* (Khovanov 1996), or a *flat braid* (Merkov 1999), on n strands is a collection of n descending arcs in the plane which connect the set of points $(1, a), \ldots, (n, a)$ with the set of points $(1, b), \ldots, (n, b)$ for some $a > b$, such that no three arcs intersect in one point. Flat braids are considered modulo horizontal deformations, vertical re-scalings and translations and the second Reidemeister move as in Figure 12.10.

Just as the usual braids, flat braids form a group with respect to concatenation. Develop the Vassiliev theory for the group of flat braids on n strands.

12.6 *Find the nilpotency class of $\mathscr{L}(n)$ for string links on two strands.

13

Gauss diagrams

In this chapter we shall show how the finite type invariants of a knot can be read off its Gauss diagram. It is not surprising that this is possible in principle, since the Gauss diagram encodes the knot completely. However, the particular method we describe, invented by Polyak and Viro and whose efficiency was proved by Goussarov, turns out to be conceptually very simple. For a given Gauss diagram, it involves only counting its subdiagrams of some particular types.

We shall prove that each finite type invariant arises in this way and describe several examples of such formulae.

13.1 The Goussarov theorem

13.1.1 A pairing on Gauss diagrams

Recall that in Chapter 12 we have constructed a universal Vassiliev invariant for the free group by sending a word to the sum of all of its subwords. A similar construction can be performed for knots if we think of a knot as being "generated by its crossings."

Let **GD** be the set of all Gauss diagrams (we shall take them to be based, or long, even though for the moment it is of little importance). Denote by $\mathbb{Z}$**GD** the set of all finite linear combinations of the elements of **GD** with integer coefficients. We define the map $I : \mathbb{Z}$**GD** $\to \mathbb{Z}$**GD** by simply sending a diagram to the sum of all its subdiagrams:

$$I(D) := \sum_{D' \subseteq D} D'$$

and continuing this definition to the whole of $\mathbb{Z}$**GD** by linearity. In other terms, the effect of this map can be described as

$$I : \quad \begin{array}{c}\text{(diagram)}\end{array} \longmapsto \begin{array}{c}\text{(diagram)}\end{array} + \begin{array}{c}\text{(diagram)}\end{array}$$

For example, we have

$$I\left(\text{} \right) = \text{} + \text{}$$

$$+ \text{} + \text{} + \text{} + 2 \text{} + \longrightarrow$$

Here all signs on the arrows are assumed to be, say, positive.

Define the pairing $\langle A, D \rangle$ of two Gauss diagrams A and D as the coefficient of A in $I(D)$:

$$I(D) = \sum_{A \in \mathbf{GD}} \langle A, D \rangle A.$$

In principle, the integer $\langle A, D \rangle$ may change if a Reidemeister move is performed on D. However, one can find invariant linear combinations of these integers. For example, in Section 3.6.6 we have proved that the Casson invariant c_2 of a knot can be expressed as

$$\langle \text{} , D \rangle - \langle \text{} , D \rangle - \langle \text{} , D \rangle + \langle \text{} , D \rangle.$$

More examples of such invariant expressions can be found in Section 13.4. In fact, as we shall now see, for each Vassiliev knot invariant there exists a formula of this type.

13.1.2 The Goussarov theorem

Each linear combination of the form

$$\sum_{A \in \mathbf{GD}} c_A \langle A, D \rangle$$

with integer coefficients, considered as a function of D, is just the composition $c \circ I$, where $c : \mathbb{Z}\mathbf{GD} \to \mathbb{Z}$ is the linear map with $c(A) = c_A$.

In what follows, usual knots will be referred to as *classical* knots, in order to distinguish them from virtual knots. Gauss diagrams that encode long classical knots, or, in other words, *realizable*, diagrams, form a subset $\mathbf{GD}^{re} \subset \mathbf{GD}$. Any integer-valued knot invariant v gives rise to a function $\mathbf{GD}^{re} \to \mathbb{Z}$ which extends by linearity to a function $\mathbb{Z}\mathbf{GD}^{re} \to \mathbb{Z}$. We also denote this extension by v. Here $\mathbb{Z}\mathbf{GD}^{re}$ is the free abelian group generated by the set $\mathbf{GD}^{re}$.

Theorem 13.1 (Goussarov). *For each integer-valued Vassiliev invariant v of classical knots of order $\leqslant n$ there exists a linear map $c : \mathbb{Z}\mathbf{GD} \to \mathbb{Z}$ such that*

$$v = c \circ I \mid_{\mathbb{Z}\mathbf{GD}^{re}}$$

and such that c is zero on each Gauss diagram with more than n arrows.

The proof of the Goussarov theorem is the main goal of this section.

13.1.3 Construction of the map c

Consider a given Vassiliev knot invariant v of order $\leqslant n$ as a linear function $v : \mathbb{Z}\mathbf{GD}^{re} \to \mathbb{Z}$. We are going to define a map $c : \mathbb{Z}\mathbf{GD} \to \mathbb{Z}$ such that the equality $c \circ I = v$ holds on $\mathbb{Z}\mathbf{GD}^{re}$.

Since the map I is an isomorphism with the inverse being

$$I^{-1}(D) = \sum_{D' \subseteq D} (-1)^{|D - D'|} D',$$

where $|D - D'|$ is the number of arrows of D not contained in D', the definition appears obvious:

$$c = v \circ I^{-1}. \tag{13.1}$$

However, for this equation to make sense we need to extend v from $\mathbb{Z}\mathbf{GD}^{re}$ to the whole of $\mathbb{Z}\mathbf{GD}$ since the image of I^{-1} contains all the subdiagrams of D and a subdiagram of a realizable diagram need not be realizable.

To make such an extension consistent, we need it to satisfy the Vassiliev skein relation. Thus we first express this relation in terms of Gauss diagrams in Section 13.1.4 introducing Gauss diagrams with undirected signed chords. It turns out (Section 13.1.6) that via Vassiliev's skein relation, an arbitrary Gauss diagram can be presented as a linear combination of some realizable Gauss diagrams (which we call *descending* and define in Section 13.1.5) plus Gauss diagrams with more than n chords. Since v is of order $\leqslant n$, it is natural to extend it by zero on Gauss diagrams with more than n chords. Also, we know the values of v on descending Gauss diagrams since they are realizable, and, thus, such a presentation gives us the desired extension. We shall complete the proof of the Goussarov theorem in Section 13.1.7.

13.1.4 Gauss diagrams with chords

Gauss diagrams can also be naturally defined for knots with double points. Apart from the arrows, these diagrams have solid undirected chords on them, each chord labelled with a sign. The *sign of a chord* is positive if in the positive resolution of the double point the overcrossing is passed first. (Recall that we are dealing with long Gauss diagrams, and that the points on a long knot are ordered.)

Gauss diagrams with at most n chords span the space $\mathbb{Z}\mathbf{GD}_n$, which is mapped to $\mathbb{Z}\mathbf{GD}$ by a version of the Vassiliev skein relation:

$$
\overset{\varepsilon}{\frown} \quad = \quad \varepsilon \quad \overset{\varepsilon}{\frown} \quad - \quad \varepsilon \quad \overset{-\varepsilon}{\frown} \qquad (13.2)
$$

Using this relation, any knot invariant, or, indeed, any function on Gauss diagrams can be extended to diagrams with chords. Note that the map $\mathbb{Z}\mathbf{GD}_n \to \mathbb{Z}\mathbf{GD}$ is not injective; in particular, changing the sign of a chord in a diagram from $\mathbf{GD}_n$ multiplies its image in $\mathbb{Z}\mathbf{GD}$ by -1. We have a commutative diagram

$$
\begin{array}{ccc}
\mathbb{Z}\mathbf{GD}_n & \xrightarrow{\text{skein (13.2)}} & \mathbb{Z}\mathbf{GD} \\
{\scriptstyle I}\downarrow & & \downarrow{\scriptstyle I} \\
\mathbb{Z}\mathbf{GD}_n & \xrightarrow{\text{skein (13.2)}} & \mathbb{Z}\mathbf{GD}
\end{array}
$$

where $I : \mathbb{Z}\mathbf{GD}_n \to \mathbb{Z}\mathbf{GD}_n$ is the isomorphism that sends a diagram to the sum of all its subdiagrams that contain the same chords.

13.1.5 Descending Gauss diagrams

We shall draw the diagrams of the long knots in the plane (x, y), assuming that the knot coincides with the x-axis outside some ball.

A diagram of a (classical) long knot is *descending* if for each crossing the over-crossing comes first. A knot whose diagram is descending is necessarily trivial. The Gauss diagram corresponding to a descending knot diagram has all its arrows pointed in the direction of the increase of the coordinate x (that is, to the right).

The notion of a descending diagram can be generalized to diagrams of knots with double points.

Definition 13.2. A Gauss diagram of a long knot with double points is called *descending* if

1. all the arrows are directed to the right;
2. no endpoint of an arrow can be followed by the left endpoint of a chord.

In other words, the following situations are forbidden:

For these two conditions to make sense the Gauss diagram with double points need not be realizable; we shall speak of descending diagrams irrespective of whether they can be realized by classical knots with double points.

Descending diagrams are useful because of the following fact.

Lemma 13.3. *Each long chord diagram with signed chords underlies a unique (up to isotopy) singular classical long knot that has a descending Gauss diagram.*

Proof The endpoints of the chords divide the line of the parameter into intervals, two of which are semi-infinite. Let us say that such an interval is *prohibited* if it is bounded from the right by a left end of a chord. Clearly, of the two semi-infinite intervals the left one is prohibited while the right one is not. If a chord diagram D underlies a descending Gauss diagram G_D, then G_D has no arrow endpoints on the prohibited intervals. We shall refer to the union of all prohibited intervals with some small neighbourhoods of the chord endpoints (which do not contain endpoints of other chords or arrows) as the *prohibited set*.

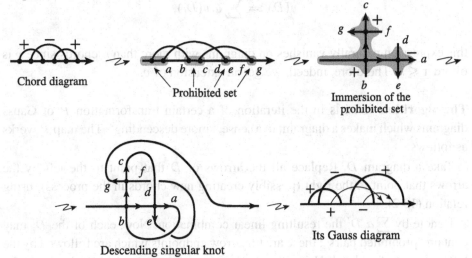

Chord diagram

Prohibited set

Immersion of the prohibited set

Descending singular knot

Its Gauss diagram

The prohibited set of a diagram can be immersed into the plane with double points corresponding to the chords, in such a way that the signs of the chords are respected. Such an immersion is uniquely defined up to isotopy. The image of the prohibited set will be an embedded tree T.

The leaves of T are numbered in the order given by the parameter along the knot. Note that given T, the rest of the plane diagram can be reconstructed as follows: the leaves of T are joined, in order, by arcs lying outside of T; these arcs only touch T at their endpoints and each arc lies below all the preceding arcs; the last arc extends to infinity. Such reconstruction is unique since the complement of T is homeomorphic to a 2-disk, so all possible choices of arcs are equivalent. $\qquad\square$

13.1.6 Extension of v

Here, using the Vassiliev skein relation, we extend v not only to singular long knots (realizable Gauss gauss diagrams with chords) but also to arbitrary Gauss diagrams with signed chords.

If D is a descending Gauss diagram with signed chords, by Lemma 13.3 there exists precisely one singular classical knot K which has a descending diagram with the same signed chords. We set $v(D) := v(K)$.

If D is an arbitrary diagram, we apply the algorithm which is described below to represent D as a linear combination $\sum a_i D_i$ of descending diagrams modulo diagrams with the number of chords $\geqslant n$. The algorithm uses the Vassiliev skein relation and has the property that it transforms a realizable Gauss diagram into a linear combination of realizable diagrams.

Now, if we set

$$v(D) := \sum a_i v(D_i),$$

this expression naturally vanishes on diagrams with more than n chords since v is of order $\leqslant n$. Therefore, indeed, we get an extension of v.

The algorithm consists in the iteration of a certain transformation P of Gauss diagrams which makes a diagram, in a sense, "more descending." The map P works as follows.

Take a diagram D. Replace all the arrows of D that point to the left by the arrows that point to the right (possibly creating new chords in the process), using relation (13.2).

Denote by $\sum a_i D_i'$ the resulting linear combination. Now, each of the D_i' may contain "prohibited pairs": these are the arrow endpoints which are followed by the left endpoint of a chord. Using the Reidemeister moves, a prohibited pair can be transformed as follows:

On a Gauss diagram this transformation can take one of the forms shown in Figure 13.1 where the arrows corresponding to the new crossings are thinner.

For each D_i' consider the leftmost prohibited pair, and replace it with the corresponding configuration of arrows and chords as in Figure 13.1; denote the

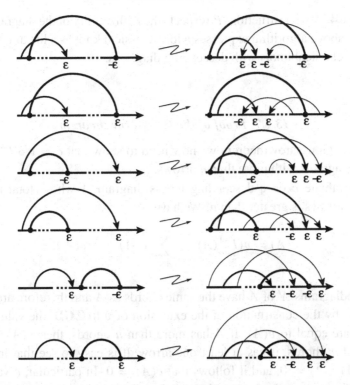

Figure 13.1 Action of the transformation P''

resulting diagram by D_i''. Set $P(D) := \sum a_i D_i''$ and extend P linearly to the whole $\mathbb{Z}\mathbf{GD}_\infty = \bigcup_n \mathbb{Z}\mathbf{GD}_n$.

If D is descending, then $P(D) = D$. We claim that applying P repeatedly to any diagram we shall eventually arrive to a linear combination of descending diagrams, modulo the diagrams with more than n chords.

Let us order the chords in a diagram by their left endpoints. We say that a diagram is *descending up to the kth chord* if the closed interval from $-\infty$ up to the left end of the kth chord contains neither endpoints of leftwards-pointing arrows, nor prohibited pairs.

If D is descending up to the kth chord, each diagram in $P(D)$ also is. Moreover, applying P either decreases the number of arrow heads to the left of the left end of the $(k+1)$st chord, or preserves it. In the latter case, it decreases the number of arrow tails in the same interval. It follows that for some finite m each diagram in $P^m(D)$ will be decreasing up to the $(k+1)$st chord. Therefore, repeating the process, we obtain after a finite number of steps a combination of diagrams descending up to the $(n+1)$st chord. Those of them that have at most n chords are descending, and the rest can be disregarded.

Remark 13.4. By construction, P respects the realizability of the diagrams. In particular, the above algorithm expresses a long classical knot as a linear combination of singular classical knots with descending diagrams.

13.1.7 Proof of the Goussarov theorem

To prove the Goussarov theorem we now need to show that $c = v \circ I^{-1}$ vanishes on Gauss diagrams with more than n arrows.

Let us evaluate c on a descending Gauss diagram A whose total number of chords and arrows is greater than n. We have

$$c(A) = v(I^{-1}(A)) = \sum_{A' \subseteq A} (-1)^{|A-A'|} v(A').$$

All the subdiagrams A' of A have the same chords as A and therefore are descending. Hence, by the construction of the extension of v to $\mathbb{Z}\mathbf{GD}$, the values of v on all the A' are equal to $v(A)$. If A has more than n chords, then $v(A) = 0$. If A has at most n chords, it has at least one arrow. It is easy to see that in this case $\sum_{A' \subseteq A} (-1)^{|A-A'|} = 0$, and it follows that $c(A) = 0$. In particular, c vanishes on all descending Gauss diagrams with more than n arrows.

In order to treat non-descending Gauss diagrams, we shall introduce an algorithm, very similar to that of Section 13.1.6 that converts any long Gauss diagram with chords into a combination of descending diagrams with at least the same total number of chords and arrows. The algorithm consists in the iteration of a certain map Q, similar to P, which also makes a diagram "more descending." We shall prove that the map Q preserves c in the sense that $c \circ Q = c$ and does not decrease the total number of chords and arrows. Then, applying Q to a Gauss diagram A enough number of times we get a linear combination of descending diagrams without altering the value of c. Then the arguments of the previous paragraph show that $c(A) = 0$, which will conclude the proof of the Goussarov theorem.

Take a Gauss diagram A. As in Section 13.1.6, we replace all the arrows of A that point leftwards by the arrows that point to the right, using relation (13.2).

Denote by $\sum a_i A_i'$ the resulting linear combination and check if the summands A_i' contain prohibited pairs. Here is where our new construction differs from the previous one. For each A_i' consider the leftmost prohibited pair, and replace it with the sum of the seven non-empty subdiagrams of the corresponding diagram from the right column of Figure 13.1 containing at least one of the three arrows. Denote the sum of these seven diagrams by A_i''. For example, if A_i' is the first diagram from the left column of Figure 13.1,

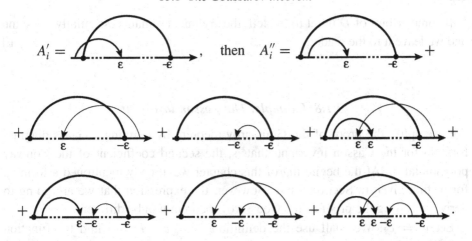

Now, set $Q(A) = \sum a_i A_i''$ and extend Q linearly to the whole $\mathbb{Z}\mathbf{GD}_\infty$.

As before, applying Q repeatedly to any diagram we shall eventually arrive to a linear combination of descending diagrams, modulo the diagrams with more than n chords. Note that Q does not decrease the total number of chords and arrows.

It remains to prove that Q preserves c. Since $I : \mathbb{Z}\mathbf{GD} \to \mathbb{Z}\mathbf{GD}$ is epimorphic, it is sufficient to check this on diagrams of the form $I(D)$. Assume that we have established that $c(Q(I(D))) = c(I(D))$ for all Gauss diagrams D with some chords and at most k arrows. If there are no arrows at all then D is descending and $Q(I(D)) = I(D)$. Now let D have $k + 1$ arrows. If D is descending, then again $Q(I(D)) = I(D)$ and there is nothing to prove. If D is not descending, then let us first assume for simplicity that all the arrows of D point to the right. Denote by l the arrow involved in the leftmost prohibited pair, and let D_l be the diagram D with l removed. We have

$$I(P(D)) = Q\big(I(D) - I(D_l)\big) + I(D_l).$$

Indeed, $P(D)$ is a diagram from the right column of Figure 13.1. Its subdiagrams fall into two categories depending on whether they contain at least one of the three arrows indicated on Figure 13.1 or none of them. The latter are subdiagrams of D_l and they are included in $I(D_l)$. The former can be represented as $Q\big(I(D) - I(D_l)\big)$.

By the induction assumption, $c(Q(I(D_l))) = c(I(D_l))$. Therefore,

$$c(Q(I(D))) = c(I(P(D))) = v(P(D)).$$

But applying P does not change the value of v because of our definition of the extension of v from Section 13.1.3. Therefore,

$$c(Q(I(D))) = v(P(D)) = v(D) = c(I(D)),$$

and, hence $c(Q(A)) = c(A)$ for any Gauss diagram A.

If some arrows of D point to the left, the argument remains essentially the same and we leave it to the reader.   □

13.1.8 Example. The Casson invariant

We exemplify the proof of the Goussarov theorem by deriving a Gauss diagram formula for the Casson invariant, that is, the second coefficient of the Conway polynomial c_2. At the beginning of this chapter we already mentioned a formula for it, first given in Section 3.6.6. However, the expression that we are going to derive following the proof of the Goussarov theorem will be different.

Let $v = c_2$. We shall use the definition $c = v \circ I^{-1}$ to find the function $c : \mathbb{Z}\mathbf{GD} \to \mathbb{Z}$.

If a Gauss diagram A has at most one arrow, then obviously $c(A) = 0$. Also, if A consists of two non-intersecting arrows, then $c(A) = 0$. So we need to consider the only situation when A consists of two intersecting arrows. There are 16 such diagrams differing by the direction of arrows and signs on them. The following table shows the values of c on all of them.

$$c(\underset{}{\overset{+}{\frown}}\overset{+}{}) = 0 \quad c(\underset{}{\overset{+}{\frown}}\overset{+}{}) = 0 \quad c(\underset{}{\overset{+}{\frown}}\overset{-}{}) = 0 \quad c(\underset{}{\overset{}{\frown}}\overset{-}{}) = 0$$

$$c(\underset{}{\overset{+}{\frown}}\overset{+}{}) = 0 \quad c(\underset{}{\overset{+}{\frown}}\overset{+}{}) = 0 \quad c(\underset{}{\overset{+}{\frown}}\overset{-}{}) = 0 \quad c(\underset{}{\overset{}{\frown}}\overset{-}{}) = 0$$

$$c(\underset{}{\overset{}{\frown}}\overset{+}{}) = 0 \quad c(\underset{}{\overset{}{\frown}}\overset{+}{}) = 0 \quad c(\underset{}{\overset{+}{\frown}}\overset{}{}) = 0 \quad c(\underset{}{\overset{}{\frown}}\overset{}{}) = 0$$

$$c(\underset{}{\overset{+}{\frown}}\overset{+}{}) = 1 \quad c(\underset{}{\overset{}{\frown}}\overset{+}{}) = -1 \quad c(\underset{}{\overset{+}{\frown}}\overset{-}{}) = -1 \quad c(\underset{}{\overset{}{\frown}}\overset{}{}) = 1$$

Let us do the calculation of some of these values in detail.

Take $A = \overset{+}{\underset{}{\frown}}\overset{-}{}$. According to the definition of I^{-1} on page 377 we have

$$c(A) = v(\longrightarrow) - v(\underset{}{\overset{+}{\frown}}) - v(\underset{}{\overset{-}{\frown}}) + v(\overset{+}{\underset{}{\frown}}\overset{-}{}).$$

The first three values vanish. Indeed, the first and third Gauss diagrams are descending, so they represent the trivial long knot, and the value of c_2 on it is equal to zero. For the second value one should use the Vassiliev skein relation (13.2)

$$v(\underset{}{\overset{+}{\frown}}) = v(\underset{}{\overset{-}{\frown}}) + v(\underset{}{\frown})$$

and then notice that both diagrams are descending. Moreover, for the second diagram with a single chord both resolutions of the corresponding double point lead to the trivial knot.

Thus we have

$$c(A) = v(\text{⌢⌢}) = v(\text{⌢⌢}) + v(\text{⌢⌢}).$$

The last two Gauss diagrams are descending. Therefore, $c(A) = 0$.

Now let us take $A = \text{⌢⌢}$. Applying I^{-1} to A we get that the value of c on the first three diagrams is equal to zero as before, and

$$c(\text{⌢⌢}) = v(\text{⌢⌢}).$$

To express the last Gauss diagram as a combination of descending diagrams, first we should reverse its right arrow using the relation (13.2):

$$\text{⌢⌢} = \text{⌢⌢} + \text{⌢⌢}.$$

The first Gauss diagram here is descending. But the second one is not, it has a prohibited pair. So we have to apply the map P from Section 13.1.6 to it. According to the first case of Figure 13.1 we have

$$\text{⌢⌢} = \text{⌢⌢⌢} = \text{⌢⌢⌢} + \text{⌢⌢⌢}.$$

In the first diagram we have to reverse one more arrow, and to the second diagram we need to apply the map P again. After that, the reversion of arrows in it would not create any problem since the additional terms would have 3 chords, and we can ignore them if we are interested in the second order invariant $v = c_2$ only. Modulo diagrams with three chords, we have

$$\text{⌢⌢} = \text{⌢⌢⌢} - \text{⌢⌢⌢} + \text{⌢⌢⌢}.$$

The first and third diagrams here are descending. But with the second one we have a problem because it has a prohibited interval with many (three) arrow ends on it. We need to apply P five times in order to make it descending modulo diagrams with three chords. The result will be a descending diagram B with two non-intersecting chords, one inside the other. So the value of v on it would be zero and we may ignore this part of the calculation (see Exercise 13.2). Nevertheless, we give the answer here so that interested readers can check their understanding of the procedure:

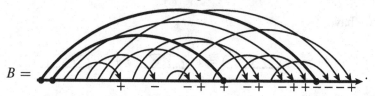

Combining all these results we have

$$\overset{+}{\underset{+}{\overbrace{\longrightarrow}}} \ = \ \overset{+}{\underset{-}{\overbrace{\longrightarrow}}} \ - B + \overset{}{\underset{+ \ +--}{\overbrace{\longrightarrow}}} \ + \overset{}{\underset{+ \ +-+--}{\overbrace{\longrightarrow}}}$$

modulo diagrams with at least three chords. The value of v on the last Gauss diagram is equal to its value on the descending knot with the same chord diagram, , namely, the knot

It is easy to see that the only resolution that gives a nontrivial knot is the positive resolution of the right double point together with the negative resolution of the left double point; the resulting knot is 4_1. The value of $v = c_2$ on it is -1 according to Table 2.1. Thus the value of v on this Gauss diagram is equal to 1. The values of v on the other three descending Gauss diagrams are zero. Therefore, we have

$$c(\overset{+}{\underset{}{\overbrace{\longrightarrow}}}{}^{+}) = 1.$$

As an exercise, the reader may wish to check all the other values of c from the table.

This table implies that the value of c_2 on a knot K with the Gauss diagram D is

$$c_2(K) = \Big\langle \ \overset{+}{\overbrace{\longrightarrow}}{}^{+} \ - \ \overset{-}{\overbrace{\longrightarrow}}{}^{+} \ - \ \overset{+}{\overbrace{\longrightarrow}}{}^{-} \ + \ \overset{-}{\overbrace{\longrightarrow}}{}^{-} \ , D \Big\rangle.$$

This formula differs from the one at the beginning of the chapter by the orientation of all its arrows.

13.2 Canonical actuality tables

As a byproduct of the proof of the Goussarov theorem, namely Lemma 13.3, we have the following refinement of the notion of an actuality table from Section 3.7.

In that section we have described a procedure of calculating a Vassiliev invariant given by an actuality table. This procedure involves some choices. First, in order to build the table, we have to choose for each chord diagram a singular knot representing it. Second, when calculating the knot invariant we have to choose repeatedly sequences of crossing changes that will express our knot as a linear combination of singular knots from the table.

It turns out that for long knots these choices can be eliminated. We shall now define something that can be described as a canonical actuality table and describe a calculation procedure for Vassiliev invariants that only depends on the initial Gauss diagram representing a knot. Strictly speaking, our "canonical actuality tables" are not actuality tables, since they contain one singular knot for each long chord diagram *with signed chords*.

A *canonical actuality table* for an invariant of order n is the set of its values on all singular long knots with descending diagrams and at most n double points.

For example, here is the canonical actuality table for the second coefficient c_2 of the Conway polynomial.

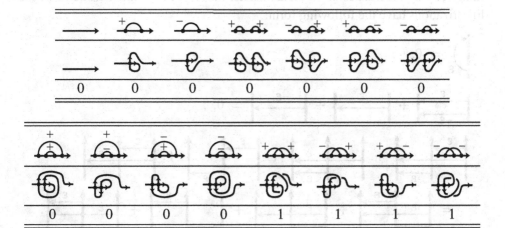

To remove the second ambiguity in the procedure of calculating a Vassiliev invariant, we use the algorithm from Section 13.1.6. It expresses an arbitrary Gauss diagram with chords as a linear combination of descending diagrams, modulo diagrams with more than n chords. The algorithm represents the value of a Vassiliev invariant as a linear combination of values from the canonical actuality table.

13.3 The Polyak algebra for virtual knots

The fact that all Vassiliev invariants can be expressed with the help of Gauss diagrams suggests that finite type invariants can be actually defined in the setup of Gauss diagrams. This is true and, moreover, there are two (inequivalent) ways to define Vassiliev invariants for virtual knots: that of Goussarov *et al.* (2000) and that of Kauffman (1999). Here we review the construction of Goussarov *et al.* (2000).

The reader should be warned that it is not known whether this definition coincides with the usual definition on classical knots. However, the logic behind it is very transparent and simple: the universal finite type invariant should send a knot to the sum of all of its "subknots." We have already seen this approach in action in Chapter 12 where the Magnus expansion of the free group was defined precisely this way.

13.3.1 The universal invariant of virtual knots

The map $I : \mathbb{Z}\mathbf{GD} \to \mathbb{Z}\mathbf{GD}$ from Section 13.1, sending a diagram to the sum of all its subdiagrams $I(D) = \sum_{D' \subseteq D} D'$, is clearly not invariant under the Reidemeister moves. However, we can make it invariant by simply taking the quotient of the image of I by the images of the Reidemeister moves, or their *linearizations*. These linearizations have the following form:

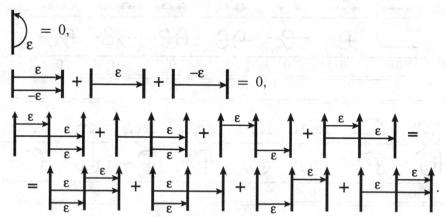

The space $\mathbb{Z}\mathbf{GD}$ modulo the linearized Reidemeister moves is called the *Polyak algebra*. The structure of an algebra comes from the connected sum of long Gauss diagrams; we shall not use it here. The Polyak algebra, which we denote by $\mathscr{P}$, looks rather different from the quotient of $\mathbb{Z}\mathbf{GD}$ by the usual Reidemeister moves, the latter being isomorphic to the free abelian group spanned by the set of all virtual knots $V\mathscr{K}$. Note, however, that by construction, the resulting invariant $I^* : \mathbb{Z}V\mathscr{K} \to \mathscr{P}$ is an isomorphism, and, therefore, contains the complete information about the virtual knot.

It is not clear how to do any calculations in $\mathscr{P}$, since the relations are not homogeneous. It may be more feasible to consider the (finite-dimensional) quotient $\mathscr{P}_n$ of $\mathscr{P}$ which is obtained by setting all the diagrams with more than n arrows equal to zero. In fact, the space $\mathscr{P}_n$ plays an important role in the theory of Vassiliev invariants for virtual knots. Namely, the map $I_n : \mathbb{Z}V\mathscr{K} \to \mathscr{P}_n$ obtained by composing I^* with the quotient map is an order n Vassiliev invariant for virtual knots, universal in the sense that any other order n invariant is obtained by composing I_n with some linear function on $\mathcal{P}_n$.

Let us now make this statement precise and define the Vassiliev invariants.

While the simplest operation on plane knot diagrams is the crossing change, for Gauss diagrams there is a similar, but even simpler manipulation: deleting/inserting of an arrow. An analogue of a knot with a double point for this operation is a diagram with a dashed arrow. A dashed arrow can be resolved by means of the following "virtual Vassiliev skein relation":

An invariant of virtual knots is said to be of finite type (or Vassiliev) of order n if it vanishes on all Gauss diagrams with more than n dashed arrows.

Observe that the effect of I on a diagram all of whose arrows are dashed consists in just making all the arrows solid. More generally, the image under I of a Gauss diagram with some dashed arrows is a sum of Gauss diagrams, all of which contain these arrows. It follows that I_n is of order n: indeed, if a Gauss diagram has more than n dashed arrows it is sent by I to a Gauss diagram with at least n arrows, which is zero in $\mathcal{P}_n$.

13.3.2 Dimensions of $\mathcal{P}_n$

The universal invariants I_n, in marked contrast with the Kontsevich integral, are defined in a simple combinatorial fashion. However, nothing comes for free: I_n takes its values in the space $\mathcal{P}_n$ which is hard to describe. For small n, the dimensions of $\mathcal{P}_n$ (over the real numbers) were calculated in Bar-Natan et al. (2009):

n	1	2	3	4
$\dim \mathcal{P}_n - \dim \mathcal{P}_{n-1}$	2	7	42	246

13.3.3 Open problems

A finite type invariant of order n for virtual knots gives rise to a finite type invariant of classical knots of at least the same order. Indeed, a crossing change can be thought of as deleting an arrow followed by inserting the same arrow with the direction reversed.

Exercise. Define the Vassiliev invariants for closed (unbased) virtual knots and show that the analogue of the space $\mathcal{P}_2$ is 0-dimensional. Deduce that the Casson knot invariant cannot be extended to a Vassiliev invariant of order 2 for closed virtual knots.

It is not clear, however, whether a finite type invariant of classical knots can be extended to an invariant of virtual *long* knots of the same order. The calculation of Goussarov *et al.* (2000) shows that this is true in orders 2 and 3.

Given that I^* is a complete invariant for virtual knots, one may hope that each virtual knot is detected by I_n for some n. It is not known whether this is the case. A positive solution to this problem would also mean that Vassiliev invariants distinguish classical knots.

It would be interesting to describe the kernel of the natural projection $\mathscr{P}_n \to \mathscr{P}_{n-1}$ which kills the diagrams with n arrows. First of all, notice that using the linearization of the second Reidemeister move, we can get rid of all signs in the diagrams in $\mathscr{P}_n$ that have exactly n arrows: changing the sign of an arrow just multiplies the diagram by -1. Now, the diagrams that have exactly n arrows satisfy the following 6T-relation in $\mathscr{P}_n$:

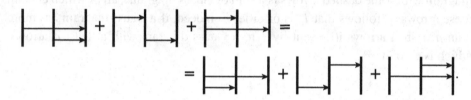

Consider the space $\vec{\mathscr{A}}_n$ of chord diagrams with n *oriented* chords, or arrows, modulo the 6T-relation. There is a map $i_n : \vec{\mathscr{A}}_n \to \mathscr{P}_n$, whose image is the kernel of the projection to $\mathscr{P}_{n-1}$. It is not clear, however, if i_n is an inclusion. The spaces $\vec{\mathscr{A}}_n$ were introduced in Polyak (2000) where their relation with usual chord diagrams is discussed. A further discussion of these spaces and their generalizations can be found in Bar-Natan (2011).

One more open problem is as follows. Among the linear combinations of Gauss diagrams of the order no greater than n, there are some that produce a well-defined invariant of degree n. Obviously, such combinations form a vector space, call it L_n. The combinations that lead to the identically zero invariant form a subspace L'_n. The quotient space L_n/L'_n is isomorphic to the space of Vassiliev invariants $\mathscr{V}_n$. The problem is to obtain a description of (or some information about) the spaces L_n and L'_n and in these terms learn something new about $\mathscr{V}_n$. For example, we have seen that the Casson invariant c_2 can be given by two different linear combinations k_1, k_2 of Gauss diagrams of order 2. It is not difficult to verify that these two combinations, together with the empty Gauss diagram k_0 that corresponds to the constant 1, span the space L_2. The subspace L'_2 is spanned by the difference $k_1 - k_2$. We see that $\dim L_2/L'_2 = 2 = \dim \mathscr{V}_2$. For degree 3 the problem is already open. We know, for instance, three linearly independent combinations of Gauss diagrams that produce the invariant j_3 (see Sections 13.4.2 and 13.4.4 below), but we do not know if their differences generate the space L'_3. Neither do we have any description of the space L_3.

13.4 Examples of Gauss diagram formulae

13.4.1 Highest part of the invariant

Let us start with one observation that will significantly simplify our formulae.

Lemma 13.5. *Let* $c : \mathbb{Z}\mathbf{GD} \to \mathbb{Z}$ *be a linear map representing an invariant of order n. If* $A_1, A_2 \in \mathbf{GD}$ *are diagrams with n arrows obtained from each other by changing the sign of one arrow, then* $c(A_1) = -c(A_2)$.

Proof As we noted before, a knot invariant c vanishes on all linearized Reidemeister moves of the form $I(R)$, where $R = 0$ is a usual Reidemeister move on realizable diagrams. Consider a linearized second Reidemeister move involving one diagram A_0 with $n + 1$ arrows and two diagrams A_1 and A_2 with n arrows. Clearly, c vanishes on A_0, and therefore $c(A_1) = -c(A_2)$. $\square$

This observation gives rise to the following notation. Let A be a Gauss diagram with n arrows *without signs, an unsigned Gauss diagram*. Given a Gauss diagram D, we denote by $\langle A, D \rangle$ the alternating sum

$$\sum_i (-1)^{\operatorname{sign} A_i} \langle A_i, D \rangle,$$

where the A_i are all possible Gauss diagrams obtained from A by putting signs on its arrows, and $\operatorname{sign} A_i$ is the number of chords of A_i whose sign is negative. Since the value of c on all the A_i coincides, up to sign, we can speak of the value of c on A.

For example, the formula for the Casson invariant of a knot K with the Gauss diagram D can be written as

$$c_2(K) = \left\langle \begin{array}{c} \text{⌢} \end{array}, D \right\rangle.$$

13.4.2 Invariants of degree 3

Apart from the Casson invariant, the simplest Vassiliev knot invariant is the coefficient $j_3(K)$ in the power series expansion of the Jones polynomial (see Section 3.6). Many formulae for $j_3(K)$ are known; the first such formula was found by M. Polyak and O. Viro in terms of unbased diagrams, see Polyak and

Viro (1994). From the results of Goussarov *et al.* (2000), the following Polyak–Viro expression for j_3 is easily derived:

$$-3\left\langle \text{} \right.$$

(In this formula a typo of Goussarov *et al.* (2000) is corrected.) Here the bracket $\langle \cdot, \cdot \rangle$ is assumed to be linear in its first argument.

S. Willerton in his thesis (Willerton 1998) found the following formula for j_3:

$$-3\left\langle \text{} \right.$$

A third Gauss diagram formula for j_3 will be given in Section 13.4.4.

Other combinatorial formulae for $c_2(K)$ and $j_3(K)$ were found earlier by J. Lannes (1993): they are not Gauss diagram formulae.

13.4.3 Coefficients of the Conway polynomial

Apart from the Gauss diagram formulae for the low degree invariants, two infinite series of such formulae are currently known: those for the coefficients of the Conway and the HOMFLY polynomials. The former can be, of course, derived from the latter, but we start from the discussion of the Conway polynomial, as it is easier. We shall follow the original exposition of Chmutov *et al.* (2009).

Definition 13.6. A chord diagram D is said to be *k-component* if after the parallel doubling of each chord as in the picture the resulting curve will have k components. We use the notation $|D| = k$. (See also Section 3.6.2.)

Example 13.7. For chord diagrams with two chords we have:

$$\left| \bigotimes \right| = 1 \Longleftarrow \bigotimes, \qquad\qquad \left| \bigcirc\!\!\!\bigcirc \right| = 3 \Longleftarrow \bigcirc\!\!\!\!\bigcirc.$$

We shall be interested in one-component diagrams only. With four chords, there are four one-component diagrams (the notation is borrowed from Table 4.1):

$$d_1^4 = \bigotimes, \quad d_5^4 = \bigotimes, \quad d_6^4 = \bigotimes, \quad \text{and} \quad d_7^4 = \bigotimes.$$

Definition 13.8. We can turn a one-component chord diagram with a base point into an arrow diagram according to the following rule. Starting from the base point we travel along the diagram with doubled chords. During this journey we pass both copies of each chord in opposite directions. Choose an arrow on each chord which corresponds to the direction of the first passage of the chord. Here is an example.

$$\bigotimes \rightsquigarrow \bigotimes \rightsquigarrow \bigotimes.$$

We call the Gauss diagram obtained in this way *ascending*.

Definition 13.9. The *Conway combination* $\mathcal{C}_{2n}$ is the sum of all based one-component ascending Gauss diagrams with $2n$ arrows. For example,

$$\mathcal{C}_2 := \bigotimes,$$

$$\mathcal{C}_4 := \bigotimes + \bigotimes + \bigotimes + \bigotimes + \bigotimes +$$

$$+ \bigotimes + \bigotimes + \bigotimes + \bigotimes + \bigotimes + \bigotimes + \bigotimes + \bigotimes +$$

$$+ \bigotimes + \bigotimes + \bigotimes + \bigotimes + \bigotimes + \bigotimes + \bigotimes + \bigotimes.$$

Note that for a given one-component chord diagram we have to consider all possible choices for the base point. However, some choices may lead to the same Gauss diagram. In $\mathcal{C}_{2n}$ we list them without repetitions. For instance, all choices of a base point for the diagram d_1^4 give the same Gauss diagram. So d_1^4 contributes only one Gauss diagram to $\mathcal{C}_4$. The diagram d_7^4 contributes four Gauss diagrams because of its symmetry, while d_5^4 and d_6^4 contribute eight Gauss diagrams each.

Theorem 13.10. *For $n \geqslant 1$, the coefficient c_{2n} of z^{2n} in the Conway polynomial of a knot K with the Gauss diagram G is equal to*

$$c_{2n} = \langle \mathfrak{C}_{2n}, G \rangle.$$

Example 13.11. Consider the knot $K := 6_2$ and its Gauss diagram $G := G(6_2)$:

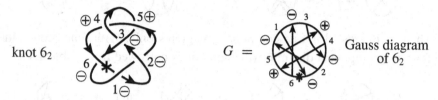

In order to compute the pairing $\langle \mathfrak{C}_4, G \rangle$ we must match the arrows of each diagram of $\mathfrak{C}_4$ with the arrows of G. One common property of all terms in $\mathfrak{C}_{2n}$ is that in each term both endpoints of the arrows that are adjacent to the base point are *arrowtails*. This follows from our construction of $\mathfrak{C}_{2n}$. Hence, the arrow $\{1\}$ of G can not participate in the matching with any diagram of $\mathfrak{C}_4$. The only candidates to match with the first arrow of a diagram of $\mathfrak{C}_4$ are the arrows $\{2\}$ and $\{4\}$ of G. If it is $\{4\}$, then $\{1, 2, 3\}$ cannot participate in the matching, and there remain only three arrows to match with the four arrows of $\mathfrak{C}_4$. Therefore, the arrow of G which matches with the first arrow of a diagram of $\mathfrak{C}_4$ must be $\{2\}$. In a similar way, we can find that the arrow of G which matches with the last arrow of a diagram of $\mathfrak{C}_4$ must be $\{6\}$. This leaves three possibilities to match with the four arrows of $\mathfrak{C}_4$: $\{2, 3, 4, 6\}$, $\{2, 3, 5, 6\}$, and $\{2, 4, 5, 6\}$. Checking them all, we find only one quadruple, $\{2, 3, 5, 6\}$, which constitute a diagram equal to the second diagram in the second row of $\mathfrak{C}_4$. The product of the local writhes of the arrows $\{2, 3, 5, 6\}$ is equal to $(-1)(-1)(+1)(-1) = -1$. Thus,

$$\langle \mathfrak{C}_4, G \rangle = \langle \bigcirc\!\!\!\!\!\!\bigstar , G \rangle = -1,$$

which coincides with the coefficient c_4 of the Conway polynomial $\nabla(K) = 1 - z^2 - z^4$.

13.4.4 Coefficients of the HOMFLY polynomial

Let $P(K)$ be the HOMFLY polynomial of the knot K. Substitute $a = e^h$ and take the Taylor expansion in h. The result will be a Laurent polynomial in z and a power series in h. Let $p_{k,l}(K)$ be the coefficient of $h^k z^l$ in that expression. The numbers $p_{0,l}$ coincide with the coefficients of the Conway polynomial, since the latter is obtained from HOMFLY by fixing $a = 1$.

Remark 13.12. It follows from Exercise 3.22 that

(1) for all nonzero terms the sum $k + l$ is non-negative;

(2) $p_{k,l}$ is a Vassiliev invariant of degree no greater than $k + l$;

(3) if l is odd, then $p_{k,l} = 0$.

We shall describe a Gauss diagram formula for $p_{k,l}$ following Chmutov and Polyak (2010).

Let A be a (based, or long) Gauss diagram, S a subset of its arrows (referred to as a *state*) and α an arrow of A. Doubling all the arrows in A that belong to S, in the same fashion as in the preceding section, we obtain a diagram consisting of one or several circles with some signed arrows attached to them. Denote by $\langle \alpha | A | S \rangle$ the expression in two variables h and z that depends on the sign of the chord α and the type of the first passage of α (starting from the basepoint) according to the following table:

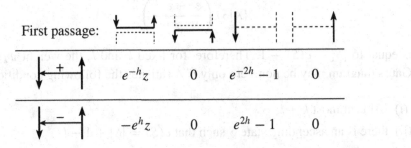

To the Gauss diagram A we then assign a power series $W(A)$ in h and z defined by

$$W(A) = \sum_{S} \langle A | S \rangle \left(\frac{e^h - e^{-h}}{z} \right)^{c(S)-1},$$

where $\langle A | S \rangle = \prod_{\alpha \in A} \langle \alpha | A | S \rangle$ and $c(S)$ is the number of components obtained after doubling all the chords in S. Denote by $w_{k,l}(A)$ the coefficient of $h^k z^l$ in this power series and consider the following linear combination of Gauss diagrams: $A_{k,l} := \sum w_{k,l}(A) \cdot A$. Note that the number $w_{k,l}(A)$ is non-zero only for a finite number of diagrams A.

Theorem 13.13. *Let G be a Gauss diagram of a knot L. Then*

$$p_{k,l}(K) = \langle A_{k,l}, G \rangle.$$

For a proof of the theorem, we refer the reader to the original paper (Chmutov and Polyak 2010). Here we only give one example. To facilitate the practical application of the theorem, we start with some general remarks.

A state S of a Gauss diagram A is called *ascending*, if in traversing the diagram with doubled arrows we approach the neighbourhood of every arrow (not only the ones in S) first at the arrow head. As follows directly from the construction, only ascending states contribute to $W(A)$.

Note that since

$$e^{\pm 2h} - 1 = \pm 2h + \text{(higher degree terms)}$$

and

$$\pm e^{\mp h} z = \pm z + \text{(higher degree terms)},$$

the power series $W(A)$ starts with terms of degree at least $|A|$, the number of arrows of A. Moreover, the z-power of

$$\langle A|S \rangle \left(\frac{e^h - e^{-h}}{z} \right)^{c(S)-1}$$

is equal to $|S| - c(S) + 1$. Therefore, for fixed k and l, the weight $w_{k,l}(A)$ of a Gauss diagram may be non-zero only if A satisfies the following conditions:

(i) $|A|$ is at most $k + l$;

(ii) there is an ascending state S such that $c(S) = |S| + 1 - l$.

For diagrams of the highest degree $|A| = k + l$, the contribution of an ascending state S to $w_{k,l}(A)$ is equal to $(-1)^{|A|-|S|} 2^k \varepsilon(A)$, where $\varepsilon(A)$ is the product of signs of all arrows in A. If two such diagrams A and A' with $|A| = k + l$ differ only by signs of the arrows, their contributions to $A_{k,l}$ differ by the sign $\varepsilon(A)\varepsilon(A')$. Thus all such diagrams may be combined to the unsigned diagram A, appearing in $A_{k,l}$ with the coefficient $\sum_S (-1)^{|A|-|S|} 2^k$ (where the summation is over all ascending states of A with $c(S) = |S| + 1 - l$).

Exercise. Prove that Gauss diagrams with isolated arrows do not contribute to $A_{k,l}$. (*Hint:* All ascending states cancel out in pairs.)

Now, by way of example, let us find an explicit formula for $A_{1,2}$. The maximal number of arrows is equal to 3. To get z^2 in $W(A)$ we need ascending states with either $|S| = 2$ and $c(S) = 1$, or $|S| = 3$ and $c(S) = 2$. In the first case the equation $c(S) = 1$ means that the two arrows of S must intersect. In the second case the equation $c(S) = 2$ does not add any restrictions on the relative position of the arrows. In the cases $|S| = |A| = 2$ or $|S| = |A| = 3$, since S is ascending, A itself must be ascending as well.

For diagrams of the highest degree $|A| = 1 + 2 = 3$, we must count ascending states of unsigned Gauss diagrams with the coefficient $(-1)^{3-|S|}2$, that is, -2 for $|S| = 2$ and $+2$ for $|S| = 3$. There are only four types of (unsigned) 3-arrow Gauss diagrams with no isolated arrows:

Diagrams of the same type differ by the directions of arrows.

For the first type, recall that the first arrow should be oriented towards the base point; this leaves 4 possibilities for the directions of the remaining two arrows. One of them, namely

does not have ascending states with $|S| = 2, 3$. The remaining possibilities, together with their ascending states, are shown in the table:

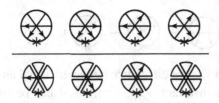

The final contribution of this type of diagram to $A_{1,2}$ is equal to

The other three types of degree 3 diagrams differ by the location of the base point. A similar consideration shows that 5 out of the total of 12 Gauss diagrams of these types, namely

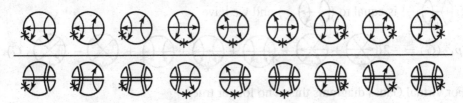

do not have ascending states with $|S| = 2, 3$. The remaining possibilities, together with their ascending states, are shown in the table:

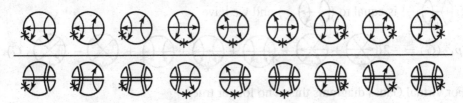

The contribution of this type of diagrams to $A_{1,2}$ is thus equal to

$$-2\ \text{⊕} - 2\ \text{⊕} - 2\ \text{⊕} + 2\ \text{⊕} - 2\ \text{⊕}.$$

Apart from diagrams of degree 3, some degree 2 diagrams contribute to $A_{1,2}$ as well. Since $|A| = 2 < k + l = 3$, contributions of 2-diagrams depend also on their signs. Such diagrams must be ascending (since $|S| = |A| = 2$) and should not have isolated arrows. There are four such diagrams: ⊗ , with all choices of the signs ε_1, ε_2 for the arrows. For each choice we have $\langle A|S \rangle = \varepsilon_1 \varepsilon_2 e^{-(\varepsilon_1 + \varepsilon_2)h} z^2$. If $\varepsilon_1 = -\varepsilon_2$, then $\langle A|S \rangle = -z^2$, so the coefficient of hz^2 vanishes and such diagrams do not occur in $A_{1,2}$. For the two remaining diagrams with $\varepsilon_1 = \varepsilon_2 = \pm$, the coefficients of hz^2 in $\langle A|S \rangle$ are equal to ∓ 2 respectively.

Combining all the above contributions, we finally get

$$A_{1,2} = -2\Big(\text{⊗} + \text{⊗} + \text{⊕} + \text{⊕} + \text{⊕} - \text{⊕}$$

$$+ \text{⊕} + \text{⊗} - \text{⊗} \Big).$$

At this point we can see the difference between virtual and classical long knots. For classical knots the invariant $\mathscr{I}_{A_{1,2}} = \langle A_{1,2}, \cdot \rangle$ can be simplified further. Note that any realizable Gauss diagram G satisfies

$$\langle \text{⊕}, G \rangle = \langle \text{⊕}, G \rangle.$$

This follows from the symmetry of the linking number. Indeed, suppose we have matched two vertical arrows (which are the same in both diagrams) with two arrows of G. Let us consider the orientation preserving smoothings of the corresponding two crossings of the link diagram D associated with G. The smoothed diagram $\widetilde{D}$ will have three components. Matchings of the horizontal arrow of our Gauss diagrams with an arrow of G both measure the linking number between the first and the third components of $\widetilde{D}$, using crossings when the first component passes over (respectively, under) the third one. Thus, as functions on classical Gauss diagrams,

$\langle , \text{⊕}, \cdot \rangle$ is equal to $\langle \text{⊕}, \cdot \rangle$ and we have

$$p_{1,2}(G) = -2\langle \text{⊗} + \text{⊗} + \text{⊕} + \text{⊕} + \text{⊕} + \text{⊗} - \text{⊗}, G \rangle.$$

For virtual Gauss diagrams this is no longer true.

In a similar manner one may check that $A_{3,0} = -4A_{1,2}$.

The obtained result implies one more formula for the invariant j_3 (compare it with the two other formulae given in Section 13.4.2). Indeed, $j_3 = -p_{3,0} - p_{1,2} = 3p_{1,2}$, therefore

$$j_3(K) = -6\langle \bigotimes + \bigotimes + \bigoplus + \bigoplus + \bigoplus + \bigotimes - \bigotimes , G\rangle.$$

13.5 The Jones polynomial via Gauss diagrams

Apart from the Gauss diagram formulae as understood in this chapter, there are many other ways to extract Vassiliev (and other) knot invariants from Gauss diagrams. Here is just one example: a description of the Jones polynomial (which is essentially a reformulation of the construction from a paper by L. Zulli (1995)). The reader should compare it to the definition of $\mathfrak{so}_N$-weight system in Section 6.1.8.

Let G be a Gauss diagram representing a knot K. Denote by $[G]$ the set of arrows of G. The sign of an arrow $c \in [G]$ can be considered as a value of the function sign : $[G] \to \{-1, +1\}$. A *state* s for G is an arbitrary function $s : [G] \to \{-1, +1\}$; in particular, for a Gauss diagram with n arrows there are 2^n states. The function $\text{sign}(\cdot)$ is one of them. With each state s we associate an immersed plane curve in the following way. Double every chord c according to the rule:

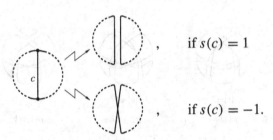

$$\text{if } s(c) = 1$$

$$\text{if } s(c) = -1.$$

Let $|s|$ denote the number of connected components of the curve obtained by doubling all the chords of G. Also, for a state s we define an integer

$$p(s) := \sum_{c \in [G]} s(c) \cdot \text{sign}(c).$$

The defining relations for the Kauffman bracket from Section 2.4.1 lead to the following expression for the Jones polynomial.

Theorem 13.14.

$$J(K) = (-1)^{w(K)} t^{3w(K)/4} \sum_s t^{-p(s)/4} \left(-t^{-1/2} - t^{1/2}\right)^{|s|-1},$$

where the sum is taken over all 2^n states for G and $w(K) = \sum_{c \in [G]} \text{sign}(c)$ is the writhe of K.

This formula can be used to extend the Jones polynomial to virtual knots.

Example 13.15. For the left trefoil knot 3_1 we have the following Gauss diagram:

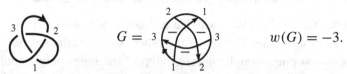

There are eight states for such a diagram. Here are the corresponding curves and numbers $|s|$, $p(s)$.

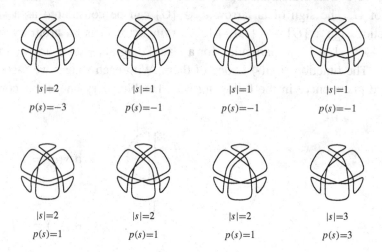

Therefore,

$$
\begin{aligned}
J(3_1) &= -t^{-9/4}\left(t^{3/4}\left(-t^{-1/2} - t^{1/2}\right) + 3t^{1/4} + 3t^{-1/4}\left(-t^{-1/2} - t^{1/2}\right)\right.\\
&\qquad\left. + t^{-3/4}\left(-t^{-1/2} - t^{1/2}\right)^2\right)\\
&= -t^{-9/4}\left(-t^{1/4} - t^{5/4} - 3t^{-3/4} + t^{-3/4}\left(t^{-1} + 2 + t\right)\right)\\
&= t^{-1} + t^{-3} - t^{-4},
\end{aligned}
$$

as we had before in Chapter 2.

Exercises

13.1   Gauss diagrams and Gauss diagram formulae may be defined for links in the same way as for knots. Prove that for a link L with two components K_1 and K_2

$$lk(K_1, K_2) = \langle\ \ \underline{\quad\quad}\ , G(L)\rangle.$$

13.2   Find a sequence of Reidemeister moves that transforms the Gauss diagram B from page 386 to the diagram

Show that this diagram is not realizable. Calculate the value of the extension, according to Section 13.1.3, of the invariant c_2 on it.

13.3   Let $\vec{\mathscr{A}}$ be the space of arrow (oriented chord) diagrams modulo the 6T relations, see page 390. Show that the map $\mathscr{A} \to \vec{\mathscr{A}}$ which sends a chord diagram to the sum of all the arrow diagrams obtained by putting the orientations on the chords is well-defined. In other words, show that the 6T relation implies the 4T relation.

13.4 * Construct analogues of the algebras of closed and open Jacobi diagrams $\mathscr{C}$ and $\mathscr{B}$ consisting of diagrams with oriented edges. (It is known how to do it in the case of closed diagrams with acyclic internal graph; see Polyak 2000.)

14

Miscellany

14.1 The Melvin–Morton Conjecture

14.1.1 Formulation

Roughly speaking, the Melvin–Morton Conjecture says that the Alexander–Conway polynomial can be read off the highest order part of the coloured Jones polynomial.

According to Exercise 6.26 (see also Melvin and Morton 1995; Bar-Natan and Garoufalidis 1996) the coefficient J_n^k of the unframed coloured Jones polynomial J^k (Section 11.2.3) is a polynomial in k, of degree at most $n + 1$ and without constant term. So we may write

$$\frac{J_n^k}{k} = \sum_{0 \leqslant j \leqslant n} b_{n,j} k^j \qquad \text{and} \qquad \frac{J^k}{k} = \sum_{n=0}^{\infty} \sum_{0 \leqslant j \leqslant n} b_{n,j} k^j h^n ,$$

where $b_{n,j}$ are Vassiliev invariants of order $\leqslant n$. The highest order part of the coloured Jones polynomial is a Vassiliev power series invariant

$$\mathrm{MM} := \sum_{n=0}^{\infty} b_{n,n} h^n .$$

The Melvin–Morton Conjecture (Melvin and Morton 1995). *The highest order part of the coloured Jones polynomial MM is inverse to the Alexander–Conway power series A defined by equations (11.1–11.2). In other words,*

$$\mathrm{MM}(K) \cdot A(K) = 1$$

for any knot K.

14.1.2 Historical remarks

H. Morton (1995) proved the conjecture for torus knots. After this L. Rozansky (1996) proved the Melvin–Morton Conjecture on the level of rigour of Witten's path integral interpretation for the Jones polynomial. The first complete proof was carried out by D. Bar-Natan and S. Garoufalidis (1996). They invented a remarkable reduction of the conjecture to a certain identity on the corresponding weight systems via canonical invariants; we shall review this reduction in Section 14.1.3. This identity was then verified by evaluating the weight systems on chord diagrams. In fact, Bar-Natan and S. Garoufalidis (1996) proved a more general theorem that relates the highest order part of an arbitrary quantum invariant to the Alexander–Conway polynomial. Following Chmutov (1997) we shall present another proof of this generalized Melvin–Morton Conjecture in Section 14.1.5. A. Kricker, B. Spence and I. Aitchison (1997) proved the Melvin–Morton Conjecture using the cabling operations. Their work was further generalized in Kricker (1997). Yet another proof of the Melvin–Morton Conjecture appeared in the paper by A. Vaintrob (1996). He used calculations on chord diagrams and the Lie superalgebra $\mathfrak{gl}(1|1)$ which gives rise to the Alexander–Conway polynomial. The idea to use the restriction of the aforementioned identity on weight systems to the primitive space was explored in Chmutov (1998) and Vaintrob (1997). We shall follow Chmutov (1998) in the direct calculation of the Alexander–Conway weight system in Section 14.1.4.

B. I. Kurpita and K. Murasugi found a different proof of the Melvin–Morton Conjecture which does not use Vassiliev invariants and weight systems (Kurpita and Murasugi 1998).

Among other things, the works on the Melvin–Morton Conjecture inspired L. Rozansky to state his Rationality Conjecture that describes the fine structure of the Kontsevich integral. This conjecture was proved by A. Kricker, and is the subject of Section 11.5.

14.1.3 Reduction to weight systems

Since both power series Vassiliev invariants MM and A are canonical, so is their product (see Exercise 11.8). The constant invariant which is identically equal to 1 on all knots is also a canonical invariant. We see that the Melvin–Morton Conjecture states that two canonical invariants are equal, and it is enough to prove that their symbols coincide.

Introduce the notation

$$S_{\mathrm{MM}} := \mathrm{symb}(\mathrm{MM}) = \sum_{n=0}^{\infty} \mathrm{symb}(b_{n,n}) ;$$

$$S_A \;:=\; \mathrm{symb}(A) = \mathrm{symb}(C) = \sum_{n=0}^{\infty} \mathrm{symb}(c_n).$$

The Melvin–Morton Conjecture is equivalent to the relation

$$S_{MM} \cdot S_A = \mathbf{I}_0 \,.$$

This is obvious in degrees 0 and 1. So, basically, we must prove that in degree $\geqslant 2$ the product $S_{MM} \cdot S_A$ equals zero. In order to show this we have to establish that $S_{MM} \cdot S_A$ vanishes on any product $p_1 \cdot \ldots \cdot p_n$ of primitive elements of degree > 1.

The weight system S_{MM} is the highest part of the weight system $\varphi_{\mathfrak{sl}_2}^{'V_k}/k$ from Exercise 6.26. The latter is multiplicative as we explained in Section 6.1.4; hence, S_{MM} is multiplicative too. Exercise 3.16 implies then that the weight system S_A is also multiplicative. In other words, both weight systems S_{MM} and S_A are group-like elements of the Hopf algebra of weight systems $\mathcal{W}$. A product of two group-like elements is group-like which shows that the weight system $S_{MM} \cdot S_A$ is multiplicative. Therefore, it is sufficient to prove that

$$S_{MM} \cdot S_A \big|_{\mathscr{P}_{>1}} = 0 \,.$$

By the definitions of the weight system product and of a primitive element

$$S_{MM} \cdot S_A(p) = (S_{MM} \otimes S_A)(\delta(p)) = S_{MM}(p) + S_A(p) \,.$$

Therefore, we have reduced the Melvin–Morton Conjecture to the equality

$$S_{MM} \big|_{\mathscr{P}_{>1}} + S_A \big|_{\mathscr{P}_{>1}} = 0 \,.$$

Now we shall exploit the filtration

$$0 = \mathscr{P}_n^1 \subseteq \mathscr{P}_n^2 \subseteq \mathscr{P}_n^3 \subseteq \cdots \subseteq \mathscr{P}_n^n = \mathscr{P}_n$$

from Section 5.5.2. Recall that the wheel $\overline{w_n}$ spans $\mathscr{P}_n^n/\mathscr{P}_n^{n-1}$ for even n and belongs to $\mathscr{P}_n^{n-1}$ for odd n.

The Melvin–Morton Conjecture is a consequence of the following theorem.

Theorem 14.1. *The weight systems S_{MM} and S_A satisfy*

1. $S_{MM}\big|_{\mathscr{P}_n^{n-1}} = S_A\big|_{\mathscr{P}_n^{n-1}} = 0;$
2. $S_{MM}(\overline{w_{2m}}) = 2, \quad S_A(\overline{w_{2m}}) = -2.$

The proof is based on several exercises from Chapter 6.

First, let us consider the weight system S_{MM}. Exercise 6.24 implies that for any $D \in \mathscr{P}_n^{n-1}$ the weight system $\varphi_{\mathfrak{sl}_2}(D)$ is a polynomial in c of degree less than or equal to $[(n-1)/2]$. The weight system of the coloured Jones polynomial is obtained from $\varphi_{\mathfrak{sl}_2}$ by fixing the representation V_k of $\mathfrak{sl}_2$ and deframing. Choosing the representation V_k means that we have substituted $c = \frac{k^2-1}{2}$; the degree of the polynomial $\varphi_{\mathfrak{sl}_2}^{V_k}(D)/k$ in k will be at most $n-1$. Therefore, its nth degree term vanishes and $S_{\mathrm{MM}}\big|_{\mathscr{P}_n^{n-1}} = 0$. According to Exercise 6.22, the highest degree term of the polynomial $\varphi_{\mathfrak{sl}_2}(\overline{w_{2m}})$ is $2^{m+1}c^m$. Again, the substitution $c = \frac{k^2-1}{2}$ (taking the trace of the corresponding operator and dividing the result by k) gives that the highest degree term of $\varphi_{\mathfrak{sl}_2}^{V_k}(\overline{w_{2m}})/k$ is $\frac{2^{m+1}k^{2m}}{2^m} = 2k^{2m}$, and, hence $S_{\mathrm{MM}}(\overline{w_{2m}}) = 2$.

In order to treat the weight system S_A we use Exercise 6.32, which contains the equality $S_A(\overline{w_{2m}}) = -2$ as a particular case. It remains to prove that $S_A\big|_{\mathscr{P}_n^{n-1}} = 0$.

14.1.4 Alexander–Conway weight system

Using the state sum formula for S_A from Exercise 6.32, we shall prove that $S_A(D) = 0$ for any closed diagram $D \in \mathscr{P}_n^{n-1}$.

First of all, note that any such $D \in \mathscr{P}_n^{n-1}$ has an internal vertex which is not connected to any leg by an edge. Indeed, each leg is connected with only one internal vertex. The diagram p has at most $n-1$ legs and $2n$ vertices in total, so there must be at least $n+1$ internal vertices, and only $n-1$ of them can be connected with legs.

Pick such a vertex connected only with other internal vertices. There are two possible cases: either all these other vertices are different or two of them coincide.

Let us start with the second, easier, case. Here we have a "bubble"

$$\!\!\!\! \multimap\!\!\mathrel{\ooalign{\hss$\circ$\hss}}\!\!\multimap \ .$$

After resolving the vertices of this fragment according to the state sum formula and erasing the curves with more than one component, we are left with the linear combination of curves

$$-2\,\succ\!\!\!\!\prec \quad = \quad + \quad 2\,\succ\!\!\!\!\prec \quad =$$

which cancel each other, so $S_A(D) = 0$.

For the first case we formulate our claim as a lemma.

Lemma 14.2. $\quad S_A\!\left(\right) = 0.$

We shall utilize the state surfaces $\Sigma_s(D)$ from Exercise 6.28. For a given state, the neighbourhoods of "+"- and "−"-vertices look on the surface like three meeting bands:

$$\tag{14.1}$$

Switching a marking (value of the state) at a vertex means reglueing the three bands along two chords on the surface:

cut along chords   interchange   glue

Proof According to Exercise 6.32, the symbol of the Conway polynomial is the coefficient of N in the polynomial $\varphi_{gl_N}^{St}$. In terms of the state surfaces this means that we only have to consider the surfaces with one boundary component. We are going to divide the set of all those states s for which the state surface $\Sigma_s(D)$ has one boundary component into pairs in such a way that the states s and s' of the same pair differ by an odd number of markings. The terms of the pairs will cancel each other and will contribute zero to $S_A(D)$.

In fact, in order to do this we shall adjust only the markings of the four vertices of the fragment pictured in the statement of the Lemma. The markings $\varepsilon_1, \ldots, \varepsilon_l$ and $\varepsilon_1', \ldots, \varepsilon_l'$ in the states s and s' will be the same except for some markings of the four vertices of the fragment. Denote the vertices by v, v_a, v_b, v_c and their markings in the state s by $\varepsilon, \varepsilon_a, \varepsilon_b, \varepsilon_c$, respectively.

Assume that $\Sigma_s(D)$ has one boundary component. Modifying the surface as in (14.1), we can suppose that the neighbourhood of the fragment has the form

Draw nine chords $a, a_1, a_2, b, b_1, b_2, c, c_1, c_2$ on our surface as shown on the picture. The chords a, b, c are located near the vertex v; a, a_1, a_2 near the vertex v_a; b, b_1, b_2 near v_b and c, c_1, c_2 near v_c.

Since the surface has only one boundary component, we can draw this boundary as a plane circle and $a, a_1, a_2, b, b_1, b_2, c, c_1, c_2$ as chords inside it. Let us consider the possible chord diagrams obtained in this way.

If two, say b and c, of three chords located near a vertex, say v, do not intersect, then the surface $\Sigma_{\ldots, -\varepsilon, \varepsilon_a, \varepsilon_b, \varepsilon_c, \ldots}(D)$ obtained by switching the marking ε to $-\varepsilon$

also has only one boundary component. Indeed, the reglueing effect along two non-intersecting chords can be seen on chord diagrams as follows:

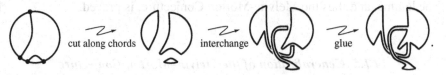

Therefore, in this case, the state $s = \{\dots, -\varepsilon, \varepsilon_a, \varepsilon_b, \varepsilon_c, \dots\}$ should be paired with $s' = \{\dots, \varepsilon, \varepsilon_a, \varepsilon_b, \varepsilon_c, \dots\}$.

We see that switching a marking at a vertex we increase the number of boundary components (so that such a marked diagram may give a non-zero contribution to $S_A(D)$) if and only if the three chords located near the vertex intersect pairwise.

Now we can suppose that any two of the three chords in each triple (a, b, c), (a, a_1, a_2), (b, b_1, b_2), (c, c_1, c_2) intersect. This leaves us with only one possible chord diagram:

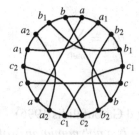

The boundary curve of the surface connects the ends of our fragment as in the left picture below.

$\Sigma_{\dots, \varepsilon, \varepsilon_a, \varepsilon_b, \varepsilon_c, \dots}(p)$ $\Sigma_{\dots, \varepsilon, -\varepsilon_a, -\varepsilon_b, -\varepsilon_c, \dots}(p)$

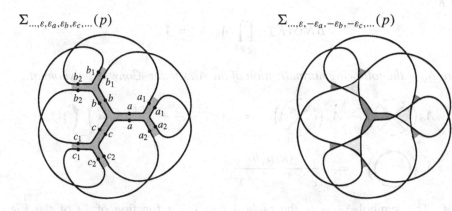

Switching markings at v_a, v_b, v_c gives a surface which also has one boundary component as in the right picture above. Pairing the state

$$s = \{\dots, \varepsilon, \varepsilon_a, \varepsilon_b, \varepsilon_c, \dots\}$$

up with $s' = \{\dots, \varepsilon, -\varepsilon_a, -\varepsilon_b, -\varepsilon_c, \dots\}$ we get the desired result.

The lemma, and thus the Melvin–Morton Conjecture, is proved. □

14.1.5 Generalization of the Melvin–Morton Conjecture to other quantum invariants

In this section we shall need some notions from the structure theory of Lie algebras; the reader can find the necessary information in Humphreys (1980).

Let $\mathfrak{g}$ be a semi-simple Lie algebra and let V_λ be an irreducible representation of $\mathfrak{g}$ of the highest weight λ. Denote by $\mathfrak{h}$ a Cartan subalgebra of $\mathfrak{g}$, by R the set of all roots and by R^+ the set of positive roots. Let $\langle \cdot, \cdot \rangle$ be the scalar product on $\mathfrak{h}^*$ induced by the Killing form. These data define the unframed quantum invariant $\theta_\mathfrak{g}^{V_\lambda}$ which, after the substitution $q = e^h$ and the expansion into a power series in h, can be written as

$$\theta_\mathfrak{g}^{V_\lambda} = \sum_{n=0}^{\infty} \theta_{\mathfrak{g},n}^{\lambda} h^n,$$

see Section 11.2.2.

Theorem 14.3. (Bar-Natan and Garoufalidis 1996)
(1) *The invariant $\theta_{\mathfrak{g},n}^{\lambda} / \dim(V_\lambda)$ is a polynomial in λ of degree at most n.*

(2) *Define the Bar-Natan–Garoufalidis function BNG as a power series in h whose coefficient at h^n is the degree n part of the polynomial $\theta_{\mathfrak{g},n}^{\lambda} / \dim(V_\lambda)$. Then for any knot K,*

$$BNG(K) \cdot \prod_{\alpha \in R^+} A_\alpha(K) = 1,$$

where A_α is the following normalization of the Alexander–Conway polynomial:

$$A_\alpha\left(\begin{array}{c}\includegraphics\end{array}\right) - A_\alpha\left(\begin{array}{c}\includegraphics\end{array}\right) = (e^{\frac{\langle \lambda, \alpha \rangle h}{2}} - e^{-\frac{\langle \lambda, \alpha \rangle h}{2}}) A_\alpha\left(\begin{array}{c}\includegraphics\end{array}\right);$$

$$A_\alpha\left(\begin{array}{c}\bigcirc\end{array}\right) = \frac{\langle \lambda, \alpha \rangle h}{e^{\frac{\langle \lambda, \alpha \rangle h}{2}} - e^{-\frac{\langle \lambda, \alpha \rangle h}{2}}}.$$

Proof The symbol S_{BNG} is the highest part (as a function of λ) of the Lie algebra weight system $\varphi_\mathfrak{g}'^{V_\lambda}$ associated with the representation V_λ. According to Exercise 11.6, the symbol of A_α in degree n equals $\langle \lambda, \alpha \rangle^n \mathrm{symb}(c_n)$.

The relation between the invariants can be reduced to the following relation between their symbols:

$$S_{BNG}\big|_{\mathscr{P}_n} + \sum_{\alpha \in R^+} \langle \lambda, \alpha \rangle^n \mathrm{symb}(c_n)\big|_{\mathscr{P}_n} = 0,$$

for $n > 1$.

As above, $S_{BNG}\big|_{\mathscr{P}_n^{n-1}} = \mathrm{symb}(c_n)\big|_{\mathscr{P}_n^{n-1}} = 0$, and $\mathrm{symb}(c_n)(\overline{w_{2m}}) = -2$. Thus it remains to prove that

$$S_{BNG}(\overline{w_{2m}}) = 2 \sum_{\alpha \in R^+} \langle \lambda, \alpha \rangle^{2m}.$$

To prove this equality we shall use the method of Section 6.2. First, we take the Weyl basis of $\mathfrak{g}$ and write the Lie bracket tensor J in this basis.

Fix the root space decomposition $\mathfrak{g} = \mathfrak{h} \oplus \left(\bigoplus_{\alpha \in R} \mathfrak{g}_\alpha \right)$. The Cartan subalgebra $\mathfrak{h}$ is orthogonal to all the $\mathfrak{g}_\alpha$'s and $\mathfrak{g}_\alpha$ is orthogonal to $\mathfrak{g}_\beta$ for $\beta \neq -\alpha$. Choose the elements $e_\alpha \in \mathfrak{g}_\alpha$ and $h_\alpha = [e_\alpha, e_{-\alpha}] \in \mathfrak{h}$ for each $\alpha \in R$ in such a way that $\langle e_\alpha, e_{-\alpha} \rangle = 2/\langle \alpha, \alpha \rangle$, and for any $\lambda \in \mathfrak{h}^*$, $\lambda(h_\alpha) = 2\langle \lambda, \alpha \rangle / \langle \alpha, \alpha \rangle$.

The elements $\{h_\beta, e_\alpha\}$, where β belongs to a basis $B(R)$ of R and $\alpha \in R$, form the Weyl basis of $\mathfrak{g}$. The Lie bracket $[\cdot, \cdot]$ as an element of $\mathfrak{g}^* \otimes \mathfrak{g}^* \otimes \mathfrak{g}$ can be written as follows:

$$[\cdot, \cdot] = \sum_{\substack{\beta \in B(R) \\ \alpha \in R}} \left(h_\beta^* \otimes e_\alpha^* \otimes \alpha(h_\beta) e_\alpha - e_\alpha^* \otimes h_\beta^* \otimes \alpha(h_\beta) e_\alpha \right)$$

$$+ \sum_{\alpha \in R} e_\alpha^* \otimes e_{-\alpha}^* \otimes h_\alpha + \sum_{\substack{\alpha, \gamma \in R \\ \alpha + \gamma \in R}} e_\alpha^* \otimes e_\gamma^* \otimes N_{\alpha, \gamma} e_{\alpha + \gamma},$$

where the stars indicate elements of the dual basis. The second sum is most important because the first and third sums give no contribution to the Bar-Natan–Garoufalidis weight system S_{BNG}.

After identification of $\mathfrak{g}^*$ and $\mathfrak{g}$ via $\langle \cdot, \cdot \rangle$ we get $e_\alpha^* = (\langle \alpha, \alpha \rangle / 2) e_{-\alpha}$. In particular, the second sum of the tensor J is

$$\sum_{\alpha \in R} \left(\langle \alpha, \alpha \rangle / 2 \right)^2 e_{-\alpha} \otimes e_\alpha \otimes h_\alpha.$$

According to Section 6.2, in order to calculate $S_{BNG}(\overline{w_{2m}})$ we must assign a copy of the tensor $-J$ to each internal vertex, perform all the contractions corresponding to internal edges and, after that, take the product $\varphi_\mathfrak{g}^{V_\lambda}(\overline{w_{2m}})$ of all the operators in V_λ corresponding to the external vertices. We have that $\varphi_\mathfrak{g}^{V_\lambda}(\overline{w_{2m}})$ is a scalar operator of multiplication by some constant. This constant is a polynomial in λ of degree at most $2m$; its part of degree $2m$ is $S_{BNG}(\overline{w_{2m}})$.

We associate the tensor $-J$ with an internal vertex in such a way that the third tensor factor of $-J$ corresponds to the edge connecting the vertex with a leg. After that we take the product of operators corresponding to these external vertices. This means that we take the product of operators corresponding to the third tensor factor of $-J$. Of course, we are interested only in those operators which are linear in λ. One can show (see, for example, Bar-Natan and Garoufalidis 1996, lemma 5.1) that it is possible to choose a basis in the space of the representation V_λ in such a way that the Cartan operators h_α and raising operators e_α ($\alpha \in R^+$) will be linear in λ while the lowering operators $e_{-\alpha}$ ($\alpha \in R^+$) will not depend on λ. So we have to take into account only those summands of $-J$ that have h_α or e_α ($\alpha \in R^+$) as the third tensor factor. Further, to calculate the multiplication constant of our product it is sufficient to act by the operator on any vector. Let us choose the highest weight vector v_0 for this. The Cartan operators h_α multiply v_0 by $\lambda(h_\alpha) = 2\langle\lambda, \alpha\rangle/\langle\alpha, \alpha\rangle$. So indeed they are linear in λ. But the raising operators e_α ($\alpha \in R^+$) send v_0 to zero. This means that we have to take into account only those summands of $-J$ whose third tensor factor is one of the h_α's. This is exactly the second sum of J with the opposite sign:

$$\sum_{\alpha \in R} \left(\langle\alpha, \alpha\rangle/2\right)^2 e_\alpha \otimes e_{-\alpha} \otimes h_\alpha.$$

Now performing all the contractions corresponding to the edges connecting the internal vertices of $\overline{w_{2m}}$ we get the tensor

$$\sum_{\alpha \in R} \left(\langle\alpha, \alpha\rangle/2\right)^{2m} \underbrace{h_\alpha \otimes \ldots \otimes h_\alpha}_{2m \text{ times}}.$$

The corresponding element of $U(\mathfrak{g})$ acts on the highest weight vector v_0 as multiplication by

$$S_{BNG}(\overline{w_{2m}}) = \sum_{\alpha \in R} \langle\lambda, \alpha\rangle^{2m} = 2 \sum_{\alpha \in R^+} \langle\lambda, \alpha\rangle^{2m}.$$

The theorem is proved. $\square$

14.2 The Goussarov–Habiro theory revisited

The term *Goussarov–Habiro theory* refers to the study of n-equivalence classes of knots (or, more generally, knotted graphs), as defined in Section 3.2.1, in terms of local moves on knot diagrams. It was first developed by M. Goussarov, who announced the main results in September 1995 at a conference in Oberwolfach, and, independently, by K. Habiro (1994, 2000). (As often happened with Goussarov's results, his publication on the subject (Goussarov 2001) appeared several years later.)

There are several different approaches to Goussarov–Habiro theory, which produce roughly the same results. In Chapter 12 we have developed the group-theoretic approach pioneered by T. Stanford (1996b, 1999), who described n-equivalence in terms of the lower central series of the pure braid groups. Habiro (1994, 2000) uses *claspers* to define local moves on knots and string links. Here we shall briefly sketch Goussarov's approach, neither giving complete proofs, nor striving for maximal generality. Other versions of theorems of the same type can be found in Conant and Teichner (2004) and Taniyama and Yasuhara (2002). A proof that the definitions of Goussarov, Habiro and Stanford are equivalent can be found in Habiro (2000).

14.2.1 Statement of the Goussarov–Habiro Theorem

In what follows we shall use the term *tangle* in the sense that is somewhat different from the rest of this book. Here, by a tangle we shall mean an oriented one-dimensional submanifold of a ball in $\mathbb{R}^3$, transversal to the boundary of the ball in its boundary points. The isotopy of tangles is understood to fix the boundary.

Theorem 14.4 (Goussarov–Habiro). *Let K_1 and K_2 be two knots. They are n-equivalent, that is, $v(K_1) = v(K_2)$ for any $\mathbb{Z}$-valued Vassiliev invariant v of order $\leqslant n$ if and only if K_1 and K_2 are related by a finite sequence of moves $\mathscr{M}_n$:*

$$\underbrace{}_{n+2\ components} \qquad\qquad \underbrace{}_{n+2\ components}$$

Denote by B_n and T_n the tangles on the left and, respectively, on the right-hand side of the move $\mathscr{M}_n$. The tangle B_n is an example of a *Brunnian tangle* characterized by the property that removing any of its components makes the remaining tangle to be isotopic to the trivial tangle T_{n-1} with $n + 1$ components.

The sequence of moves $\mathscr{M}_n$ starts with $n = 0$:

$$\mathscr{M}_0: \quad B_0 = \; \text{⚇} \quad \text{∿} \quad \text{⊃⊂} \; = T_0.$$

In terms of knot diagrams, $\mathcal{M}_0$ consists of a crossing change followed by a second Reidemeister move.

The move $\mathcal{M}_1$ looks like

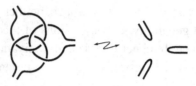

$$\mathcal{M}_1: \quad B_1 = \qquad\qquad\qquad\qquad\qquad = T_1$$

It is also known as the *Borromean move*

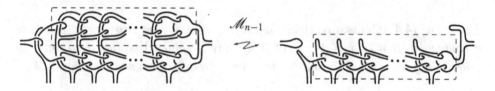

Since there are no invariants of order $\leqslant 1$ except constants (see page 64), the Goussarov–Habiro theorem implies that any knot can be transformed to the unknot by a finite sequence of Borromean moves $\mathcal{M}_1$; in other words, $\mathcal{M}_1$ is an *unknotting operation*.

Remark 14.5. The coincidence of all Vassiliev invariants of order $\leqslant n$ implies the coincidence of all Vassiliev invariants of order $\leqslant n - 1$. This means that one can accomplish a move $\mathcal{M}_n$ by a sequence of moves $\mathcal{M}_{n-1}$. Indeed, let us draw the tangle B_n as shown below on the left:

(In order to see that the tangle on the left is indeed B_n, untangle the components one by one, working from right to left.) The tangle in the dashed rectangle is B_{n-1}. To perform the move $\mathcal{M}_{n-1}$ we must replace it with T_{n-1}. This gives us the tangle on the right also containing B_{n-1}. Now performing once more the move $\mathcal{M}_{n-1}$ we obtain the trivial tangle T_n.

14.2.2 Reformulation of the Goussarov–Habiro theorem

Recall some notation from Sections 1.5 and 3.2.1. We denote by $\mathcal{K}$ the set of all isotopy classes of knots, $\mathbb{Z}\mathcal{K}$ is the free $\mathbb{Z}$-module (even an algebra) consisting of all finite formal $\mathbb{Z}$-linear combinations of knots and $\mathcal{K}_n$ stands for the nth term of the singular knot filtration in $\mathbb{Z}\mathcal{K}$. Using the moves $\mathcal{M}_n$, we can define another filtration in the module $\mathbb{Z}\mathcal{K}$.

Let $\mathcal{H}_n$ be the $\mathbb{Z}$-submodule of $\mathbb{Z}\mathcal{K}$ spanned by the differences of two knots obtained one from another by a single move $\mathcal{M}_n$. For example, the difference $3_1 - 6_3$ belongs to $\mathcal{H}_2$. The Goussarov–Habiro theorem can be restated as follows.

Theorem 14.6. *For all n the submodules $\mathcal{K}_{n+1}$ and $\mathcal{H}_n$ coincide.*

Proof of the equivalence of the two statements. As we have seen in Section 3.2.1, the values of any Vassiliev invariant of order $\leqslant n$ are the same on the knots K and K' if and only if the difference $K - K'$ belongs to $\mathcal{K}_{n+1}$. On the other hand, K can be obtained from K' by a sequence of $\mathcal{M}_n$-moves if and only if $K - K'$ belongs to $\mathcal{H}_n$. Indeed, if $K - K' \in \mathcal{H}_n$, we can write

$$K - K' = \sum_i a_i (K_i - K_i')$$

where a_i are positive integers and each K_i differs from K_i' by a single $\mathcal{M}_n$ move. Since all the knots in this sum apart from K and K' cancel each other, we can write it as $\sum_{i=1}^n K_i - K_{i-1}$ with $K_n = K$, $K_0 = K'$ and where each K_i differs from K_{i-1} by a single $\mathcal{M}_n$ move. Then K is obtained by from K' by a sequence of $\mathcal{M}_n$ moves that change consecutively K_{i-1} into K_i. $\qquad\square$

Exercise. Prove that two knots are related by a sequence of $\mathcal{M}_n$-moves if and only if they are γ_{n+1}-equivalent (see section 12.6.1).

The Goussarov–Habiro theorem is a corollary of this exercise and Theorem 12.22. Nevertheless, we shall verify one part of the Goussarov–Habiro theorem directly, in order to give the reader some feel of the Goussarov–Habiro theory. Namely, let us show that $\mathcal{H}_n \subseteq \mathcal{K}_{n+1}$. (The inclusion $\mathcal{K}_{n+1} \subseteq \mathcal{H}_n$ is rather more difficult to prove.)

14.2.3 Proof that $\mathcal{H}_n$ is contained in $\mathcal{K}_{n+1}$

In order to prove that $\mathcal{H}_n \subseteq \mathcal{K}_{n+1}$ it is sufficient to represent the difference $B_n - T_n$ as a linear combination of singular tangles with $n + 1$ double points each. Let us choose the orientations of the components of our tangles as shown. Using the Vassiliev skein relation, we shall gradually transform the difference $B_n - T_n$ into the required form.

$$B_n - T_n \; = \;$$

$$= \;$$

But the difference of the last two tangles can be expressed as a singular tangle:

$$= \;$$

We got a presentation of $B_n - T_n$ as a linear combination of two tangles with one double point on the first two components. Now we add and subtract isotopic singular tangles with one double point:

$$B_n - T_n \; = \;$$

Then using the Vassiliev skein relation we can see that the difference in the first pair of parentheses is equal to

Similarly, the difference in the second pair of parentheses would be equal to

Now we have represented $B_n - T_n$ as a linear combination of four singular tangles with two double points each; in each tangle one double point lies on the first and on the second components and the other double point, on the second and on the third components:

$$B_n - T_n = \quad - \quad \cdots \quad + \quad \cdots$$

$$+ \quad \cdots \quad - \quad \cdots$$

Continuing in the same way we arrive to a linear combination of 2^n tangles with $n + 1$ double points each; one double point for every pair of consecutive components. It is easy to see that if we change the orientations of arbitrary k components of our tangles B_n and T_n, then the whole linear combination will be multiplied by $(-1)^k$.

Example 14.7.

$$B_2 - T_2 = \qquad - \qquad - \qquad + $$

Example 14.8. There is only one (up to multiplication by a scalar and adding a constant) nontrivial Vassiliev invariant of order $\leqslant 2$, namely It is the coefficient c_2 of the Conway polynomial.

Consider two knots

$$3_1 \quad = \qquad , \qquad 6_3 \quad = \qquad .$$

We choose the orientations as indicated. Their Conway polynomials

$$C(3_1) = 1 + t^2, \qquad C(6_3) = 1 + t^2 + t^4$$

have equal coefficients of t^2. Therefore, for any Vassiliev invariant v of order $\leqslant 2$ we have $v(3_1) = v(6_3)$. In this case the Goussarov–Habiro theorem states that it is possible to obtain the knot 6_3 from the knot 3_1 by moves $\mathcal{M}_2 : B_2 \rightsquigarrow T_2$

$\mathcal{M}_2$:

Let us show this. We start with the standard diagram of 3_1, and then transform it in order to obtain B_2 as a subtangle.

$$3_1 = \;\cong\;\cong\;\cong\;\cong$$

$$\cong\;\cong\;\cong\;\cong$$

$$\cong\;\cong\;\cong$$

$$\cong\;\cong$$

Now we have the tangle B_2 in the dashed oval. Perform the move $\mathcal{M}_2$ replacing B_2 with the trivial tangle T_2:

$$\cong\;\cong\;\cong\;\cong\; = 6_3$$

14.2.4 Vassiliev invariants and local moves

The *mod* 2 reduction of c_2 is called the *Arf invariant* of a knot. A description of the Arf invariant similar to the Goussarov–Habiro description of c_2 was obtained by L. Kauffman.

Theorem (Kauffman 1983, 1985). *K_1 and K_2 have the same Arf invariant if and only if K_1 can be obtained from K_2 by a finite number of so called pass moves:*

The orientations are important. Allowing pass moves with arbitrary orientations we obtain an unknotting operation (see Kawauchi 1996).

Actually, one can develop the whole theory of Vassiliev invariants using the pass move instead of the crossing change in the Vassiliev skein relation. It turns out, however, that all primitive finite type invariants with respect to the pass move of order n coincide with primitive Vassiliev invariants of order n for all $n \geqslant 1$. The Arf invariant is the unique finite type invariant of order 0 with respect to the pass move (Conant *et al.* 2010).

More generally, in the definition of the finite type invariants, one can replace a crossing change with an arbitrary local move, that is, a modification of a knot that replaces a subtangle of some fixed type with another subtangle. For a wide class of moves, one obtains theories of finite type invariants for which the Goussarov–Habiro theorem holds, see Taniyama and Yasuhara (2002) and Conant *et al.* (2010).

One such move is the *doubled-delta move*:

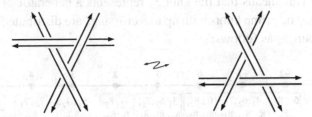

S. Naik and T. Stanford (2003) have shown that two knots can be transformed into each other by doubled-delta moves if and only if they are *S-equivalent*, that is, if they have a common Seifert matrix, see Kawauchi (1996). The theory of finite type invariants based on the doubled-delta move appears to be rather rich. In particular, for each n there are an infinite number of independent invariants of order $2n$ which are not of order $2n - 1$. We refer the reader to Conant *et al.* (2010) for more details.

14.2.5 The Goussarov groups of knots

There are two main results in the Goussarov–Habiro theory. One is what we called the Goussarov–Habiro theorem. The other result says that classes of knots (more generally, string links) related by $\mathcal{M}_n$-moves form groups under the connected sum operation.

Modulo the exercise on page 413, we have proved this in Chapter 12, see Theorem 12.22. There we were mostly interested in applying the technique of braid closures and the theory of nilpotent groups. Here let us give some concrete examples.

We shall denote by $\mathcal{G}_n$ the nth *Goussarov group*, that is, the set $\mathcal{K} / \Gamma_{n+1}\mathcal{K}$ of n-equivalence classes of knots with the connected sum operation. A *j-inverse* for a knot K is a knot K' such that $K\#K'$ is j-trivial. An n-inverse for K provides an inverse for the class of K in $\mathcal{G}_n$.

Since there are no Vassiliev invariants of order $\leqslant 1$ except constants, the zeroth and the first Goussarov groups are trivial.

14.2.6 The second Goussarov group $\mathcal{G}_2$.

Consider the coefficient c_2 of the Conway polynomial $C(K)$. According to Exercise 2.6, $C(K)$ is a multiplicative invariant of the form $C(K) = 1 + c_2(K)t^2 + \ldots$. This implies that $c_2(K_1\#K_2) = c_2(K_1) + c_2(K_2)$, and, hence, c_2 is a homomorphism of $\mathcal{G}_2$ into $\mathbb{Z}$. Since c_2 is the only nontrivial invariant of order $\leqslant 2$ and there are knots on which it takes value 1, the homomorphism $c_2 : \mathcal{G}_2 \to \mathbb{Z}$ is, in fact, an isomorphism and $\mathcal{G}_2 \cong \mathbb{Z}$. From Table 2.1 we can see that $c_2(3_1) = 1$ and $c_2(4_1) = -1$. This means that the knot 3_1 represents a generator of $\mathcal{G}_2$, and 4_1 is 2-inverse of 3_1. The prime knots with up to 8 crossings are distributed in the second Goussarov group $\mathcal{G}_2$ as follows:

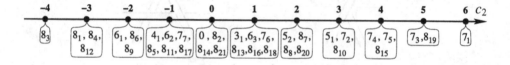

14.2.7 The third Goussarov group $\mathcal{G}_3$.

In order 3 we have one more Vassiliev invariant; namely, j_3, the coefficient at h^3 in the power series expansion of the Jones polynomial with the substitution $t = e^h$. The Jones polynomial is multiplicative, $J(K_1\#K_2) = J(K_1) \cdot J(K_2)$ (see Exercise 2.7) and its expansion has the form $J(K) = 1 + j_2(K)h^2 + j_3(K)h^3 + \ldots$ (see Section 3.6). Thus we can write

$$J(K_1\#K_2) = 1 + (j_2(K_1) + j_2(K_2))\, h^2 + (j_3(K_1) + j_3(K_2))\, h^3 + \ldots .$$

In particular, $j_3(K_1\#K_2) = j_3(K_1) + j_3(K_2)$. According to Exercise 3.6, j_3 is divisible by 6. Then $j_3/6$ is a homomorphism from $\mathcal{G}_3$ to $\mathbb{Z}$. The direct sum with c_2 gives the isomorphism

$$\mathcal{G}_3 \cong \mathbb{Z} \oplus \mathbb{Z} = \mathbb{Z}^2; \qquad K \mapsto (c_2(K), j_3(K)/6).$$

Let us identify $\mathscr{G}_3$ with the integral lattice on a plane. The distribution of prime knots on this lattice is shown in Figure 14.1; recall that $\overline{K}$ is the mirror reflection of K.

In particular, the 3-inverse of the trefoil 3_1 can be represented by 6_2, or by 7_7. Also, we can see that $3_1\#4_1$ is 3-equivalent to $\overline{8_2}$. Therefore $3_1\#4_1\#8_2$ is 3-equivalent to the unknot, and $4_1\#8_2$ also represents the 3-inverse to 3_1. The knots 6_3 and $\overline{8_2}$ represent the standard generators of $\mathscr{G}_3$.

Open problem. Is there any torsion in the group $\mathscr{G}_n$?

14.3 Willerton's fish and bounds for c_2 and j_3

Willerton's fish is a graph where the Vassiliev invariant c_2 is plotted against the invariant j_3 for all prime knots of a given crossing number (Willerton 2002b). The shape of this graph, at least for the small values of the crossing number ($\leqslant 14$) where there is enough data to construct it, is reminiscent of a fish, hence the name. (This shape is already discernible on Figure 14.1, which shows all prime knots up to 8 crossings.)

A plausible explanation for the strange shape of these graphs could involve some inequality on c_2, j_3 and the crossing number c. At the moment, no such inequality is known. However, there are several results relating the above knot invariants.

Theorem 14.9 (Polyak and Viro 2001). *For any knot K $|c_2(K)| \leqslant \left[\frac{c(K)^2}{8}\right]$.*

Proof Recall the Gauss diagram formulae for c_2 on pages 75 and 386 which we can write as follows:

$$c_2(K) = \left\langle \text{⌢⌢} , D \right\rangle = \left\langle \text{⌢⌢} , D \right\rangle,$$

where D is a based Gauss diagram with n arrows representing the knot K. Let C^+ be the set of arrows of D that point forward (this makes sense since D is based) and let C^- be the set of backwards-pointing arrows. If C^+ consists of k elements, then C^- has $n - k$ elements.

Now, assume that the diagram

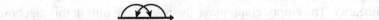

appears n_1 times as a subdiagram of D and the diagram

appears n_2 times. Each of these diagrams contains one arrow from C^+ and one from C^-. Therefore, we have

$$|c_2(K)| \leqslant \min(n_1, n_2) \leqslant \frac{k(n-k)}{2} \leqslant \left[\frac{n^2}{8}\right].$$

Now, the smallest possible n in this formula, that is, the minimal number of arrows in a Gauss diagram representing K, is, by definition, nothing else but the crossing number $c(K)$.   □

Similar inequalities exist for all Vassiliev invariants. Indeed, each invariant of order n can be represented by a Gauss diagram formula (see Chapter 13). This means that its value on a knot can be calculated by representing this knot by a Gauss diagram D and counting subdiagrams of D of certain types, all with at most n arrows. The number of such subdiagrams grows as $(\deg D)^n$, so for each invariant of degree n there is a bound by a polynomial of degree n in the crossing number. In particular, S. Willerton found the following bound (unpublished):

$$|j_3(K)| \leqslant \frac{3}{2} \cdot c(K)(c(K)-1)(c(K)-2).$$

One particular family of knots for which c_2 and j_3 are related by explicit inequalities are the torus knots (Willerton 2002b). We have

$$24c_2(K)^3 + 12c_2(K)^2 \leqslant j_3(K)^2 \leqslant 32c_2(K)^3 + 4c_2(K)^2$$

for any torus knot K. These bounds are obtained from the explicit expressions for c_2 and j_3 for torus knots obtained in Alvarez and Labastida (1996).

14.4 Bialgebra of graphs

It turns out that the natural mapping that assigns to every chord diagram its intersection graph, can be converted into a homomorphism of bialgebras $\gamma : \mathscr{A} \to \mathscr{L}$, where $\mathscr{A}$ is the algebra of chord diagrams and $\mathscr{L}$ is an algebra generated by graphs modulo certain relations, introduced by S. Lando (2002). Here is his construction.

Let $\mathfrak{G}$ be the graded vector space spanned by all simple graphs (without loops or multiple edges) as free generators:

$$\mathfrak{G} = \mathfrak{G}_0 \oplus \mathfrak{G}_1 \oplus \mathfrak{G}_2 \oplus \dots.$$

It is graded by the order (the number of vertices) of a graph. This space is easily turned into a bialgebra:

(1) The product is defined as the disjoint union of graphs, then extended by linearity. The empty graph plays the role of the unit in this algebra.

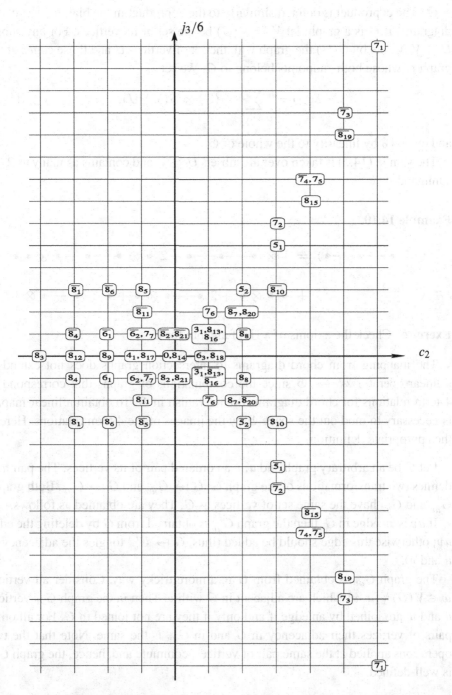

Figure 14.1 Values of Vassiliev invariants c_2 and j_3 on prime knots with up to 8 crossings. The mirror images of 8_{14} and of 8_{17}, not shown, have the same invariants as the original knots.

(2) The coproduct is defined similarly to the coproduct in the bialgebra of chord diagrams. If G is a graph, let $V = V(G)$ be the set of its vertices. For any subset $U \subset V$ denote by $G(U)$ the graph with the set of vertices U and those edges of the graph G whose both endpoints belong to G. We set

$$\delta(G) = \sum_{U \subseteq V(G)} G(U) \otimes G(V \setminus U), \tag{14.2}$$

and extend δ by linearity to the whole of $\mathfrak{G}$.

The sum in (14.2) is taken over all subsets $U \subset V$ and contains as many as $2^{\#(V)}$ summands.

Example 14.10.

$$\delta(\,\bullet\!\!-\!\!\!-\!\!\bullet\!\!\!-\!\!\bullet\,) = 1 \otimes \bullet\!\!-\!\!\!-\!\!\bullet\!\!\!-\!\!\bullet + 2\,\bullet \otimes \bullet\!\!-\!\!\!-\!\!\bullet + \bullet \otimes \bullet\,\bullet$$

$$+ \,\bullet\,\bullet \otimes \bullet + 2\,\bullet\!\!-\!\!\!-\!\!\bullet \otimes \bullet + \bullet\!\!-\!\!\!-\!\!\bullet\!\!\!-\!\!\bullet \otimes 1$$

Exercise. Check the axioms of a Hopf algebra for $\mathfrak{G}$.

The mapping from chord diagrams to intersection graphs does not extend to a linear operator $\mathscr{A} \to \mathfrak{G}$ since the combinations of graphs that correspond to 4-term relations for chord diagrams do not vanish in $\mathfrak{G}$. To obtain a linear map, it is necessary to mod out the space $\mathfrak{G}$ by the images of the 4 term relations. Here is the appropriate definition.

Let G be an arbitrary graph and u, v an ordered pair of its vertices. The pair u, v defines two transformations of the graph G: $G \mapsto G'_{uv}$ and $G \mapsto \tilde{G}_{uv}$. Both graphs G'_{uv} and $\tilde{G}_{uv}$ have the same set of vertices as G. They are obtained as follows.

If uv is an edge in G, then the graph G'_{uv} is obtained from G by deleting the edge uv; otherwise this edge should be added (thus, $G \mapsto G'_{uv}$ toggles the adjacency of u and v).

The graph $\tilde{G}_{uv}$ is obtained from G in a more tricky way. Consider all vertices $w \in V(G) \setminus \{u, v\}$ which are adjacent in G with v. Then in the graph $\tilde{G}_{uv}$ vertices u and w are joined by an edge if and only if they are not joined in G. For all other pairs of vertices their adjacency in G and in $\tilde{G}_{uv}$ is the same. Note that the two operations applied at the same pair of vertices, commute and, hence, the graph G'_{uv} is well-defined.

Definition 14.11. A four-term relation for graphs is

$$G - G'_{uv} = \tilde{G}_{uv} - \tilde{G}'_{uv}. \tag{14.3}$$

Example 14.12.

$$\cdots - \cdots = \cdots - \cdots$$

$$u \quad v \qquad u \quad v \qquad u \quad v \qquad u \quad v$$

Exercises.

(1) Check that, passing to intersection graphs, the four-term relation for chord diagram carries over exactly into this four-term relation for graphs.

(2) Find the four-term relation of chord diagrams which is the preimage of the relation shown in the example above.

Definition 14.13. The *graph bialgebra of Lando* $\mathscr{L}$ is the quotient of the graph algebra $\mathfrak{G}$ by the ideal generated by all 4-term relations (14.3).

Theorem 14.14. *The product and the coproduct defined above induce a bialgebra structure in the quotient space* $\mathscr{L}$.

Proof The only thing that needs checking is that both the product and the coproduct respect the 4-term relation (14.3). For the product, which is the disjoint union of graphs, this statement is obvious. In order to verify it for the coproduct, it is sufficient to consider two cases. Namely, let $u, v \in V(G)$ be two distinct vertices of a graph G. The right-hand side summands in the formula (14.2) for the coproduct split into two groups: the summands where both vertices u and v belong either to the subset $U \subset V(G)$ or to its complement $V(G) \setminus U$, and those where u and v belong to different subsets. By cleverly grouping the terms of the first kind for the coproduct $\delta(G - G'_{uv} - \widetilde{G}_{uv} + \widetilde{G}'_{uv})$ we can see that they all cancel out in pairs. The terms of the second kind cancel out in pairs already within each of the two summands $\delta(G - G'_{uv})$ and $\delta(\widetilde{G}_{uv} - \widetilde{G}'_{uv})$. $\square$

Relations (14.3) are homogeneous with respect to the number of vertices; therefore $\mathscr{L}$ is a graded algebra. By the Milnor–Moore theorem (Theorem A.32), the algebra $\mathscr{L}$ is polynomial with respect to its space of primitive elements.

Now we have a well-defined bialgebra homomorphism

$$\gamma : \mathscr{A} \to \mathscr{L}$$

which extends the assignment of the intersection graph to a chord diagram. It is defined by the linear mapping between the corresponding primitive spaces $\mathscr{P}\mathscr{A} \to \mathscr{P}\mathscr{L}$.

According to S. Lando (2000), the dimensions of the homogeneous components of $\mathscr{P}\mathscr{L}$ are known up to degree 7. It turns out that the homomorphism γ is an isomorphism in degrees up to 6, while the map $\gamma : \mathscr{P}_7\mathscr{A} \to \mathscr{P}_7\mathscr{L}$ has a one-dimensional kernel. See Lando (2000) for further details and open problems related to the algebra $\mathscr{L}$.

14.5 Estimates for the number of Vassiliev knot invariants

Knowing the dimensions of the primitive subspaces $\mathscr{P}_i = \mathscr{P}\mathscr{A}_i'$ for $i \leqslant n$ is equivalent to knowing $\dim \mathscr{A}_i'$ or $\dim \mathscr{V}_i$ for $i \leqslant n$. These numbers have been calculated only for small values of n and, at present, their exact asymptotic behaviour as n tends to infinity is not known. Below we give a summary of all available results on these dimensions.

14.5.1 Historical remarks: exact results

The precise dimensions of the spaces related to Vassiliev invariants are known up to $n = 12$, and are listed in the table below. They were obtained by V. Vassiliev for $n \leqslant 4$ in 1990 (Vassiliev 1990b), then by D. Bar-Natan for $n \leqslant 9$ in 1993 (Bar-Natan 1995a) and by J. Kneissler, for $n = 10, 11, 12$, in 1997 (Kneissler 1997). Vassiliev used manual calculations. Bar-Natan wrote a computer program that implemented the direct algorithm to solve the system of linear equations coming from one-term and four-term relations. Kneissler obtained a lower bound using the marked surfaces (Bar-Natan 1990b) and an upper bound using the action of Vogel's algebra Λ on the primitive space $\mathscr{P}$: miraculously, these two bounds coincided for $n \leqslant 12$.

n	0	1	2	3	4	5	6	7	8	9	10	11	12
$\dim \mathscr{P}_n$	0	0	1	1	2	3	5	8	12	18	27	39	55
$\dim \mathscr{A}_n'$	1	0	1	1	3	4	9	14	27	44	80	132	232
$\dim \mathscr{V}_n$	1	1	2	3	6	10	19	33	60	104	184	316	548

The splitting of the numbers $\dim \mathscr{P}_n$ for $n \leqslant 12$ according to the second grading in $\mathscr{P}$ is given in the table on page 127.

Exercise. Prove that $\dim \mathscr{A}_n = \dim \mathscr{V}_n$ for all n.

14.5.2 Historical remarks: upper bounds

A priori it was obvious that $\dim \mathscr{A}_n' < (2n - 1)!! = 1 \cdot 3 \cdots (2n - 1)$, since this is the total number of linear chord diagrams.

Then, there appeared five papers where this estimate was successively improved:

1. (1993) Chmutov and Duzhin (1994) proved that $\dim \mathscr{A}_n' < (n - 1)!$
2. (1995) K. Ng (1998) replaced $(n - 1)!$ by $(n - 2)!/2$.

3. (1996) A. Stoimenow (1998a) proved that dim $\mathscr{A}'_n$ grows slower than $n!/a^n$, where $a = 1.1$.

4. (2000) B. Bollobás and O. Riordan (2000) obtained the asymptotical bound $n!/(2\ln(2) + o(1))^n$ (approximately $n!/1.38^n$).

5. (2001) D. Zagier (2001) improved the last result to $\frac{6^n \sqrt{n} \cdot n!}{\pi^{2n}}$, which is asymptotically smaller than $n!/a^n$ for any constant $a < \pi^2/6 = 1.644....$

For the proofs of these results, we refer the interested reader to the original papers, and only mention here the methods used to get these estimates. Chmutov and Duzhin proved that the space $\mathscr{A}'_n$ is spanned by the *spine chord diagrams*, that is, diagrams containing a chord that intersects all other chords, and estimating the number of such diagrams. Stoimenow did the same with *regular linearized diagrams*; Zagier gave a better estimate for the number of such diagrams.

14.5.3 Historical remarks: lower bounds

In the story of lower bounds for the number of Vassiliev knot invariants there is an amusing episode. The first paper by Kontsevich about Vassiliev invariants (Kontsevich 1993, section 3) contains the following passage:

"Using this construction,[1] one can obtain the estimate

$$\dim(\mathscr{V}_n) > e^{c\sqrt{n}}, \quad n \to +\infty$$

for any positive constant $c < \pi\sqrt{2/3}$ (see Bar-Natan (1992), exercise 6.14)."

Here $\mathscr{V}_n$ is a slip of the pen, instead of $\mathscr{P}_n$, because of the reference to Exercise 6.14 where primitive elements are considered. Exercise 6.14 was present, however, only in the first edition of Bar-Natan's preprint and eliminated in the following editions as well as in the final published version of his text (1995a). In Bar-Natan (1992) it reads as follows (page 43):

"*Exercise* 6.14. (Kontsevich, [24]) Let $P_{\geqslant 2}(m)$ denote the number of partitions of an integer m into a sum of integers bigger than or equal to 2. Show that $\dim \mathscr{P}_m \geqslant P_{\geqslant 2}(m + 1)$.

Hint 6.15. Use a correspondence like

$$\longleftrightarrow \quad 10 + 1 = 4 + 3 + 2 + 2,$$

and ..."

---

[1] Of Lie algebra weight systems.

The reference [24] was to "M. Kontsevich. Private communication."! Thus, both authors referred to each other, and none of them gave any proof. Later, however, Kontsevich explained what he had in mind (see item 5 below).

Arranged by the date, the history of world records in asymptotic lower bounds for the dimension of the primitive space $\mathscr{P}_n$ looks as follows.

1. (1994) dim $\mathscr{P}_n \geqslant 1$ ("forest elements" found by Chmutov, Duzhin and Lando 1994c).
2. (1995) dim $\mathscr{P}_n \geqslant [n/2]$ (given by coloured Jones function – see Melvin–Morton 1995 and Chmutov–Varchenko 1997).
3. (1996) dim $\mathscr{P}_n \gtrsim n^2/96$ (see Duzhin 1996).
4. (1997) dim $\mathscr{P}_n \gtrsim n^{\log_b n}$ for any $b > 4$, i.e. the growth is faster than any polynomial (Chmutov–Duzhin 1999).
5. (1997) dim $\mathscr{P}_n \gtrsim e^{\pi\sqrt{n/3}}$ (Kontsevich 1997).
6. (1997) dim $\mathscr{P}_n \gtrsim e^{c\sqrt{n}}$ for any constant $c < \pi\sqrt{2/3}$ (Dasbach 2000).

Each lower bound for the dimensions of the primitive space $p_n = \dim \mathscr{P}_n$ implies a certain lower bound for the dimensions of the whole algebra $a_n = \dim \mathscr{A}'_n$.

Proposition 14.15. *The lower bound of Dasbach implies that* $a_n \gtrsim e^{n/\log_b n}$ *for any constant* $b < \pi^2/6$.

Sketch of the proof   Fix a basis in each $\mathscr{P}_k$, suppose that $n = km$ and consider the elements of $\mathscr{A}'_n$ which are products of m basis elements of $\mathscr{P}_k$. Finding the maximum of this number over k with fixed n, we get the desired lower bound. $\square$

Note that the best known upper and lower bounds on the dimensions of $\mathscr{A}'_n$ are very far apart. Indeed, using the relation between the generating functions

$$\sum_{n=0}^{\infty} a_n t^n = \prod_{k=1}^{\infty} (1 - t^k)^{-p_k} = \exp \sum_{n=1}^{\infty} \Big(\sum_{k|n} p_k \Big) t^n,$$

one can easily prove (see Stoimenow 1998b) that any subexponential lower bound on p_n can only lead to a subexponential lower bound on a_n, while the existing upper bound is essentially factorial, that is, much greater than exponential.

14.5.4 Proof of the lower bound

We will sketch the proof of the lower bound for the number of Vassiliev knot invariants, following Chmutov and Duzhin (1999) and then explain how O. Dasbach (2000), using the same method, managed to improve the estimate and establish the bound which is still (in 2011) the best.

The idea of the proof is simple: we construct a large family of open diagrams whose linear independence in the algebra $\mathscr{B}$ follows from the linear independence of the values on these diagrams of a certain polynomial invariant P, which is obtained by simplifying the universal $\mathfrak{gl}_N$ invariant.

As we know from Chapter 6, the $\mathfrak{gl}_N$ invariant $\rho_{\mathfrak{gl}_N}$, evaluated on an open diagram, is a polynomial in the generalized Casimir elements $x_0, x_1, \ldots, x_N$. This polynomial is homogeneous in the sense of the grading defined by setting $\deg x_m = m$. However, in general, it is not homogeneous if the x_m are considered as variables of degree 1.

Definition 14.16. The polynomial invariant $P : \mathscr{B} \to \mathbb{Z}[x_0, \ldots, x_N]$ is the highest degree part of $\rho_{\mathfrak{gl}_N}$ if all the variables are taken with degree 1.

For example, if we had $\rho_{\mathfrak{gl}_N}(C) = x_0^2 x_2 - x_1^2$, then we would have $P(C) = x_0^2 x_2$.

Now we introduce the family of primitive open diagrams whose linear independence we shall prove.

Definition 14.17. The *baguette diagram* $B_{n_1,\ldots,n_k}$ is

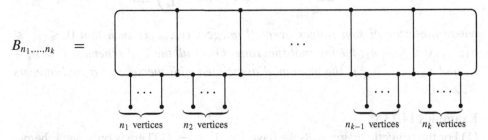

$$B_{n_1,\ldots,n_k} \quad = $$

n_1 vertices $\quad$ n_2 vertices $\quad\quad$ n_{k-1} vertices $\quad$ n_k vertices

It has a total of $2(n_1 + \cdots + n_k + k - 1)$ vertices, out of which $n_1 + \cdots + n_k$ are univalent. It is a particular case of a caterpillar diagram; see Exercise 5.17.

To write down the formula for the value $P(B_{n_1,\ldots,n_k})$, we shall need the following definitions.

Definition 14.18. Consider k pairs of points arranged in two rows like $\begin{smallmatrix}\bullet & \bullet & & \bullet \\ \bullet & \bullet & \cdots & \bullet\end{smallmatrix}$. Choose one of the 2^{k-1} subsets of the set $\{1, \ldots, k-1\}$. If a number s belongs to the chosen subset, then we connect the lower points of sth and $(s+1)$th pairs, otherwise we connect the upper points. The resulting combinatorial object is called a *two-line scheme* of order k.

Example 14.19. Here is the scheme corresponding to $k = 5$ and the subset $\{2, 3\}$:

The number of connected components in a scheme of order k is $k + 1$.

Definition 14.20. Let σ be a scheme; $i_1, \ldots, i_k$ be non-negative integers: $0 \leqslant i_1 \leqslant n_1, \ldots, 0 \leqslant i_k \leqslant n_k$. We assign i_s to the lower vertex of the sth pair of σ and $j_s = n_s - i_s$ to the upper vertex. For example:

$$\begin{matrix} j_1 & j_2 & j_3 & j_4 & j_5 \\ \bullet\!\!-\!\!\bullet & \bullet & \bullet\!\!-\!\!\bullet & \bullet & \bullet \\ i_1 & i_2 & i_3 & i_4 & i_5 \end{matrix}.$$

Then the *monomial corresponding to* σ is $x_{\sigma_0} x_{\sigma_1} \ldots x_{\sigma_k}$ where σ_t is the sum of integers assigned to the vertices of tth connected component of σ.

Example 14.21. For the above weighted scheme we get the monomial

$$x_{i_1} x_{j_1+j_2} x_{i_2+i_3+i_4} x_{j_3} x_{j_4+j_5} x_{i_5} .$$

Now the formula for P can be stated as follows.

Proposition 14.22. *If $N > n_1 + \cdots + n_k$ then*

$$P_{gl_N}(B_{n_1,\ldots,n_k}) = \sum_{i_1,\ldots,i_k} (-1)^{j_1+\cdots+j_k} \binom{n_1}{i_1} \cdots \binom{n_k}{i_k} \sum_{\sigma} x_{\sigma_0} x_{\sigma_1} \ldots x_{\sigma_k},$$

where the external sum ranges over all integers $i_1, \ldots, i_k$ such that $0 \leqslant i_1 \leqslant n_1, \ldots, 0 \leqslant i_k \leqslant n_k$; the internal sum ranges over all the 2^{k-1} schemes, $j_s = n_s - i_s$, and $x_{\sigma_0} x_{\sigma_1} \ldots x_{\sigma_k}$ is the monomial associated with the scheme σ and integers $i_1, \ldots, i_k$.

Examples 14.23.
(1) For the baguette diagram B_2 we have $k = 1, n_1 = 2$. There is only one scheme: $\bullet\!\!-\!\!\bullet$. The corresponding monomial is $x_{i_1} x_{j_1}$, and

$$\begin{aligned} P_{gl_N}(B_2) &= \sum_{i_1=0}^{2} (-1)^{j_1} \binom{2}{i_1} x_{i_1} x_{j_1} \\ &= x_0 x_2 - 2x_1 x_1 + x_2 x_0 = 2(x_0 x_2 - x_1^2) \end{aligned}$$

which agrees with Example 6.31.

(2) For the diagram $B_{1,1}$ we have $k = 2, n_1 = n_2 = 1$. There are two schemes: $\bullet\!\!-\!\!\bullet$ and $\begin{matrix}\bullet\ \bullet\\ \bullet\!\!-\!\!\bullet\end{matrix}$. The corresponding monomial are $x_{i_1} x_{i_2} x_{j_1+j_2}$ and $x_{i_1+i_2} x_{j_1} x_{j_2}$. We have

$$P_{gl_N}(B_{1,1}) = \sum_{i_1=0}^{1}\sum_{i_2=0}^{1} (-1)^{j_1+j_2} x_{i_1} x_{i_2} x_{j_1+j_2} + \sum_{i_1=0}^{1}\sum_{i_2=0}^{1} (-1)^{j_1+j_2} x_{i_1+i_2} x_{j_1} x_{j_2}$$
$$= x_0 x_0 x_2 - x_0 x_1 x_1 - x_1 x_0 x_1 + x_1 x_1 x_0 + x_0 x_1 x_1 - x_1 x_0 x_1 - x_1 x_1 x_0 + x_2 x_0 x_0$$
$$= 2(x_0^2 x_2 - x_0 x_1^2).$$

Sketch of the proof The diagram $B_{n_1,...,n_k}$ has k parts separated by $k - 1$ walls. Each wall is an edge connecting trivalent vertices to which we shall refer as *wall vertices*. The sth part has n_s outgoing legs. We shall refer to the corresponding trivalent vertices as *leg vertices*.

The proof consists of three steps.

Recall that in order to evaluate the universal $\mathfrak{gl}_N$ weight system on a diagram we can use the graphical procedure of "resolving" the trivalent vertices of a diagram and associating a tensor to each of these resolutions, see Sections 6.2.4 and 6.3.2. At the first step we study the effect of resolutions of the wall vertices. We prove that the monomial obtained by certain resolutions of these vertices has the maximal possible degree if and only if for each wall both resolutions of its vertices have the same sign. These signs are related to the above defined schemes in the following way. If we take the positive resolutions at both endpoints of the wall number s, then we connect the lower vertices of the sth and the $(s + 1)$st pairs in the scheme. If we take the negative resolutions, then we connect the upper vertices.

At the second step we study the effect of resolutions of leg vertices. We show that the result depends only on the numbers of positive resolutions of leg vertices in each part and does not depend on which vertices in a part were resolved positively and which negatively. We denote by i_s the number of positive resolutions in part s. This yields the binomial coefficients $\binom{n_s}{i_s}$ in the statement of the proposition. The total number $j_1 + \cdots + j_k$ of negative resolutions of leg vertices gives the sign $(-1)^{j_1 + \cdots + j_k}$.

The first two steps allow us to consider only those cases where the resolutions of the left i_s leg vertices in the part s are positive, the rest j_s resolutions are negative and both resolutions at the ends of each wall have the same sign. At the third step we prove that such resolutions of wall vertices lead to monomials associated with corresponding schemes, according to Definition 14.20.

We shall make some comments only about the first step, because it is exactly at this step where Dasbach found an improvement of the original argument of Chmutov and Duzhin (1999).

Let us fix certain resolutions of all trivalent vertices of $B_{n_1,...,n_k}$. We denote the obtained diagram of $n = n_1 + \cdots + n_k$ pairs of points and n arrows (see page 175) by T. After a suitable permutation of the pairs T will look like a disjoint union of certain x_m's. Hence it defines a monomial in x_m's which we denote by $m(T)$.

Let us close up all arrows in the diagram by connecting the two points in every pair with an additional short line. We obtain a number of closed curves, and we can draw them in such a way that they have 3 intersection points in the vicinity of each negative resolution and do not have other intersections. Each variable x_m

gives precisely one closed curve. Thus the degree of $m(T)$ is equal to the number of these closed curves.

Consider an oriented surface S which has our family of curves as its boundary (the Seifert surface):

The degree of $m(T)$ is equal to the number of boundary components b of S. The whole surface S consists of an annulus corresponding to the big circle in $B_{n_1,\ldots,n_k}$ and $k - 1$ bands corresponding to the walls. Here is an example:

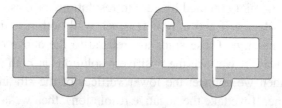

where each of the two walls on the left has the same resolutions at its endpoints, while the two walls on the right have different resolutions at their endpoints. The resolutions of the leg vertices do not influence the surface S.

The Euler characteristic χ of S can be easily computed. The surface S is contractible to a circle with $k - 1$ chords, thus $\chi = -k + 1$. On the other hand $\chi = 2 - 2g - b$, where g and b are the genus and the number of boundary components of S. Hence $b = k + 1 - 2g$. Therefore, the degree of $m(T)$, equal to b, attains its maximal value $k + 1$ if and only if the surface S has genus 0.

We claim that if there exists a wall whose ends are resolved with the opposite signs then the genus of S is not zero. Indeed, in this case we can draw a closed curve in S which does not separate the surface (independently of the remaining resolutions):

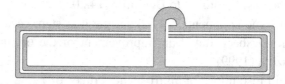

Hence the contribution to $P(B_{n_1,\ldots,n_k})$ is given by only those monomials which come from equal resolutions at the ends of each wall.   □

Now we can prove the following result.

Theorem 14.24. *Let $n = n_1 + \cdots + n_k$ and $d = n + k - 1$. Baguette diagrams $B_{n_1,\ldots,n_k}$ are linearly independent in $\mathscr{B}$ if $n_1, \ldots, n_k$ are all even and satisfy the following conditions:*

$$n_1 < n_2$$
$$n_1 + n_2 < n_3$$
$$n_1 + n_2 + n_3 < n_4$$
$$\dots\dots\dots\dots\dots\dots\dots$$
$$n_1 + n_2 + \cdots + n_{k-2} < n_{k-1}$$
$$n_1 + n_2 + \cdots + n_{k-2} + n_{k-1} < n/3.$$

The proof is based on the study of the *supports* of polynomials $P(B_{n_1,\dots,n_k})$ — the subsets of $\mathbb{Z}^k$ corresponding to non-zero terms of the polynomial.

Counting the number of elements described by the theorem, one arrives at the lower bound $n^{\log(n)}$ for the dimension of the primitive subspace $\mathscr{P}_n$ of $\mathscr{B}$.

The main difficulty in the above proof is the necessity to consider the 2^k resolutions for the wall vertices of a baguette diagram that correspond to the zero genus Seifert surface. O. Dasbach (2000) avoided this difficulty by considering a different family of open diagrams for which there are only two ways of resolution of the wall vertices leading to the surface of minimum genus. These are the *Pont-Neuf diagrams*:

$$PN_{a_1,\dots,a_k,b} \quad = \quad$$

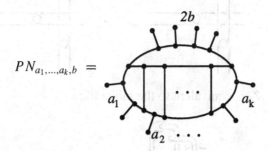

(the numbers $a_1, \dots, a_k, 2b$ refer to the number of legs attached to the corresponding edge of the inner diagram).

The reader may wish to check the above property of Pont-Neuf diagrams by way of exercise. It is remarkable that Pont-Neuf diagrams not only lead to simpler considerations, but they are more numerous, too, and thus lead to a much better asymptotic estimate for dim $\mathscr{P}_n$. The exact statement of Dasbach's theorem is as follows.

Theorem 14.25. *For fixed n and k, the diagrams* $PN_{a_1,\dots,a_k,b}$ *with* $0 \leqslant a_1 \leqslant \dots \leqslant a_k \leqslant b$, $a_1 + \dots + a_k + 2b = 2n$ *are linearly independent.*

Counting the number of such partitions of $2n$, we obtain precisely the estimate announced by Kontsevich (1993).

Corollary 14.26. dim $\mathscr{P}_n$ *is asymptotically greater than* $e^{c\sqrt{n}}$ *for any constant* $c < \pi\sqrt{2/3}$.

Exercises

14.1 Show that the move $\mathscr{M}_1$ is equivalent to the Δ *move* in the sense that, modulo Reidemeister moves, the $\mathscr{M}_1$ move can be  accomplished by Δ moves and vice versa. The fact that Δ is an unknotting operation was proved in Matveev (1987) and Murakami and Nakanishi (1989).

14.2 Prove that $\mathscr{M}_1$ is equivalent to the move

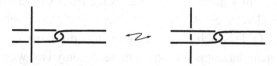

14.3 Prove that $\mathscr{M}_2$ is equivalent to the so-called *clasp-pass* move

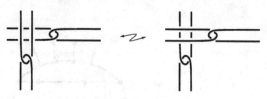

14.4 Prove that $\mathscr{M}_n$ is equivalent to the move $\mathscr{C}_n$:

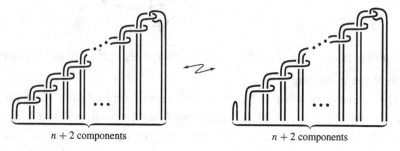

$n + 2$ components $n + 2$ components

14.5 Find the inverse element of the knot 3_1 in the group $\mathscr{G}_4$.

14.6 1.*(Kauffman) Find a set of moves relating the knots with the same c_2 modulo n, for $n = 3, 4,\ldots$.

2.*Find a set of moves relating any two knots with the same Vassiliev invariants modulo 2 (3, 4, ...) up to the order n.

3.*Find a set of moves relating any two knots with the same Conway polynomial.

14.7 (Lando). Let N be a formal variable. Prove that $N^{\operatorname{corank} A(G)}$ defines an algebra homomorphism $\mathscr{L} \to \mathbb{Z}[N]$, where $\mathscr{L}$ is the graph algebra of Lando and $A(G)$ stands for the adjacency matrix of the graph G considered over the field $\mathbb{F}_2$ of two elements.

14.8 *Let $\lambda : \mathscr{A} \to \mathscr{L}$ be the natural homomorphism from the algebra of chord diagrams into the graph algebra of Lando.

- Find $\ker \lambda$ (unknown in degrees greater than 7).
- Find $\operatorname{im} \lambda$ (unknown in degrees greater than 7).
- Describe the primitive space $P(\mathscr{L})$.
- $\mathscr{L}$ is the analog of the algebra of chord diagrams in the case of intersection graphs. Are there any counterparts of the algebras $\mathscr{C}$ and $\mathscr{B}$?

15

The space of all knots

Throughout this book we used the definition of finite type invariants based on the Vassiliev skein relation. This definition is justified by the richness of the theory based on it, but it may appear to be somewhat *ad hoc*. In fact, in Vassiliev's original approach the skein relation is a consequence of a rather sophisticated construction, which we are going to review briefly in this chapter.

One basic idea behind Vassiliev's work is that knots, considered as smooth embeddings $\mathbb{R}^1 \to \mathbb{R}^3$, form a topological space K. An isotopy of a knot can be thought of as a continuous path in this space. Knot invariants are the locally constant functions on K; therefore, the vector space of $\mathscr{R}$-valued invariants, where $\mathscr{R}$ is a ring, is the cohomology group $H^0(K, \mathscr{R})$. We see that the problem of describing all knot invariants can be generalized to the following:

Problem. *Find the cohomology ring $H^*(K, \mathscr{R})$.*

There are several approaches to this problem. Vassiliev replaces the study of knots by the study of singular knots with the help of Alexander duality and then uses simplicial resolutions for the spaces of singular knots. This method produces a spectral sequence which can be explicitly described. It is not clear how much information about the cohomology of the space of knots is contained in it, but the zero-dimensional classes coming from this spectral sequence are precisely the Vassiliev invariants.

The second approach is an attempt to build the space of knots out of the configuration spaces of points in $\mathbb{R}^3$. We are not going to discuss it here; an instructive explanation of this construction is given in Sinha (2009). Both points of view lead to a new description of the chord diagram algebra $\mathscr{A}'$. It turns out that $\mathscr{A}'$ is a part of an algebraic object which, in a way, is more fundamental than knots, namely, the *Hochshild homology of the Poisson operad*.

It is inevitable that the prerequisites for this chapter include rather advanced material such as spectral sequences; at the same time we delve into less detail. Our goal here is to give a brief introduction into the subject, after which the reader is encouraged to consult the original sources.

15.1 The space of all knots

15.1.1 Long curves and long knots

First of all, let us give precise definitions.

Definition 15.1. A *long curve* is a smooth curve $f : \mathbb{R} \to \mathbb{R}^3$ which at infinity tends to the diagonal embedding of $\mathbb{R}$ into $\mathbb{R}^3$:

$$\lim_{t \to \pm\infty} |f(t) - (t, t, t)| + \lim_{t \to \pm\infty} |f'(t) - (1, 1, 1)| = 0.$$

Here, of course, we could have chosen any fixed linear embedding of $\mathbb{R}$ into $\mathbb{R}^3$ instead of the diagonal.

There are many ways to organize long curves into a topological space. For example, one can introduce the C^1-metric on the set V of all long curves with the distance between f and g defined as

$$d(f, g) = \max_{t \in \mathbb{R}} |f(t) - g(t)| + \max_{t \in \mathbb{R}} |f'(t) - g'(t)|.$$

Alternatively, let V_n be the set of long curves of the form

$$f(t) = \frac{(P_x(t), P_y(t), P_z(t))}{(1 + t^2)^n},$$

where P_x, P_y and P_z are polynomials of the form

$$t^{2n+1} + a_{2n-1}t^{2n-1} + a_{2n-2}t^{2n-2} + \ldots + a_1 t + a_0$$

and $n > 0$. (Note the absence of the term of degree $2n$.) We can consider V_n as a Euclidean space with the coefficients of the polynomials as coordinates. The space V_n can be identified with the subspace of V_{n+1} corresponding to the triples of polynomials divisible by $1 + t^2$. Write

$$V_\infty = \bigcup_{n=1}^{\infty} V_n$$

with the topology of the union (weak topology). We can think of V_∞ as of the space of all *polynomial curves*.

Exercise. 1. Show that any long curve can be uniformly approximated by polynomial curves.

2. Is the weak topology on V_∞ equivalent to the topology given by the C^1-metric?

Definition 15.2. The *space of knots* K is the subset of V consisting of non-singular curves (smooth embeddings). Similarly, the space K_n, for $n \leqslant \infty$ is the subspace of smooth embeddings in V_n.

Clearly, K and K_∞ are just two of the possible definitions of the space of all knots.

Exercise. Show that the natural map from K_∞ to K is a weak homotopy equivalence. In other words, prove that this map induces a bijection on the set of connected components and an isomorphism in homotopy groups for each component. Note that this implies that the cohomology rings of K_∞ and K are the same for all coefficients.

We shall refer to K as *the space of long knots*. In the first chapter we defined long knots as string links on one string. It is not hard to see that any reasonable definition of a topology on the space of such string links produces a space that is weakly homotopy equivalent to K.

15.1.2 A remark on the definition of the knot space

One essential choice that we have made in the definition of the knot spaces is to consider long curves. As we know, the invariants of knots in $\mathbb{R}^3$ are the same as those of knots in S^3, or those of long knots. This, however, is no longer true for the higher invariants of the knot spaces. For example, the component of the trivial knot in K is contractible, while in the space of usual knots $S^1 \to \mathbb{R}^3$ it is not simply connected, see Hatcher (1983).

The space of long knots K has many advantages over the other types of knot spaces. An important feature of K is a natural product

$$K \times K \to K$$

given by the connected sum of long knots. Indeed, the sum of long knots is defined simply as a (suitably scaled) concatenation, and is well-defined not just for isotopy classes but for knots as geometric objects. The connected sum of usual knots, on the contrary, depends on many choices and is only well-defined as an isotopy class.

Exercise. Give a precise definition of the product on K and show that it is commutative up to homotopy. Show that the trivial knot is a unit for this product, up to homotopy. (The homotopy commutativity of the product on K is not entirely trivial, see Budney (2007)).

15.2 Complements of discriminants

In this section we shall describe the technical tools necessary for the construction of the Vassiliev spectral sequence for the space of knots. This machinery is very general and can be applied in many situations that are not related to knots in any way; we refer the reader to Vassiliev's book (1994) (or its more complete Russian version (1997)) for details.

The space K, whose cohomology we are after, is the complement in the space of all long curves of the closed set whose points correspond to long curves that fail to be embeddings. In other words, K is a complement of a *discriminant* in the space of curves.

The term "discriminant" usually denotes the subspace of singular maps in the space of all maps between two geometric objects, say, manifolds. For the discussion that follows, the word "discriminant" will simply mean "a closed subvariety in an affine space."

Vassiliev's construction involves three general technical tools: Alexander duality, simplicial resolutions and stabilization. Let us describe them in this order.

15.2.1 Alexander duality and the spectral sequence

If Σ is a discriminant, that is, a closed possibly singular subvariety of an N-dimensional real vector space V, the Alexander duality theorem states that

$$\tilde{H}^q(V - \Sigma, \mathbb{Z}) \simeq \tilde{H}_{N-q-1}(\Sigma^\bullet),$$

where $0 \leqslant q < N$, the tilde indicates reduced (co)homology and $\Sigma^\bullet$ is the one-point compactification of Σ. The geometric meaning of this isomorphism is as follows: given a cycle c of dimension $N - q - 1$ in $\Sigma^\bullet$ we assign to each q-dimensional cycle in $V - \Sigma$ its linking number with c in the sphere $S^N := V \cup \{\infty\}$. This is a q-dimensional cocyle representing the cohomology class dual to the class of c. (Here the integer coefficients can be replaced by coefficients in any abelian group.)

Now, suppose that the discriminant Σ is filtered by closed subspaces

$$\Sigma_1 \subseteq \Sigma_2 \subseteq \ldots \subseteq \Sigma_k = \Sigma.$$

Taking one-point compactifications of all terms, we get the following filtration:

$$\Sigma_{-1}^\bullet \subseteq \Sigma_0^\bullet \subseteq \Sigma_1^\bullet \subseteq \Sigma_2^\bullet \subseteq \ldots \subseteq \Sigma_k^\bullet = \Sigma^\bullet,$$

with $\Sigma_{-1}^\bullet = \emptyset$ and $\Sigma_0^\bullet = \{\infty\}$ being the added point. Then the homology of $\Sigma^\bullet$ can be studied using the *spectral sequence* arising from this filtration. (We refer the reader to Hatcher (2004) or Weibel (1994) for basics on spectral sequences.)

The term $E^1_{p,q}$ of this spectral sequence is isomorphic to $H_{p+q}(\Sigma_p^\bullet, \Sigma_{p-1}^\bullet)$ and the E^∞ term

$$E^\infty_m = \bigoplus_{p+q=m} E^\infty_{p,q}$$

is associated with $\widetilde{H}_m(\Sigma^\bullet)$ in the following sense: let $_{(i)}\widetilde{H}_m(\Sigma^\bullet)$ be the image of $\widetilde{H}_m(\Sigma_i^\bullet)$ in $\widetilde{H}_m(\Sigma^\bullet)$. Then

$$E^\infty_{i,m-i} = {}_{(i)}\widetilde{H}_m(\Sigma^\bullet)/{}_{(i-1)}\widetilde{H}_m(\Sigma^\bullet).$$

Let us define the cohomological spectral sequence $E^{p,q}_r$ by setting

$$E^{p,q}_r = E^r_{-p,N-q-1}$$

and defining the differentials correspondingly, by renaming the differentials in the homological spectral sequence. According to Alexander duality, the term E_∞ of this sequence is associated with the cohomology of $V - \Sigma$. All non-zero entries of this sequence lie in the region $p < 0$, $p + q \geqslant 0$.

The functions on the connected components of $V - \Sigma$ can be identified with the elements of $H^0(V - \Sigma, \mathbb{Z})$. The information about this group is contained in the anti-diagonal entries $E^{-i,i}_\infty$ with positive i. Namely, let $_{(i)}H^0(V - \Sigma, \mathbb{Z})$ be the subgroup of H^0 consisting of the functions that are obtained as linking numbers with cycles of maximal dimension contained in the one-point compactification of Σ_i; as we shall soon see, these classes can be thought of as Vassiliev invariants of order i.

Remark 15.3. The spectral sequence that we just described was first defined by V. Arnol'd (1970), who studied with its help the cohomology of the braid groups. A similar method was later used by G. Segal (1979) to describe the topology of the spaces of rational functions.

15.2.2 Simplicial resolutions

Assume that $f : X \to Y$ is a finite-to-one proper surjective map of topological spaces. Then Y is obtained from X by identifying the preimages of each point $y \in Y$. Assume (for simplicity) that there exists a constant R such that for any point $y \in Y$ the preimage $f^{-1}(y)$ consists of at most R points, and that X is embedded in some Euclidean space V in such a way that any $k + 1$ distinct points of X span a non-degenerate k-simplex for all $k < 2R + 1$. Then we can form the space $\widetilde{Y}$ as the union, over all points $y \in Y$, of the convex hulls of the sets $\{f^{-1}(y)\}$ in V.

We have a map $\tilde{f} : \widetilde{Y} \to Y$ which assigns to a point in the convex hull of the set $\{f^{-1}(y)\}$ the point $y \in Y$. This map is proper and its fibres are simplices, possibly, of different dimensions. It can be deduced that under mild assumptions on Y the map $\tilde{f}$ is a homotopy equivalence; it is called a *simplicial resolution of Y associated with f*. We shall refer to the space $\tilde{Y}$ as the *space of the simplicial resolution $\tilde{f}$*, or, abusing the terminology, as the simplicial resolution of Y.

Example 15.4. The map of a circle onto the figure eight which identifies two points has the following simplicial resolution:

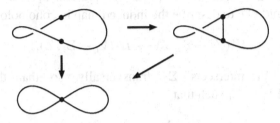

Here the space of the resolution is shown on the right.

Exercise. Describe the simplicial resolution associated with the double cover of a circle by itself.

Since we are interested only in calculating the homology groups, we lose nothing by replacing a space by the space of its simplicial resolution. On the other hand, simplicial resolutions often have interesting filtrations on them. For instance, since the space $\tilde{Y}$ is a union of simplices, it is natural to consider the subspaces $\tilde{Y}_i$ of $\tilde{Y}$ which are the unions of the $i-1$-skeleta of these simplices. In the case of the discriminant in the space of long curves we shall consider another geometrically natural filtration, see Section 15.3.3.

Exercise. Check that in the preceding section instead of a filtration on a discriminant, one can use a filtration on its simplicial resolution.

Simplicial resolutions are especially useful for studying spaces of functions with singularities of some kind. In such a situation Y is taken to be the space of functions with singularities and X is the space of all pairs (ϕ, x) where $\phi \in Y$ is a function and x is a point in the domain of ϕ where ϕ is singular; the map $X \to Y$ simply forgets the second element of the pair (that is, the singular point). Various examples of this kind are described in Vassiliev's book (1994, 1997). While many of the ingredients of Vassiliev's approach to knot spaces were well-known before Vassiliev, the simplicial resolutions are the main innovation of his work.

15.2.3 Stabilization

Strictly speaking, Alexander duality and, as a consequence, all the foregoing constructions, only makes sense in finite-dimensional spaces. However, in the case of knots the space V is infinite-dimensional. This problem can be circumvented by using finite-dimensional approximations to the space of long curves. For this we have to understand first how complements of discriminants behave with respect to inclusions.

Consider two discriminants Σ_1 and Σ_2 inside the Euclidean spaces V_1 and V_2 respectively. If $V_1 \subset V_2$ and $\Sigma_1 = V_1 \cap \Sigma_2$ we see that $V_1 - \Sigma_1$ is a subspace of $V_2 - \Sigma_2$. We would like to describe the induced map in cohomology

$$H^i(V_2 - \Sigma_2, \mathbb{Z}) \to H^i(V_1 - \Sigma_1, \mathbb{Z}). \tag{15.1}$$

Assume that V_1 intersects Σ_2 transversally, so that there exists an ε-neighbourhood V_ε of V_1 such that

$$V_\varepsilon \cap \Sigma_2 = \Sigma_1 \times \mathbb{R}^s,$$

where $s = \dim V_2 - \dim V_1$. There is a homomorphism of reduced homology groups

$$\widetilde{H}_i(\Sigma_2^\bullet) \to \widetilde{H}_{i-s}(\Sigma_1^\bullet)$$

where $X^\bullet$, as before, denotes the one-point compactification of X. This homomorphism is known as the *Pontrjagin–Thom homomorphism* and is constructed in two steps. First, we collapse the part of $\Sigma_2^\bullet$ which lies outside of $\Sigma_2 \cap V_\varepsilon$ to one point and take the induced homomorphism in homology. Then notice that the quotient space with respect to this collapsing map is precisely the s-fold suspension of $\Sigma_1^\bullet$, so we can apply the suspension isomorphism which decreases the degree by s and lands in the reduced homology of $\Sigma_1^\bullet$.

Since Alexander duality is defined by taking linking numbers, it follows from this construction that the cohomology map (15.1) is dual to the Pontrjagin–Thom homomorphism.

Now, let us consider the situation when both discriminants Σ_1 and Σ_2 are filtered, the inclusion $V_1 \to V_2$ is transversal to the filtration and $V_\varepsilon \cap (\Sigma_2)_j = (\Sigma_1)_j \times \mathbb{R}^s$ for all j. Then we have relative Pontrjagin–Thom maps $H_i((\Sigma_2^\bullet)_j, (\Sigma_2^\bullet)_{j-1}) \to H_{i-s}((\Sigma_1^\bullet)_j, (\Sigma_1^\bullet)_{j-1})$.

Proposition 15.5. *If, in the above notation and under the above assumptions, the relative Pontrjagin–Thom maps are isomorphisms for all $j \leqslant P$ and $i > \dim V - Q + j$, for some positive P and Q, the terms $E_1^{p,q}$ of the Vassiliev spectral sequences for the cohomology of $V_1 - \Sigma_1$ and $V_2 - \Sigma_2$ coincide in the region $-P \leqslant p$ and $q \leqslant Q$.*

The proof consists in combining Alexander duality with the definition of the spectral sequence.

The above proposition will allow us to work in infinite-dimensional Euclidean spaces as if they had finite dimension, see Section 15.3.3.

15.2.4 Vassiliev invariants

Suppose that we want to enumerate the connected components of the complement of a discriminant Σ in a vector space V; in other words, we would like to calculate $H^0(V - \Sigma, \mathscr{R})$, the space of $\mathscr{R}$-valued functions on the set of connected components of $V - \Sigma$. If Σ is filtered by closed subspaces Σ_i, we can define *Vassiliev invariants* for the connected components of $V - \Sigma$ as follows.

Definition 15.6. A Vassiliev invariant of degree i is an element of $H^0(V - \Sigma, \mathscr{R})$ defined as the linking number with a cycle in $H_{\dim V - 1}(\Sigma_i^{\,\bullet}, \mathscr{R})$.

This definition also makes sense when we only have the filtration on the homology of Σ, rather than on the space Σ itself. Such a situation arises when we consider a filtration on a simplicial resolution of Σ. Let us consider the following rather special situation where the Vassiliev invariants have a transparent geometric interpretation.

Let Σ be the image of a smooth manifold X immersed in a finite-dimensional vector space V, and assume that each point in V, where Σ has a singularity, is a point of transversal k-fold self-intersection for some k. Without loss of generality we can suppose that Σ is of codimension 1, since its complement would be connected otherwise. Locally, Σ looks like $T^k \times \mathbb{R}^{\dim V - k}$ where T^k is the union of all coordinate hyperplanes in $\mathbb{R}^k$. We shall also assume that Σ is *co-oriented*, that is, that there is a continuous field of unit normal vectors (*co-orientation*) at the smooth points of Σ which extends to the self-intersection points as a multivalued vector field.

Consider the simplicial resolution $\sigma \to \Sigma$ associated with the map $X \to \Sigma$, and the filtration σ_i on σ by the $i - 1$-skeleta of the inverse images of points of Σ. Then we have the following criterion for an element $f \in H^0(V - \Sigma, \mathscr{R})$ to be a Vassiliev invariant of order n.

Grouping together the points of the discriminant Σ according to the multiplicity of self-intersection at each point, we get a decomposition of Σ into a union of open strata $\Sigma_{(i)}$, with $\Sigma_{(i)}$ consisting of points of i-fold self-intersection (that is, of $i + 1$ sheets of X) and having codimension i in Σ. The function f can be extended from $V - \Sigma$ to a locally constant function on each stratum of Σ. If x is a point on the maximal stratum $\Sigma_{(0)}$ which consists of the points where Σ is smooth, let x_+ and x_- be two points in $V - \Sigma$ obtained by shifting x by $\pm\varepsilon$ in the direction of the co-orientation, where ε is small. Then, we set $f(x) = f(x_+) - f(x_-)$. For $x \in \Sigma_{(1)}$

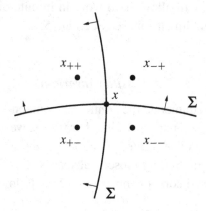

Figure 15.1 The neighbourhood of a generic self-intersection point of the discriminant Σ

we take four points x_{++}, x_{+-}, x_{-+} and x_{--} obtained from x by shifting it to each of the four quadrants in $V - \Sigma$, see Figure 15.1.

For such x we set $f(x) = f(x_{++}) - f(x_{+-}) - f(x_{-+}) + f(x_{--})$. It is clear how to continue: at a point $x \in \Sigma_{(k)}$ the value of f is the alternating sum of its values at the 2^{k+1} points obtained by shifting x to the 2^{k+1} adjacent quadrants of $V - \Sigma$.

Proposition 15.7. *An element* $f \in H^0(V - \Sigma, \mathscr{R})$ *is a Vassiliev invariant of order* $\leqslant n$ *if and only if its extension to the stratum* $\Sigma_{(n)}$ *of n-fold self-intersection points of Σ is identically equal to zero.*

This second characterization of the Vassiliev invariants in terms of their extensions to the strata of the discriminant is the definition that we used throughout the book. Indeed, a generic point of the discriminant in the space of long curves is a singular knot with one simple double point. Knots with two simple double points correspond to transversal self-intersections of the discriminant, etc. Note, however, that this proposition, as stated, does not apply directly to the space of knots, since the discriminant in this case has singularities more complicated than transversal self-intersections. It turns out that these singularities have no influence on the homology of the discriminant in the relevant dimensions, and can be omitted from consideration.

Exercise. Let Σ be the union of the coordinate hyperplanes $x_i = 0$ in $V = \mathbb{R}^n$, X be the disjoint union of these hyperplanes and $X \to \Sigma$ be the natural projection. Describe the cycles that represent classes in $H_{n-1}(\sigma_i{}^\bullet)$ and the space of Vassiliev invariants of degree i. Show that the Vassiliev invariants distinguish the connected components of $V - \Sigma$.

Sketch of the proof. We are interested in the cycles of top dimension on $\Sigma^\bullet$, and these are linear combinations of the (closures of) connected components of $\Sigma_{(1)}$, that is, the top-dimensional stratum consisting of smooth points.

By construction of the simplicial resolution, any cycle that locally is diffeomorphic to the boundary of the "$k-1$-corner"

$$\{(x_1, \ldots, x_n) \mid x_1 > 0, \ldots, x_k > 0\}$$

in $\mathbb{R}^n$ defines a homology class in $H_{n-1}(\sigma_i{}^\bullet, \mathscr{R})$, where $i \geqslant k$. Conversely, any cycle in $H_{n-1}(\sigma_i{}^\bullet, \mathscr{R})$ after projection to $\Sigma^\bullet$ locally looks like a linear combination of $k-1$-corners with $k \leqslant i$.

Now it remains to observe that, locally, the linking number with the boundary of a $(k-1)$-corner vanishes on the strata of dimension $k+1$ and, moreover, this property defines linear combinations of cycles that locally look like j-corners with $j < k$. □

15.3 The space of singular knots and Vassiliev invariants

We want to relate the topology of the space of knots to the structure of the discriminant in the space of long curves. In order to use one of our main tools, namely, Alexander duality, we need to use finite-dimensional approximations to the space of long curves. Spaces V_n of polynomial curves of bounded degree (Section 15.1.1) provide such an approximation; however, it cannot be used directly for the following reason:

Exercise. Denote by $\Sigma(V_n)$ the discriminant in the space V_n consisting of non-embeddings. Then the intersection of V_{n-1} with $\Sigma(V_n)$ inside V_n is not transversal.

As a consequence, we cannot apply the stabilization procedure described in Section 15.2.3, since it requires transversality in an essential way. Nevertheless, this is only a minor technical problem.

15.3.1 Good approximations to the space of long curves

Let

$$U_1 \subset U_2 \subset \ldots \subset U_n \subset \ldots \subset V_\infty$$

be a sequence of finite-dimensional affine subspaces in the space of all polynomial long curves. Note that each U_j is contained in a subspace V_k for some finite k that depends on j. We say that the sequence (U_j) is a *good approximation* to the space of long curves if

- each finite-dimensional compact family of long curves can be uniformly approximated by a continuous family of curves from (U_j) for some j;

- for each U_j and each V_n that contains U_j the intersection of U_j with $\Sigma(V_n)$ is transversal inside V_n, where $\Sigma(V_n)$ is the subspace in V_n consisting of non-embeddings.

Write K'_j for the (topological) subspace of U_j that consists of knots. The first condition in this definition guarantees that the union of all the spaces K'_j has the same homotopy, and, hence, cohomology groups as the space of all knots K. Indeed, it allows us to approximate homotopy classes of maps $S^n \to K$ and homotopies among them by maps and homotopies whose images are contained in the K'_j. The second condition is to ensure that we can use the stability criterion from Section 15.2.3.

A general position argument gives the following

Proposition 15.8 (Vassiliev 1994, 1997). *Good approximations to the space of long curves exist.*

The precise form of a good approximation will be unimportant for us. One crucial property of good approximations is the following:

Exercise. Show that good approximations only contain long curves with a finite number of singular points (that is, points where the tangent vector to the curve vanishes) and self-intersections.

Hint: Show that long curves with an infinite number of self-intersections and singular points form a subset of infinite codimension in V_∞.

In what follows by the "space of long curves" we shall mean the union U_∞ of all the U_j from a good approximation to the space of long curves and by the "space of knots" we shall understand the space $K'_\infty = \cup_j K'_j$ constructed with the help of this approximation.

15.3.2 Degenerate chord diagrams

The discriminant in the space of long curves consists of various parts (*strata*) that correspond to various types of *singular knots*, that is, long curves with self-intersections and singular points.

In our definition of Vassiliev invariants in Chapter 3, we associated a chord diagram with n chords to a knot with n double points. In fact, we saw that chord diagrams are precisely the equivalence classes of knots with double points modulo isotopies and crossing changes. If we consider knots with more complicated self-intersections and with singular points we must generalize the notion of chord diagram.

A *degenerate chord diagram* is a set of distinct pairs (x_k, y_k) of real numbers (called *vertices*) with $x_k \leqslant y_k$. These pairs can be thought of as chords on $\mathbb{R}$, with

x_k and y_k being the left and right endpoints of the chords respectively. If all the x_k and y_k are distinct, we have a usual linear chord diagram.

The "degeneracy" of a degenerate chord diagram can be of two kinds: one chord can degenerate into a singular point ($x_k = y_k$) or two chords can glue together and share an endpoint. Two degenerate chord diagrams are *combinatorially equivalent* if there is a self-homeomorphism of $\mathbb{R}$ that preserves the orientation and sends one diagram to the other.

The vertices of a degenerate chord diagram are of two types: the *singularity* vertices which participate in chords with $x_k = y_k$, and *self-intersection* vertices which participate in chords with $x_k < y_k$. The same vertex can be a singularity vertex and a self-intersection vertex at the same time; in this case we shall count it twice and say that a singularity vertex coincides with a self-intersection vertex. As with usual chord diagrams, one can speak of the *internal graph* of a degenerate diagram: this is the abstract graph formed by the chords whose ends are distinct. The self-intersection vertices are divided into *groups*: two vertices belong to the same group if and only if they belong to the same connected component of the internal graph. Here is an example of a degenerate chord diagram with two groups of self-intersection vertices; the singularity vertices are indicated by hollow dots:

Let us say that two degenerate chord diagrams D_1 and D_2 are *equivalent* if D_1 is combinatorially equivalent to a diagram that has the same set of singularity vertices and the same groups of self-intersection vertices as D_2. For instance, the following diagrams are equivalent:

In Vassiliev's terminology, equivalence classes of degenerate chord diagrams are called (A, b)-*configurations*.

Define the *complexity* of a degenerate chord diagram as the total number of its vertices minus the number of groups of its self-intersection vertices. This number only depends on the equivalence class of the diagram. The complexity of a usual chord diagram is, clearly, equal to its degree.

Remark 15.9. Note that the number of chords of a degenerate chord diagram is not mentioned in the definitions of equivalence and complexity. This is only natural, of course, since equivalent diagrams can have different numbers of chords.

An arbitrary singular knot with a finite number of singular points and self-intersections defines an equivalence class of degenerate chord diagrams: each group of self-intersection vertices is a preimage of a self-intersection and the singularity vertices are the preimages of the singularities. We shall say that a singular knot $f : \mathbb{R} \to \mathbb{R}^3$ *respects* a degenerate chord diagram D if it glues together all points within each group of self-intersection vertices of D and its tangent vector is zero at each singularity vertex of D.

Exercise. Show that equivalence classes of degenerate chord diagrams coincide with the equivalence classes of singular knots with a finite number of singular points and self-intersections under isotopies and crossing changes.

15.3.3 The discriminant

The discriminant in the space of long curves U_∞ is a complicated set. Its strata can be enumerated: they correspond to equivalence classes of degenerate chord diagrams. However, the structure of these strata is not easy to describe, since they can (and do) have self-intersections. The most convenient tool for studying the discriminant are the simplicial resolutions described in Section 15.2.2.

In order to tame the multitude of indices, let us write simply U for the approximating space U_j, $N = N_j$ for its dimension and $\Sigma = \Sigma(U_j)$ for the discriminant. Write $Sym^2(\mathbb{R})$ for the space of all unordered pairs of points in $\mathbb{R}$; this space can be thought of as the subset of $\mathbb{R}^2$ defined by the inequality $x \leqslant y$.

In the product space $\Sigma \times Sym^2(\mathbb{R})$ consider the subspace $\widetilde{\Sigma}$ consisting of pairs $(f, (x, y))$ such that either $x \neq y$ and $f(x) = f(y)$ or $x = y$ and $f'(x) = 0$. Forgetting the pair (x, y) gives a map $\widetilde{\Sigma} \to \Sigma$ which is finite-to-one and proper, so we can associate a simplicial resolution with it. (Strictly speaking, in order to define a simplicial resolution, we must embed $\widetilde{\Sigma}$ in a Euclidean space in a particular way, but let us sweep this issue under the carpet and refer to Vassiliev (1994).)

Denote the space of this simplicial resolution by σ. A point in σ is uniquely described by a collection

$$\left(f, (x_0, y_0), \ldots, (x_k, y_k), \tau \right),$$

where f is a singular knot with $f(x_j) = f(y_j)$ whenever $x_j \neq y_j$ and $f'(x_j) = 0$ when $x_j = y_j$, all pairs (x_j, y_j) are distinct and τ is a point in the interior of a k-simplex with vertices labelled by the points (x_j, y_j). In other words, a point in σ is a triple consisting of a singular knot f, a degenerate chord diagram D which f respects, and a point τ in a simplex whose vertices are labelled by the chords of D. Here k can be arbitrary.

Let σ_i be the closed subspace of σ consisting of triples (f, D, τ) with D of complexity at most i. The *cohomological Vassiliev spectral sequence for the space of knots* is the spectral sequence that comes from the filtration of σ by the σ_i (see Section 15.2.1). We have

$$E_1^{p,q} = E_{-p,N-q-1}^1 = H_{N-(p+q+1)}(\sigma_p^\bullet, \sigma_{p-1}^\bullet) = \widetilde{H}_{N-(p+q+1)}(\sigma_p^\bullet/\sigma_{p-1}^\bullet).$$

Note that $\sigma_p^\bullet/\sigma_{p-1}^\bullet$ is homeomorphic to the one-point compactification of $\sigma_p - \sigma_{p-1}$, and this space can be described rather explicitly, at least when the dimension of U is sufficiently large.

Indeed, the condition that a singular knot $f \in U \subset V_d$ respects a degenerate chord diagram D produces several linear constraints on the coefficients of the polynomials P_1, P_2 and P_3 which determine f. Namely, if (x_j, y_j) is a chord of D, the polynomials satisfy the conditions

$$\frac{P_\alpha(x_j)}{(1 + x_j^2)^d} = \frac{P_\alpha(y_j)}{(1 + y_j^2)^d}$$

when $x_j < y_j$, and

$$\left(\frac{P_\alpha(x_j)}{(1 + x_j^2)^d}\right)' = 0$$

if $x_j = y_j$. Each of these conditions with x_j and y_j fixed gives one linear equation on the coefficients of each of the polynomials P_1, P_2 and P_3.

In general, these linear equations may be linearly dependent. However, the rank of this system of equations can be explicitly calculated when the dimension of U is large.

Exercise. Show that for a given degenerate chord diagram of complexity p there exists N_0 such that for $N > N_0$ the number of linearly independent conditions on the coefficients of the P_α is equal to exactly $3p$.

Hint. For any set of distinct real numbers $x_1, \ldots, x_k$ there exists d such that the vectors $(1, x_i, x_i^2, \ldots, x_i^d)$ are linearly independent.

This exercise shows that, for N sufficiently big, the forgetful map that sends a triple $(f, D, \tau) \in \sigma_p - \sigma_{p-1}$ to the pair (D, τ) is an affine bundle with the fibre of dimension $N - 3p$ over a base W_p which only depends on p. In particular, we have the Thom isomorphism

$$\widetilde{H}_{N-s}((\sigma_p - \sigma_{p-1})^\bullet) = H_{3p-s}(W_p^\bullet).$$

As a consequence, when $p > 0$ and $q \geqslant p$, the first term of the Vassiliev spectral sequence has the entries

$$E_1^{-p,q} = E_{p,N-q-1}^1 = \widetilde{H}_{N+p-q-1}((\sigma_p - \sigma_{p-1})^\bullet) = H_{4p-q-1}(W_p^\bullet),$$

and for all other values of p and q the corresponding entry is zero. The space $W_p{}^\bullet$ whose homology is, therefore, so important for the theory of Vassiliev invariants will be called here *the diagram complex*.

15.4 Topology of the diagram complex

The diagram complexes $W_p{}^\bullet$ are constructed out of simplices and spaces of degenerate chord diagrams of complexity p.

15.4.1 A cell decomposition for the diagram complex

The space $W_p{}^\bullet$ has a cell decomposition with cells indexed by degenerate chord diagrams (more precisely, combinatorial equivalence classes of such diagrams) of complexity p. The cell $[D]$ is a product of an open simplex Δ_D whose vertices are indexed by the chords of D, and the space E_D of all diagrams combinatorially equivalent to D. This latter space is also an open simplex, of dimension k, where k is the number of geometrically distinct vertices of D. Indeed, it is homeomorphic to the configuration space of k distinct points in an open interval.

The boundaries in these cell complexes can also be explicitly described. Since

$$[D] = \Delta_D \times E_D$$

is a product, its boundary consists of two parts. The first part consists of the cells that come from $\partial \Delta_D \times E_D$. These are of the form $[D']$, where D' is obtained from D by removing a number of chords. The second part comes from $\Delta_D \times \partial E_D$. The diagram D' of a cell $[D']$ of this kind is obtained from D by collapsing to zero the distance between two adjacent vertices. Note that by removing chords or glueing together two adjacent vertices we can decrease the complexity of a diagram; in this case the corresponding part of $\partial[D]$ is glued to the base point in $W_p{}^\bullet$.

If we are interested in the homology of $W_p{}^\bullet$, we need to describe the boundaries in the corresponding cellular chain complex and this involves only those cells $[D'] \subset \partial[D]$ with $\dim[D'] = \dim[D] - 1$. For the dimension of $[D]$ we have the formula

$$\dim[D] = \text{no. of geometrically distinct vertices} + \text{no. of chords} - 1$$

and the complexity $c(D)$ is given by the expression

$$c(D) = \text{total no. of vertices} - \text{no. of groups of self-intersection vertices}.$$

These formulae show that if by removing chords of D we obtain a diagram D' with $c(D') = c(D)$ and $\dim[D'] = \dim[D] - 1$, then D' is obtained from D by removing one chord, and the endpoints of the removed chord belong to the same group of self-intersection vertices in D'.

Now, suppose that by collapsing two adjacent vertices of D we obtain a diagram D' with $c(D') = c(D)$ and $\dim[D'] = \dim[D] - 1$. There are several possibilities for this:

1. Both vertices are self-intersection vertices and belong to different groups, and at most one of the two vertices is also a singularity vertex:

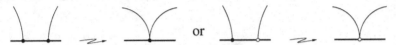

2. Both vertices are endpoints of the same chord and of no other chord:

3. One vertex is a self-intersection vertex only and the other is a singularity vertex only:

Each of the cells $[D']$ in the above list appears in the boundary of $[D]$ exactly once. This describes the cellular chain complex for $W_p^{\bullet}$ up to signs. This is sufficient if we work modulo 2. In order to calculate the integral homology, we need to fix the orientations for each $[D]$ and work out the signs.

Recall that the cell $[D]$ is a product of two simplices; therefore, its orientation can be specified by ordering the vertices of the factors. E_D is the configuration space of k points in an interval and its vertices are naturally ordered: the ith vertex is a configuration with $k - i$ points in the left end of the interval and i points in the right end. (Note that the vertices belong to closure of E_D, but not to E_D itself, and the corresponding configurations may have coinciding points.) The vertices of Δ_D are the chords of D. In order to order them, we first order the chords within each group of self-intersection vertices: a chord is smaller than another chord if its left endpoint is smaller; if both chords have the same left endpoint, the one with the smaller right endpoint is smaller. Next, we order the groups: a group is smaller if its leftmost vertex is smaller. It is convenient to consider in this context each singularity vertex which does not coincide with any self-intersection vertex as a separate group consisting of one "degenerate" chord; the leftmost vertex of such a group is, of course, the singularity vertex itself. Finally, we list the chords lexicographically: first, all the chords from the first group, then the chords from the second group, and so on.

Now, it is clear how to assign the signs in the boundaries.

Example 15.10. In the cellular chain complex for $W_2{}^\bullet$ we have

$$d\left(\text{⌢⌢}\right) = \text{⌢}\ -\ \text{⌢⌢}\ +\ \text{⌢},$$

$$d\left(\text{⌢⌢}\right) = -\ \text{∘⌢}\ +\ \text{⌢⌢}\ -\ \text{⌢∘},$$

and

$$d\left(\text{⌢⌢}\right) = -\ \text{⌢}\ +\ \text{⌢⌢}\ -\ \text{⌢}.$$

Note that our convention for the orientations of the cells is different from that of Vassiliev (1994, 1997).

Exercise. Formulate the general rule for the signs in the boundary of a cell $[D]$ for an arbitrary D. Calculate the boundary for the 5-dimensional cell of $W_2{}^\bullet$ not shown in Example 15.10.

15.4.2 The filtration on $W_p{}^\bullet$ by the number of vertices

In principle, the homology of $W_p{}^\bullet$ can be calculated directly from the cellular chain complex that we have just described. There is, however, a better way to calculate this homology.

Let $W_p{}^\bullet(k)$ be the subspace of $W_p{}^\bullet$ consisting of all the cells $[D]$ where D has at most k geometrically distinct vertices, together with the added basepoint. The smallest k for which $W_p{}^\bullet(k)$ is nontrivial is equal to $[p/2] + 1$ where $[\cdot]$ denotes the integer part; this corresponds to the diagrams all of whose singularity vertices are combined with self-intersection vertices and the latter are joined into only one group. The maximal number of distinct vertices is $k = 2p$; it is achieved for chord diagrams of degree p. We get the increasing filtration

$$* = W_p{}^\bullet([p/2]) \subset W_p{}^\bullet([p/2] + 1) \subset \ldots \subset W_p{}^\bullet(2p) = W_p{}^\bullet$$

by the number of vertices; in Vassiliev (1994, 1997) it is called the *auxiliary filtration*.

The successive quotients in this filtration are bouquets of certain spaces indexed by equivalence classes of degenerate diagrams:

$$W_p{}^\bullet(k)/W_p{}^\bullet(k-1) = \bigvee_{\mathfrak{D}} [\mathfrak{D}],$$

where $\mathfrak{D}$ runs over all equivalence classes of diagrams with k distinct vertices and $[\mathfrak{D}]$ is the union of all the cells $[D]$ such that the equivalence class of D is $\mathfrak{D}$ (the basepoint counted as one of such cells). In turn, each $[\mathfrak{D}]$ can be constructed out of several standard pieces.

For a positive integer a define the *complex of connected graphs* $\Delta^1(a)$ as follows. Given a set A of a points, consider the simplex of dimension $a(a-1)/2-1$ whose vertices are indexed by chords connecting pairs of points of A. Each face of this simplex corresponds to a graph whose set of vertices is A. Collapsing the union of all those faces that correspond to non-connected graphs to a point, we get the complex of connected graphs $\Delta^1(a)$.

The proof of the following statement can be found in Vassiliev (1994, 1997):

Lemma 15.11. $H_i(\Delta^1(a)) = 0$ *unless* $i = 0, a-2$, *and* $H_{a-2}(\Delta^1(a)) = \mathbb{Z}^{(a-1)!}$.

The spaces $[\mathfrak{D}]$ can now be described as follows:

Lemma 15.12. *Let $\mathfrak{D}$ be an equivalence class of diagrams with m groups of self-intersection vertices consisting of $a_1, \ldots, a_m$ vertices respectively, b singularity vertices, and k geometrically distinct vertices in total. Then*

$$[\mathfrak{D}] = \Delta^1(a_1) \wedge \ldots \wedge \Delta^1(a_m) \wedge S^{k+m+b-1}.$$

Here $A \wedge B$ is the "smash" product of topological spaces: $A \wedge B = A \times B / A \vee B$.

Proof It is instructive to verify first the case when the diagrams in $\mathfrak{D}$ have k self-intersection vertices only, all in one group. The space of all possible sets of vertices for such diagrams is an open k-dimensional simplex. Over each point of this space, we have the $k(k+1)/2 - 1$-dimensional simplex without the faces corresponding to non-connected graphs, that is, $\Delta^1(k)$ minus the basepoint. $[\mathfrak{D}]$ is the one-point compactification of this product:

$$[\mathfrak{D}] = S^k \wedge \Delta^1(k)$$

and in this case $m = 1$ and $k = a_1$.

In the general case the space of all possible sets of vertices of a diagram is still a k-dimensional simplex. Now, over each set of vertices we have the interior of the join of all $\Delta^1(a_i)$, taken without their basepoints, and b singularity points. This is nothing but the product

$$(\Delta^1(a_1) - *) \times \ldots \times (\Delta^1(a_m) - *) \times \mathbb{R}^{m+b-1}.$$

Taking one-point compactification we get the statement of the lemma. □

The two above lemmas imply that the homology of $[\mathfrak{D}]$ vanishes in all dimensions apart from 0 and $p+k-1$. Now, using the homology exact sequences, or, which is the same, the spectral sequence associated with the filtration $W_p{}^\bullet(k)$, we arrive at the following:

Lemma 15.13.

$$H_{3p-1}(W_p{}^\bullet) = H_{3p-1}(W_p{}^\bullet/W_p{}^\bullet(2p-2)).$$

15.4.3 Chord diagrams and 4T relations

Cohomology classes of dimension zero, that is, knot invariants, produced by the Vassiliev spectral sequence correspond to elements of the groups $E_\infty^{-p,p}$ obtained from $E_1^{-p,p}$ as quotients.

A consequence of last lemma is the following description of the group $E_1^{-p,p} = H_{3p-1}(W_p{}^\bullet)$:

Proposition 15.14 (Vassiliev 1994, 1997). *For any ring $\mathscr{R}$ of coefficients, $E_1^{-p,p}$ is isomorphic to the group $\mathscr{W}_p$ of $\mathscr{R}$-valued weight systems, that is, functions on chord diagrams that vanish on the 1T and 4T relations.*

Note that, according to the Fundamental Theorem of Section 4.2.1, over the rational numbers the group $\mathscr{W}_p$ is isomorphic to the space $E_\infty^{-p,p}$ of Vassiliev invariants of order $\leqslant p$, modulo those of order $\leqslant p-1$.

Proof The proof uses the cell decomposition of $H_{3p-1}(W_p{}^\bullet)$. The $3p-1$-dimensional cells that are not contained in $W_p{}^\bullet(2p-2)$ are of the form $[D]$ where D is either

- a non-degenerate chord diagram of order p;
- a degenerate chord diagram with $2p-1$ self-intersection vertices and $p-1$ groups, of which $p-2$ are pairs and one is a group with 3 vertices connected by 3 chords.

None of these cells is contained in the boundary of a $3p$-dimensional cell.

The $3p-2$-dimensional cells that are not contained in $W_p{}^\bullet(2p-2)$ are of the form $[D]$ where D is either

(1T) a diagram with $2p-1$ distinct vertices one of which is a singularity vertex and the rest are self-intersection vertices grouped into pairs;

(4T) a degenerate chord diagram with $2p-1$ self-intersection vertices and $p-1$ groups, of which $p-2$ are pairs and one is a group with 3 vertices connected by 2 chords.

The cellular chain complex consists of free modules, so the kernel of the boundary on the $3p-1$-cells, is isomorphic to the dual of the cokernel for the coboundary on the $3p-2$-cells. Unlike the boundary, the coboundary is easy to calculate.

To be precise, let $d_{3p-1} : C_{3p-1} \to C_{3p-2}$ be the boundary in the chain complex for $W_p{}^\bullet / W_p{}^\bullet (2p-2)$. The dual modules $\mathrm{Hom}(C_i, \mathscr{R})$ can be identified with C_i; in particular, they are generated by the same degenerate chord diagrams. The dual homomorphism $d_{3p-1}^* : C_{3p-2} \to C_{3p-1}$ sends a diagram of the type (1T) to a diagram which has a chord with adjacent vertices; moreover, every diagram with such a chord is in the image of d_{3p-1}^*.

At this point it will be convenient to modify our convention on the orientation of the cells. Let us change the orientation of the cells that correspond to non-degenerate chord diagrams by multiplying it by $(-1)^r$, where r is the number of intersections among the chords of the diagram, or, in a slightly fancier language, the number of edges of its intersection graph. Diagrams of the type (4T) are sent by d_{3p-1}^* to linear combinations of three diagrams:

$$d_{3p-1}^* \left(\vcenter{\hbox{\includegraphics{}}} \right) = \vcenter{\hbox{}} + (-1)^s \vcenter{\hbox{}} - (-1)^s \vcenter{\hbox{}},$$

$$d_{3p-1}^* \left(\vcenter{\hbox{}} \right) = - \vcenter{\hbox{}} - (-1)^s \vcenter{\hbox{}} + (-1)^s \vcenter{\hbox{}},$$

and

$$d_{3p-1}^* \left(\vcenter{\hbox{}} \right) = \vcenter{\hbox{}} + (-1)^s \vcenter{\hbox{}} - (-1)^s \vcenter{\hbox{}},$$

where s is the same number in all cases.

Exercise. Find an expression for s and verify the above formulae.

It follows that each 4T relation is in the image of d_{3p-1}^*.

Now, let C'_{3p-1} be the subspace of C_{3p-1} spanned by non-degenerate chord diagrams. Each functional f on $C_{3p-1}/d_{3p-1}^* C_{3p-2}$ can be uniquely reconstructed from its value on $C'_{3p-1}/\langle 1T, 4T \rangle$ since

$$f \left(\vcenter{\hbox{}} \right) = -(-1)^s f \left(\vcenter{\hbox{}} \right) + (-1)^s f \left(\vcenter{\hbox{}} \right)$$

where s is as above, and, hence, we see that $H_{3p-1}(W_p{}^\bullet / W_p{}^\bullet (2p-2))$ consists precisely of the $\mathscr{R}$-valued weight systems. $\qquad \Box$

The reader who has survived to this point may note that in Vassiliev's original approach, the road to the combinatorial description of the weight systems has been long and winding, especially if compared to the method presented in the first chapters of this book. We stress, however, that while the present approach to the 0-dimensional cohomology classes can be dramatically simplified, there are no low-tech solutions for classes of higher dimensions.

15.5 Homology of the space of knots and Poisson algebras

The same methods that we have used in this chapter to study the cohomology of the space of knots can be applied to describe its homology. In particular, we get a homological spectral sequence whose first term consists of the cohomology groups of the diagram complexes $W_p^\bullet$ and can be described completely in terms of degenerate chord diagrams. The bialgebra of chord diagrams $\mathscr{A}'$ forms a part of this spectral sequence; namely $\mathscr{A}'_p$ is isomorphic over a ring $\mathscr{R}$ to the diagonal entry $E^1_{-p,p} = \widetilde{H}^{3p-1}(W_p^\bullet, \mathscr{R})$.

It is very interesting to note that the first term of this spectral sequence has another interpretation, which, at first, seems to be completely unrelated to knots. Namely, as discovered by V. Turchin (2004), it is closely related to the *Hochschild homology of the Poisson algebras operad.*

The details of Turchin's work are outside the scope of this book. Let us just give a rough explanation of how Poisson algebras appear in the homological Vassiliev spectral sequence.

Recall that a *Poisson algebra* has two bilinear operations: a commutative and associative product, and an antisymmetric bracket satisfying the Jacobi identity. The two operations are related by the Leibniz rule

$$[ab, c] = a[b, c] + b[a, c].$$

Using the Leibniz rule, one can re-write any composition of products and brackets as a linear combination of products of iterated brackets, which we call *Poisson monomials.*

In order to describe the cohomology of the diagram complex $W_p^\bullet$ one can use the auxiliary filtration on it by the number of vertices of a diagram, see Section 15.4.2. The successive quotients in this filtration are built out of certain standard pieces indexed by equivalence classes of degenerate diagrams. As pointed out in in Turchin (2004), these equivalence classes give rise to Poisson monomials in the following fashion.

Let us restrict our attention to degenerate chord diagrams without singularity vertices. Label the vertices of such a diagram by numbers from 1 to n according to their natural order on the real line. The equivalence class of the diagram is then determined by a partition of the set $1, \ldots, n$ into several subsets with at least two elements each. For every such subset $i_1, \ldots, i_k$ form an iterated bracket

$[\dots[[x_{i_1}, x_{i_2}]\dots, x_{i_k}]$ and take the product of these brackets over all the subsets. For instance, the equivalence class of the diagram

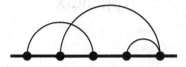

gives rise to the monomial $[x_1, x_3][[x_2, x_4], x_5]$. A chord diagram of degree p gives a product of p simple brackets.

Poisson monomials of this type appear in the Hochschild complex for the Poisson algebra operad. V. Turchin proves the following result:

Theorem 15.15. *The first term of the Vassiliev spectral sequence for the homology of the space of knots coincides with the Hochschild homology bialgebra for the operad of Poisson algebras with the Poisson bracket of degree 2, taken modulo two explicit relations.*

We refer to Turchin (2004) for the basics on operads and their Hochschild homology, and the precise form of this statement. A further reference is the paper by D. Sinha (2006) where the relationship between the Vassiliev spectral sequence and the Hochschild complex is explained "on the level of spaces."

Appendix

A.1 Lie algebras and their representations

A.1.1 Lie algebras

A Lie algebra $\mathfrak{g}$ over a field $\mathbb{F}$ of characteristic zero is a vector space equipped with a bilinear operation (*Lie bracket*) $(x, y) \mapsto [x, y]$ subject to the identities

$$[x, y] = -[y, x],$$
$$[x, [y, z]] + [y, [z, x]] + [z, [x, y]] = 0.$$

In this section we shall only consider finite-dimensional Lie algebras over $\mathbb{C}$.

An *abelian* Lie algebra is a vector space with the bracket which is identically 0: $[x, y] = 0$ for all $x, y \in \mathfrak{g}$. Any vector space with the zero bracket is an abelian Lie algebra.

Considering the Lie bracket as a product, one may speak about homomorphisms of Lie algebras, Lie subalgebras, and so on. In particular, an *ideal* in a Lie algebra is a vector subspace stable under taking the bracket with an arbitrary element of the whole algebra. A Lie algebra is called *simple* if it is not abelian and does not contain any proper ideal. Simple Lie algebras are classified (see, for example, Fulton and Harris 1991; Humphreys 1980). Over the field of complex numbers $\mathbb{C}$ there are four families of *classical algebras*:

Type	$\mathfrak{g}$	$\dim \mathfrak{g}$	description
A_n	$\mathfrak{sl}_{n+1}$	$n^2 + 2n$	$(n + 1) \times (n + 1)$ matrices with zero trace, $(n \geqslant 1)$
B_n	$\mathfrak{so}_{2n+1}$	$2n^2 + n$	skew-symmetric $(2n + 1) \times (2n + 1)$ matrices, $(n \geqslant 2)$
C_n	$\mathfrak{sp}_{2n}$	$2n^2 + n$	$2n \times 2n$ matrices X satisfying the relation $X^t \cdot M + M \cdot X = 0$, where M is the standard $2n \times 2n$ skew-symmetric matrix $M = \begin{pmatrix} O & Id_n \\ -Id_n & 0 \end{pmatrix}$, $(n \geqslant 3)$
D_n	$\mathfrak{so}_{2n}$	$2n^2 - n$	skew-symmetric $2n \times 2n$ matrices, $(n \geqslant 4)$

and five *exceptional algebras*:

Type	E_6	E_7	E_8	F_4	G_2
dim $\mathfrak{g}$	78	133	248	52	14

The Lie bracket in the matrix Lie algebras above is the commutator of matrices.
Apart from the low-dimensional isomorphisms

$$\mathfrak{sp}_2 \cong \mathfrak{so}_3 \cong \mathfrak{sl}_2; \qquad \mathfrak{sp}_4 \cong \mathfrak{so}_5; \qquad \mathfrak{so}_4 \cong \mathfrak{sl}_2 \oplus \mathfrak{sl}_2; \qquad \mathfrak{so}_6 \cong \mathfrak{sl}_4,$$

all the Lie algebras in the list above are different. The Lie algebra $\mathfrak{gl}_N$ of all $N \times N$ matrices is isomorphic to the direct sum of $\mathfrak{sl}_N$ and the abelian one-dimensional Lie algebra $\mathbb{C}$.

A.1.2 Metrized Lie algebras

For $x \in \mathfrak{g}$ write ad_x for the linear map $\mathfrak{g} \to \mathfrak{g}$ given by $\mathrm{ad}_x(y) = [x, y]$.
The *Killing form* on a Lie algebra $\mathfrak{g}$ is defined by the equality

$$\langle x, y \rangle^K = \mathrm{Tr}(\mathrm{ad}_x \mathrm{ad}_y).$$

Cartan's criterion says that this bilinear form is non-degenerate if and only if the algebra is *semi-simple*, that is, isomorphic to a direct sum of simple Lie algebras.

Exercise. Prove that the Killing form is ad-invariant in the sense of the following definition.

Definition A.1. A bilinear form $\langle \cdot, \cdot \rangle : \mathfrak{g} \otimes \mathfrak{g} \to \mathbb{C}$ is said to be *ad-invariant* if it satisfies the identity

$$\langle \mathrm{ad}_z(x), y \rangle + \langle x, \mathrm{ad}_z(y) \rangle = 0,$$

or, equivalently,

$$\langle [x, z], y \rangle = \langle x, [z, y] \rangle. \tag{A.1}$$

for all $x, y, z \in \mathfrak{g}$.

This definition is justified by the fact described in the following exercise.

Exercise. Let $\mathfrak{G}$ be the connected Lie group corresponding to the Lie algebra $\mathfrak{g}$ and let $\mathrm{Ad}_g : \mathfrak{g} \to \mathfrak{g}$ be its adjoint representation (see, for instance, Adams 1969). Then the ad-invariance of a bilinear form is equivalent to its Ad-invariance defined by the natural rule

$$\langle \mathrm{Ad}_g(x), \mathrm{Ad}_g(y) \rangle = \langle x, y \rangle$$

for all $x, y \in \mathfrak{g}$ and $g \in \mathfrak{G}$.

A Lie algebra is said to be *metrized* if it is equipped with an ad-invariant symmetric non-degenerate bilinear form $\langle \cdot, \cdot \rangle : \mathfrak{g} \otimes \mathfrak{g} \to \mathbb{C}$. The class of metrized algebras contains simple Lie algebras with (a multiple of) the Killing form, abelian Lie algebras with an arbitrary non-degenerate bilinear form and their direct sums. If a Lie algebra is simple, all ad-invariant symmetric non-degenerate bilinear forms are multiples of each other.

For the classical simple Lie algebras which consist of matrices, it is often more convenient to use, instead of the Killing form, a different bilinear form $\langle x, y \rangle = \mathrm{Tr}(xy)$, which is proportional to the Killing form with the coefficient $\frac{1}{2N}$ for $\mathfrak{sl}_N$, $\frac{1}{N-2}$ for $\mathfrak{so}_N$, and $\frac{1}{N+2}$ for $\mathfrak{sp}_N$.

Exercise. Prove that for the Lie algebra $\mathfrak{gl}_N$ the Killing form $\langle x, y \rangle^K = \mathrm{Tr}(\mathrm{ad}_x \cdot \mathrm{ad}_y)$ is degenerate with defect 1 and can be expressed as follows:

$$\langle x, y \rangle^K = 2N\mathrm{Tr}(xy) - 2\mathrm{Tr}(x)\mathrm{Tr}(y).$$

Exercise. Prove that the form $\mathrm{Tr}(xy)$ on $\mathfrak{gl}_N$ is non-degenerate and ad-invariant.

The bilinear form $\langle \cdot, \cdot \rangle$ is an element of $\mathfrak{g}^* \otimes \mathfrak{g}^*$. It identifies $\mathfrak{g}$ with $\mathfrak{g}^*$ and, hence, can be considered as an element $c \in \mathfrak{g} \otimes \mathfrak{g}$, called the *quadratic Casimir tensor*. If $\{e_i\}$ is a basis for $\mathfrak{g}$ and $\{e_i^*\}$ is the dual basis, the Casimir tensor can be written as

$$c = \sum_i e_i \otimes e_i^*.$$

The quadratic Casimir tensor is ad-invariant in the sense that for any $x \in \mathfrak{g}$

$$\mathrm{ad}_x c := \sum_i \mathrm{ad}_x e_i \otimes e_i^* + \sum_i e_i \otimes \mathrm{ad}_x e_i^* = 0.$$

A.1.3 Structure constants

Given a basis $\{e_i\}$ for the Lie algebra $\mathfrak{g}$ of dimension d, the Lie brackets of the basis elements can be written as

$$[e_i, e_j] = \sum_{k=1}^{d} c_{ijk} e_k.$$

The numbers c_{ijk} are called the *structure constants of $\mathfrak{g}$ with respect to $\{e_i\}$*.

Lemma A.2. *Let c_{ijk} be the structure constants of a metrized Lie algebra in a basis $\{e_i\}$, orthonormal with respect to an ad-invariant bilinear form. Then the constants c_{ijk} are antisymmetric with respect to the permutations of the indices i, j and k.*

Proof The equality $c_{ijk} = -c_{jik}$ is the coordinate expression of the fact that the commutator is antisymmetric: $[x, y] = -[y, x]$. It remains to prove that $c_{ijk} = c_{jki}$.

This follows immediately from Equation (A.1), simply by setting $x = e_i$, $y = e_k$, $z = e_j$. ∎

The Lie bracket, being a bilinear map $\mathfrak{g} \otimes \mathfrak{g} \to \mathfrak{g}$, can be considered as an element of $\mathfrak{g}^* \otimes \mathfrak{g}^* \otimes \mathfrak{g}$. The metric defines an isomorphism $\mathfrak{g} \simeq \mathfrak{g}^*$ and, hence, the Lie bracket of a metrized Lie algebra produces an element in $J \in \mathfrak{g}^{\otimes 3}$ called the *structure tensor* of $\mathfrak{g}$.

Corollary A.3. *The structure tensor J of a metrized Lie algebra $\mathfrak{g}$ is totally antisymmetric: $J \in \wedge^3 \mathfrak{g}$.*

A.1.4 Representations of Lie algebras

A *representation* of a Lie algebra $\mathfrak{g}$ in a vector space V is a Lie algebra homomorphism of $\mathfrak{g}$ into the Lie algebra $\mathfrak{gl}(V)$ of linear operators in V, that is, a map $\rho : \mathfrak{g} \to \mathfrak{gl}(V)$ such that

$$\rho([x, y]) = \rho(x)\rho(y) - \rho(y)\rho(x).$$

It is also said that V is a $\mathfrak{g}$-*module* and that $\mathfrak{g}$ *acts on* V by ρ. When ρ is understood from the context, the element $\rho(x)(v)$ can be written as $x(v)$. The *invariants* of an action of $\mathfrak{g}$ on V are the elements of V that lie in the kernel of $\rho(x)$ for all $x \in \mathfrak{g}$. The space of all invariants in V is denoted by $V^{\mathfrak{g}}$.

The *standard representation St* of a matrix Lie algebra, such as $\mathfrak{gl}_N$ or $\mathfrak{sl}_N$, is the representation in $\mathbb{C}^N$ given by the identity map.

The *adjoint representation* is the action ad of $\mathfrak{g}$ on itself according to the rule

$$x \mapsto \mathrm{ad}_x \in \mathrm{Hom}(\mathfrak{g}, \mathfrak{g}), \quad \mathrm{ad}_x(y) = [x, y].$$

It is indeed a representation, since $\mathrm{ad}_{[x,y]} = \mathrm{ad}_x \cdot \mathrm{ad}_y - \mathrm{ad}_y \cdot \mathrm{ad}_x = [\mathrm{ad}_x, \mathrm{ad}_y]$.

A representation $\rho : \mathfrak{g} \to \mathfrak{gl}(V)$ is *reducible* if there exist $\rho_1 : \mathfrak{g} \to \mathfrak{gl}(V_1)$ and $\rho_2 : \mathfrak{g} \to \mathfrak{gl}(V_2)$ with $V_i \neq 0$ and $V = V_1 \oplus V_2$, such that $\rho = \rho_1 \oplus \rho_2$. A representation that is not reducible is *irreducible*.

Example A.4. The algebra $\mathfrak{sl}_2$ of 2×2-matrices with zero trace has precisely one irreducible representation of dimension $n + 1$ for each positive n. Denote this representation by V_n. There exist a basis $e_0, \ldots, e_n$ for V_n in which the matrices

$$H = \begin{pmatrix} 1 & 0 \\ 0 & -1 \end{pmatrix}, \qquad E = \begin{pmatrix} 0 & 1 \\ 0 & 0 \end{pmatrix}, \qquad F = \begin{pmatrix} 0 & 0 \\ 1 & 0 \end{pmatrix}$$

that span $\mathfrak{sl}_2$ act as follows:

$$H(e_i) = (n - 2i)e_i, \qquad E(e_i) = (n - i + 1)e_{i-1}, \qquad F(e_i) = (i + 1)e_{i+1},$$

where it is assumed that $e_{-1} = e_{n+1} = 0$.

The *Casimir element* of a representation ρ of a metrized Lie algebra is the matrix

$$c(\rho) = \sum_i \rho e_i \rho e_i^*$$

for some basis $\{e_i\}$ of $\mathfrak{g}$. If ρ is finite-dimensional, the trace of the Casimir element of ρ is well-defined; it is called the *Casimir number* of ρ.

Exercise. Show that $c(\rho)$ is well-defined and commutes with the image of ρ.

For further information on representations of Lie algebras, we refer the reader to Humphreys (1980).

A.1.5 Tensor algebras

Let $\mathfrak{g}$ be a vector space over a field of characteristic zero. A *tensor* is an element of a tensor product of several copies of $\mathfrak{g}$ and its dual space $\mathfrak{g}^*$. The number of factors in this product is called the *rank* of the tensor. The canonical map $\mathfrak{g} \otimes \mathfrak{g}^* \to \mathbb{C}$ induces maps

$$\mathfrak{g}^{\otimes p} \otimes (\mathfrak{g}^*)^{\otimes q} \to \mathfrak{g}^{\otimes p-1} \otimes (\mathfrak{g}^*)^{\otimes q-1}$$

called *contractions*, defined for any pair of factors $\mathfrak{g}$ and $\mathfrak{g}^*$ in the tensor product.

Denote by

$$T(\mathfrak{g}) = \bigoplus_{n \geqslant 0} \mathfrak{g}^{\otimes n}$$

the *tensor algebra* of the vector space $\mathfrak{g}$, whose multiplication is given by the tensor product. In particular, $\mathfrak{g} \subset T(\mathfrak{g})$ is the subspace spanned by the generators of $T(\mathfrak{g})$.

The *symmetric algebra* of $\mathfrak{g}$, denoted by $S(\mathfrak{g})$, is the quotient of $T(\mathfrak{g})$ by the two-sided ideal generated by all the elements $x \otimes y - y \otimes x$. The symmetric algebra decomposes as

$$S(\mathfrak{g}) = \bigoplus_{n \geqslant 0} S^n(\mathfrak{g}),$$

where the *nth symmetric power* $S^n(\mathfrak{g})$ is the images of $\mathfrak{g}^{\otimes n}$.

Let $\{e_i\}$ be a basis of $\mathfrak{g}$. Then $T(\mathfrak{g})$ can be identified with the free algebra on the generators e_i, and $S(\mathfrak{g})$ with the free commutative algebra on the e_i. In particular, elements of $S(e_i)$ can be thought of as polynomials in the e_i and products $e_{j_1} e_{j_2} \ldots e_{j_m}$ such that $m \geqslant 0$ and $j_1 \leqslant j_2 \leqslant \ldots \leqslant j_m$ form an additive basis for $S(\mathfrak{g})$.

The symmetric algebra is a quotient, rather than a subalgebra, of the tensor algebra. However, it can be identified with the subspace of *symmetric tensors* in $T(\mathfrak{g})$. Namely, the image of the linear map $S^n(\mathfrak{g}) \to \mathfrak{g}^{\otimes n}$ given by

$$e_{j_1} e_{j_2} \dots e_{j_m} \rightarrow \frac{1}{m!} \sum_{\sigma \in S_m} e_{\sigma(j_1)} \otimes e_{\sigma(j_2)} \otimes \dots \otimes e_{\sigma(j_m)}$$

consists of the tensors invariant under all permutations of the factors in $\mathfrak{g}^{\otimes m}$.

A.1.6 Universal enveloping and symmetric algebras

Any associative algebra can be considered as a Lie algebra whose Lie bracket is the commutator

$$[a, b] = ab - ba.$$

While not every Lie algebra is of this form, each Lie algebra is contained in an associative algebra as a subspace closed under the commutator.

The *universal enveloping algebra* of $\mathfrak{g}$, denoted by $U(\mathfrak{g})$, is the quotient of $T(\mathfrak{g})$ by the two-sided ideal generated by all the elements

$$x \otimes y - y \otimes x - [x, y],$$

$x, y \in \mathfrak{g}$. In other words, we force the commutator of two elements of $\mathfrak{g} \subset T(\mathfrak{g})$ to be equal to their Lie bracket in $\mathfrak{g}$. An example of a universal enveloping algebra is the symmetric algebra $S(\mathfrak{g})$: one can think of it as the universal enveloping algebra of the abelian Lie algebra obtained from $\mathfrak{g}$ by endowing it with the zero bracket.

The universal enveloping algebra of $\mathfrak{g}$ is always infinite-dimensional. A basis of $\mathfrak{g}$ gives rise to an explicit additive basis of $U(\mathfrak{g})$:

Theorem A.5 (Poincaré–Birkhoff–Witt). *Let $\{e_i\}$ be a basis of the Lie algebra $\mathfrak{g}$. Then all the products $e_{j_1} e_{j_2} \dots e_{j_m}$ such that $m \geq 0$ and $j_1 \leq j_2 \leq \dots \leq j_m$ form an additive basis for $U(\mathfrak{g})$.*

Corollary A.6. *The map $\mathfrak{g} \rightarrow T(\mathfrak{g}) \rightarrow U(\mathfrak{g})$ is an inclusion; the restriction to $\mathfrak{g}$ of the commutator on $U(\mathfrak{g})$ coincides with the Lie bracket.*

Exercise. Show that $U(\mathfrak{g})$ has the following universal property: for each homomorphism f of $\mathfrak{g}$ into a commutator algebra of an associative algebra A there exists the unique homomorphism of associative algebras $U(\mathfrak{g}) \rightarrow A$ whose restriction to $\mathfrak{g}$ is f.

The basis given by the Poincaré–Birkhoff–Witt theorem does not depend on the Lie bracket of $\mathfrak{g}$. In particular, we see that

$$U(\mathfrak{g}) \simeq S(\mathfrak{g})$$

as vector spaces.

Further, both $S(\mathfrak{g})$ and $U(\mathfrak{g})$ are $\mathfrak{g}$-modules: the adjoint representation of $\mathfrak{g}$ can be extended to $S(\mathfrak{g})$ or $U(\mathfrak{g})$ by the condition that $\mathfrak{g}$ acts by *derivations*:

$$\mathrm{ad}_x(yz) = \mathrm{ad}_x(y)z + y\,\mathrm{ad}_x(z)$$

for all y, z in $S(\mathfrak{g})$ or $U(\mathfrak{g})$. Note that in the case of $U(\mathfrak{g})$ the element x simply acts by taking the commutator with x. In particular, the Casimir element for this action is simply the image of the Casimir tensor under the map $T(\mathfrak{g}) \to U(\mathfrak{g})$.

The *Poincaré–Birkhoff–Witt isomorphism* is the map $S(\mathfrak{g}) \to U(\mathfrak{g})$ defined by

$$e_{j_1}e_{j_2}\cdots e_{j_m} \to \frac{1}{m!}\sum_{\sigma \in S_m} e_{\sigma(j_1)}e_{\sigma(j_2)}\cdots e_{\sigma(j_m)}.$$

Exercise. Show that this is a vector space isomorphism.

It follows from the definition that the Poincaré–Birkhoff–Witt isomorphism is an *isomorphism of $\mathfrak{g}$-modules*, that is, it commutes with the action of $\mathfrak{g}$. In fact, it is also an isomorphism of coalgebras, see Exercise A.18. (Clearly, it is not an algebra isomorphism, since $S(\mathfrak{g})$ is commutative and $U(\mathfrak{g})$ is not, unless $\mathfrak{g}$ is abelian.)

A.1.7 Duflo isomorphism

Since the universal enveloping algebra and the symmetric algebra of a Lie algebra $\mathfrak{g}$ are isomorphic as $\mathfrak{g}$-modules, we have an isomorphism of vector spaces

$$S(\mathfrak{g})^{\mathfrak{g}} \simeq U(\mathfrak{g})^{\mathfrak{g}} = Z(U(\mathfrak{g}))$$

between the subalgebra of invariants in the symmetric algebra and the centre of the universal enveloping algebra. This map does not respect the product, but it turns out that $S(\mathfrak{g})^{\mathfrak{g}}$ and $Z(U(\mathfrak{g}))$ are actually isomorphic as commutative algebras. The isomorphism between them is given by the *Duflo–Kirillov map*, defined as follows.

A *differential operator* $S(\mathfrak{g}) \to S(\mathfrak{g})$ is just an element of the symmetric algebra $S(\mathfrak{g}^*)$. The action of $S(\mathfrak{g}^*)$ on $S(\mathfrak{g})$ is obtained by extending the pairing of $\mathfrak{g}^*$ and $\mathfrak{g}$: we postulate that

$$x(ab) = x(a) \cdot b + a \cdot x(b)$$

for any $x \in \mathfrak{g}^*$ and $a, b \in S(\mathfrak{g})$, and that

$$(xy)(a) = x(y(a))$$

for $x, y \in S(\mathfrak{g}^*)$ and $a \in S(\mathfrak{g})$. An element of $S^k(\mathfrak{g}^*)$ is a differential operator of order k: it sends $S^m(\mathfrak{g})$ to $S^{m-k}(\mathfrak{g})$. We can also speak of differential operators of infinite order; these are elements of the graded completion of $S(\mathfrak{g}^*)$.

If $\mathfrak{g}$ is a metrized Lie algebra, its bilinear form gives an isomorphism between $S(\mathfrak{g})$ and $S(\mathfrak{g}^*)$, which sends elements of $S(\mathfrak{g})$ to differential operators. Explicitly, if we think of elements of $S(\mathfrak{g})$ as symmetric tensors, for $a \in S(\mathfrak{g})$ the operator $\partial_a \in S(\mathfrak{g}^*)$ is obtained by taking the sum of all possible contractions with a.

Let $j(x)$ be a formal power series with $x \in \mathfrak{g}$ given by

$$j(x) = \det\left(\frac{\sinh \frac{1}{2}\mathrm{ad}\, x}{\frac{1}{2}\mathrm{ad}\, x}\right).$$

This power series starts with the identity, so we can take the square root $\sqrt{j}$. It is an element of the graded completion of $S(\mathfrak{g}^*)$ so it can be considered as a differential operator of infinite order, called the Duflo–Kirillov map:

$$\sqrt{j} : S(\mathfrak{g}) \to S(\mathfrak{g}).$$

Theorem A.7 (Duflo 1970; see also Alekseev and Torossian 2009; Bar-Natan *et al.* 2003; Kontsevich 2003). *The composition of the Duflo–Kirillov map with the Poincaré–Birkhoff–Witt isomorphism defines an isomorphism of commutative algebras $S(\mathfrak{g})^{\mathfrak{g}} \to Z(U(\mathfrak{g}))$.*

This isomorphism is known as the *Duflo isomorphism*.

A.1.8 Lie superalgebras

A *super vector space*, or a $\mathbb{Z}_2$-*graded vector space*, is a vector space, decomposed as a direct sum

$$V = V_0 \oplus V_1.$$

The indices (or *degrees*) 0 and 1 are thought of as elements of $\mathbb{Z}_2$; V_0 is called the *even* part of V and V_1 is the *odd* part of V. An element $x \in V$ is *homogeneous* if it belongs to either V_0 or V_1. For x homogeneous we write $|x|$ for the degree of x. The (super) dimension of V is the pair $(\dim V_0 \mid \dim V_1)$ also written as $\dim V_0 + \dim V_1$.

An endomorphism f of a super vector space V is a sum of four linear maps $f_{ij} : V_i \to V_j$. If V is finite-dimensional the *supertrace* of f is defined as

$$\mathrm{sTr}\, f = \mathrm{Tr}\, f_{00} - \mathrm{Tr}\, f_{11}.$$

A *superalgebra* is a super vector space A together with a bilinear product which respects the degree:

$$|xy| = |x| + |y|$$

for all homogeneous x and y in A. The *supercommutator* in a superalgebra A is a bilinear operation defined on homogeneous $x, y \in A$ by

$$[x, y] = xy - (-1)^{|x||y|}yx.$$

The elements of A whose supercommutator with the whole of A is zero form the *super centre* of A.

The supercommutator satisfies the following identities:

$$|[x, y]| = |x| + |y|,$$

$$[x, y] = -(-1)^{|x||y|}[y, x]$$

and

$$(-1)^{|z||x|}[x, [y, z]] + (-1)^{|y||z|}[z, [x, y]] + (-1)^{|x||y|}[y, [z, x]] = 0,$$

where x, y, z are homogeneous. A super vector space with a bilinear bracket satisfying these identities is called a *Lie superalgebra*.

Each Lie superalgebra $\mathfrak{g}$ can be thought of as a subspace of its *universal enveloping superalgebra* $U(\mathfrak{g})$ defined as the quotient of the tensor algebra on $\mathfrak{g}$ by the ideal generated by

$$x \otimes y - (-1)^{|x||y|}y \otimes x - [x, y],$$

where x and y are arbitrary homogeneous elements of $\mathfrak{g}$; the supercommutator in $U(\mathfrak{g})$ induces the bracket of $\mathfrak{g}$.

The theory of Lie superalgebras was developed by V. Kac (1977a, 1977b); it closely parallels the usual Lie theory.

Example A.8 (Figueroa-O'Farrill *et al.* 1997). The Lie superalgebra $\mathfrak{gl}(1|1)$ consists of the endomorphisms of the super vector space of dimension $1+1$ with the bracket being the supercommutator of endomorphisms. The supertrace gives a bilinear form on $\mathfrak{gl}(1|1)$

$$\langle x, y \rangle = \mathrm{sTr}(xy),$$

which is non-degenerate and ad-invariant in the same sense as for Lie algebras:

$$\langle [x, z], y \rangle = \langle x, [z, y] \rangle.$$

Take a basis in the $1 + 1$ - dimensional space whose first vector is even and the second vector is odd. Then the even part of $\mathfrak{gl}(1|1)$ is spanned by the matrices

$$H = \begin{pmatrix} 1 & 0 \\ 0 & 1 \end{pmatrix}, \qquad G = \begin{pmatrix} 0 & 0 \\ 0 & 1 \end{pmatrix},$$

and the odd part by

$$Q_+ = \begin{pmatrix} 0 & 0 \\ 1 & 0 \end{pmatrix}, \qquad Q_- = \begin{pmatrix} 0 & 1 \\ 0 & 0 \end{pmatrix}.$$

The Lie bracket of H with any element vanishes and we have

$$[G, Q_\pm] = \pm Q_\pm \qquad \text{and} \qquad [Q_+, Q_-] = H.$$

The quadratic Casimir tensor for $\mathfrak{gl}(1|1)$ is

$$H \otimes G + G \otimes H - Q_+ Q_- + Q_- Q_+.$$

Its image c in the universal enveloping algebra $U(\mathfrak{gl}(1|1))$ together with the image of H under the inclusion of $\mathfrak{gl}(1|1)$, which we denote by h, generate a polynomial subalgebra of $U(\mathfrak{gl}(1|1))$ which coincides with the super centre of $U(\mathfrak{gl}(1|1))$.

A.2 Bialgebras and Hopf algebras

Here we give a brief summary of necessary information about bialgebras and Hopf algebras. More details can be found in Abe (1980); Cartier (2007); and Milnor and Moore (1965).

A.2.1 Coalgebras and bialgebras

In what follows, all vector spaces and algebras will be considered over a field $\mathbb{F}$ of characteristic zero. First, let us recall the definition of an algebra in the language of commutative diagrams.

Definition A.9. A *product*, or a *multiplication*, on a vector space A is a linear map $\mu : A \otimes A \to A$. The product μ on A is *associative* if the diagram

$$
\begin{array}{ccc}
A \otimes A \otimes A & \xrightarrow{\mu \otimes \mathrm{id}} & A \otimes A \\
{\scriptstyle \mathrm{id} \otimes \mu} \downarrow & & \downarrow {\scriptstyle \mu} \\
A \otimes A & \xrightarrow{\mu} & A
\end{array}
$$

commutes. A *unit* for μ is a linear map $\iota : \mathbb{F} \to A$ (uniquely defined by the element $\iota(1) \in A$) that makes commutative the diagram

$$
\begin{array}{ccc}
\mathbb{F} \otimes A & \xrightarrow{\iota \otimes \mathrm{id}} & A \otimes A \\
\uparrow & & \downarrow {\scriptstyle \mu} \\
A & =\!=\!=\!= & A
\end{array}
$$

where the upward arrow is the natural isomorphism. A vector space with an associative product is called an (associative) *algebra*.

The unit in an algebra, if it exists, is always unique. We shall only consider associative algebras with a unit.

Reversing the arrows in the above definition, we arrive to the notion of a *coalgebra*.

Definition A.10. A coalgebra is a vector space A equipped with a linear map $\delta : A \to A \otimes A$, referred to as *comultiplication*, or *coproduct*, and a linear map $\varepsilon : A \to \mathbb{F}$, called the *counit*, such that the following two diagrams commute:

$$
\begin{array}{ccc}
A \otimes A \otimes A & \xleftarrow{\ \delta \otimes \mathrm{id}\ } & A \otimes A \\
{\scriptstyle \mathrm{id} \otimes \delta} \uparrow & & \uparrow {\scriptstyle \delta} \\
A \otimes A & \xleftarrow{\ \delta\ } & A
\end{array}
\qquad
\begin{array}{ccc}
\mathbb{F} \otimes A & \xleftarrow{\ \varepsilon \otimes \mathrm{id}\ } & A \otimes A \\
\downarrow & & \uparrow {\scriptstyle \delta} \\
A & =\!=\!= & A
\end{array}
$$

Algebras (coalgebras) may possess an additional property of commutativity (respectively, cocommutativity), defined via the following commutative diagrams:

$$
\begin{array}{ccc}
A \otimes A & \xrightarrow{\ \mu\ } & A \\
{\scriptstyle \tau} \uparrow & & \| \\
A \otimes A & \xrightarrow{\ \mu\ } & A
\end{array}
\qquad
\begin{array}{ccc}
A \otimes A & \xleftarrow{\ \delta\ } & A \\
{\scriptstyle \tau} \downarrow & & \| \\
A \otimes A & \xleftarrow{\ \delta\ } & A
\end{array}
$$

where $\tau : A \otimes A \to A \otimes A$ is the permutation of the tensor factors:

$$\tau(a \otimes b) = b \otimes a.$$

Definition A.11. A *bialgebra* is a vector space A with the structure of an algebra given by μ, ι and the structure of a coalgebra given by δ, ε which agree in the sense that the following identities hold:

1. $\varepsilon(1) = 1$;
2. $\delta(1) = 1 \otimes 1$;
3. $\varepsilon(ab) = \varepsilon(a)\varepsilon(b)$;
4. $\delta(ab) = \delta(a)\delta(b)$.

Here μ is written as a usual product and in the last equation $\delta(a)\delta(b)$ denotes the component-wise product in $A \otimes A$ induced by the product μ in A.

Note that these conditions, taken in pairs, have the following meaning:

- $(1,3) \Leftrightarrow \varepsilon$ is a homomorphism of unital algebras.
- $(2,4) \Leftrightarrow \delta$ is a homomorphism of unital algebras.

- (1,2) $\Leftrightarrow \iota$ is a homomorphism of coalgebras.
- (3,4) $\Leftrightarrow \mu$ is a homomorphism of coalgebras.

The coherence of the two structures in the definition of a bialgebra can thus be stated in either of the two equivalent ways:

- ε and δ are algebra homomorphisms,
- μ and i are coalgebra homomorphisms.

Example A.12. The group algebra $\mathbb{F}G$ of a group G over the field $\mathbb{F}$ consists of finite formal linear combinations $\sum_{x \in G} \lambda_x x$ where $\lambda_x \in \mathbb{F}$ with the product defined on the basis elements by the group multiplication in G. The coproduct is defined as $\delta(x) = x \otimes x$ for $x \in G$ and then extended by linearity. Instead of a group G, in this example one can, actually, take a monoid, that is, a semigroup with a unit.

Example A.13. The algebra $\mathbb{F}^G$ of $\mathbb{F}$-valued functions on a finite group G with pointwise multiplication

$$(fg)(x) = f(x)g(x)$$

and the comultiplication defined by

$$\delta(f)(x, y) = f(xy)$$

where the element $\delta(f) \in \mathbb{F}^G \otimes \mathbb{F}^G$ is understood as a function on $G \times G$ via the natural isomorphism $\mathbb{F}^G \otimes \mathbb{F}^G \cong \mathbb{F}^{G \times G}$.

Example A.14. The symmetric algebra $S(V)$ of a vector space V is a bialgebra with the coproduct defined on the elements $x \in V = S^1(V)$ by setting $\delta(x) = 1 \otimes x + x \otimes 1$ and then extended as an algebra homomorphism to the entire $S(V)$.

Example A.15. The *completed* symmetric algebra

$$\widehat{S}(\mathfrak{g}) = \prod_{n \geqslant 0} S^n(\mathfrak{g}),$$

of a vector space V, whose elements are formal power series in the coordinates of V, is a bialgebra whose coproduct extends that of the symmetric algebra.

Example A.16. Let $U(\mathfrak{g})$ be the universal enveloping algebra of a Lie algebra $\mathfrak{g}$ (see Section A.1.6). Define $\delta(g) = 1 \otimes g + g \otimes 1$ for $g \in \mathfrak{g}$ and extend it to all of A by the axioms of bialgebra. If $\mathfrak{g}$ is abelian, this example reduces to that of the symmetric algebra.

Exercise A.17. Define the appropriate unit and counit in each of the above examples.

Exercise A.18. Show that the Poincaré–Birkhoff–Witt isomorphism is an isomorphism of coalgebras (that is, commutes with the counit and the comultiplication).

A.2.2 Primitive and group-like elements

In bialgebras there are two remarkable classes of elements: primitive elements and group-like elements.

Definition A.19. An element $a \in A$ of a bialgebra A is said to be *primitive* if

$$\delta(a) = 1 \otimes a + a \otimes 1.$$

The set of all primitive elements forms a vector subspace $\mathcal{P}(A)$ called the *primitive subspace* of the bialgebra A. The primitive subspace is closed under the commutator $[a, b] = ab - ba$, and, hence, forms a Lie algebra (which is abelian, if A is commutative). Indeed, since δ is a homomorphism, the fact that a and b are primitive implies

$$\delta(ab) = \delta(a)\delta(b) = 1 \otimes ab + a \otimes b + b \otimes a + ab \otimes 1,$$
$$\delta(ba) = \delta(b)\delta(a) = 1 \otimes ba + b \otimes a + a \otimes b + ba \otimes 1$$

and, therefore,

$$\delta([a, b]) = 1 \otimes [a, b] + [a, b] \otimes 1.$$

Definition A.20. An element $a \in A$ is said to be *semigroup-like* if

$$\delta(a) = a \otimes a.$$

If, in addition, a is invertible, then it is called *group-like*.

The set of all semigroup-like elements in a bialgebra is closed under multiplication. It follows that the set of all group-like elements $\mathcal{G}(A)$ of a bialgebra A forms a multiplicative group.

Among the examples of bialgebras given above, the notions of the primitive and group-like elements are especially transparent in the case $A = \mathbb{F}^G$. As follows from the definitions, primitive elements are the *additive* functions ($f(xy) = f(x) + f(y)$) while group-like elements are the *multiplicative* functions ($f(xy) = f(x)f(y)$).

In the example of the symmetric algebra, there is an isomorphism

$$S(V) \otimes S(V) \cong S(V \oplus V)$$

which allows to rewrite the definition of the coproduct as $\delta(x) = (x, x) \in V \oplus V$ for $x \in V$. It can be even more suggestive to view the elements of the symmetric algebra $S(V)$ as polynomial functions on the dual space V^* (where homogeneous subspaces $S^0(V)$, $S^1(V)$, $S^2(V)$ and so on correspond to constants, linear functions, quadratic functions, etc. on V^*). In these terms, the product in $S(V)$ corresponds to the usual (pointwise) multiplication of functions, while the coproduct $\delta : S(V) \to S(V \oplus V)$ acts according to the rule

$$\delta(f)(\xi, \eta) = f(\xi + \eta), \quad \xi, \eta \in X^*.$$

Under the same identifications,

$$(f \otimes g)(\xi, \eta) = f(\xi)g(\eta),$$

in particular,

$$(f \otimes 1)(\xi, \eta) = f(\xi),$$
$$(1 \otimes f)(\xi, \eta) = f(\eta).$$

We see that an element of $S(V)$, considered as a function on V^*, is primitive (group-like) if and only if this function is additive (multiplicative):

$$f(\xi, \eta) = f(\xi) + f(\eta),$$
$$f(\xi, \eta) = f(\xi)f(\eta).$$

The first condition means that f is a linear function on V^*, that is, it corresponds to an element of V itself; therefore,

$$\mathscr{P}(S(V)) = V.$$

Over a field of characteristic zero, the second condition cannot hold for polynomial functions except for the constant function equal to 1; thus

$$\mathscr{G}(S(V)) = \{1\}.$$

The completed symmetric algebra $\widehat{S}(V)$, in contrast with $S(V)$, has a lot of group-like elements. Namely,

$$\mathscr{G}(\widehat{S}(V)) = \{\exp(x) \mid x \in V\},$$

where $\exp(x)$ is defined as a formal power series $1 + x + x^2/2! + \ldots$, see Section A.2.9.

Exercise. Describe the primitive and group-like elements in $\mathbb{F}G$ and in $U(\mathfrak{g})$.

Answer. In $\mathbb{F}G$ we have $\mathscr{P} = 0, \mathscr{G} = G$; in $U(\mathfrak{g})$ we have $\mathscr{P} = \mathfrak{g}, \mathscr{G} = \{1\}$.

A.2.3 Filtrations and gradings

A decreasing filtration on a vector space A is a sequence of subspaces A_i, $i = 0, 1, 2, \ldots$ such that

$$A = A_0 \supseteq A_1 \supseteq A_2 \supseteq \ldots$$

The *factors* of a decreasing filtration are the quotient spaces $\mathrm{gr}_i A = A_i / A_{i+1}$.

An increasing filtration on a vector space A is a sequence of subspaces A_i, $i = 0, 1, 2, \ldots$ such that

$$A_0 \subseteq A_1 \subseteq A_2 \subseteq \cdots \subseteq A.$$

The factors of an increasing filtration are the quotient spaces $\mathrm{gr}_i A = A_i / A_{i-1}$, where by definition $A_{-1} = 0$.

A filtration (either decreasing or increasing) is said to be of *finite type* if all its factors are finite-dimensional. Note that in each case the whole space has a (possibly infinite-dimensional) "part" not covered by the factors, namely $\cap_{i=1}^{\infty} A_i$ for a decreasing filtration and $A / \cup_{i=1}^{\infty} A_i$ for an increasing filtration.

A vector space is said to be *graded* if it is represented as a direct sum of its subspaces

$$A = \bigoplus_{i=0}^{\infty} A_i.$$

A graded space A has a canonical increasing filtration by the subspaces $\oplus_{i=0}^{k} A_i$ and a canonical decreasing filtration by $\oplus_{i=k}^{\infty} A_i$.

With a filtered vector space A one can associate a graded vector space $G(A)$ setting

$$\mathrm{gr} A = \bigoplus_{i=0}^{\infty} G_i A = \bigoplus_{i=0}^{\infty} A_i / A_{i+1}$$

in case of a decreasing filtration and

$$\mathrm{gr} A = \bigoplus_{i=0}^{\infty} G_i A = \bigoplus_{i=0}^{\infty} A_i / A_{i-1}$$

in case of an increasing filtration.

If A is a filtered space of finite type, then the homogeneous components $G_i A$ are also finite-dimensional; their dimensions have a compact description in terms of the *Poincaré series*

$$\sum_{k=0}^{\infty} \dim(\mathrm{gr}_i A) \, t^k,$$

where t is an auxiliary formal variable.

Example A.21. The Poincaré series of the algebra of polynomials in one variable is

$$1 + t + t^2 + \dots = \frac{1}{1 - t}.$$

Exercise. Find the Poincaré series of the polynomial algebra with n independent variables.

One can also speak of filtered and graded algebras, coalgebras and bialgebras: these are filtered (graded) vector spaces with operations that respect the corresponding filtrations (gradings).

Definition A.22. We say that an algebra A is *filtered* if its underlying vector space has a filtration by subspaces A_i compatible with the product in the sense that

$$A_p A_q \subset A_{p+q} \quad \text{for} \quad p, q \geqslant 0.$$

A coalgebra A is filtered if it is filtered as a vector space and

$$\delta(A_n) \subset \sum_{p+q=n} A_p \otimes A_q \quad \text{for} \quad n \geqslant 0.$$

Finally, a bialgebra is filtered if it is filtered both as an algebra and as a coalgebra, with respect to the same filtration.

Definition A.23. A *graded algebra* A is a graded vector space with a product satisfying

$$A_p A_q \subset A_{p+q} \quad \text{for} \quad p, q \geqslant 0 \quad \text{and} \quad 1 \in A_0.$$

A *graded coalgebra* A is a graded vector space with a coproduct satisfying

$$\delta(A_n) \subset \sum_{p+q=n} A_p \otimes A_q \quad \text{for} \quad n \geqslant 0 \quad \text{and} \quad \varepsilon|_{A_k} = 0 \quad \text{for} \quad k > 0.$$

A *graded bialgebra* is a graded vector space which is graded both as an algebra and as a coalgebra.

The operations on filtered vector spaces descend to the associated graded spaces.

Proposition A.24. *The graded vector space associated to a filtered algebra (coalgebra, bialgebra) has a natural structure of a graded algebra (respectively, coalgebra, bialgebra) except, possibly, for the existence of the unit and the counit.*

Exercise. Find the conditions that a filtered algebra (coalgebra) has to satisfy so that the associated graded vector space is a graded algebra (respectively, coalgebra).

Definition A.25. The *graded completion* of a graded vector space $A = \oplus_{i=0}^{\infty} A_i$ is the vector space $\widehat{A} = \prod_{i=0}^{\infty} A_i$.

For instance, the graded completion of the vector space of polynomials in n variables is the space of formal power series in the same variables. Note that a priori there is no non-trivial grading on the graded completion of a graded space.

Note that the product in a graded algebra extends uniquely to its graded completion; the same is true for the coproduct in a graded coalgebra.

A.2.4 Dual filtered bialgebra

Let A be a filtered bialgebra with a decreasing filtration A_k of finite type. For each $k \geqslant 0$ define W_k to be the the subspace of A^* consisting of all the linear functions on A that vanish on A_{k+1}. Then W_k is contained in W_{k+1} and the union

$$W = \bigcup_{k \geqslant 0} W_k$$

is a filtered vector space with the increasing filtration by the W_k.

Proposition A.26. *W is a bialgebra with an increasing filtration, with the operations induced by duality by those of A.*

We say that W is a bialgebra *dual* to A. Note that $W \subseteq A^*$ and the equality holds if and only if A is finite-dimensional.

Proof If μ and δ are the product and the coproduct in A, respectively, with ι the unit and ε the counit, the operations in W are as follows:

$$\delta^* : \sum_{k+l=n} W_k \otimes W_l \to W_n \quad \text{is the product in } W,$$

$$\mu^* : W_n \to \sum_{k+l=n} W_k \otimes W_l \quad \text{is the coproduct in } W,$$

$$\iota^* : W \to \mathbb{F} \qquad\qquad\qquad \text{is the counit in } W,$$

$$\varepsilon^* : \mathbb{F} \to W \qquad\qquad\qquad \text{is the unit in } W.$$

First, let us see that δ^* is indeed a product which agrees with the filtration. $W \otimes W$ is a subspace of $A^* \otimes A^*$ which, in turn, is a subspace of $(A \otimes A)^*$. (The three spaces coincide if and only if A is finite-dimensional.) We need to show that the image of the composition

$$W_k \otimes W_l \hookrightarrow (A \otimes A)^* \xrightarrow{\delta^*} A^*$$

lies in W_{k+l}.

Take $w_1 \in W_k$ and $w_2 \in W_l$. The product of these elements is the composition

$$A \xrightarrow{\delta} A \otimes A \xrightarrow{w_1 \otimes w_2} \mathbb{F}.$$

If $a \in A_{k+l+1}$ then

$$\delta(a) = \sum_i b_i \otimes c_i,$$

where for each i we have $b_i \in A_p$ and $c_i \in A_q$ with $p + q = k + l + 1$. As a consequence, either $p > k$ or $q > l$ which implies that $(w_1 \otimes w_2)(b_i \otimes c_i) = 0$ for all i and, hence, $\delta^*(w_1 \otimes w_2) \in W_{k+l}$.

In order to see that μ^* gives a coproduct on W which respects the filtration, we have to verify that the image of the map

$$W_k \hookrightarrow A^* \xrightarrow{\mu^*} (A \otimes A)^*$$

lies in $\sum_{p+q=k} W_p \otimes W_q$.

Take $w \in W_k$ and consider the composition

$$A \otimes A \xrightarrow{\mu} A \xrightarrow{w} \mathbb{F}.$$

Since w vanishes on A_{k+1}, the composition $w \circ \mu$ is equal to zero on the subspace $\sum_{p+q=k+1} A_p \otimes A_q$ and thus may be considered as a linear function on the quotient vector space

$$A \otimes A / \sum_{p+q=k+1} A_p \otimes A_q.$$

Since the filtration A_i is of finite type, this quotient does not change if we replace A with the finite-dimensional vector space A/A_{k+1}. Now, for any finite-dimensional vector space A with a descending filtration and for all k the subspaces

$$\left(A \otimes A / \sum_{p+q=k+1} A_p \otimes A_q \right)^*$$

and

$$\sum_{p+q=k+1} (A/A_{p+1})^* \otimes (A/A_{q+1})^*$$

of $A^* \otimes A^*$ coincide. This implies that $\mu^*(w) \in \sum_{p+q=k} W_p \otimes W_q$.

We leave checking the bialgebra axioms to the reader. □

Example A.27. The bialgebra of $\mathbb{F}$-valued functions on a finite group G is dual to the bialgebra $\mathbb{F}G$. Here the filtration of $\mathbb{F}G$ is concentrated in degree 0: $(\mathbb{F}G)_k = 0$ for $k > 0$.

The fact the dual is defined only for filtered bialgebras of finite type and not for bialgebras in general is explained by the following observation. If the vector space A is infinite-dimensional, the inclusion

$$A^* \otimes A^* \subset (A \otimes A)^*$$

is strict. The dual to a coproduct $A \to A \otimes A$ is a map $(A \otimes A)^* \to A^*$ which restricts to a product $A^* \otimes A^* \to A^*$, and, hence, the dual of a coalgebra is an algebra. However, the dual to a product on A is a map $A^* \to (A \otimes A)^*$, whose image does not necessarily lie in $A^* \otimes A^*$. As a consequence, the dual of an algebra may fail to be a coalgebra.

Exercise. Give an example of a bialgebra whose product does not induce a coproduct on the dual space.

A.2.5 Group-like and primitive elements in the dual bialgebra

Primitive and group-like elements in the dual bialgebra have a very transparent meaning.

Proposition A.28. *Primitive (respectively, group-like) elements in the dual of a filtered bialgebra A are those linear functions which are additive (respectively, multiplicative), that is, satisfy the respective identities*

$$f(ab) = f(a) + f(b),$$
$$f(ab) = f(a)f(b)$$

for all $a, b \in A$.

Proof An element f is primitive if $\delta(f) = 1 \otimes f + f \otimes 1$. Evaluating this on an arbitrary tensor product $a \otimes b$ with $a, b \in A$, we obtain

$$f(ab) = f(a) + f(b).$$

An element f is group-like if $\delta(f) = f \otimes f$. Evaluating this on an arbitrary tensor product $a \otimes b$, we obtain

$$f(ab) = f(a)f(b).$$

In the same way, the additivity (multiplicativity) of a linear function implies that it defines a primitive (respectively, group-like) element. $\square$

A.2.6 Hopf algebras

Definition A.29. A *Hopf algebra* is a connected graded bialgebra of finite type. This means that A is a graded vector space A, with the grading by non-negative integers

$$A = \bigoplus_{k \geqslant 0} A_k$$

and the grading is compatible with the operations μ, ι, δ, ε in the following sense:

$$\mu : A_m \otimes A_n \to A_{m+n},$$

$$\delta : A_n \to \bigoplus_{k+l=n} A_k \otimes A_l,$$

$$1 \in A_0,$$

$$\varepsilon|_{A_k} = 0 \text{ for } k > 0.$$

A graded algebra A is said to be of *finite type*, if all its homogeneous components A_n are finite-dimensional. It is said to be *connected*, if $\iota : \mathbb{F} \to A$ is an isomorphism of $\mathbb{F}$ onto $A_0 \subset A$.

Remark A.30. The above definition follows the classical paper Milnor and Moore (1965). Nowadays a Hopf algebra is usually defined as a not necessarily graded bialgebra with an additional operation, called *antipode*, which is a linear map $S : A \to A$ such that

$$\mu \circ (S \otimes 1) \circ \delta = \mu \circ (1 \otimes S) \circ \delta = \iota \circ \varepsilon.$$

The bialgebras of interest for us (those that satisfy the premises of the Milnor–Moore theorem below) always have an antipode.

Example A.31. Recall that, given a basis of a vector space V, the symmetric algebra $S(V)$ is spanned by commutative monomials in the elements of this basis. If V is a graded vector space, and the basis is chosen to consist of homogeneous elements of V, we define the degree of a monomial to be the sum of the degrees of its factors. With this grading $S(V)$ is a Hopf algebra.

A.2.7 Dual Hopf algebra

If A is a Hopf algebra let $W_k = A_k^*$ and

$$W = \oplus_{k \geqslant 0} W_k.$$

The space W is also a Hopf algebra; its operations are dual to those of A:

$$\mu^* : W_n \to \bigoplus_{k+l=n} (A_k \otimes A_l)^* \cong \bigoplus_{k+l=n} W_k \otimes W_l \quad \text{is the coproduct in } W$$

$$\delta^* : W_n \otimes W_m \to W_{m+n} \qquad\qquad\qquad \text{is the product in } W$$

$$\iota^* : W \to \mathbb{F} \qquad\qquad\qquad\qquad\qquad \text{is the counit in } W$$

$$\varepsilon^* : \mathbb{F} \to W \qquad\qquad\qquad\qquad\qquad \text{is the unit in } W$$

The Hopf algebra W is called the *dual* of A.

Exercise. Check that this definition is a particular case of the dual of a filtered bialgebra.

Exercise. Show that the dual of the dual of a Hopf algebra A is canonically isomorphic to A.

A.2.8 Structure theorem for Hopf algebras

Is it easy to see that in a Hopf algebra the primitive subspace $\mathscr{P} = \mathscr{P}(A) \subset A$ is the direct sum of its homogeneous components: $\mathscr{P} = \bigoplus_{n \geqslant 0} \mathscr{P} \cap A_n$.

Theorem A.32 (Milnor–Moore 1965). *Any commutative cocommutative Hopf algebra is canonically isomorphic to the symmetric algebra on its primitive subspace:*

$$A = S(\mathscr{P}(A)).$$

This isomorphism sends a polynomial in the primitive elements of A into its value in A.

In other words, if a linear basis is chosen in every homogeneous component $\mathscr{P}_n = \mathscr{P} \cap A_n$, then each element of A can be written uniquely as a polynomial in these variables.

Proof There are two assertions to prove:

(1) every element of A can expressed as a polynomial, that is, as a sum of products, of primitive elements;
(2) the value of a nonzero polynomial on a set of linearly independent homogeneous primitive elements cannot vanish in A.

First, let us prove assertion (1) for the homogeneous elements of A by induction on their degree.

Note that under our assumptions the coproduct of a homogeneous element $x \in A_n$ has the form

$$\delta(x) = 1 \otimes x + \cdots + x \otimes 1, \tag{A.2}$$

where the dots stand for an element of $A_1 \otimes A_{n-1} + \cdots + A_{n-1} \otimes A_1$. Indeed, we can always write $\delta(x) = 1 \otimes y + \cdots + z \otimes 1$. By cocommutativity, $y = z$. Then, $x = (\varepsilon \otimes \mathrm{id})(\delta(x)) = y + 0 + \cdots + 0 = y$.

In particular, for any element $x \in A_1$, Equation (A.2) ensures that $\delta(x) = 1 \otimes x + x \otimes 1$, so that $A_1 = \mathscr{P}_1$. (It may happen that $A_1 = 0$, but this does not interfere with the subsequent argument!)

Take an element $x \in A_2$. We have

$$\delta(x) = 1 \otimes x + \sum \lambda_{ij} p_i^1 \otimes p_j^1 + x \otimes 1,$$

where p_i^1 constitute a basis of $A_1 = \mathscr{P}_1$ and λ_{ij} is a symmetric matrix over the ground field. Let

$$x' = \frac{1}{2} \sum \lambda_{ij} p_i^1 p_j^1.$$

Then

$$\delta(x') = 1 \otimes x' + \sum \lambda_{ij} p_i^1 \otimes p_j^1 + x' \otimes 1.$$

It follows that

$$\delta(x - x') = 1 \otimes x' + x' \otimes 1,$$

that is, $x - x'$ is primitive, and x is expressed via primitive elements as $(x - x') + x'$, which is a polynomial, linear in $\mathscr{P}_2$ and quadratic in $\mathscr{P}_1$.

Proceeding in this way, assertion (1) can be proved in degrees 3, 4, and so on. We omit the formal inductive argument.

Now, assume that there exists a polynomial in the basis elements of $\mathscr{P}(A)$ which is equal to zero in A. Among all such expressions there exists one, which we denote by w, of the smallest degree; we can assume that it is homogeneous (lies in A_n for some n). In particular, all monomials of degree smaller than n are linearly independent. (We remind the reader that we are working in a graded algebra, so the degree of a polynomial is calculated taking into the account the degrees of the variables. In particular, a linear monomial has the degree equal to the degree of the corresponding variable.)

Let a be a basis primitive element which appears in w as a factor in at least one of the summands; we can write

$$w = a^k f_k + a^{k-1} f_{k-1} + \ldots f_0,$$

where the f_i for $i > 0$ are polynomials in the basis primitive elements of degree smaller than n. Now, $\delta(w) - (1 \otimes w + w \otimes 1)$ lies in the sum of the terms $A_p \otimes A_q$ with $p + q = n$ and $p, q > 0$; the sum of these terms has a basis consisting of expressions $m_p \otimes m_q$ where m_p, m_q are monomials in the basis elements of $\mathscr{P}(A)$ of degrees p, q respectively. Inspection shows that the terms in $\delta(w) - (1 \otimes w + w \otimes 1)$ with $m_p = a^k$ add up to $a^k \otimes f_k$. Since $\delta(w)$ must be zero in $A \otimes A$ this implies that $f_k = 0$ in A, which gives a contradiction since the degree of f_k is smaller than n.

This completes the proof. $\square$

Corollary A.33. *An algebra A satisfying the assumptions of the theorem*

(1) has no zero divisors,

(2) has the antipode S defined on primitive elements by

$$S(p) = -p \, .$$

(3) splits as a direct sum of vector spaces $A_k = P_k \oplus R_k$, where R_k is spanned by products of elements of non-zero degrees.

The space $R = \oplus R_k$ is called the space of *decomposable* elements.

Remark A.34. In fact, there is a more general version of the Milnor–Moore theorem which describes the structure of a cocommutative but not necessarily commutative Hopf algebra. The primitive subspace of such a Hopf algebra is a graded Lie algebra; a cocommutative Hopf algebra A is canonically isomorphic to the universal enveloping algebra of $\mathscr{P}(A)$ (see Etingof and Schiffmann 1998).

A.2.9 Primitive and group-like elements in Hopf algebras

As the Milnor–Moore theorem shows, a nontrivial cocommutative Hopf algebra always has a non-empty primitive subspace. However, the only group-like element in such a Hopf algebra is the identity. (In the case of commutative algebras, which are all isomorphic to symmetric algebras, this was noted in Section A.2.2.) As we shall now see, all these Hopf algebras acquire a wealth of group-like elements *after completion.*

Let $\widehat{A}$ be the graded completion of a Hopf algebra A. We remind that while any element of A can be written as a finite sum $\sum_{i<N} x_i$ with $x_i \in A_i$, elements of $\widehat{A}$ are represented by infinite sums $\sum_i x_i$ with $x_i \in A_i$. The operations on A extend to $\widehat{A}$ uniquely; note, however, that a priori $\widehat{A}$ comes with no nontrivial grading.

Lemma A.35. *For the graded completion $\widehat{A}$ of a connected Hopf algebra A the functions $\exp$ and $\log$, defined by the usual power series, establish a one-to-one correspondence between the set of primitive elements $\mathscr{P}(\widehat{A})$ and the set of group-like elements $\mathscr{G}(\widehat{A})$.*

Proof Let $p \in \mathscr{P}(\widehat{A})$. Then

$$\delta(p^n) = (1 \otimes p + p \otimes 1)^n = \sum_{k+l=n} \frac{n!}{k!l!} p^k \otimes p^l$$

and therefore

$$\delta(e^p) = \delta\left(\sum_{n=0}^{\infty} \frac{p^n}{n!}\right) = \sum_{k=0}^{\infty} \sum_{l=0}^{\infty} \frac{1}{k!l!} p^k \otimes p^l = \sum_{k=0}^{\infty} \frac{1}{k!} p^k \otimes \sum_{l=0}^{\infty} \frac{1}{l!} p^l = e^p \otimes e^p$$

which means that $e^p \in \mathscr{G}(\widehat{A})$.

Vice versa, assuming that $g \in \mathscr{G}(\widehat{A})$, we want to prove that $\log(g) \in \mathscr{P}(\widehat{A})$. By assumption, our Hopf algebra A is connected which implies that the graded component $g_0 \in A_0 \cong \mathbb{F}$ is equal to 1. Therefore we can write $g = 1 + h$ where $h \in \prod_{k>0} A_k$. The condition that g is group-like transcribes as

$$\delta(h) = 1 \otimes h + h \otimes 1 + h \otimes h. \tag{A.3}$$

Now,

$$p = \log(g) = \log(1 + h) = \sum_{k=1}^{\infty} \frac{(-1)^{k-1}}{k} h^k$$

and an exercise in power series combinatorics shows that Equation (A.3) implies the required property

$$\delta(p) = 1 \otimes p + p \otimes 1.$$

$\square$

Exercise (Schmitt 1994; Lando 1997). Define the convolution product of two vector space endomorphisms of a commutative and cocommutative Hopf algebra A by

$$(f * g)(a) = \sum_{\delta(a)=\sum a_i' \otimes a_i''} f(a_i') g(a_i'').$$

Let $I : A \to A$ be the operator defined as zero on A_0 and as the identity on each A_i with $i > 0$. Show that the map

$$I - \frac{1}{2} I * I + \frac{1}{3} I * I * I - \frac{1}{4} I * I * I * I + \dots$$

is the projector of A onto the subspace of primitives P parallel to the subspace R of decomposable elements.

A.3 Free algebras and free Lie algebras

Here we briefly mention the definitions and basic properties of the the free associative and free Lie algebras. For a detailed treatment see, for example, Reutenauer (1993).

A.3.1 Free algebras

The *free algebra* $\mathscr{R}\langle x_1, \ldots, x_n\rangle$ over a commutative unital ring $\mathscr{R}$ is the associative algebra of non-commutative polynomials in the x_i with coefficients in $\mathscr{R}$. If $\mathscr{R} = \mathbb{F}$ and V is the vector space spanned by the symbols $x_1, \ldots, x_n$ then the free algebra on the x_i is isomorphic to the tensor algebra $T(V)$.

Example A.36. The algebra $\mathscr{R}\langle x_1, x_2\rangle$ consists of finite linear combinations of the form $c + c_1 x_1 + c_2 x_2 + c_{11} x_1^2 + c_{12} x_1 x_2 + c_{21} x_2 x_1 + c_{22} x_2^2 + \ldots$, $c_\alpha \in \mathscr{R}$, with natural addition and multiplication.

The free algebra $\mathscr{R}\langle x_1, \ldots, x_n\rangle$ is characterized by the following universal property: given an $\mathscr{R}$-algebra A and a set of elements $a_1, \ldots, a_n$ in A there exists a unique map

$$\mathscr{R}\langle x_1, \ldots, x_n\rangle \to A$$

which sends x_i to a_i for all i. As a consequence, every $\mathscr{R}$-algebra generated by n elements is a quotient of the free algebra $\mathscr{R}\langle x_1, \ldots, x_n\rangle$.

The word *free* refers to the above universal property, which is analogous to the universal property of free groups or Lie algebras (see below). This property amounts to the fact that the only identities that hold in $\mathscr{R}\langle x_1, \ldots, x_n\rangle$ are those that follow from the axioms of an algebra, such as $(x_1 + x_2)^2 = x_1^2 + x_1 x_2 + x_2 x_1 + x_2^2$.

The algebra $\mathscr{R}\langle x_1, \ldots, x_n\rangle$ is graded by the degree of the monomials; its homogeneous component of degree k has dimension n^k, and its Poincaré series is $1/(1 - nt)$. The graded completion of $\mathscr{R}\langle x_1, \ldots, x_n\rangle$ is denoted by $\mathscr{R}\langle\langle x_1, \ldots, x_n\rangle\rangle$.

The free algebra $\mathscr{R}\langle x_1, \ldots, x_n\rangle$ has a coproduct δ defined by the condition that the generators x_i are primitive:

$$\delta(x_i) = x_i \otimes 1 + 1 \otimes x_i.$$

This condition determines δ completely since the coproduct is an algebra homomorphism. There also exists a counit: it sends a non-commutative polynomial to its constant term.

Proposition A.37. *The free algebra* $\mathbb{F}\langle x_1, \ldots, x_n\rangle$ *is a cocommutative Hopf algebra.*

The proof is immediate.

A.3.2 Free Lie algebras

Recall that the space of primitive elements in a bialgebra is a Lie algebra whose Lie bracket is the algebra commutator $[a, b] = ab - ba$. Let $L(x_1, \ldots, x_n)$ be the Lie algebra of primitive elements in $\mathbb{F}\langle x_1, \ldots, x_n \rangle$. Note that the x_i belong to $L(x_1, \ldots, x_n)$.

Proposition A.38. *The Lie algebra* $L(x_1, \ldots, x_n)$ *has the following universal property: given a Lie algebra* $\mathfrak{g}$ *and a set of elements* $a_1, \ldots, a_n \in \mathfrak{g}$ *there exists the unique Lie algebra homomorphism* $L(x_1, \ldots, x_n) \to \mathfrak{g}$ *sending each* x_i *to* a_i.

Indeed, since $\mathbb{F}\langle x_1, \ldots, x_n \rangle$ is free, there exists a unique algebra homomorphism

$$\mathbb{F}\langle x_1, \ldots, x_n \rangle \to U(\mathfrak{g})$$

sending the x_i to the a_i. Passing to the primitive spaces we recover the proposition.

Definition A.39. The Lie algebra $L(x_1, \ldots, x_n)$ is called *the free Lie algebra on* $x_1, \ldots, x_n$.

The explicit construction of $L(x_1, \ldots, x_n)$ uses *Lie monomials*, which are defined inductively as follows. A Lie monomial of degree 1 in $x_1, \ldots, x_n$ is simply one of these symbols. A Lie monomial of degree d is an expression of the form $[a, b]$ where a and b are Lie monomials the sum of whose degrees is d.

The Lie algebra $L(x_1, \ldots, x_n)$ as a vector space is spanned by all Lie monomials in $x_1, \ldots, x_n$, modulo the subspace spanned by all expressions of the form

$$[a, b] - [b, a]$$

and

$$[[a, b], c] + [[b, c], a] + [[c, a], b]$$

where a, b, c are Lie monomials. The Lie bracket is the linear extension of the operation $[\ ,\ ]$ on Lie monomials. Note that as a vector space a free Lie algebra is graded by the degree of Lie monomials. Understanding the bracket as the commutator, we get an embedding of $L(x_1, \ldots, x_n)$ constructed in this way into $\mathbb{F}\langle x_1, \ldots, x_n \rangle$ as the primitive subspace.

Finding a good basis for a free Lie algebra is a nontrivial problem; it is discussed in detail in Reutenauer (1993). One explicit basis, the so-called *Lyndon basis*, is constructed with the help of *Lyndon words*. The Lyndon words can be defined as follows. Take an aperiodic necklace (see page 185) and choose the lexicographically smallest among all its cyclic shifts. Replacing each bead with the label i by x_i we get a non-commutative monomial (Lyndon word) in the x_i. A Lyndon word w gives rise to an iterated commutator by means of the following recurrent procedure.

First, $w = x_i x_j$ is declared to produce the commutator $[x_i, x_j]$. If w is of degree more than two, among all decompositions of w into a nontrivial product $w = uv$ choose the decomposition with lexicographically the smallest possible v, and take the commutator of the (possibly iterated) commutators that correspond to u and v.

Shown below is the Lyndon basis for the free Lie algebra $L(x, y)$ in small degrees (up to signs).

m	dim $L(x, y)_m$	basis
1	2	x, y
2	1	$[x, y]$
3	2	$[x, [x, y]]$ $[y, [x, y]]$
4	3	$[x, [x, [x, y]]]$ $[y, [x, [x, y]]]$ $[y, [y, [x, y]]]$
5	6	$[x, [x, [x, [x, y]]]]$ $[y, [x, [x, [x, y]]]]$ $[y, [y, [x, [x, y]]]]$
		$[y, [y, [y, [x, y]]]]$ $[[x, y], [x, [x, y]]]$ $[[x, y], [y, [x, y]]]$

Exercise: Prove that in any Lie algebra $[a, [b, [a, b]]] = [b, [a, [a, b]]]$.

References

Abe, E. (1980). *Hopf Algebras*. Cambridge University Press, Cambridge–New York.

Adams, C. C. (1994). *The Knot Book*. W. H. Freeman and Company, New York.

Adams, J. F. (1969). *Lectures on Lie Groups*. W. A. Benjamin Inc., New York–Amsterdam.

Adyan, S. I. (1984). *Fragments of the Word Δ in the Braid Group*. Math. Notes, 36, Princeton University Press, Princeton, NJ.

Akbulut, S. and J. McCarthy. (1990). *Casson's Invariant for Oriented Homology 3-spheres. An Exposition*. Math. Notes, 36, Princeton University Press, Princeton, NJ.

Alekseev, A. and C. Torossian. (2009). On Triviality of the Kashiwara–Vergne Problem for Quadratic Lie Algebras. *C. R. Math. Acad. Sci. Paris* **347**: 1231–1236.

Alexander, J. W. (1923). A Lemma on Systems of Knotted Curves. *Proc. Nat. Acad. Sci. USA* **9**: 93–95.

Alexander, J. W. (1928). Topological Invariants of Knots and Links. *Trans. Amer. Math. Soc.* **30**: 275–306.

Alexander, J. W. and G. B. Briggs. (1926/1927). On Types of Knotted Curves. *Ann. of Math. (2)* **28**: 562–586.

Alvarez, M. and J. M. F. Labastida. (1996). Vassiliev Invariants for Torus Knots. *J. Knot Theory Ramifications* **5**: 779–803.

Arnol'd, V. I. (1969). The Cohomology Ring of the Group of Dyed Braids. *Mat. Zametki* **5**: 227–231.

Arnol'd, V. I. (1970). On Some Topological Invariants of Algebraic Functions. *Trudy MMO* **21**: 27–46; English translation: *Transact. Moscow Math. Soc.* **21** (1970) 30–52.

Arnold, V. I. (1992). *The Vassiliev Theory of Discriminants and Knots*. First European Congress of Mathematics, Vol. I (Paris), 3–29, Birkhäuser, Basel–Boston–Berlin.

Arnold, V. I. (1994). *Topological Invariants of Plane Curves and Caustics*. University Lecture Series 5, American Mathematical Society, Providence, RI.

Atiyah, M. F. (1990). *The Geometry and Physics of Knots*. Cambridge University Press, Cambridge.

Bar-Natan, D. (1991a). *Weights of Feynman Diagrams and the Vassiliev Knot Invariants*. preprint. http://www.math.toronto.edu/~drorbn/papers.

Bar-Natan, D. (1991b). *Perturbative Aspects of the Chern-Simons Topological Quantum Field Theory*. PhD thesis, Harvard University. http://www.math.toronto.edu/ drorbn/papers.

Bar-Natan, D. (1992). On the Vassiliev Knot Invariants. Draft version of Bar-Natan (1995a).

Bar-Natan, D. (1995a). On the Vassiliev Knot Invariants. *Topology* **34**: 423–472 (an updated version available at `http://www.math.toronto.edu/~drorbn/papers`).

Bar-Natan, D. (1995b). Vassiliev Homotopy String Link Invariants, *J. Knot Theory Ramifications* **4**: 13–32.

Bar-Natan, D. and S. Garoufalidis. (1996). On the Melvin–Morton–Rozansky conjecture. *Invent. Math.*, **125**: 103–133.

Bar-Natan, D. (1996a). *Computer Programs and Data Files*, available online from `http://www.math.toronto.edu/~drorbn/papers/OnVassiliev` and `http://www.math.toronto.edu/~drorbn/papers/nat/natmath.html`.

Bar-Natan, D. (1996b). *Some Computations Related to Vassiliev Invariants*, last updated May 5, 1996. `http://www.math.toronto.edu/~drorbn/papers`.

Bar-Natan, D. (1996c). Vassiliev and Quantum Invariants of Braids, In *the Interface of Knots and Physics* (San Francisco, CA, 1995) 129–144, Proc. Sympos. Appl. Math., 51, Amer. Math. Soc., Providence, RI.

Bar-Natan, D. (1997a). Non-Associative Tangles. In *Geometric Topology* (Athens, GA, 1993) 139–183, Amer. Math. Soc., Providence, RI.

Bar-Natan, D. (1997b). Lie Algebras and the Four Color Theorem. *Combinatorica* **17**: 43–52.

Bar-Natan, D. (1998). On Associators and the Grothendieck-Teichmuller Group. I, *Selecta Math. (N.S.)*, **4**: 183–212.

Bar-Natan, D., S. Garoufalidis, L. Rozansky, and D. Thurston. (2002). The Århus Integral of Rational Homology 3-Spheres II:Invariance and Universality. *Selecta Math, N.S.* **8**: 341–371.

Bar-Natan, D., T. Q. T. Le and D. P. Thurston. (2003). Two Applications of Elementary Knot Theory to Lie Algebras and Vassiliev invariants. *Geom. Topol.* **7**: 1–31.

Bar-Natan, D. and R. Lawrence. (2004). A Rational Surgery Formula for the LMO Invariant. *Israel J. Math.* **140**: 29–60.

Bar-Natan, D., I. Halacheva, L. Leung, and F. Roukema. (2009). *Some Dimensions of Spaces of Finite Type Invariants of Virtual Knots.* `arXiv:0909.5169v1 [math.GT]`.

Bar-Natan, D. (2011). *Finite Type Invariants of W-Knotted Objects: From Alexander to Kashiwara and Vergne*, in preparation, available online at `http://www.math.toronto.edu/drorbn/papers/WKO/`

Bigelow, S. (1999). The Burau Representation is not Faithful for $n = 5$. *Geom. Topol.* **3**: 397–404.

Bigelow, S. (2001). Braid Groups are Linear. *J. Amer. Math. Soc.* **14**: 471–486.

Birman, J. (1976). On the Stable Equivalence of Plat Representations of Knots and Links. *Canad. J. Math.* **28**: 264–290.

Birman, J. S. (1974). *Braids, Links and Mapping Class Groups.* Princeton University Press, Princeton, NJ; University of Tokyo Press, Tokyo.

Birman, J. S. (1993). New Points of View in Knot Theory. *Bull. Amer. Math. Soc. (N.S.)* **28**: 253–287.

Birman, J. S and X.-S. Lin. (1993). Knot Polynomials and Vassiliev's Invariants. *Invent. Math.* **111**: 225–270.

Bollobás, B. and O. Riordan. (2000). Linearized chord Diagrams and an Upper Bound for Vassiliev Invariants. *J. Knot Theory Ramifications* **9**: 847–853.

Borwein, J., D. Bradley, D. Broadhurst and P. Lisoněk. (1998). Combinatorial Aspects of Multiple Zeta Values, *Electron. J. Combin.* **5**: Research Paper 38, 12 pp.

Bouchet, A. (1987). Reducing Prime Graphs and Recognizing Circle Graphs. *Combinatorica* **7**: 243–254.

Bouchet, A. (1994). Circle Graph Obstructions. *J. Combin. Theory Ser. B*, **60**: 107–144.

Broadhurst, D. (1997). *Conjectured Enumeration of Vassiliev Invariants.* arXiv:q-alg/9709031.

Budney, R. (2007). Little Cubes and Long Knots. *Topology* **46**: 1–27.

Burde, G. and H. Zieschang. (2003). *Knots.* Second edition. Walter de Gruyter, Berlin–New York.

Cairns, G. and D. M. Elton. (1993). The Planarity Problem for Signed Gauss Words. *J. Knot Theory Ramifications* **2**: 359–367.

Cartier, P. (1993). Construction Combinatoire des Invariants de Vassiliev–Kontsevich des Nœuds. *C. R. Acad. Sci. Paris Sér. I* **316**: 1205–1210.

Cartier, P. (1999). *Values of the Zeta Function.* Notes of survey lectures for undergraduate students, Independent University of Moscow. http://www.mccme.ru/ium/stcht.html (in Russian).

Cartier, P. (2007). *A Primer of Hopf Algebras.* Frontiers in number theory, physics, and geometry II. On conformal field theories, discrete groups and renormalization. Papers from the meeting, Les Houches, France, March 9–21, 2003, 537–615. Springer, Berlin.

Chen, K.-T. (1977a). Extension of C^∞ Function Algebra by Integrals and Malcev Completion of π_1. *Advances in Math.* **23**: 181–210.

Chen, K.-T. (1977b). Iterated Path Integrals. *Bull. Amer. Math. Soc.* **83**: 831–879.

Chmutov, S. (1997). Combinatorial Analog of the Melvin–Morton Conjecture. *In Proceedings of KNOTS '96 (Tokyo)*, 257–266. World Scientific Publishing Co., River Edge, NJ.

Chmutov, S. (1998). A Proof of Melvin–Morton Conjecture and Feynman Diagrams. *J. Knot Theory Ramifications* **7**: 23–40.

Chmutov, S. and S. Duzhin. (1994). An Upper Bound for the Number of Vassiliev Knot Invariants. *J. Knot Theory Ramifications* **3**: 141–151.

Chmutov, S. and S. Duzhin. (1999). A Lower Bound for the Number of Vassiliev Knot Invariants. *Topology Appl.* **92**: 201–223.

Chmutov, S. and S. Duzhin. (2001). The Kontsevich Integral. *Acta Appl. Math.* **66**: 155–190.

Chmutov, S., S. Duzhin and S. Lando. (1994a). *Vassiliev Knot Invariants I. Introduction, Singularities and bifurcations*, Adv. Soviet Math. 21: 117–126, Amer. Math. Soc., Providence, RI.

Chmutov, S., S. Duzhin and S. Lando. (1994b). *Vassiliev Knot Invariants II. Intersection Graph Conjecture for Trees. Singularities and Bifurcations*, Adv. Soviet Math. 21: 127–134, Amer. Math. Soc., Providence, RI.

Chmutov, S., S. Duzhin and S. Lando. (1994c). *Vassiliev Knot Invariants III. Forest Algebra and Weighted Graphs. Singularities and Bifurcations*, Adv. Soviet Math. 21: 135–145, Amer. Math. Soc., Providence, RI.

Chmutov, S., M. Khoury and A. Rossi. (2009). Polyak-Viro Formulas for Coefficients of the Conway Polynomial. *J. Knot Theory Ramifications*, **18**: 773–783.

Chmutov, S. and S. Lando. (2007). Mutant Knots and Intersection Graphs. *Algebr. Geom. Topol.* **7**: 1579–1598.

Chmutov, S. and M. Polyak. (2010). Elementary Combinatorics of the HOMFLYPT Polynomial. *Int. Math. Res. Not. IMRN* 480–495.

Chmutov, S. and A. Varchenko. (1997). Remarks on the Vassiliev Knot Invariants Coming from sl_2. *Topology* **36**: 153–178.

Conant, J., J. Mostovoy and T. Stanford. (2010). Finite Type Invariants Based on the Band-Pass and Doubled Delta Moves. *J. Knot Theory Ramifications* **19**: 355–384.

Conant, J. and P. Teichner. (2004). Grope Cobordism of Classical Knots, *Topology* **43**: 119–156.

Conway, J. H. (1970). An Enumeration of Knots and Links and Some of Their Algebraic Properties. In *Computational Problems in Abstract Algebra*, 329–358. Pergamon Press, Oxford.

Crowell, R. and R. Fox. (1963). *Introduction to Knot Theory*. Ginn and Co., Boston.

Cromwell, P. (2004). *Knots and Links*. Cambridge University Press, Cambridge.

Cunningham, W. H. (1982). Decomposition of Directed Graphs. *SIAM J. Algebraic Discrete Methods*, **3**: 214–228.

Dasbach, O. (1996). *On Subspaces of the Space of Vassiliev Invariants*. PhD thesis. Düsseldorf, August.

Dasbach, O. (1997). Private Communication. July.

Dasbach, O. (1998). On the Combinatorial Structure of Primitive Vassiliev Invariants. II. *J. Combin. Theory Ser. A* **81**: 127–139.

Dasbach, O. (2000). On the Combinatorial Structure of Primitive Vassiliev Invariants. III. A Lower Bound. *Commun. Contemp. Math.* **2**: 579–590.

Dold, A. (1995). *Lectures on Algebraic Topology*. Classics in Mathematics, Springer-Verlag, Berlin.

Donkin, S. (1993). Invariant Functions on Matrices. *Math. Proc. Cambridge Philos. Soc.* **113**: 23–43.

Drinfeld, V. G. (1985). Hopf Algebras and the Quantum Yang–Baxter Equation. *Soviet Math. Dokl.* **32**: 254–258.

Drinfeld, V. G. (1987). Quantum Groups. In *Proceedings of the International Congress of Mathematicians (Berkeley, 1986)*, 798–820, Amer. Math. Soc., Providence, RI.

Drinfeld, V. G. (1989). Quasi-Hopf Algebras. *Algebra i Analiz*, **1**: 114–148. English translation: Leningrad Math. J. **1** (1990): 1419–1457.

Drinfeld, V. G. (1990). On Quasi-Triangular Quasi-Hopf Algebras and a Group Closely Connected with Gal($\overline{\mathbb{Q}}/\mathbb{Q}$). *Algebra i Analiz*, **2**, no. 4: 149–181. English translation: Leningrad Math. J., **2** (1991): 829–860.

Duflo, M. (1970). Caractères des Groupes et des Algèbres de Lie Résolubles. *Ann. Sci. Éc. Norm. Sup. (4)* **3**: 23–74.

Duzhin, S. (1996). A Quadratic Lower Bound for the Number of Primitive Vassiliev Invariants. Extended abstract, *KNOT'96 Conference/Workshop Report*, 52–54, Waseda University, Tokyo, July.

Duzhin, S. (2002). *Lectures on Vassiliev Knot Invariants*. Lectures in Mathematical Sciences, vol. 19, University of Tokyo.

Duzhin, S. (2009). *Expansion of the Logarithm of the KZ Drinfeld Associator Over Lyndon Basis up to Degree 12*, online at http://www.pdmi.ras.ru/~arnsem/dataprog/.

Duzhin, S. (2010). *Conway Polynomial and Magnus Expansion*. arXiv:1001.2500v2 [math.GT].

Duzhin, S. and S. Chmutov. (1999). Knots and their Invariants. *Matematicheskoe Prosveschenie* **3**: 59–93 (in Russian).

Duzhin, S., A. Kaishev and S. Chmutov. (1998). The Algebra of 3-Graphs. *Proc. Steklov Inst. Math.* **221**: 157–186.

Duzhin, S. V. and M. V. Karev. (2007). Determination of the Orientation of String Links using Finite-Type Invariants. *Funct. Anal. Appl.* **41**: 208–216.

Dynnikov, I. A. (2006). Arc-Presentations of Links: Monotonic Simplification. *Fund. Math.* **190**: 29–76.

El-Rifai, E. and H. Morton. (1994). Algorithms for Positive Braids. *Quart. J. Math. Oxford* **45**: 479–497.

Etingof, P., I. Frenkel and A. Kirillov Jr. (1998). *Lectures on Representation Theory and Knizhnik–Zamolodchikov Equations*. Mathematical Surveys and Monographs 58, Amer. Math. Soc., Providence, RI.

Etingof, P. and O. Schiffmann. (1998). *Lectures on Quantum Groups*. International Press, Boston.

Euler, L. (1775). Meditationes Circa Singulare Serierum Genus. *Novi Commentarii Academiae Scientiarum Petropolitanae* **20**: 140–186. Reprinted in: *Opera Omnia*, Ser.1, XV, Teubner, Leipzig–Berlin (1927): 217–267.

Falk, M. and R. Randell. (1985). The Lower Central Series of a Fiber-Type Arrangement. *Invent. Math.* **82**: 77–88.

Fiedler, T. (2001). *Gauss Diagram Invariants for Knots and Links*. Kluwer Academic Publishers, Dordrecht.

Figueroa-O'Farrill, J. M., T. Kimura and A. Vaintrob. (1997). The Universal Vassiliev Invariant for the Lie Superalgebra gl(1|1). *Comm. Math. Phys.* **185**: no. 1, 93–127.

Freyd, P., D. Yetter, J. Hoste, W. B. R. Lickorish, K. Millett and A. Ocneanu. (1985). A New Polynomial Invariant of Knots and Links. *Bull. Amer. Math. Soc. (N.S.)* **12**: 239–246.

Fox, R- H. (1953). Free Differential Calculus. I. Derivation in the Free Group Ring. *Ann. of Math.* (2) **57**: 547–560.

Fulton, W. and J. Harris. (1991). *Representation Theory, A First Course*. Graduate Texts in Mathematics 129, Springer-Verlag, New York.

Gabor, C. P., K. J. Supowit and W.-L. Hsu. (1989). Recognizing Circle Graphs in Polynomial Time. *J. Assoc. Comput. Mach.* **36**: 435–473.

Garside, F. (1969). The Braid Group and Other Groups. *Quart. J. Math. Oxford* **20**: 235–254.

Garoufalidis, S. and A. Kricker. (2004). A Rational Noncommutative Invariant of Boundary Links. *Geom. Topol.* **8**: 115–204.

Goeritz, L. (1934). Bemerkungen zur Knotentheorie. *Abh. Math. Sem. Univ. Hamburg.* **10**: 201–210.

Goncharov, A. (1998). Multiple Polylogarithms, Cyclotomy and Modular Complexes. *Math. Res. Lett.* **5**: 497–516.

Goncharov, A. (2002). *Periods and Mixed Motives.* arXiv:math/0202154.

Goryunov, V. (1998). Vassiliev Type Invariants in Arnold's J^+-Theory of Plane Curves without Direct Self-Tangencies. *Topology* **37**: 603–620.

Goryunov, V. (1999). *Vassiliev Invariants of Knots in $\mathbb{R}^3$ and in a Solid Torus. Differential and Symplectic Topology of Knots and Curves*. Amer. Math. Soc. Transl. Ser. 2 190: 37–59, Amer. Math. Soc., Providence, RI.

Gorin, E. A. and V. Ya. Lin. (1969). Algebraic Equations with Continuous Coefficients and Some Problems of the Algebraic Theory of Braids. *Math. USSR Sb.* **7** (1969) 569–596; translation from *Mat. Sb., N. Ser.* **78(120)**: 579–610.

Goussarov, M. (1991). A New Form of the Conway-Jones Polynomial of Oriented Links. *Zap. Nauchn. Sem. Leningrad. Otdel. Mat. Inst. Steklov (LOMI)* **193**: 4–9. English translation: *Topology of manifolds and varieties* (O. Viro, editor), Adv. Soviet Math. 18 (1994) 167–172, Amer. Math. Soc., Providence, RI.

Goussarov, M. (1993). On *n*-Equivalence of Knots and Invariants of Finite Degree. *Zap. Nauchn. Sem. S-Peterburg. Otdel. Mat. Inst. Steklov (POMI)* **208**: 152–173. English translation: Topology of manifolds and varieties (O. Viro, editor), *Adv. Soviet Math.* **18** (1994) 173–192, Amer. Math. Soc., Providence, RI.

Goussarov, M. (1998a). *Finite Type Invariants are Presented by Gauss Diagram Formulas*, preprint (translated from Russian by O. Viro). http://www.math.toronto.edu/~drorbn/~Goussarov/.

Goussarov, M. (1998b). Interdependent Modifications of Links and Invariants of Finite Degree. *Topology* **37**: 595–602.

Goussarov, M. (2001). Variations of Knotted Graphs. The Geometric Technique of N-equivalence. *St. Petersburg Math. J.* **12**: 569–604.

Goussarov, M., M. Polyak and O. Viro. (2000). Finite Type Invariants of Classical and Virtual Knots. *Topology* **39**: 1045–1068.

Habegger, N. and X.-S. Lin. (1990). Classification of links up to link homotopy. *J. Amer. Math. Soc.* **3**: 389–419.

Habegger, N. and G. Masbaum. (2000). The Kontsevich Integral and Milnor's Invariants. *Topology* **39**: 1253–1289.

Habiro, K. (1994). *Aru Musubime no Kyokusyo Sousa no Zoku ni Tuite*. Master's thesis at Tokyo University (in Japanese).

Habiro, K. (2000). Claspers and Finite Type Invariants of Links. *Geom. Topol.* **4**: 1–83.

Hall, P. (1957). *Nilpotent Groups*. Canadian Mathematical Congress, University of Alberta, Edmonton, 1957; also published as *Queen Mary College Mathematics Notes*, Queen Mary College, London, 1969, and in *The Collected Works of Philip Hall*, Oxford University Press, New York, 1988.

Harary, F. (1969). *Graph Theory*. Addison Wesley, Reading, MA–Menlo Park, CA–London.

Harris, B. (2004). *Iterated Integrals and Cycles on Algebraic Manifolds*. Nankai Tracts in Mathematics 7. World Scientific, River Edge, NJ.

Hatcher, A. (1983). A Proof of the Smale Conjecture. *Ann. of Math.* **177**: 553–607.

Hatcher, A. (2004). *Spectral Sequences in Algebraic Topology*, http://www.math.cornell.edu/~hatcher/SSAT/SSATpage.html.

Hinich, V. and A. Vaintrob. (2002). Cyclic Operads and Algebra of Chord Diagrams. *Selecta Math. (N.S.)* **8**: 237–282.

Hoffman, M. (1992). Multiple Harmonic Series. *Pacific J. Math.* **152**: 275–290.

Hoffman, M. and Y. Ohno. (2003). Relations of Multiple Zeta Values and their Algebraic Expression. *J. Algebra* **262**: 332–347.

Hoste, J. and M. Thistlethwaite. (1999). *Knotscape*, computer program available from www.math.utk.edu/~morwen/knotscape.html.

Hoste, J., M. Thistlethwaite and J. Weeks. (1998). The first 1,701,936 knots. *Math. Intelligencer* **20**: 33–48.

Humphreys, J. (1980). *Introduction to Lie Algebras and Representation Theory*. Graduate Texts in Mathematics 9, Springer-Verlag, New York–Heidelberg–Berlin.

Jantzen, J. (1996). *Lectures on Quantum Groups*. Graduate Studies in Math. 6, Amer. Math. Soc., Providence, RI.

Jennings, S. A. (1955). The Group Ring of a Class of Infinite Nilpotent Groups. *Can. J. Math.* **7**: 169–187.

Jimbo, M. (1985). A *q*-difference Analogue of $U(\mathfrak{g})$ and the Yang–Baxter Equation. *Lett. Math. Phys.* **10**: 63–69.

Jones, V. (1985). A Polynomial Invariant for Knots Via von Neumann Algebras, *Bull. Amer. Math. Soc. (N.S.)* **12**: 103–111.

Jones, V. (1987). Hecke Algebra Representations of Braid Groups and Link Polynomials. *Math. Ann.* **126**: 335–388.

Jones, V. (1989). On Knot Invariants Related to some Statistical Mechanics Models. *Pacific J. Math.* **137**: 311–334.

Kac, V. (1977a). A Sketch of Lie Superalgebra Theory. *Comm. Math. Phys.* **53**: 31–64.

Kac, V. (1977b). Lie Superalgebras. *Adv. Math.* **26**: 8–96.

Kaishev, A. (2000). *A Program System for the Study of Combinatorial Algebraic Invariants of Topological Objects of Small Dimension*. PhD thesis, Program Systems Institute of the Russian Academy of Sciences, Pereslavl-Zalessky.

Kalfagianni, E. (1998). Finite Type Invariants for Knots in Three-Manifolds. *Topology* **37**: 673–707.

Kanenobu, T. and H. Murakami. (1986). Two-bridge with Unknotting Number 1. *Proc. Amer. Math. Soc.* **98**: 499–501.

Kassel, C. (1995). *Quantum Groups*. Graduate Texts in Math. 155, Springer-Verlag, New York–Heidelberg–Berlin.

Kassel, C., M. Rosso and V. Turaev. (1997). *Quantum Groups and Knot Invariants*. Panoramas et Synthèses, 5, Société Mathématique de France, Paris.

Kauffman, L. (1983). *Formal Knot Theory*. Math. Notes 30, Princeton University Press, Princeton, NJ.

Kauffman, L. (1985). The Arf Invariant of Classical Knots. *Contemp. Math.* **44**: 101–106.

Kauffman, L. (1987a). *On Knots*. Annals of Mathematics Studies 115, Princeton University Press, Princeton, NJ.

Kauffman, L. (1987b). State Models and the Jones Polynomial. *Topology* **26**: 395–407.

Kauffman, L. (1990). An Invariant of Regular Isotopy. *Trans. Amer. Math. Soc.* **318**: 417–471.

Kauffman, L. (1993). *Knots and Physics*. 3rd edition. World Scientific Publishing.

Kauffman, L. (1999). Virtual Knot Theory. *European J. Combin.* **20**: 663–690.

Kawauchi, A. (1979). The Invertibility Problem on Amphicheiral Excellent Knots. *Proc. Japan Acad. Ser. A Math. Sci.* **55**: 399–402.

Kawauchi, A. (1996). *A Survey of Knot Theory*. Birkhäuser, Basel.

Khovanov, M. (1996). Real $K(\pi, 1)$ Arrangements from Finite Root Systems. *Math. Res. Lett.* **3**: 261–274.

Kirillov, A. N. (1995). Dilogarithm Identities. *Progress of Theor. Phys. Suppl.* **118**: 61–142.

Kneissler, J. (1997). *The Number of Primitive Vassiliev Invariants up to Degree Twelve*. arXiv:math.QA/9706022.

Kneissler, J. (2000). On Spaces of Connected Graphs. I. Properties of Ladders. In *Knots in Hellas '98*, 252–273. World Sci. Publishing, River Edge, NJ.

Kneissler, J. (2001a). On Spaces of Connected Graphs. II. Relations in the Algebra Λ. *J. Knot Theory Ramifications* **10**: 667–674.

Kneissler, J. (2001b). On Spaces of Connected Graphs. III. The Ladder Filtration. *J. Knot Theory Ramifications* **10**: 675–686.

Knizhnik, V. and A. Zamolodchikov. (1984). Current Algebra and the Wess–Zumino Model in Two Dimensions. *Nucl. Phys.* **B 247**: 83–103.

Knot Atlas. http://katlas.org.

Kobayashi, S. and K. Nomizu. (1996). *Foundations of Differential Geometry*. John Wiley & Sons, New York.

Kohno, T. (1985). Série de Poincaré-Koszul Associée Aux Groupes de Tresses Pures. *Invent. Math.* **82**: 57–75.

Kohno, T. (1987). Monodromy Representations of Braid Groups and Yang–Baxter Equations. *Ann. Inst. Fourier* **37**: 139–160.

Kohno, T. (1994). Vassiliev Invariants and the de Rham Complex on the Space of Knots. *Contemp. Math.* **179**: 123–138.

Kohno, T. (2002). *Conformal Field Theory and Topology*, Translations of Mathematical Monographs, 210, Amer. Math. Soc., Providence, RI.

Kontsevich, M. (1993). *Vassiliev's Knot Invariants. I. M. Gelfand seminar. Part 2*, Adv. Soviet Math. 16 137–150, Amer. Math. Soc., Providence, RI.

Kontsevich, M. (1997). Fax to S. Chmutov.

Kontsevich, M. (2003). Deformation Quantization of Poisson Manifolds, I. *Lett. Math. Phys.* **66**: 157–216.

Krammer, D. (2002). Braid Groups are Linear. *Ann. of Math.* (2) **155**: 131–156.

Kreimer, D. (2000). *Knots and Feynman Diagrams*. Cambridge Lecture Notes in Physics, 13. Cambridge University Press, Cambridge.

Kricker, A. (1997). Alexander–Conway Limits of Many Vassiliev Weight Systems. *J. Knot Theory Ramifications* **6**: 687–714.

Kricker, A. (2000). *The Lines of the Kontsevich Integral and Rozansky's Rationality Conjecture.* arXiv:math.GT/0005284.

Kricker, A., B. Spence and I. Aitchison. (1997). Cabling the Vassiliev Invariants. *J. Knot Theory Ramifications* **6**: 327–358.

Kurlin, V. (2004). *Explicit Description of Compressed Logarithms of All Drinfeld Associators.* math.GT/0408398.

Kurpita, B. I. and K. Murasugi. (1998). A Graphical Approach to the Melvin–Morton Conjecture, Part I. *Topology Appl.* **82**: 297–316.

Labastida, J. M. F. (1999). Chern-Simons Gauge Theory: Ten Years After. In *Trends in Theoretical Physics II*, American Institute of Physics, AIP Conf. Proc. 484: 1–40.

Lando, S. (1997). On Primitive Elements in the Bialgebra of Chord Diagrams. In *Topics in Singularity Theory*, 167–174, Amer. Math. Soc. Transl. Ser. 2, 180, Amer. Math. Soc., Providence, RI.

Lando, S. (2000). On a Hopf Algebra in Graph Theory. *J. Combin. Theory Ser. B* **80**: 104–121.

Lando, S. and A. Zvonkin. (2004). *Graphs on Surfaces and their Applications*. Springer-Verlag, Berlin.

Lannes, J. (1993). Sur les Invariants de Vassiliev de Degré Inférieur ou égal à 3. *Enseign. Math.* **39**: 295–316.

Lewin, L. (1981). *Polylogarithms and Associated Functions*. Elsevier North Holland, New York, Oxford.

Le, T. Q. T. (1994). Private communication, November.

Le, T. Q. T. and J. Murakami. (1995a). Kontsevich's Integral for the Homfly Polynomial and Relations Between Values of Multiple Zeta functions. *Topology Appl.* **62**: 193–206.

Le, T. Q. T. and J. Murakami. (1995b). Representation of the Category of Tangles by Kontsevich's Iterated Integral. *Comm. Math. Phys.* **168**: 535–562.

Le, T. Q. T. and J. Murakami. (1996a). The Universal Vassiliev–Kontsevich Invariant for Framed Oriented Links. *Compositio Math.* **102**: 41–64.

Le, T. Q. T and J. Murakami. (1996b). The Kontsevich Integral for the Kauffman Polynomial. *Nagoya Math. J.* **142**: 39–65.

Le, T. Q. T. and J. Murakami. (1997). Parallel Version of the Universal Kontsevich–Vassiliev Invariant. *J. Pure Appl. Algebra* **212**: 271–291.

Le, T. Q. T., J. Murakami and T. Ohtsuki. (1998). On a Universal Perturbative Invariant of 3-Manifolds. *Topology* **37**: 539–574.

Lescop, C. (1999). *Introduction to the Kontsevich Integral of Framed Tangles*. CNRS Institut Fourier preprint. (PostScript file available online at `http://www-fourier.ujf-grenoble.fr/~lescop/publi.html`.)

Lickorish, W. B. R. (1997). *An Introduction to Knot Theory*. Graduate Texts in Math. 175, Springer-Verlag, New York.

Lieberum, J. (1999). On Vassiliev Invariants Not Coming from Semisimple Lie Algebras, *J. Knot Theory Ramifications* **8**: 659–666.

MacLane, S. (1988). *Categories for the Working Mathematician*. Graduate Texts in Math. 5, Springer-Verlag, New York.

Magnus, W., A. Karrass and D. Solitar. (1976). *Combinatorial Group Theory*, Revised Edition, Dover Publications, New York.

Marché, J. (2004). A Computation of Kontsevich Integral of Torus Knots. *Algebr. Geom. Topol.* **4**: 1155–1175.

Markov, A. A. (1935). Über Die Freie Aquivalenz Geschlossener Zöpfe. *Recueil Mathematique Moscou* **1**: 73–78.

Massuyeau, G. (2007). Finite-Type Invariants of 3-Manifolds and the Dimension Subgroup Problem. *J. Lond. Math. Soc.* **75**: 791–811.

Matveev, S. (1987). Generalized Surgeries of Three-Dimensional Manifolds and Representations of Homology spheres. *Mat. Zametki*, **42** (1987) 268–278; English translation: *Math. Notes*, **42**: 651–656.

Mellor, B. (2000). The Intersection Graph Conjecture for Loop Diagrams. *J. Knot Theory Ramifications* **9**: 187–211.

Mellor, B. (2003). A Few Weight Systems Arising from Intersection Graphs. *Michigan Math. J.* **51**: 509–536.

Melvin, P. M. and H. R. Morton. (1995). The Coloured Jones Function. *Comm. Math. Phys.* **169**: 501–520.

Merkov, A. B. (1999). *Vassiliev Invariants Classify Flat Braids. Differential and Symplectic Topology of Knots and Curves*. Amer. Math. Soc. Transl. Ser. 2, 190 83–102, Amer. Math. Soc., Providence, RI.

Mikhailov, R. and I. B. S. Passi. (2009). *Lower Central and Dimension Series of Groups*. Lecture Notes in Mathematics 1952, Springer-Verlag, Berlin.

Milnor, J. and J. Moore. (1965). On the Structure of Hopf Algebras. *Ann. of Math. (2)* **81**: 211–264.

Minh, H. N., M. Petitot and J. Van Der Hoeven. (2000). Shuffle Algebra and Polylogarithms. Formal power series and algebraic combinatorics (Toronto '98), *Discrete Math.* **225**: 217–230.

Moran, G. (1984). Chords in a Circle and Linear Algebra over $GF(2)$. *J. Combin. Theory Ser. A* **37**: 239–247.

Morton, H. R. (1995). The Coloured Jones Function and Alexander Polynomial for Torus Knots. *Math. Proc. Cambridge Philos. Soc.* **117**: 129–135.

Morton, H. R. and P. R. Cromwell. (1996). Distinguishing Mutants by Knot Polynomials. *J. Knot Theory Ramifications* **5**: 225–238.

Mostovoy, J. and T. Stanford. (2003). On a Map from Pure Braids to Knots. *J. Knot Theory Ramifications* **12**: 417–425.

Mostovoy, J. and S. Willerton. (2002). Free Groups and Finite-Type Invariants of Pure Braids. *Math. Proc. Camb. Philos. Soc.* **132**: 117–130.

Murakami, J. (2000). Finite Type Invariants Detecting the Mutant Knots. *Knot Theory*. A volume dedicated to Professor Kunio Murasugi for his 70th birthday. Editors: M. Sakuma et al., Osaka University. Preprint available at `http://www.f.waseda.jp/murakami/papers/finitetype.pdf`.

Murakami, H. and Y. Nakanishi. (1989). On Certain Moves Generating Link-Homology, *Math. Ann.* **284**: 75–89.

Murasugi, K. (1987). Jones Polynomials and Classical Conjectures in Knot Theory. *Topology* **26**: 187–194.

Murasugi, K. (1996). *Knot Theory and its Applications*. Birkhäuser, Basel.

Naik, S. and Stanford, T. (2003). A Move on Diagrams that Generates *S*-Equivalence of Knots. *J. Knot Theory Ramifications* **12**: 717–724.

Ng, K. Y. (1998). Groups of Ribbon Knots. *Topology* **37**: 441–458.

Ohtsuki, T. (2002). *Quantum Invariants. A Study of Knots, 3-manifolds and their Sets.* Series on Knots and Everything, 29, World Scientific.

Ohtsuki, T. (2004a). A Cabling Formula for the 2-Loop Polynomial of Knots. *Publ. Res. Inst. Math. Sci.* **40**: 949–971.

Ohtsuki, T. (ed.). (2004b). *Problems on Invariants of Knots and 3-Manifolds. Invariants of Knots and 3-Manifolds* (Kyoto 2001), 377–572, Geom. Topol. Monogr. 4, Geom. Topol. Publ., Coventry.

Okuda, J. and Ueno, K. (2004). Relations for Multiple Zeta Values and Mellin Transforms of Multiple Polylogarithms. *Publ. Res. Inst. Math. Sci.* **40**: 537–564.

Östlund, O.-P. (2001). Invariants of Knot Diagrams and Relations Among Reidemeister Moves. *J. Knot Theory Ramifications* **10**: 1215–1227.

Papadima, S. (2002). The Universal Finite-Type Invariant for Braids, with Integer Coefficients. *Topology Appl.* **118**: 169–185.

Passi, I. B. S. (1979). *Group Rings and their Augmentation Ideals*, Lecture Notes in Mathematics 715, Springer-Verlag, Berlin.

Passman, D. S. (1977). *The Algebraic Structure of Group Rings*, Interscience, New York.

Penrose, R. (1971). Applications of Negative Dimensional Tensors. In *Combinatorial Mathematics and its Applications* (ed. D. J. A. Welsh). New York, London: Academic Press.

Perko, K. A. (1973). On the Classification of Knots. *Notices Amer. Math. Soc.* **20**: A453–454.

Perko, K. A. (2002). Letter to the Authors, November 15.

Petitot, M. *Web site*, http://www2.lifl.fr/~petitot/. Last visited in 2011.

Piunikhin, S. (1995). Combinatorial Expression for Universal Vassiliev Link Invariant. *Comm. Math. Phys.* **168**: 1–22.

Poincaré, A. (1884). Sur les Groupes d'équations Linéaires. *Acta Math.* **4**: 201–311.

Polyak, M. (2000). On the Algebra of Arrow Diagrams. *Let. Math. Phys.* **51**: 275–291.

Polyak, M. (2005). Feynman Diagrams for Pedestrians and Mathematicians. *Proc. Sympos. Pure Math.* **73**: 15–42.

Polyak, M. (2010). Minimal Sets of Reidemeister Moves. *Quantum Topology* **1**: 399–411.

Polyak, M. and O. Viro. (1994). Gauss Diagram Formulas for Vassiliev Invariants. *Int. Math. Res. Notes* **11**: 445–454.

Polyak, M. and O. Viro. (2001). On the Casson Knot Invariant. Knots in Hellas '98 Vol 3 (Delphi), *J. Knot Theory Ramifications*, **10**: 711–738.

Prasolov, V. V. and A. B. Sossinsky. (1997). *Knots, Links, Braids and 3-Manifolds.* Translations of Mathematical Monographs 154, Amer. Math. Soc., Providence, RI.

Przytycki, J. and P. Traczyk. (1988). Invariants of Links of the Conway Type. *Kobe J. Math.* **4**: 115–139.

Quillen, D. (1968). On the Associated Graded Ring of a Group Ring. *J. Algebra* **10**: 411–418.

Reidemeister, K. (1948). *Knot Theory*. Chelsea Publ., New York.

Reshetikhin, N. and V. Turaev. (1990). Ribbon Graphs and their Invariants Derived from Quantum Groups. *Comm. Math. Phys.* **127**: 1–26.

Reshetikhin, N. and V. Turaev. (1991). Invariants of 3-Manifolds via Link Polynomials and Quantum Groups. *Invent. Math.* **103**: 547–597.

Reutenauer, C. (1993). *Free Lie algebras.* Clarendon Press, Oxford.

Rips, E. (1972). On the Fourth Integer Dimension Subgroup. *Israel J. Math.* **12**: 342–346.

Rolfsen, D. (1976). *Knots and Links.* Publish or Perish, Berkeley.

Roos, G. (2003). *Functional Equations of the Dilogarithm and the Kontsevich Integral,* preprint. St. Petersburg.

Rozansky, L. (1996). A Contribution of the Trivial Connection to the Jones Polynomial and Witten's Invariant of 3d Manifolds. *Comm. Math. Phys.* **175**: 267–296.

Rozansky, L. (2003). A Rationality Conjecture about Kontsevich Integral of Knots and its Implications to the Structure of the Colored Jones Polynomial. *Topology Appl.* **127**: 47–76.

Sawon, J. (2006). *Perturbative Expansion of Chern-Simons Theory.* Geometry and Topology Monographs 8: 145–166.

Schmitt, W. R. (1994). Incidence Hopf Algebras. *J. Pure Appl. Algebra* **96**: 299–330.

Segal, G. (1979). The Topology of Spaces of Rational Functions. *Acta Math.* **143**: 39–72.

Sinha, D. (2006). Operads and Knot Spaces. *J. Amer. Math. Soc.* **19**: 461–486.

Sinha, D. (2009). The Topology of Spaces of Knots: Cosimplicial Models. *Am. J. Math.* **131**: 945–980.

Sossinsky, A. (2002). *Knots.* Harvard University Press, Cambridge.

Spivak, M. (1979). *A Comprehensive Introduction to Differential Geometry.* 2nd edition. Publish or Perish, Berkeley.

Stallings, J. (1965). Homology and Central Series of Groups. *J. Algebra* **20**: 170–181.

Stanford, T. (1996a). Finite Type Invariants of Knots, Links, and Graphs. *Topology* **35**: 1027–1050.

Stanford, T. (1996b). Braid Commutators and Vassiliev Invariants. *Pacific J. Math.* **174**: 269–276.

Stanford, T. (1999). *Vassiliev Invariants and Knots Modulo Pure Braid Subgroups.* arXiv:math.GT/9907071.

Stanford, T. (2002). *Some Computational Results on Mod 2 Finite-type Invariants of Knots and String Links.* Invariants of Knots and 3-manifolds (Kyoto 2001), Geom. Topol. Monogr. 4 363–376, Geom. Topol. Publ., Coventry.

Stoimenow, A. (1998a). Enumeration of Chord Diagrams and an Upper Bound for Vassiliev Invariants. *J. Knot Theory Ramifications* **7**: 94–114.

Stoimenow, A. (1998b). *On Enumeration of Chord Diagrams and Asymptotics of Vassiliev Invariants.* PhD dissertation, online at http://www.diss.fu-berlin.de/diss/receive/FUDISS_thesis_000000000198.

Stoimenow, A. *Knot Data Tables,* http://stoimenov.net/stoimeno/homepage/.

Taniyama, K. and A. Yasuhara. (2002). Band Description of Knots and Vassiliev Invariants. *Math. Proc. Cambridge Philos. Soc.* **133**: 325–343.

Terasoma, T. (2002). Mixed Tate Motives and Multiple Zeta Values. *Invent. Math.* **149**: 339–369.

Thistlethwaite, M. (1987). A Spanning Tree Expansion for the Jones Polynomial. *Topology* **26**: 297–309.

Trotter, H. F. (1964). Non-Invertible Knots Exist. *Topology* **2**: 275–280.

Turaev, V. (1987). A Simple Proof of the Murasugi and Kauffman Theorems on Alternating Links. *Enseign. Math.* **33**: 203–225.

Turaev, V. (1988). The Yang–Baxter Equation and Invariants of Links. *Invent. Math.* **92**: 527–553.

Turaev, V. (1990). Operator Invariants of Tangles, and R-Matrices. *Math. USSR Izvestiya* **35**: 411–444.

Tourtchine, V. (2004). *On the Homology of the Spaces of Long Knots.* Advances in topological quantum field theory, NATO Science Series II: Mathematics, Physics and Chemistry 179, 23–52, Kluwer Academic Publishers.

Tourtchine, V. (2007). On the Other Side of the Bialgebra of Chord Diagrams. *J. Knot Theory Ramifications* **16**: 575–629.

Turchin, V. (2010). Hodge-Type Decomposition in the Homology of Long Knots. *J. Topol.* **3**: 487–534.

Vaintrob, A. (1994). Vassiliev Knot Invariants and Lie S-Algebras. *Math. Res. Lett.* **1**: 579–595.

Vaintrob, A. (1996). Universal Weight Systems and the Melvin–Morton Expansion of the Colored Jones Knot Invariant. *J. Math. Sci.* **82**: 3240–3254.

Vaintrob, A. (1997). Melvin–Morton Conjecture and Primitive Feynman Diagrams. *Internat. J. Math.* **8**(4): 537–553.

Varchenko, A. (2003). *Special Functions, KZ Type Equations and Representation Theory.* CBMS Regional Conference Series in Mathematics 98, Amer. Math. Soc., Providence, RI.

Vassiliev, V. A. (1990a). *Cohomology of Knot Spaces. Theory of singularities and its applications* (ed. V. I. Arnold). Adv. Soviet Math. 1: 23–69.

Vassiliev, V. A. (1990b). *Homological Invariants of Knots: Algorithm and Calculations.* Preprint Inst. of Applied Math. (Moscow), no. 90: 23 pp. (in Russian).

Vassiliev, V. A. (1994). *Complements of Discriminants of Smooth Maps: Topology and Applications.* revised ed., Transl. of Math. Monographs 98, Amer. Math. Soc., Providence, RI.

Vassiliev, V. A. (1997). *Topologiya Dopolnenii k Diskriminantam* (Russian). [Topology of complements of discriminants], Izdatel'stvo FAZIS, Moscow.

Vassiliev, V. A. (1998). On Invariants and Homology of Spaces of Knots in Arbitrary Manifolds. In *Topics in Quantum Groups and Finite-type Invariants*, (B. Feigin and V. Vassiliev, eds.) Amer. Math. Soc. Translations Ser. 2, 185 155–182, Amer. Math. Soc. Providence, RI.

Vassiliev, V. A. (2001). On Combinatorial Formulas for Cohomology of Spaces of Knots. *Mosc. Math. J.* **1**: 91–123.

Vassiliev, V. A. (2005). Combinatorial Computation of Combinatorial Formulas for Knot Invariants. *Trans. Moscow Math. Soc.* 1–83.

Vermaseren, J. *The Multiple Zeta Value Data Mine*, http://www.nikhef. nl/~form/datamine/datamine.html.

Vogel, P. (1997). *Algebraic Structures on Modules of Diagrams.* Institut de Mathématiques de Jussieu, Prépublication 32, August 1995, Revised in 1997. Online at http://www.math.jussieu.fr/~vogel/.

Vogel, P. (2006). Email to S. Duzhin. March.

Walter, W. (1998). *Ordinary Differential Equations.* Graduate Texts in Math. 192, Springer-Verlag, Berlin.

Weeks, J. et al. (2010). *SnapPea,* Computer program for investigation of hyperbolic 3-manifolds, knots and links. http://geometrygames.org/SnapPea/.

Weibel, C. (1994). *An Introduction to Homological Algebra.* Cambridge University Press, Cambridge.

Willerton, S. (1996). Vassiliev Invariants and the Hopf Algebra of Chord Diagrams. *Math. Proc. Cambridge Philos. Soc.*, **119**: 55–65.

Willerton, S. (1998). *On the Vassiliev Invariants for Knots and for Pure Braids*. PhD thesis, University of Edinburgh.

Willerton, S. (2000). *The Kontsevich Integral and Algebraic Structures on the Space of Diagrams*. Knots in Hellas '98. Series on Knots and Everything, vol. 24, World Scientific, 530–546.

Willerton, S. (2002a). An Almost Integral Universal Vassiliev Invariant of Knots, *Algebr. Geom. Topol.* **2**: 649–664.

Willerton, S. (2002b). On the First Two Vassiliev Invariants. *Experiment. Math.* **11**: 289–296.

Witten, E. (1989). Quantum Field Theory and the Jones Polynomial. *Comm. Math. Phys.* **121**: 351–399.

Witten, E. (1995). *Chern-Simons Gauge Theory as a String Theory. The Floer memorial volume*, 637–678, Progr. Math. 133, Birkhäuser, Basel.

Zagier, D. (1988). The Remarkable Dilogarithm. *J. Math. Phys. Sci.* **22**: 131–145.

Zagier, D. (1994). *Values of Zeta Functions and their Applications*. First European Congress of Mathematics, Progr. in Math. 120, Birkhauser, Basel, II: 497–512.

Zagier, D. (2001). Vassiliev Invariants and a Strange Identity Related to the Dedekind Eta-function. *Topology* **40**: 945–960.

Zhelobenko, D. P. (1973). *Compact Lie Groups and their Representations*. Moscow, Nauka, 1970. AMS translation, Providence, RI.

Zulli, L. (1995). A Matrix for Computing the Jones Polynomial of a Knot. *Topology* **34**: 717–729.

Notations

$\mathbb{Z}, \mathbb{Q}, \mathbb{R}, \mathbb{C}$ — rings of integer, rational, real and complex numbers.

$\mathscr{A}'$ — algebra of unframed chord diagrams on the circle, p. 97.

$\mathscr{A}$ — algebra of framed chord diagrams on the circle, p. 94.

$\mathscr{A}'_n$ — space of unframed chord diagrams of degree n, p. 93.

$\mathscr{A}_n$ — space of framed chord diagrams of degree n, p. 93.

$\mathscr{A}(n)$ — algebra of chord diagrams on n lines, p. 152.

$\mathscr{A}^h(n)$ — algebra of horizontal chord diagrams, p. 150.

$\widehat{\mathscr{A}}$ — graded completion of the algebra of chord diagrams, p. 219.

$\mathbf{A}_n$ — set of chord diagrams of degree n, p. 66.

A — Alexander-Conway power series invariant, p. 316.

α_n — map from $\mathscr{V}_n$ to $\mathscr{R}\mathbf{A}_n$, symbol of an invariant, p. 67.

$\mathscr{B}$ — algebra of open Jacobi diagrams, p. 131.

$\mathscr{B}(m)$ — space of m-coloured open Jacobi diagrams, p. 145.

$\mathscr{B}^\circ$ — enlarged algebra $\mathscr{B}$, p. 319.

$\mathbf{B}_n$ — set of open Jacobi diagrams of degree n, p. 130.

BNG — the Bar-Natan–Garoufalidis function, p. 408.

$\mathscr{C}$ — space of closed Jacobi diagrams, p. 116.

$\mathscr{C}_n$ — space of closed Jacobi diagrams of degree n, p. 116.

C_n — Goussarov–Habiro moves, p. 432.

$\mathscr{C}(x_1, \ldots, x_n \mid y_1, \ldots, y_m)$ — space of mixed Jacobi diagrams, p. 145.

C — Conway polynomial, p. 30.

$\mathfrak{C}_{2n}$ — Conway combination of Gauss diagrams, p. 393.

$\mathbf{C}_n$ — set of closed diagrams of degree n, p. 123.

c_n — n-th coefficient of the Conway polynomial, p. 32.

∂_C — diagrammatic differential operator on $\mathscr{B}$, p. 317.

∂_C° — diagrammatic differential operator on $\mathscr{B}^\circ$, p. 320.

∂_Ω — wheeling map, p. 318.

δ — coproduct in a coalgebra, p. 466; in particular, for the bialgebra $\mathscr{A}$, see p. 96.

$\Delta^{n_1,\ldots,n_k}$ — operation $\mathscr{A}(k) \to \mathscr{A}(n_1 + \ldots + n_k)$, p. 264.

ε — counit in a coalgebra, p. 466; in particular, for the bialgebra $\mathscr{A}$, see p. 97.

$F(L)$ — unframed two-variable Kauffman polynomial, p. 45.

Φ — map $\mathscr{B}(y) \to \mathscr{C}(x)$, p. 323.

Φ_0 — map $\mathscr{B} \to \mathscr{C}$, p. 326.

Φ_2 — map $\mathscr{B}(y_1, y_2) \to \mathscr{C}(x)$, p. 324.

$\varphi_\mathfrak{g}$ — universal Lie algebra weight system, p. 158.

$\varphi_\mathfrak{g}^T$ — Lie algebra weight system associated with the representation, p. 164.

$\mathcal{G}_n$ — Goussarov group, p. 418.

$\mathfrak{G}$ — bialgebra of graphs, p. 420.

Γ — algebra of 3-graphs, p. 196.

$\Gamma(D)$ — intersection graph of a chord diagram D, p. 104.

$H = \begin{pmatrix} 1 & 0 \\ 0 & -1 \end{pmatrix}$ — element of the Lie algebra $\mathfrak{sl}_2$, p. 161.

H — hump unknot, p. 222.

$\mathbf{I}_n$ — constant 1 weight system on $\mathscr{A}_n$, p. 100.

I — a map of Gauss diagrams to arrow diagrams, p. 375.

$I(K)$ — final Kontsevich integral, p. 222.

$\mathscr{I}$ — algebra of knot invariants, p. 35.

ι — unit in an algebra, p. 465; in particular, for the bialgebra $\mathscr{A}$, see p. 97.

j_n — n-th coefficient of the modified Jones polynomial, p. 70.

$\mathscr{K}$ — set of (equivalence classes of) knots, p. 10.

Li_2 — Euler's dilogarithm, p. 245.

$\mathscr{L}$ — bialgebra of Lando, p. 423.

$\mathscr{L}_m$ — the monoid of string links on m strings, p. 358.

Λ — Vogel's algebra, p. 212.

$\Lambda(L)$ — framed two-variable Kauffman polynomial, p. 45.

∇ — difference operator for Vassiliev invariants, p. 61.

$\mathscr{M}_n$ — Goussarov–Habiro moves, p. 411.

MM — highest order part of the coloured Jones polynomial, p. 402.

M_T — mutation of a knot with respect to a tangle T, p. 310.

μ — product in an algebra, p. 465; in particular, for the bialgebra $\mathscr{A}$, see p. 94.

$\mathscr{P}$ — Polyak algebra, p. 388.

$\mathscr{P}_n$ — primitive subspace of the algebra of chord diagrams, p. 102.

P — HOMFLY polynomial, p. 43.

P^{fr} — framed HOMFLY polynomial, p. 56.

$p_{k,l}(L)$ — k, l-th coefficient of the modified HOMFLY polynomial, p. 83.

$\rho_\mathfrak{g}$ — universal Lie algebra weight system on $\mathscr{B}$, p. 181.

$\psi_{\mathbf{n}}$ — n-th cabling of a chord diagram, p. 255.

$\mathscr{R}$ — ground ring (usually $\mathbb{Q}$ or $\mathbb{C}$), p. 59.

$\mathscr{R}(\mathbf{A}_n)$ — $\mathscr{R}$-valued functions on chord diagrams, p. 67.

R — R-matrix, p. 37.

R, R^{-1} — Kontsevich integrals of two braided strings, p. 225.

S_A — symbol of the Alexander–Conway invariant A, p. 404.

S_{MM} — symbol of the Melvin–Morton invariant MM, p. 404.

S_i — operation on tangle (chord) diagrams, p. 247.

$\mathrm{symb}(v)$ — symbol of the Vasiliev invariant v, p. 68.

σ — mirror reflection of knots, p. 7.

τ — changing the orientation of a knot, p. 7.

τ — reversing the orientation of the Wilson loop, p. 128.

τ — inverse of $\chi : \mathscr{B} \to \mathscr{C}$, p. 135.

Θ — the chord diagram with one chord, , p. 97.

θ^{fr} — quantum invariant, p. 38.

θ^{fr} — $\mathfrak{sl}_2$-quantum invariant, p. 40.

θ — unframed quantum invariant, p. 42.

$\theta^{fr,St}_{\mathfrak{sl}_N}$ — $\mathfrak{sl}_N$-quantum invariant, p. 54.

$\mathscr{V}$ — space of Vassiliev (finite type) invariants, p. 59.

$\mathscr{V}_n$ — space of unframed Vassiliev knot invariants of degree $\leqslant n$, p. 59.

$\mathscr{V}_n^{fr}$ — space of framed Vassiliev knot invariants of degree $\leqslant n$, p. 68.

$\mathscr{V}_{\bullet}$ — space of polynomial Vassiliev invariants, p. 64.

$\widehat{\mathscr{V}}_{\bullet}$ — space of power series invariants, graded completion of $\mathscr{V}_{\bullet}$, p. 64.

$\mathscr{W}_n$ — space of unframed weight systems of degree n, p. 87.

$\mathscr{W}_n^{fr}$ — space of framed weight systems of degree n, p. 87.

$\widehat{\mathscr{W}}^{fr}$ — graded completion of the algebra of weight systems, p. 99.

$\mathscr{Z}(\phi)$ — Kontsevich integral of ϕ in algebra $\mathscr{B}(y)$, p. 323.

$\mathscr{Z}(\phi)$ — Kontsevich integral of ϕ in algebra $\mathscr{B}(y_1, y_2)$, p. 324.

$\mathscr{Z}_i(\phi)$ — i-th part of the Kontsevich integral $\mathscr{Z}(\phi)$, p. 325.

$Z(K)$ — Kontsevich integral, p. 220.

$\mathbb{Z}\mathscr{K}$ — algebra of knots, p. 10.

χ — symmetrization map $\mathscr{B} \to \mathscr{C}$, p. 135.

χ_{y_m} — map $\mathscr{C}(X \mid y_1, \ldots, y_m) \to \mathscr{C}(X, y_m \mid y_1, \ldots, y_{m-1})$, p. 145.

Ω' — part of $\mathscr{Z}_0(\phi)$ containing wheels, p. 326.

$\langle \, , \, \rangle_y$ — pairing $\mathscr{C}(x \mid y) \otimes \mathscr{B}(y) \to \mathscr{C}(x)$, p. 147.

ϕ — open Hopf link, p. 323.

ϕ — doubled open Hopf link, p. 324.

∞ — closed Hopf link, p. 332.

$\#$ — connected sum of two knots, p. 9 or two diagrams, p. 96; also the action of $\mathscr{C}$ on tangle diagrams, p. 148.

Index

Printed in the United States
by Baker & Taylor Publisher Services